AF688723

# Principles and Applications of Electrical Engineering

# Principles and Applications of Electrical Engineering

Editor: Ryan Wright

MURPHY & MOORE
www.murphy-moorepublishing.com

Published by Murphy & Moore Publishing
1 Rockefeller Plaza,
New York City, NY 10020, USA
www.murphy-moorepublishing.com

Principles and Applications of Electrical Engineering
Edited by Ryan Wright

© 2022 Murphy & Moore Publishing

International Standard Book Number: 978-1-63987-450-7 (Hardback)

This book contains information obtained from authentic and highly regarded sources. Copyright for all individual chapters remain with the respective authors as indicated. All chapters are published with permission under the Creative Commons Attribution License or equivalent. A wide variety of references are listed. Permission and sources are indicated; for detailed attributions, please refer to the permissions page and list of contributors. Reasonable efforts have been made to publish reliable data and information, but the authors, editors and publisher cannot assume any responsibility for the validity of all materials or the consequences of their use.

**Trademark Notice:** Registered trademark of products or corporate names are used only for explanation and identification without intent to infringe.

**Cataloging-in-Publication Data**

Principles and applications of electrical engineering / edited by Ryan Wright.
p. cm.
Includes bibliographical references and index.
ISBN 978-1-63987-450-7
1. Electrical engineering. 2. Engineering. I. Wright, Ryan.
TK153 .P75 2022
621.3--dc23

# Contents

# Preface

The world is advancing at a fast pace like never before. Therefore, the need is to keep up with the latest developments. This book was an idea that came to fruition when the specialists in the area realized the need to coordinate together and document essential themes in the subject. That's when I was requested to be the editor. Editing this book has been an honour as it brings together diverse authors researching on different streams of the field. The book collates essential materials contributed by veterans in the area which can be utilized by students and researchers alike.

Electrical engineering is the engineering discipline which encompasses the use of electronics, electricity and electromagnetism for driving innovations. It is divided into other significant sub-disciplines such as power engineering, electronics, computer engineering, signal processing, microelectronics, etc. Power engineering is concerned with the production, transmission and distribution of electricity. It also deals with the design of electric generators, transformers, electric motors, etc. Electronic circuits are designed using capacitors, resistors, inductors, transistors and diodes to achieve a particular function. The design and testing of such circuits is under the scope of electronic engineering. Signal processing deals with the manipulation and analysis of signals. It has applications in communications, control, broadcast engineering, radar, power electronics, etc. This book covers in detail some existent theories and innovative concepts revolving around electrical engineering. The various sub-fields of this discipline along with technological progress that have future implications are glanced at in this book. It is a vital tool for all researching or studying electrical engineering as it gives incredible insights into emerging trends.

Each chapter is a sole-standing publication that reflects each author´s interpretation. Thus, the book displays a multi-facetted picture of our current understanding of application, resources and aspects of the field. I would like to thank the contributors of this book and my family for their endless support.

**Editor**

1

# A Novel Electric Bicycle Battery Monitoring System based on Android Client

**Chuanxue Song,[1] Yulong Shao,[1] Shixin Song,[2] Silun Peng,[1] and Feng Xiao[1]**

[1] *State Key Laboratory of Automotive Simulation and Control, Jilin University, Changchun 130022, China*
[2] *School of Mechanical Science and Engineering, Jilin University, Changchun, China*

Correspondence should be addressed to Shixin Song; ssx@jlu.edu.cn

Academic Editor: Yuezhong Meng

The battery monitoring system (BMS) plays a crucial role in maintaining the safe operation of the lithium battery electric bicycle and prolonging the life of the battery pack. This paper designed a set of new battery monitoring systems based on the Android system and ARM single-chip microcomputer to enable direct management of the lithium battery pack and convenient monitoring of the state of the battery pack. The BMS realizes the goal of monitoring the voltage, current, and ambient temperature of lithium batteries, estimating the state of charge (SOC) and state of health (SOH), protecting the battery from abuse during charging or discharging, and ensuring the consistency of the batteries by integrating the passive equalization circuit. The BMS was proven effective and feasible through several tests, including charging/discharging, estimation accuracy, and communication tests. The results indicated that the BMS could be used in the design and application of the electric bicycle.

## 1. Introduction

Social progress and economic development have resulted in the continued gradual aggravation of environmental pollution and the decrease of fossil energy reserves. Developing energy-saving and environmentally friendly vehicles has become the new target. In recent years, electric bicycles have gained wide attention because of its characteristics, such as convenience, low noise, and less pollution [1–5]. The popularity rate of electric bicycles has also been expanding gradually. In many countries and regions with a dense population, such as California in the United States, Central and Southeast China, and Japan and other Southeast Asian countries, the electric bicycle is generally used in urban traffic as a form of short-range transportation [6–8].

However, despite the increasing popularity of electric bicycles, current technology still has the following limitations: first, the energy density of the bulky lead-acid batteries is extremely low. Second, lead is harmful to human health during the making; furthermore, scrap batteries have the potential to pollute the environment. Finally, the power battery pack is a key component in electric bicycles and the battery status information is generally displayed through the micro dashboard in front of the electric bicycle, making the checking of the battery capacity and status difficult. The accuracy of state of charge (SOC) is extremely poor, and estimating the SOH (state of health) of the battery can be difficult. The lithium battery has been considered one of the major viable energy storage media in electric vehicles [9]. Lithium batteries have high energy density and power density and hence are the most promising power battery. Therefore, replacing the lead-acid batteries with lithium-ion batteries can reduce the volume/weight of the battery pack and achieve a qualitative leap in travel distance. However, because of the characteristics of lithium battery, the battery pack is prone to damage due to overcharging, overdischarging, and overheating [10, 11]. Therefore, monitoring battery status and managing the cells according to the characteristics of the lithium battery are meaningful for battery safety and maintenance effectiveness [12]. This function has utmost significance for the stable operation of the bicycle. Most of the electric bicycles now use lead-acid battery, which does not need extra management. While current research on the battery management system is mostly focused on electric vehicles, studies on electric bicycles BMS are relatively limited. In the replacement of lead-acid batteries with lithium-ion batteries, research on electric bicycle BMS is highly necessary. Reference [13] designed a low-cost intelligent Li-ion BMS based on Atmega16 and used multichannel analog switch and differential amplifier to detect the voltage on each cell. Therefore, the voltages of the cells were not collected at the same time, and the current fluctuation during driving can lead to voltage changes. The voltages collected in this way may not be sufficiently accurate. Reference [14] presents the development of an over current protection scheme for BMS that protects the battery and other system components from damage. Reference [15] presents a fast charging scheme for the Li-FePO4 battery pack used in electric bicycles. The BMS adjusts the charging current to the maximum possible limit that corresponds to the voltage of the cell that has the highest voltage in the pack so that no individual cell suffers overvoltage during charging. Reference [16] presents a direct measure voltage of each individual cell to avoid overcharge. The measurement has a resolution as good as 7.8 mV, which is not sufficient, because the required voltage measurement accuracy is 1 mv in this paper. What is more, the current electric bike does not have enough display interfaces to display all the serial cell information; and most electric bicycles are not able to accurately tell the owner how many miles are left during the current drive. This ability can be achieved by the battery management system in this article through the Android client. The user can control the charging and discharging process, as well as provide a timely charging reminder through a mobile phone. Further, the mobile client BMS can be transplanted in the application of electric kart, home energy reserve system, and other fields that require power systems. Therefore, designing the lithium battery monitoring system based on the Android platform and ARM microcontroller is necessary. The novelty of the designed BMS in this paper is the monitoring method using Android client. And the system has higher voltage measurement accuracy; what is more, it is able to balance the cells and estimate SOH.

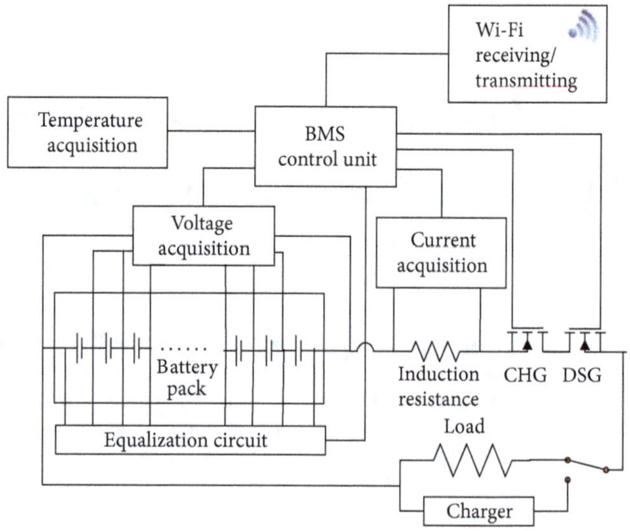


FIGURE 1: Architecture of the BMS (lower machine).

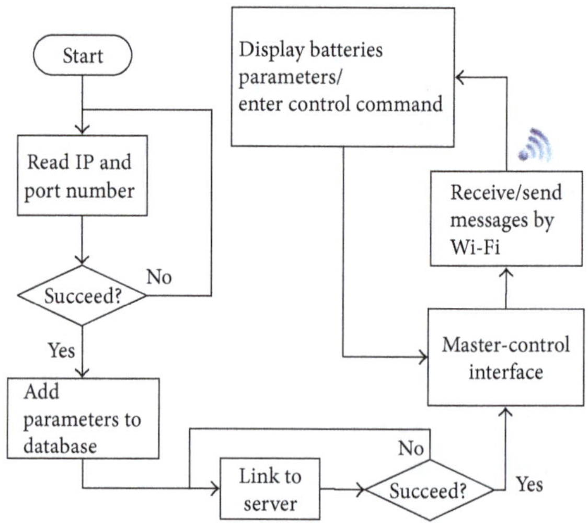


FIGURE 2: Architecture of the BMS (upper machine).

## 2. Composition and Function of the Battery Monitoring System

The power battery pack used in electric bicycle is originally the 48 V, 20 Ah, lead-acid battery. This paper will use the pack of NCR18650B (3.7 V, 3250 mAh) cells to replace lead-acid battery. The lithium battery pack consists of 13 elements in a series, each consisting of eight cells in parallel. Compared with the lead-acid batteries, this pack's nominal voltage is 48.1 V, the total capacity increased by 28%, and the volume and weight reduced to approximately 63% and 80%, respectively. Specifically, the lithium battery pack significantly improves the bicycle's performance and simultaneously reduces the volume and total mass of the bicycle.

The battery monitoring system in this paper consists of the following two parts: the upper and lower machines. Figure 1 is the architecture of the lower machine installed in the electric bike; the lower machine obtains the signals, such as voltage, current, and temperature. The master control unit utilizes these messages to estimate the batteries' SOC and SOH (state of health). The back-to-back MOSFET controls the charging and discharging procession, thereby ensuring the security of the circuit. The equalization circuit can improve the consistency of series connected cells. Scientific and technological progress has brought the smart phones to numerous households allowing for the monitoring of information of electric bicycle batteries by mobile phone [17]. The data exchange between cell phones and lower machine can be achieved via Wi-Fi. Accordingly, a Wi-Fi sending and receiving module is added to the system as follows: the battery status information was sent to mobile phones via Wi-Fi, and the lower machine could obtain the commands sent by mobile phones.

Figure 2 shows the architecture of client-monitoring software installed on the Android phone (upper machine). First, in the upper machine, the Wi-Fi module reads the IP and port number before it establishes a connection between the lower machine and the upper machine after adding the parameters to the database and enters the main control interface, which processes the received data and displays related messages. The user can send a command to control the procession of discharging/charging through the Wi-Fi.

Information displayed by the upper machine includes all cells' voltage, equalization status, battery pack's voltage, current, temperatures of the environment and inside the battery pack, the SOCs, and capacities of the cells, and apart from displaying the messages, the upper machine can also record information for playback analysis.

*2.1. Signal Acquisition.* Voltage, current, and temperature are the foundation of estimation, protection, and control. The voltages of the pack and cells were utilized to protect the cells from overcharging and overdischarging and can also be used to estimate the cells' capacity when the voltages are adequately stable. Current was used to calculate the SOC. The temperature of multiple points was utilized to correct temperature-related battery characteristics, such as internal resistance and capacity.

*2.2. Lithium Battery Protection.* Because of the characteristics of lithium battery, the battery pack is prone to damage by overcharging, overdischarging, and overheating. Thus, cutting the power from the circuit is important when the voltages, current, or temperature reaches the thresholds and provides the corresponding warning for malfunction. Thus, the back-to-back MOSFET was employed to control the on/off of the main circuit. Battery charging control is an important part of BMS. Panasonic recommended the Constant Current-Constant Voltage charging method, and the standard charging current is 1.625 A. As the experimental electric bicycle uses eight batteries in parallel as a group, thus, the battery pack's standard charging current is 13 A. However, the charge current of the commonly used charger for electric bicycle is only from 1.5 A to 3 A, which is safe in the constant current charging stage. When the charge voltage reaches 54 V, it will go to the constant voltage charging mode. During the entire process of charging, the main control unit monitors the voltage of each cell in real-time. When any of the cell voltages reaches 4.2 V, the system will immediately control the MOSFET to turn off the power bus to suspend charging. When cell voltages are equalized, the control unit will control the MOSFET to turn on the power bus to go on charging until the discharge current reaches below 520 mA (65 mA∗8). Similarly, the discharge phase needs to be controlled as well. The NCR18650B datasheet recommends that the end of discharge voltage is 2.5 V, but to extend battery life, the undervoltage protection is set to 2.7 V. Hence, during the discharge process, when any one of the battery voltage reaches or falls below 2.7 V, the system will immediately control the MOSFET to turn off the power bus and remind the user to charge the battery through the mobile client.

To extend battery life, the maximum charge current is set as 15 A. After calculation, in normal working conditions, the maximum discharge current can reach 30 A. To prevent short circuit and ensure the safety of the battery, we set the maximum discharge current to 40 A. When the circuit current exceeds threshold, the MOSFET will automatically turn off the power bus.

*2.3. SOC and SOH Estimation.* The SOC of a pack of batteries is analogous to the fuel gauge of a conventional motorcycle. SOC can be used to predict travel distance in the current load conditions and is highly dependent on individual cell performance. SOC was calculated in Coulomb counting using Formula (1), which is a simple technique that integrates the current with time [18]. However, the process actually suffers from drift and requires periodic recalibration [19]. SOC $(t_0)$ was recalibrated by SOC–OCV (open circuit voltage) curve when the voltage is adequately stable.

The SOH of a battery cell used to describe its physical condition is characterized by a cell parameter correlated with its aging, which in turn is correlated with the battery's ability to store energy [20].

SOH was calculated using Formulas (2) and (3) as follows:

$$\mathrm{SOC}(t) = \mathrm{SOC}(t_0) - \frac{\left(\int I(t)\,dt\right)}{C_M}, \quad (1)$$

$$C_M = \frac{(\text{Passed Charge})}{(\mathrm{SOC}_2 - \mathrm{SOC}_1)}, \quad (2)$$

$$\mathrm{SOH} = \frac{C_M}{C_N} \times 100\%, \quad (3)$$

where SOC$(t_0)$ is the SOC at the beginning of the discharge or charge; $I(t)$ is the working current; $C_M$ is the current capacity of the battery; $C_N$ is the brand new capacity of the battery; Passed Charge is the total capacity flow out of the battery; and $\mathrm{SOC}_1$ and $\mathrm{SOC}_2$ are the SOCs of the battery before and after discharge, respectively, and they are both obtained by using the OCV method according to the OCV-SOC curve.

*2.4. Passive Equalization.* Differences are observed in terms of capacity, internal resistance, and voltage among different batteries, even the same type and batch of batteries. Second, the change in battery temperature in the battery pack, ventilation conditions, self-discharging factors, and electrolyte density difference and the inconsistency of cells will increase along with the increasing working time. The pack's capacity will also be affected by the poor cells. Thus, the equalization circuit is required. The passive equalization circuit has a simple structure, with low-cost, easy-to-implement, and high-reliability features, suitable for the lithium battery pack in the electric bicycle.

*2.5. Android Client-Monitoring Application.* The screen in front of the electric bicycle is extremely simple, which cannot display all necessary information of the lithium-ion battery pack. An Android client app can solve this problem; the user can use a smart phone to check the status (voltage, current, SOC, and SOH) of the battery pack. Thus, client interface display is a great supplement for the screen in the electric bicycle.

The Android system is designed for the mobile Internet system, which is supported by several Internet links. As an open access intelligent operating system, Android has the most obvious advantage because it can support users' software with comprehensive functionality. Each application can design its unique function module and when the module is set up in accordance with the rules of the application framework, it can gain the support of the Android system and interact with other function modules. Based on the advantages of Android [17], the smart phone's efficient processor and large high-resolution screens can efficiently monitor the lithium battery pack of electric bicycle using smart phones [21].

SOC characterizes the proportion between the residual battery capacity in its present state and total capacity after completely charging the battery. However, SOC is unable to clearly tell the user how many miles the bike can run, which is a major concern to the user. The remaining mileage can be calculated by (4) as follows.

$$M_L = \text{Voltage}_1 * \text{SOC} * C_M * M_E$$
$$M_E = \frac{\text{Trip}}{(\Delta\text{SOC} * C_M * \text{Voltage}_2)} \tag{4}$$
$$\text{Trip} = \int_0^{\Delta t} \frac{2 \times \pi \times R_{pm} \times R}{I_r} dt,$$

where $M_L$ is the left miles, $C_M$ is the current capacity of the battery, Trip is the driven distance during time $\Delta t$; $\Delta$SOC is SOC change during $\Delta t$; $R_{pm}$ is the motor speed ($r$/min), $I_r$ is the transmission ratio between motor and wheels, and $R$ represents the radius of the drive wheels; $\text{Voltage}_1$ is the equivalent voltage related to the current SOC and $\text{Voltage}_2$ is the equivalent voltage during time $\Delta t$. Equivalent voltage related to the current SOC is the current nominal voltage. It can be calculated by dividing the remaining energy (Wh) by the remaining capacity (Ah) of the battery.

Then the remaining mileage is sent to the mobile client to display through the Wi-Fi module.

## 3. Design of the Hardware System

*3.1. The Control Unit.* To improve accuracy of SOC estimation, we collected current and voltage data with high accuracy. The microcontroller collects battery voltage, temperature, and current. Further, the microcontroller calculates the SOC by Formula (1) and the SOH by Formula (3). Other features include the ability to manage and control serial-connected 13 cells, obtain motor speed from motor controller through controller area network (CAN) bus, and communicate with mobile client through the WIFI model. Given that the workload is very large and we want to apply further the system to the electric kart BMS, thus, we used a microcontroller with stronger performance. This paper uses low power consumption, 32-bit STM32F401RCT6 single-chip microcomputer as the main control unit's microprocessor and has the following features: frequency up to 180 MHz, 512 kB of flash memory, 128 KB of SRAM, up to 20 communication interfaces, and up to 24 channels of 2.4 MSPS ADC. An external 256 KB of EEPROM chip is used to save the last SOC and SOH when the BMS suddenly loses power value. The main chips and peripheral circuit structure are shown in Figure 3.

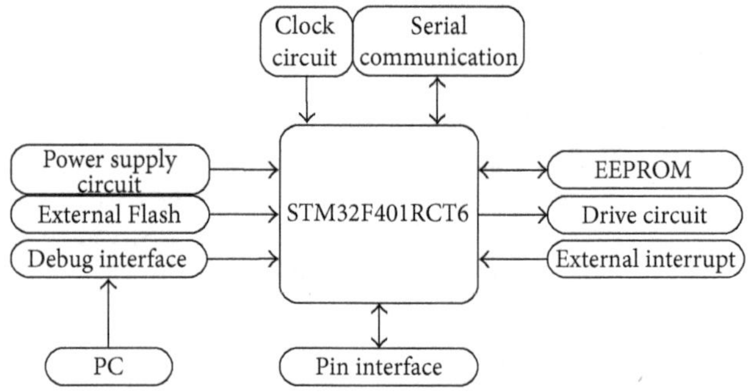


FIGURE 3: Structure of the main control unit.

*3.2. Voltage Acquisition.* The lithium battery pack contains 13 cells in series and thus, 13 voltages need to be acquired. Finding a chip to collect the voltages of 13 cells simultaneously can be difficult. LTC6804-2 is one of the few voltage collection chips which can simultaneously collect the voltages of the 12 cells. Its measurement error is also less than 1.2 mV, and the measurement time is less than 290 $\mu$s. Therefore, the former 12-battery voltage is measured with the chip, and for the 13th battery, this paper used the continuously self-calibrating analog to digital (A/D) converter (ADS1100) to sample voltage. The device has high precision and can communicate with a single-chip microcomputer through IIC interface. Given the long-term connection between the battery management system and the 13th cell, we designed a voltage follower based on the op amp, which can improve the input impedance and reduce leakage current, to avoid excessive power consumption. This configuration does not seem to be as strong as two LTC6804s, but we have tested the setup to meet the requirement of the system (the measurement error of the 13th cell's voltage is 0.018%.). This proposed system can save one LTC6804 chip, which could greatly reduce costs.

*3.3. Current Acquisition.* Circuit current is not only an important parameter of battery protection system but also crucial to the estimation of the SOC.

This design adopts shunt resistance to measure the current through the ADS1246, which is a highly integrated precision 24-bit analog to digital converter (ADC), to ensure accurate current measurement. The resistance of the shunt resistor is 1 mR and accuracy is 1%, with ±100 PPM/°C. We used the open-loop voltage and the ampere-hour counting method to calculate the SOC. The dynamic range of current is quite large, which is almost up to 30 A, but typically, the current value is below 5 A. To reduce energy loss and impact of temperature drift, we used this small resistance measurement resistor. The planned accuracy for the current measurement is 1 mA, which needs a high-resolution sampling method. Thus, we used the AD1246 chip, which is a 24-bit resolution, with integrated gain and differential input A/D, and has up to 2 kSPS conversion chips. In this design, the current is integrated every 5 ms. To obtain the highest possible accuracy, we set the sampling frequency as 200 Sps.

*3.4. Temperature Acquisition.* Temperature is gathered by the LTC6804 temperature acquisition function. Using a 10 K NTC measuring resistance, the B25/50 value is 3950. The temperature acquisition frequency is 1 Hz. The planned accuracy for the temperature measurement is 1°C. The function

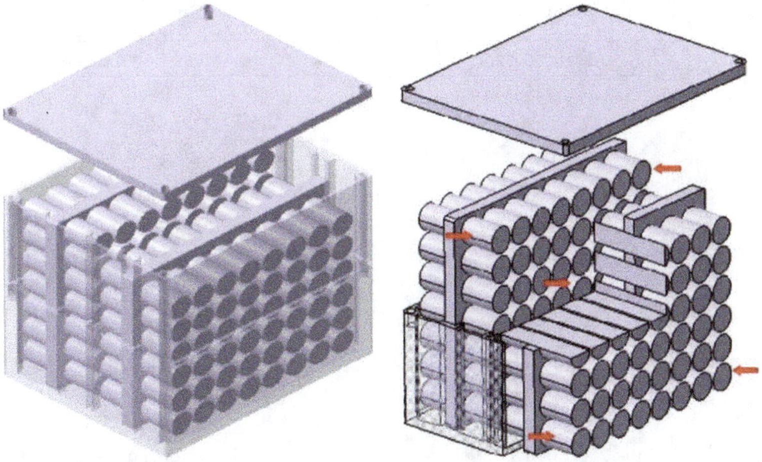

(a) Battery pack 3D model (b) Sectional view

FIGURE 4: Positions of the temperature sensors in the battery pack.

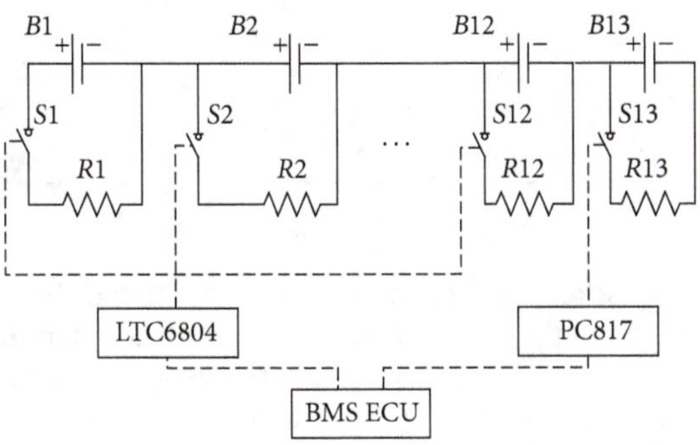


FIGURE 5: Principle of passive equalization.

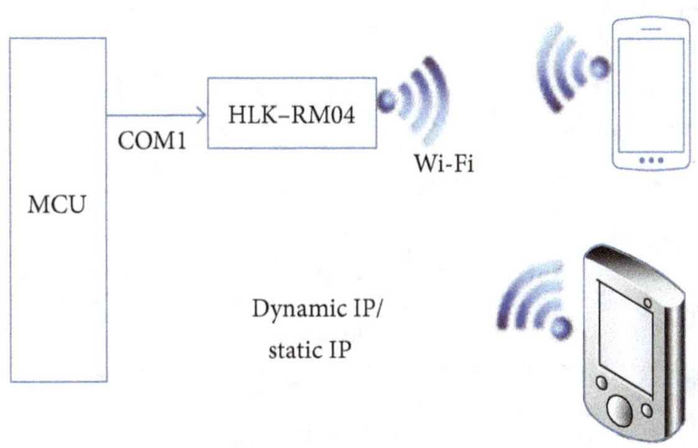


FIGURE 6: Schematic diagram of communication between the upper monitor and the lower computer.

of the temperature sensor is to ensure that the temperature of the battery does not exceed the temperature specified in the manual (charge −10~+45°C, discharge −20~+60°C). Thus, the five temperature sensors are evenly placed inside the battery packs and stick on the surface of the cells. We have roughly marked the position of the temperature sensors with red arrow in Figure 4(b).

*3.5. Equalization Function.* Passive equalization is used for its simple structure, low-cost, easy-to-implement, and high-reliability features. Figure 5 shows the principle of passive equalization. The bypass resistance is 100 Ω. The maximum current that the equalization resistor can withstand is 60 mA. The power of the equalization resistor is 1/3 W. We used the LTC6804 chip to control the first 12 equalization resistors and used PMOSFET to control the 13th equalization resistor in LTC6804's control principle. The PMOSFET is controlled by the control unit using opt coupler (PC817) isolation technology. During the charging process, if the voltages of some cells were higher than others by 50 mV, the electronic control unit (ECU) in the lower machine will control the bypass switches to deplete the excess energy of the higher voltage cells.

*3.6. Wi-Fi Sending and Receiving.* Figure 6 shows that the Wi-Fi module selects the UART ETH WIFI chip HLK-RM04. The chip is based on universal serial interface and abides by the TCP/IP protocol. It can also be used in a port equipment to transmit data through the network. The Android smart phone devices typically have Wi-Fi communication functions, which can communicate with the lower machine through Wi-Fi.

## 4. Design of Software System

*4.1. Software Design of Lower Machine.* Figure 7 shows that the control module obtains the voltages of 13 cells through the voltage collection module and calculates the total voltage. The control model supervises the charging and discharging processes according to the current acquired by the current module and the temperature acquired by the temperature module. The SOC and SOH of batteries were estimated through Formulas (1) and (2), and the serial cells were balanced through the passive equalizer module. The control model communicates with the Android client to send the batteries' message or receive the user's command simultaneously.

TABLE 1: Comparison of acquisition value with measured value.

| Parameters/unit | Acquired data | Measured data | Error (%) |
|---|---|---|---|
| Voltage of cell 1/V | 3.9852 | 3.9860 | −0.020 |
| Voltage of cell 2/V | 3.9832 | 3.9831 | 0.003 |
| Voltage of cell 3/V | 3.9574 | 3.9579 | −0.013 |
| Voltage of cell 4/V | 3.9637 | 3.9630 | 0.018 |
| Voltage of cell 5/V | 3.9845 | 3.9850 | −0.013 |
| Voltage of cell 6/V | 3.9802 | 3.9807 | −0.013 |
| Voltage of cell 7/V | 3.9795 | 3.9788 | 0.018 |
| Voltage of cell 8/V | 3.9832 | 3.9839 | −0.018 |
| Voltage of cell 9/V | 3.9808 | 3.9817 | −0.023 |
| Voltage of cell 10/V | 3.9809 | 3.9815 | −0.015 |
| Voltage of cell 11/V | 3.9810 | 3.9816 | −0.015 |
| Voltage of cell 12/V | 3.9797 | 3.9789 | 0.020 |
| Voltage of cell 13/V | 3.9763 | 3.9770 | −0.018 |
| Current/A | 1.621 | 1.62242 | −0.088 |
| Current/A | −12.136 | −12.11489 | −0.174 |
| Temperature 1/°C | 19.9 | 19.8 | 0.505 |
| Temperature 2/°C | 19.9 | 19.9 | 0.000 |
| Temperature 3/°C | 19.8 | 19.9 | −0.503 |
| Temperature 4/°C | 19.4 | 19.7 | −1.523 |
| Temperature 5/°C | 19.7 | 19.6 | 0.510 |

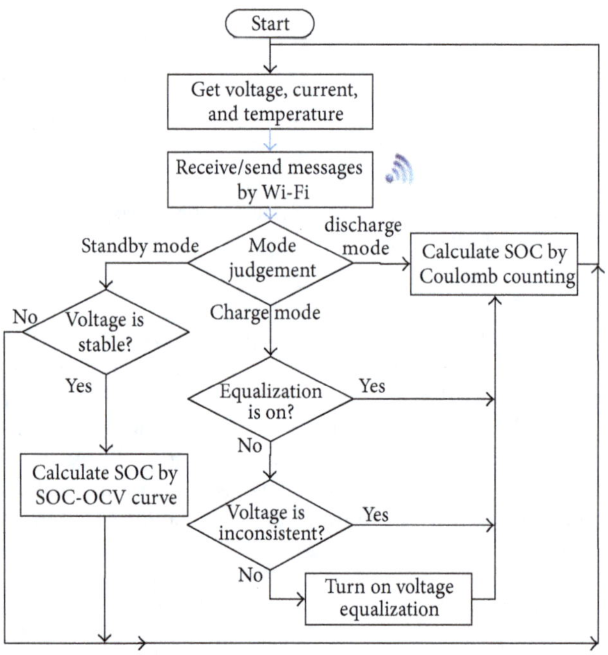


FIGURE 7: Control flow of the lower computer.

*4.2. Android Client Design.* A Samsung GALAXY NOTE 3 N9006, Android version 4.3, was used as a platform to run the upper machine software and test the Android client system. The development environment for the software is Android SDK, Java JDK6, and Eclipse. The flowchart of the operating program is shown in Figure 8. The lower machine can communicate simultaneously with multiple Android clients by assigning a port number for each client. When the users log in to the system by using the port number and IP address, the Android client receives the message posted by the lower machine and then analyzes and assigns parameters to the monitor interface. The actual process is shown in Figure 8.

## 5. Validations of System Functions

An experimental circuit board was designed according to the lower machine function and installed on the experimental electric bicycle to verify the feasibility and practicability of the battery monitoring system. A BMS app (based on Android 4.4) was designed according to the upper machine function. After installing the app on a Samsung GALAXY NOTE 3 and connecting to the Wi-Fi network launched by the lower machine, the NOTE 3 could receive and display messages from the battery pack. The experimental tools and monitoring system are shown in Figure 9.

*5.1. Verification of Data Acquisition Accuracy.* Verifications of voltage and current acquisition are obtained by comparing voltage and current values collected from the measurement module and measured by Agilent 34450A (Figure 9(c)), which is a 5.5 Digit Multimeter. After calibration, the fabricated circuit can obtain almost the same measurement results as that of the 34450A. The accuracy of temperature is verified by the Yokogawa MV2000 (Figure 9(b)), which has accuracy of 0.15%. The battery monitoring system uploads data to the Android monitoring software via Wi-Fi for monitoring and storage while simultaneously measuring the voltages of the battery cells, the current in the circuit, and ambient temperature through a multimeter. Table 1 shows that the

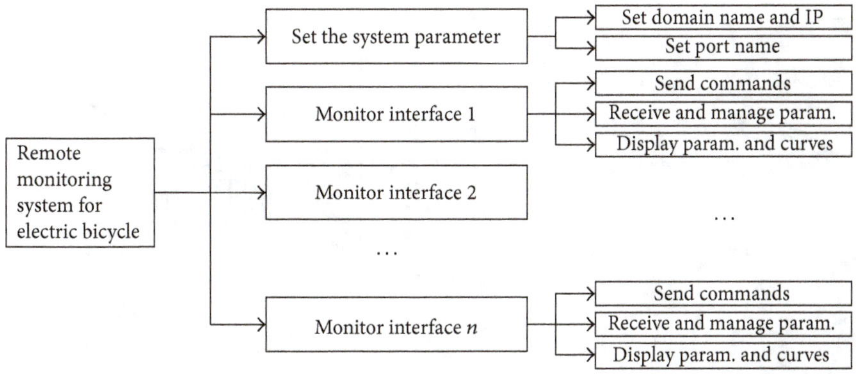


FIGURE 8: Framework of Android client.

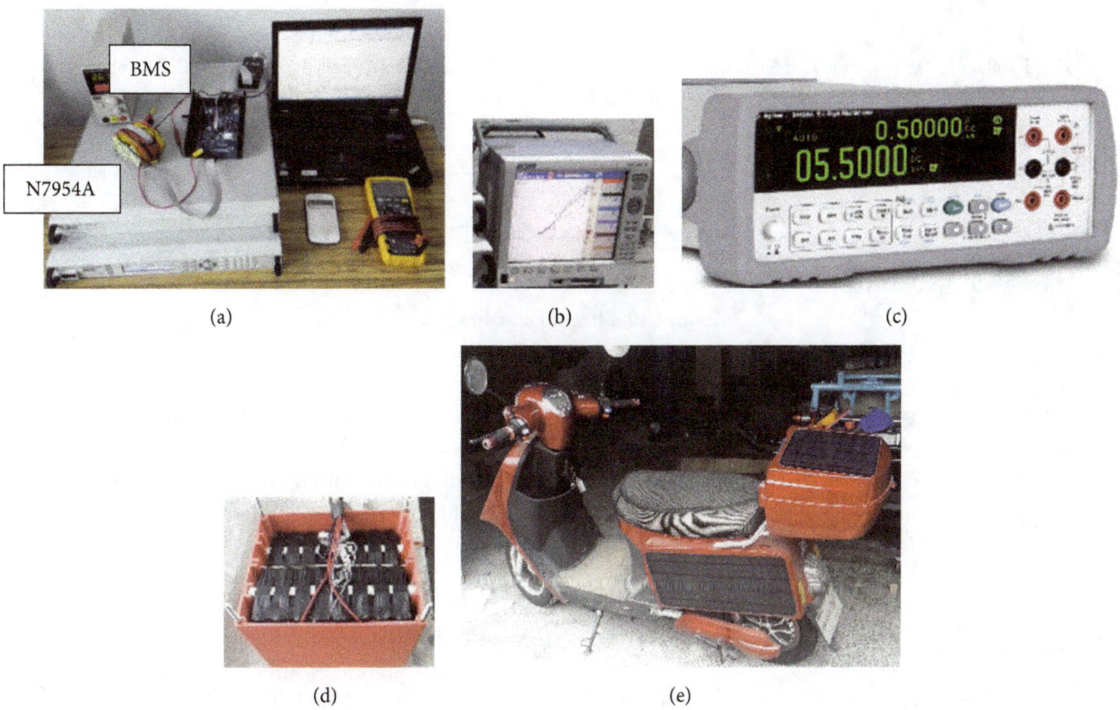


FIGURE 9: Pictures of the laboratory equipment (a, b, c), battery pack (d), and e-bike (e).

errors of voltage and current were within 0.2% and the errors of temperature were within 2%, which were sufficiently precise for application in the electric bicycle.

*5.2. Verification of Battery Protection Function.* Verification of battery protection function is focused mainly on whether the circuit can automatically disconnect when the batteries encounter too low or too high voltage, as well as extraordinary current, and temperature. Battery protection is verified by this concept: MOSFET switches off when any one of the battery voltages reaches below 2.7 V or higher than 4.2 V (the range of 2.7 V to 4.2 V is safe). The NCR18650B datasheet recommends that the end of discharge voltage is 2.5 V, but to extend battery life, the undervoltage protection is set to 2.7 V. The test shows that the back-to-back MOSFET can guarantee the batteries to work safely in any condition.

*5.3. Verification of Equalization Function.* Verification of balance function is centered on the charging stage and should ensure that all series cells can simultaneously achieve the fully charged stage. Figure 10 shows the equalization effect of each cell. The results indicate that the voltage difference among all battery cells was below 50 mV at the end of the charge.

*5.4. Verification of SOC Estimation Function.* Verification of SOC estimation is conducted by controlling the discharge current through a programmable power supply and comparing the estimated SOC value with the actual tested SOC value. When the voltage is stable, we used the OCV-SOC lookup table to estimate the SOC. In the discharge/charge process, SOC is calculated by ampere-hour counting method, and thus, SOC's accuracy also depends on current measurement accuracy. Figure 11 is used to compare the BMS estimated and

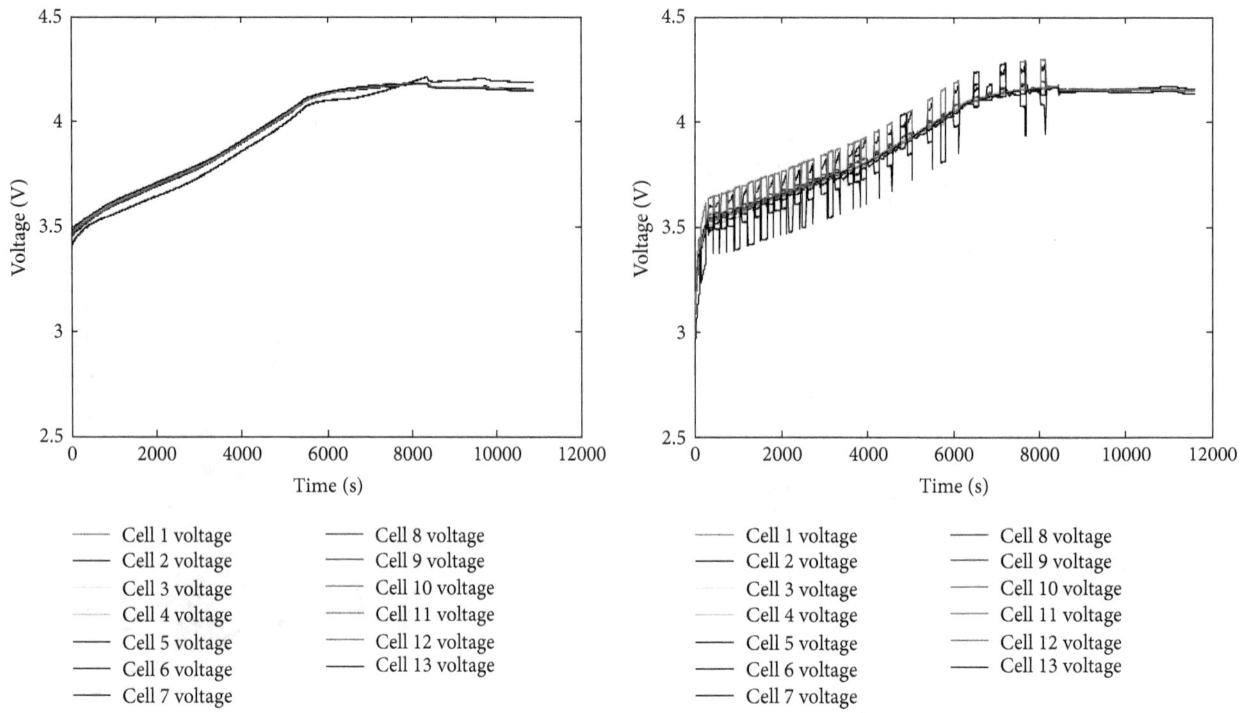


(a) Charge without equalization

(b) Charge with equalization

FIGURE 10: Effect of equalization.

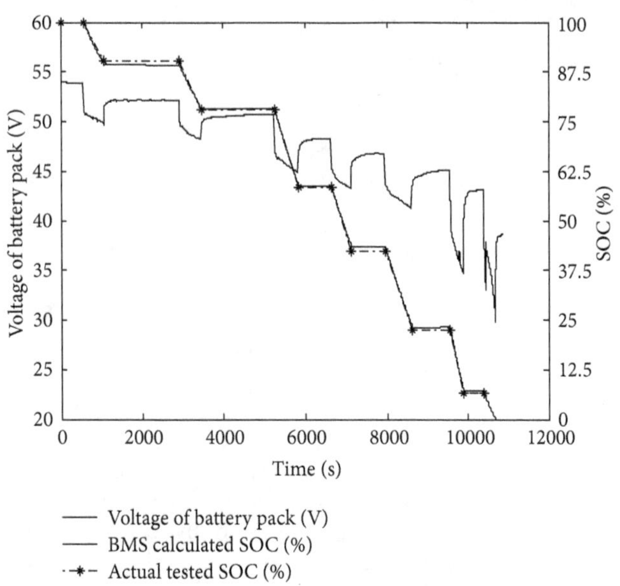


FIGURE 11: SOC comparison between the calculated and the estimated.

actually tested SOC. The actually tested SOC is calculated using the programmable power N7954A (Figure 9(a)), and the difference between the two is within 3%, which is higher than the SOC precision requirement of electric bicycles.

*5.5. Verification of SOH Estimation Function.* To verify that the SOH estimation can automatically correct, we set up three different cell capacities. The first group's initial capacity was set to 4000 mAh, which was far greater than the actual value; the second group's initial capacity was set to 3250 mAh, which was equal to the actual value; and the third group's initial capacity was set to 2000 mAh, which was far lower than the actual value. Then, we controlled battery discharge intermittently by using programmable power supply N7954A. Figure 12(b) shows battery voltage. When the voltage of the battery pack stabilizes after discharge and meets the SOH update conditions (the SOH updates only when the voltages become stable before and after a charge or discharge process, and the amount of passed charge is greater than 30% of the battery capacity; Formula (2) will be used when these conditions are met), the cell capacity will be automatically updated in accordance with Formula (2). As shown in Figure 12(a), the cell capacity can be automatically corrected by Formula (2), which results in a gradually close solution to the ideal value, regardless of the initial value. The estimated SOH then can be calculated by Formula (3) and can be corrected automatically to nearly the actual value, after the multiple discharge cycles.

*5.6. Verification of Communication System.* The verification communication system was created by connecting the mobile phone to the Wi-Fi network delivered by the low computer. Figure 13 shows that after connecting successfully, the monitoring software begins to receive the data and display them on the interface. The Android client can provide a more flexible interface, as shown in Figures 13(a), 13(b), and 13(c), without adding too much cost or upgrading the existing front panel.

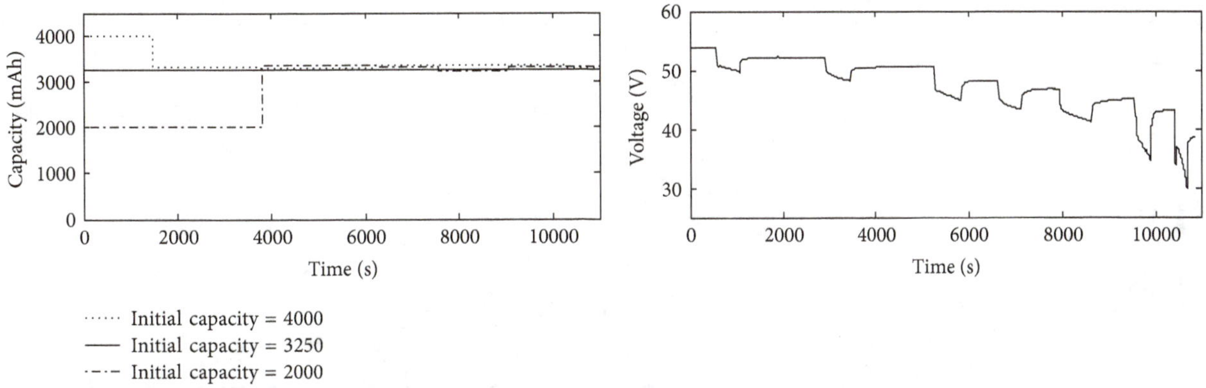


(a) Cell capacity update

(b) Battery pack voltage

FIGURE 12: Automatic correction of the SOH.

(a) (b) (c) (d)

FIGURE 13: Monitor interface (a, b), data replay interface (c), and command send interface (d) on the Android client.

The Android client displays each cell's voltage, balance state, the pack's SOC and SOH, and total power to the user. In addition, the program reports how many miles the battery can provide, which allows the users to understand their e-bike better. Figure 13(d) shows the commands send interface. By the power on or power off button, we can send command to the lower machine to control the MOSFET's on or off. When we turn the MOSFET off, the pack cannot charge or discharge any more. Data with time stamping will be recorded into the new notepad for further experimental analysis and data playback.

*5.7. Verification of Protecting Battery Life.* Reference [22] shows that when the cut-off voltage decreases from 4.2 v to 4.1 v, the capacity degradation rate will be reduced by 38%, and that when the charging current rate is greater than 1 C, the increase rate of 1 s resistance will be accelerated rapidly, which is a key parameter to identify the battery life. Reference [23] provides some principles to prolong battery life, such as charging the battery just before using and charging only when the capacity for next use is insufficient. The aging test in [24] shows that the cycle count is 4,200 when 20% capacity is consumed by cycling in the SOC range 0.4–0.7, and the cycle count is reduced to 550 when 20% capacity is consumed by cycling in the SOC range 0-1.

To prolong the battery life, this paper monitors the battery pack through the app running on the mobile phone. When the SOC is less than 20%, the phone will remind the user to charge in time. When the SOC is higher than 80%, it will remind the user to stop charging, and when the SOC reaches 95%, the app will send a message to the lower machine to stop charging. In our experiment, we collected data every 100 cycles. Figure 14(a) shows the capacity fading when the BMS does not have charge management, the battery is cycling in the SOC range 0-1, and the capacity fades quickly. The battery capacity will be reduced by 20% when the cycle count is 600. Figure 14(b) shows that the capacity fading is very slow when the BMS has charge management; the battery is cycling in the SOC range 0.2–0.8; and the capacity is reduced by only 3% when the cycle count is 600.

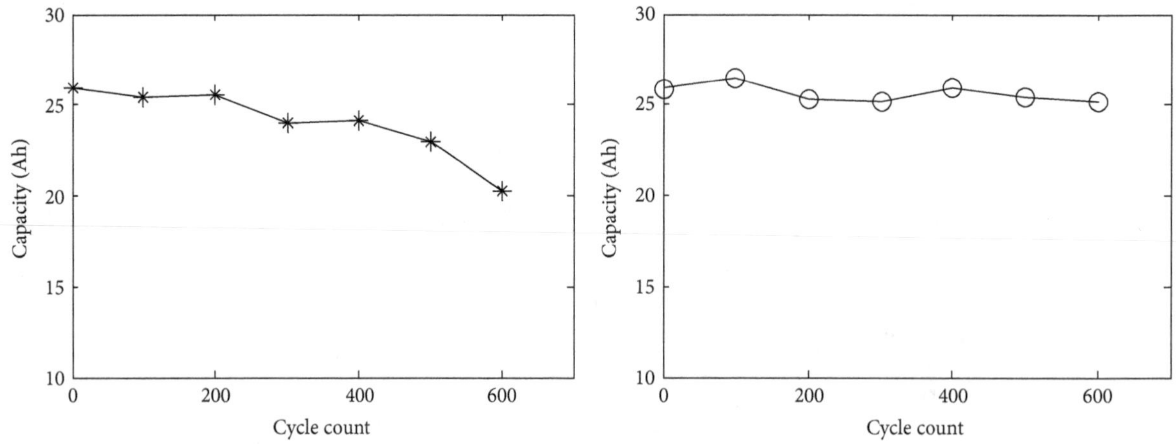


(a) Without charging management

(b) With charging management

FIGURE 14: Effect of the charging management.

Next step, we will specifically develop a matching charger. When we enter the travel plan through a mobile phone, the BMS will automatically command the charger to charge the battery to a reasonable SOC a few hours just before the trip. This approach will prolong the battery life further.

## 6. Conclusions

This paper designed a battery monitoring system based on Android client and the ARM microcontroller for electric bicycle, which can not only meet the requirements of the conventional battery monitoring functions effectively but also facilitate easy monitoring of the state of the battery pack. The battery monitoring system with the Android client system has been successfully applied in electric bicycles, and the experimental schemes in this paper have been found to be feasible and replicable.

The main conclusions are as follows:

(i) The battery monitoring system in this paper can effectively manage the lithium battery pack in the electric bicycle and replacing lead-acid battery with the lithium battery in the electric bicycle can be considered.

(ii) Monitoring the lithium battery pack through the Android client app is feasible and convenient.

## Authors' Contributions

Chuanxue Song conceived and designed the experiments; Yulong Shao and Shixin Song performed the experiments; Silun Peng and Feng Xiao analyzed the data; Chuanxue Song contributed analysis tools; Yulong Shao and Shixin Song wrote the paper.

## Acknowledgments

This research is supported by Graduate Innovation Fund of Jilin University (Project 2017012) and also is funded by the Key Tackling Item in Science and Technology Department of Jilin Province (Grant no. 20150204017GX).

## References

[1] I. Tal, B. Ciubotaru, and G.-M. Muntean, "Vehicular-Communications-Based Speed Advisory System for Electric Bicycles," *IEEE Transactions on Vehicular Technology*, vol. 65, no. 6, pp. 4129–4143, 2016.

[2] T. Jones, L. Harms, and E. Heinen, "Motives, perceptions and experiences of electric bicycle owners and implications for health, wellbeing and mobility," *Journal of Transport Geography*, vol. 53, pp. 41–49, 2016.

[3] P. C. Baptista, C. M. Silva, J. A. Peças Lopes, F. J. Soares, and P. R. Almeida, "Evaluation of the benefits of the introduction of electricity powered vehicles in an island," *Energy Conversion and Management*, vol. 76, pp. 541–553, 2013.

[4] C. Song, Y. Shao, S. Song et al., "Energy management of parallel-connected cells in electric vehicles based on fuzzy logic control," *Energies*, vol. 10, no. 3, p. 404, 2017.

[5] C. Song, Y. Shao, S. Song et al., "Insulation resistance monitoring algorithm for battery pack in electric vehicle based on extended kalman filtering," *Energies*, vol. 10, no. 6, p. 714, 2017.

[6] H. B. Zhou, B. Long, and B. G. Cao, "New energy recovery H∞ robust controller for electric bicycles," *International Journal of Automotive Technology*, vol. 14, no. 2, pp. 283–289, 2013.

[7] Y.-P. Yang and J.-M. Jiang, "Optimal design of an axial-flux permanent-magnet middle motor integrated in a cycloidal reducer for a pedal electric cycle," *Energies*, vol. 8, no. 12, pp. 14151–14167, 2015.

[8] B. Gao, W. Zhang, Y. Tang, M. Hu, M. Zhu, and H. Zhan, "Game-theoretic energy management for residential users with dischargeable plug-in electric vehicles," *Energies*, vol. 7, no. 11, pp. 7499–7518, 2014.

[9] Z. Xin, Y. Ge, W. Bo, Y. Zhang, Y. Li, and J. Yang, "Design of microcontroller-based battery management system for pure

electric vehicle," *Transactions of the Chinese Society of Agricultural Engineering*, vol. 30, no. 12, pp. 163–170, 2014 (Chinese).

[10] C. Sen and N. C. Kar, "Battery pack modeling for the analysis of battery management system of a hybrid electric vehicle," in *Proceedings of the 5th IEEE Vehicle Power and Propulsion Conference, VPPC '09*, pp. 207–212, Dearborn, Mich, USA, September 2009.

[11] S. Lee, J. Kim, J. Lee, and B. H. Cho, "State-of-charge and capacity estimation of lithium-ion battery using a new open-circuit voltage versus state-of-charge," *Journal of Power Sources*, vol. 185, no. 2, pp. 1367–1373, 2008.

[12] P. Singh, C. Fennie Jr., and D. Reisner, "Fuzzy logic modelling of state-of-charge and available capacity of nickel/metal hydride batteries," *Journal of Power Sources*, vol. 136, no. 2, pp. 322–333, 2004.

[13] G. H. Hu, J. M. Xu, J. J. Zhou, and X. Hu, "Design of Li-ion battery management system based on Atmega16," in *Information Technology Applications in Industry*, J. Zhang, Z. J. Wang, S. R. Zhu, and X. M. Meng, Eds., vol. 1-42013, pp. 335–338, 2012.

[14] W. Puviwatnangkurn, B. Tanboonjit, and N. H. Fuengwarodsakul, "Overcurrent protection scheme of BMS for Li-Ion battery used in electric bicycles," in *Proceedings of the 2013 10th International Conference on Electrical Engineering/Electronics, Computer, Telecommunications and Information Technology, ECTI-CON 2013*, Krabi, Thailand, May 2013.

[15] B. Tanboonjit and N. H. Fuengwarodsakul, "Implementation of charger and battery management system for fast charging technique of Li-FePO4 battery in electric bicycles," in *Proceedings of the 2014 9th International Conference on Ecological Vehicles and Renewable Energies, EVER 2014*, Monte-Carlo, Monaco, March 2014.

[16] Y-F. Chou, MW. Cheng, and K-K. Pang, "High precision cell-voltage measurement by a low cost micro-controller for li-ion battery packs used in electric bikes," in *Proceeding of the 22nd International Battery, Hybrid and Fuel Cell Electric Vehicle Symposium and Exposition, EVS*, pp. 583–591, Yokohama, Japan, October 2006.

[17] H. Ben Salem, T. Damarla, K. Sudusinghe, W. Stechele, and S. S. Bhattacharyya, "Adaptive tracking of people and vehicles using mobile platforms," *Eurasip Journal on Advances in Signal Processing*, vol. 2016, no. 1, article 65, 2016.

[18] Z. Chen, X. Li, J. Shen, W. Yan, and R. Xiao, "A novel state of charge estimation algorithm for lithium-ion battery packs of electric vehicles," *Energies*, vol. 9, no. 9, 2016.

[19] J. Li, L. Wang, C. Lyu, W. Luo, K. Ma, and L. Zhang, "A method of remaining capacity estimation for lithium-ion battery," *Advances in Mechanical Engineering*, vol. 2013, Article ID 154831, 2013.

[20] Y. B. MingYu and V. Michael, "Theory and Implementation of Impedance Track™ Battery Fuel-Gauging Algorithm in bq2750x Family," Texas Instruments Application Note (SLUA450), 2008.

[21] Y. Yang, B. Chen, L. Su, and D. Qin, "Research and development of hybrid electric vehicles can-bus data monitor and diagnostic system through OBD-II and android-based smartphones," *Advances in Mechanical Engineering*, vol. 2013, article 741240, 2013.

[22] Y. Gao, J. Jiang, C. Zhang, W. Zhang, Z. Ma, and Y. Jiang, "Lithium-ion battery aging mechanisms and life model under different charging stresses," *Journal of Power Sources*, vol. 356, pp. 103–114, 2017.

[23] G. Lacey and G. Putrus, "Controlling EV Charging Schedules: Supporting the Grid and Protecting Battery Life," in *Proceedings of the 12th IEEE Vehicle Power and Propulsion Conference, VPPC 2015*, Montreal, Canada, October 2015.

[24] Y. Zhu, F. Yan, J. Kang, C. Du, C. Zhang, and R. F. Turkson, "Fading analysis of the Li(NiCoMn)O2 battery under different SOC cycle intervals," *Ionics*, vol. 23, no. 6, pp. 1383–1390, 2017.

# 2

# Effect of Microstructure and Exothermic Reaction on the Thermal Convection in an Enclosure of Nanoliquid with Continuous and Discontinuous Heating from below

**Abeer Alhashash**

*Department of Mathematics, Faculty of Science, Jouf University, 24241 Sakaka Aljawf, Saudi Arabia*

Correspondence should be addressed to Abeer Alhashash; dr.abeeralhashash@gmail.com

Academic Editor: Sergio Nardini

Effect of continuous and discontinuous external heating and internal exothermic reaction on thermal convection of micropolar nanoliquid is studied in the present work. The liquid in the enclosure is a water-based nanoliquid containing Cu nanoparticles. The governing equations are solved numerically using the iterative finite difference method (FDM). The studied parameters are the material viscosity ($0 \leq K \leq 6$), nanoparticles volume fraction ($0.0 \leq \phi \leq 0.2$), and the internal heating ($0.0 \leq G \leq 2.0$). It is found that the convective flow acceleration by adding nanoparticles is retarded by the microrotation and the suppression has a great impact on the weak exothermic reaction for both cases. Increasing the internal reaction decreases the heat transfer rate at the hot wall but increases the heat transfer rate at the cool wall for both cases, Newtonian or micropolar nanoliquid.

## 1. Introduction

Today electrical machines are commonly used in the industry. It is desirable to have machines which are as small, efficient, and long lasting. To be able to make the machines smaller without reducing the power, it is important to get a better cooling of the machine, preferably by using natural convection. The convection describes the exchange of thermal energy between physical systems depending on the thermal, velocity, and pressure, by dissipating heat. Heat transfer rate could be improved by modifying thermophysical properties of the liquid. Using solid nanoparticles having very high thermal conductivities dispersed in liquid would substantially modify the liquid properties. This technique is proposed by Choi [1] and the new liquid is called nanoliquid. Nanoparticles are basically metal, oxides, and some other compounds such as graphene. This engineered liquids have potential applications, for example, heat exchanger, building construction, micro electromechanical systems, nuclear reactors, geothermal power, solar cells, electronics cooling, and so forth. Khanafer et al. [2] and Jou and Tzeng [3] obtained 25% augmentation in the heat transfer rate with 20% weight fractions of the suspended Cu nanoparticles. The Ag, Cu, CuO, $Al_2O_3$, or $TiO_2$ nanoparticles are utilized by Öğüt [4]. They reported 155% augmentation in the heat transfer rate utilizing 20% weight fractions of the Ag nanoparticles. Rashmi et al. [5] filled the enclosure with $Al_2O_3$ nanoparticle. They reported a reduction in heat transfer rate with an increment percentage of the nanoparticle for a particular heating parameter. Sheikhzadeh et al. [6] considered the Brownian and thermophoresis diffusions and found the heat transfer reduction by increasing the bulk volume fraction of nanoparticles. Boulahia et al. [7] observed that the heat transfer rate increases with decreasing the nanoparticle diameter and the highest values of the heat transfer rate occur at 25nm diameter. Motlaghc and Soltanipour [8] reported about 26% augmentation in the heat transfer rate with 4% weight fractions of the solid nanoparticles. Liao [9] investigated systematically the influence of the Rayleigh number on the heat transfer behavior with increasing $Al_2O_3$ nanoparticle volume fraction to 6%. He built a correlation equation for reproducing the critical Rayleigh number with the averaged temperature.

Heat transfer performance strongly depends on the media, i.e., liquid. Aydin and Pop [10] and Saleem et al. [11] showed that micropolar liquid give lower heat transfer values than those of the Newtonian liquid. Micropolar liquid is a subset of the non-Newtonian liquid. It is composed of dumb-bell structural molecules or small and stiff cylindrical components, for example, liquid mixtures, polymer liquid, animal blood, and engine oil. The earlier study of micropolar liquid in enclosures was conducted by Jena and Bhattacharyya [12]. They compared convection stability for several values of micropolar liquid parameters. Wang and Hsu [13] studied the influence of material parameter, geometry aspect ratio, and geometry orientation for the enclosure filled with micropolar liquid at unsteady and stationary conditions. They found that angles of inclination at the maximum values of the heat transfer were coincident for various micropolar liquid in the range of aspect ratio 1.75 to 0.75. Hsu and Chen [14] concluded that thermal performance of a micropolar liquid reduces with the vortex viscosity enhancement and stability of micropolar liquid is higher than that of the stability of Newtonian liquid. A heat sources effect was later included by Hsu et al. [15]. Hsu and Hong [16] investigated the microstructure in an open cavity and found that increasing the Grashof number increases both heat transfer rate and liquid circulation. Gibanov et al. [17] found that an increase in the vortex viscosity parameter leads to attenuation of the convective flow and heat transfer inside a trapezoidal enclosure. Later, Miroshnichenko et al. [18] analyzed the effects of Rayleigh number, Prandtl number, vortex viscosity parameter, and the heater location on streamlines, isotherms, and vorticity profile. Sheremet et al. [19] studied a right-angled wavy triangular enclosure and obtained an essential heat transfer reduction and liquid flow attenuation with vortex viscosity parameter. Turk and Tezer-Sezgin [20] observed that the streamlines and microrotation contours are similar to altering magnitudes. Recently, Ali et al. [21] observed that the expansion of isotherms toward the top boundary surface for greater values of the micropolar parameter and the Nusselt numbers decrease with change in the behavior of the liquid from Newtonian to micropolar.

The vehicle of the current investigation is to study a natural convection heat transfer in a square enclosure filled with micropolar nanoliquid when the bottom boundary is continuously and discontinuously heated at $T_h$ temperature. The top boundary is adiabatic, while the side boundaries walls have constant $T_c$ temperature where $T_c < T_h$. The liquid in the enclosure is a water-based nanoliquid containing Cu nanoparticles. Quadratic heat profile is assumed to be generated internally by the exothermic reaction. An exothermic reaction is a chemical reaction that releases energy by light or heat and the typical application of this process occurs in chemical industry. Similar research conducted by Bourantas and Loukopoulos [22] for continuously heating left wall showed that the microrotation of the nanoparticles decreases heat transfer and should not therefore be neglected when computing heat and liquid flow of micropolar liquid, as nanoliquid. Hashemi et al. [23] and Izadi et al. [24] investigated the copper-water micropolar nanoliquid inside a porous enclosure with continuous heating left wall. The nanoparticles are translating, rotating, and choosing a discontinuous thermal condition along the bottom wall which could have a great impact on the heat transfer rate. Systematical comparison between continuous and discontinuous heating cases is also carried out.

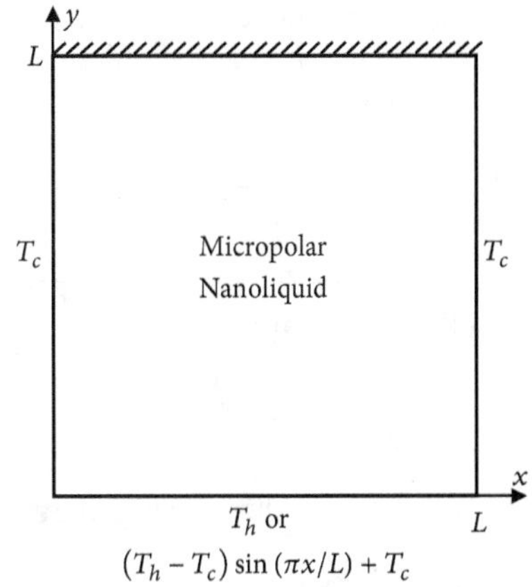


FIGURE 1: Schematic representation of the model.

## 2. Mathematical Formulation

A schematic diagram of a square enclosure with micropolar nanoliquid is shown in Figure 1. The liquid in the enclosure is a water-based nanoliquid containing Cu nanoparticles. Quadratic heat profile is generated internally by the exothermic reaction. The bottom boundary is continuously and discontinuously heated while the top boundary is adiabatic and side boundaries walls have constant low temperature. The governing equation is based on conservation laws of mass, momentum, and energy with appropriate rheological models and equations. For micropolar nanoliquid flow the continuity equation, linear momentum equation, angular momentum equation and energy equations are given as follows:

$$\nabla^2 u = -\frac{\partial \omega}{\partial y} \tag{1}$$

$$\nabla^2 v = -\frac{\partial \omega}{\partial x} \tag{2}$$

$$\rho_{nl}(\mathbf{u}\cdot\nabla\omega) = (\mu_{nl}+\kappa)\nabla^2\omega - \kappa\nabla^2 N + g(\rho\beta)_{nl}\frac{\partial(T-T_h)}{\partial x} \tag{3}$$

$$\rho_{nl} j(u\cdot\nabla N) = \gamma_{nl}\nabla^2 N - 2\kappa N + \kappa\omega \tag{4}$$

$$u\cdot\nabla T = \alpha_{nl}\nabla^2 T + \frac{\lambda}{(\rho Cp)_{nl}}(T-T_c)^2 \tag{5}$$

where subscript $nl$ is nanoliquid, $u$ and $v$ are the velocity components along $x$ and $y$ axes, $T$ is the liquid temperature, $N$ is the component of the microrotation vector normal to the $xy$ plane, $g$ is the magnitude of the acceleration due to gravity, $\rho$ is the density, $\mu$ is the dynamic viscosity, $\kappa$ is the vortex viscosity, $\gamma$ is the spin-gradient viscosity, $j$ is the microinertia density, and $n$ is a constant with values $0 \le n \le 1$, with $n = 0$, called strong concentration of microelements. Further, we assume that $\gamma_{nl}$ has the following form:

$$\gamma_{nl} = \left(\mu_{nl} + \frac{\kappa}{2}\right) \tag{6}$$

The viscosity of the nanoliquid can be approximated as viscosity of a base liquid if it contains dilute suspension of fine spherical particles, which is given by Brinkman [25] as

$$\frac{\mu_{nl}}{\mu_{bl}} = \frac{1}{(1-\phi)^{2.5}} \quad (7)$$

where $\phi$ is the solid volume fraction of nanoparticles. Thermal diffusivity of the nanoliquid is

$$\alpha_{nl} = \frac{k_{nl}}{(\rho Cp)_{nl}} \quad (8)$$

where the heat capacitance of the nanoliquid given is

$$(\rho Cp)_{nl} = (1-\phi)(\rho Cp)_{bl} + \phi(\rho Cp)_{sp} \quad (9)$$

and $k_{nl}$ stands for the effective thermal conductivity of nanoliquid restricted to spherical nanoparticles is approximated by the Maxwell-Garnetss (MG), Öğüt [4] model:

$$\frac{k_{nl}}{k_{bl}} = \frac{k_{sp} + 2k_{bl} - 2\phi(k_{bl} - k_{sp})}{k_{sp} + 2k_{bl} + \phi(k_{bl} - k_{sp})} \quad (10)$$

The thermophysical properties of liquid and the solid copper phases are given by Khanafer et al. [2]. The appropriate the boundary conditions are as follows:

$$u = v = 0 \quad \text{and} \quad \frac{\partial T}{\partial y} = 0 \quad \text{at } y = \ell \quad (11)$$

$$u = v = 0 \quad \text{and} \quad T = T_c \quad \text{at } x = 0 \quad (12)$$

$$u = v = 0 \quad \text{and} \quad T = T_c \quad \text{at } x = \ell \quad (13)$$

$$u = v = 0 \quad \text{and} \quad T = T_h \quad \text{or} \quad T = (T_h - T_c)\sin\left(\frac{\pi x}{\ell}\right) + T_c \quad \text{at } y = 0 \quad (14)$$

The above equations can be cast in nondimensional form by incorporating the following dimensionless parameters:

$$X = \frac{x}{\ell}, \quad Y = \frac{y}{\ell}, \quad U = \frac{u\ell}{\nu_{bl}}, \quad V = \frac{V\ell}{\nu_{bl}}, \quad \Theta = \frac{T - T_c}{T_h - T_c}, \quad Pr_{bl} = \frac{\nu_{bl}}{\alpha_{bl}}, \quad Ra_{bl} = \frac{g\beta_{bl}(T_h - T_c)\ell^3}{\nu_{bl}\alpha_{bl}}, \quad G = \frac{\lambda(T_h - T_c)l}{\alpha_{bl}(\rho Cp)_{bl}} \quad (15)$$

Writing in stream function–vorticity formulation and performing nondimensionalization, the dimensionless form of the governing equations is expressed as follows:

$$\nabla^2\Psi = -\Omega \quad (16)$$

$$\frac{\partial\Psi}{\partial Y}\frac{\partial\Omega}{\partial X} - \frac{\partial\Psi}{\partial X}\frac{\partial\Omega}{\partial Y} = \left(\frac{\mu_{nl}}{\mu_{bl}} + K\right)\frac{\rho_{bl}}{\rho_{nl}}\nabla^2\Omega - K\frac{\rho_{bl}}{\rho_{nl}}\nabla^2 N + \frac{Ra_{bl}}{Pr_{bl}}\frac{(\rho\beta)_{nl}}{(\rho\beta)_{bl}}\frac{\rho_{bl}}{\rho_{nl}}\frac{\partial\Theta}{\partial X} \quad (17)$$

$$\frac{\partial\Psi}{\partial Y}\frac{\partial N}{\partial X} - \frac{\partial\Psi}{\partial X}\frac{\partial N}{\partial Y} = \left(\frac{\mu_{nl}}{\mu_{bl}} + \frac{K}{2}\right)\frac{\rho_{bl}}{\rho_{nl}}\nabla^2 N - 2K\frac{\rho_{bl}}{\rho_{nl}}N + K\frac{\rho_{bl}}{\rho_{nl}}\Omega \quad (18)$$

$$\frac{k_{nl}}{k_{bl}}\frac{(\rho Cp)_{bl}}{(\rho Cp)_{nl}}\nabla^2\Theta = Pr_{bl}\left(\frac{\partial\Psi}{\partial Y}\frac{\partial\Theta}{\partial X} - \frac{\partial\Psi}{\partial X}\frac{\partial\Theta}{\partial Y}\right) - G\Theta^2 \quad (19)$$

The dimensionless boundary conditions are as follows:

$$\Psi = 0 \quad \text{and} \quad \frac{\partial\Theta}{\partial Y} = 0 \quad \text{at } Y = 1 \quad (20)$$

$$\Psi = 0 \quad \text{and} \quad \Theta = 0 \quad \text{at } X = 0 \quad (21)$$

$$\Psi = 0 \quad \text{and} \quad \Theta = 0 \quad \text{at } X = 1 \quad (22)$$

$$\Psi = 0 \quad \text{and} \quad \Theta = 1 \quad \text{or} \quad \Theta = \sin(\pi X) \quad \text{at } Y = 0 \quad (23)$$

## 3. Numerical Method and Validation

The governing equations are categorized as elliptical partial differential equation; one of the well-known methods to solve these equations is using iterative finite difference method. Central difference method is utilized for discretizing the first and second derivative of the governing equations while backward difference method is utilized for discretizing the insulated boundary. The finite difference form of equation relating the energy equation (19) is

$$\left(\frac{k_{nl}}{k_{bl}}\frac{(\rho Cp)_{bl}}{(\rho Cp)_{nl}}\right)\left[\frac{\Theta_{i+1,j} - 2\Theta_{i,j} + \Theta_{i-1,j}}{\Delta X^2} + \frac{\Theta_{i,j+1} - 2\Theta_{i,j} + \Theta_{i,j-1}}{\Delta Y^2}\right] = S_{i,j} \quad (24)$$

This can be simplified as

$$\Theta_{i,j} = \frac{1}{2\left(1 + \Delta X^2/\Delta Y^2\right)} \left[ \Theta_{i+1,j} + \Theta_{i-1,j} + \frac{\Delta X^2}{\Delta Y^2}\left(\Theta_{i,j+1} + \Theta_{i,j-1}\right) + \Delta X^2 \left( \frac{k_{bl}}{k_{nl}} \frac{(\rho Cp)_{nl}}{(\rho Cp)_{bl}} \right) S_{i,j} \right] \tag{25}$$

where

$$S_{i,j} = Pr_{bl}\left[ \left(\frac{\Psi_{i,j+1} - \Psi_{i,j-1}}{2\Delta Y}\right)\left(\frac{\Theta_{i+1,j} - \Theta_{i-1,j}}{2\Delta X}\right) - \left(\frac{\Psi_{i+1,j} - \Psi_{i-1,j}}{2\Delta X}\right) \times \left(\frac{\Theta_{i,j+1} - \Theta_{i,j-1}}{2\Delta Y}\right) - G\left(\frac{\Theta_{i+1,j} + \Theta_{i-1,j} + \Theta_{i,j+1} + \Theta_{i,j-1}}{4}\right)^2 \right] \tag{26}$$

The computation is assumed to move through the grid points from the east to the west and the south to the north. The finite difference form of equation relating the stream function (16), vorticity (17), and microrotation (18) could be treated in the same way. Good approximation of the vorticity at the boundaries is the most critical step in the stream function-vorticity formulation. The vorticity is calculated by

$$\Omega = -\frac{\left(8\Psi_{1,j} - \Psi_{2,j}\right)}{2\left(\Delta Y\right)^2} \tag{27}$$

Similar expressions are written for the left and right walls. Next, the solutions of the algebraic equations are performed using Gaussian SOR iteration. The unknowns $\Psi$, $\Omega$, $\Theta$, and $N$ are calculated until the following criterium of convergence is fulfilled:

$$\max\left|\frac{\zeta_{i,j}^{n+1} - \zeta_{i,j}^{n}}{\zeta_{i,j}^{n}}\right| \le \epsilon \tag{28}$$

where $\zeta$ is either $\Psi$, $\Omega$, $\Theta$, or $N$ and $n$ represents the iteration number and $\epsilon$ is the convergence criterion. Finally, the integration of average Nusselt number is done by using the second-order Simpson method. In this study, the convergence criterion is set at $\epsilon = 10^{-6}$. Regular and uniform grid distribution is used for the whole enclosure. The effect of grid resolution was examined in order to select the appropriate grid density as demonstrated in Figure 2 for $K = 4$, $\phi = 0.02$, $Ra = 10^5$, $G = 1.0$ for continuous case 1. Considering the accuracy and the computation time, the results indicate that an $121 \times 121$ mesh can be used in the final computations. As a validation, the results (left) of streamlines are compared with that obtained by Aydin and Pop [10] (right) for the continuously heated left wall at $\phi = 0$, $Ra = 10^6$, $Pr = 0.71$, $K = 0, 2$ as shown in Figure 3.

## 4. Results and Discussion

The reported results are considered for water and the Rayleigh number is set at $Ra = 10^5$. The microrotation contour is not presented in this work. The analysis in the undergoing numerical investigation are performed in the following range of the associated dimensionless groups: the material viscosity ($0 \le K \le 6$), the solid particle concentration ($0.0 \le \phi \le 0.2$), and the internal heating parameter, $0.0 \le G \le 2.0$.

Figure 4 exhibits the influence of the viscosity parameter, $K$, for continuous heating case and nanoliquid (solid lines) with volume fraction 10% and pure fluid (dashed lines) for $G = 1.0$ on the streamlines and the isotherms. $K$ affects the liquid temperatures and the flow characteristics. An increase in the vortex viscosity parameter reduces the mean velocity, due to increasing of the vortex viscosity on the global viscous effect. The strength of nanoliquid flow circulation is stronger than the strength of base liquid flow circulation at $K = 2$ and $K = 4$. Later, at $K = 6$ the flow acceleration by adding nanoparticles is then retarded by the microrotation effect. The isotherms are focused near the corners of bottom wall for base liquid and nanoliquid irrespective to viscosity parameter. This happened because of the effect of the microstructures nanoliquid; the mean velocity decelerates with the vortex viscosity enhancement. The concentration of microconstituents is enhanced with the raise in $K$. Thus, big amount of the energy inside the system is consumed in processing rotational speed of the micropolar nanoliquid, and, as a result, the liquid movement from the lower to the top portion is retarded. The chamber in the middle, top surface, is obtained for base liquid at $K = 2, 4, 6$. This is related to the quadratic internal heating from the exothermic reaction.

Figure 5 exhibits the influence of the viscosity parameter, $K$, for discontinuous heating case 2 and nanoliquid (solid lines) with volume fraction 10% and pure liquid (dashed lines) where $G = 1.0$ on the streamlines and the isotherms. The strength of nanoliquid flow circulation is higher than the strength of base liquid flow circulation at $K = 2$. Later, at $K = 4$, the flow acceleration by adding nanoparticles is then slightly retarded by the microrotation. Afterwards, suppression of the flow circulation by the increasing the vortex viscosity is more pronounced at $K = 6$. The maximum flow strength of continuous heating case depicted in the previous figure is slightly higher than the maximum flow strength of discontinuous heating case for the considered viscosity parameters. The isotherms of base liquid and nanoliquid are found to be distributed sinusoidally for the all material parameter. The streamlines of micropolar base liquid and micropolar nanoliquid are similar and the isotherms of micropolar base liquid and micropolar nanoliquid are deviated in the upper part of enclosure. There are no chambers in the middle, top surface for the base liquid or nanoliquid at $K = 2, 4, 6$. It seems the sinusoidal external heating effect is superior to the exothermic reaction effect.

Tables 1 and 2 present the maximum values of clockwise and counter clockwise flow circulations against internal heating parameter with different $K$ for continuous case 1 and discontinuous case 2, respectively. It is seen from both tables that, for given values of internal heating parameter, the

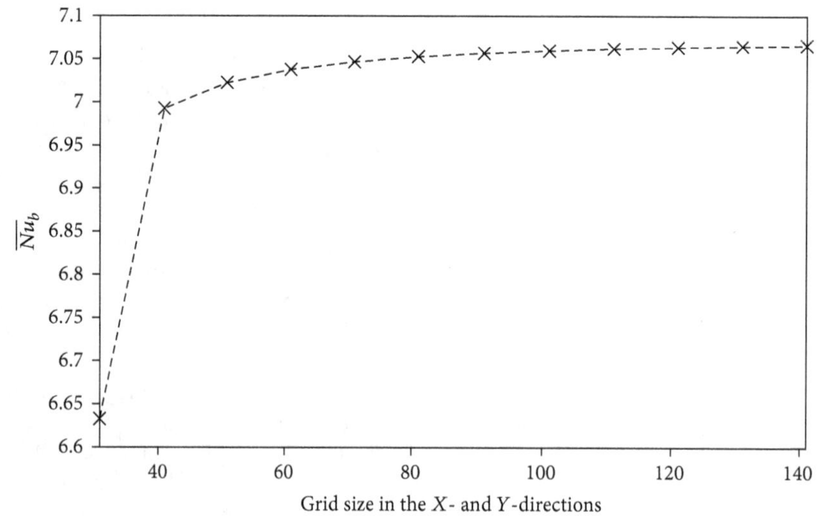


FIGURE 2: Grid independency study: $\overline{Nu_b}$ versus number of grid points.

TABLE 1: $|\Psi_{\min}^{\max}|$ against internal heating parameter with different $K$ for continuous case at $\phi = 0.1$.

| $G$ | $K = 0$ | $K = 2$ | $K = 4$ | $K = 6$ |
|---|---|---|---|---|
| 0.0 | 2.3522 | 1.2945 | 0.9082 | 0.6891 |
| 0.5 | 2.3519 | 1.3061 | 0.9269 | 0.7130 |
| 1.0 | 2.3482 | 1.3161 | 0.9448 | 0.7357 |
| 1.5 | 2.3401 | 1.3234 | 0.9610 | 0.7587 |
| 2.0 | 2.3233 | 1.3240 | 0.9723 | 0.7789 |

decrease of the strength convective flow circulation is with the increase of $K$. The reducing is more significant for lower values of $G$. The strengths of the convective flow circulation between Newtonian and micropolar liquid are different. The maximum flow magnitude decreases by increasing the internal heating parameter for the Newtonian liquid. However, maximum flow magnitude increases by increasing the internal heating parameter for the micropolar liquid ($K = 2, 4, 6$). The maximum flow strength of continuous heating case is slightly higher than the maximum flow strength of discontinuous heating case for the considered viscosity and internal heating parameters.

Figure 6 shows the variations of average Nusselt number at bottom (left) and average Nusselt number at side (right) against $K$ by varying $\phi$ for continuous case 1 (top) and discontinuous heating case 2 (bottom) at $G = 1.0$. Apparently, as the nanoparticles concentration increases, the average Nusselt number at the bottom and side increases, revealing the significant role of the improved thermal properties of the nanoliquid in heat transport. The nanoparticles concentration modifies the heat transfer properties of the nanoliquid significantly. This is due to the fact that the thermal conductivity of the base liquid is weak and the addition of nanoparticles enhances the thermal conductivity of the nanoliquid. The microrotation effect is boosted in the low solid concentration and gets weaker as the concentration of the nanoparticles rises. Here it is doubted how the suppressing reaction of the microstructure on the thermal performance varies with the solid volume fraction. For each nanoparticles concentration, the effect of the material parameter was observed to suppress the thermal performance along the bottom and side wall for both cases, since an enhanced vortex viscosity increases the global viscosity of the liquid circulation, thus reducing the heat transfer rate. Indeed, the quadratic internal heating modifies the convective flow especially at the upper region.

An increase in material parameter reduces the average Nusselt number for cases 1 and 2 as shown in Figure 7. For both cases and fixed $K$, increasing the internal reaction decreases the heat transfer rate at the bottom but increases the heat transfer rate at the cool side. The continuous heating is observed to gain better thermal performance than the discontinuous heating case. This is due to the fact that the continuous thermal condition generates uniform heat along the heated plate that support the buoyancy. At the side cool wall, for $G = 0.5$, the heat transfer rate decreases by $K$, then after reaching a sufficiently large K value the heat transfer rate tends to be stagnant. This interesting result indicates that average Nusselt number at cool wall tends to be identical between Newtonian and micropolar nanoliquid for the weak internal heating at discontinuous heating case. This is due to the weak internal convective flow and nonuniform external convective flow being unable to produce the buoyancy.

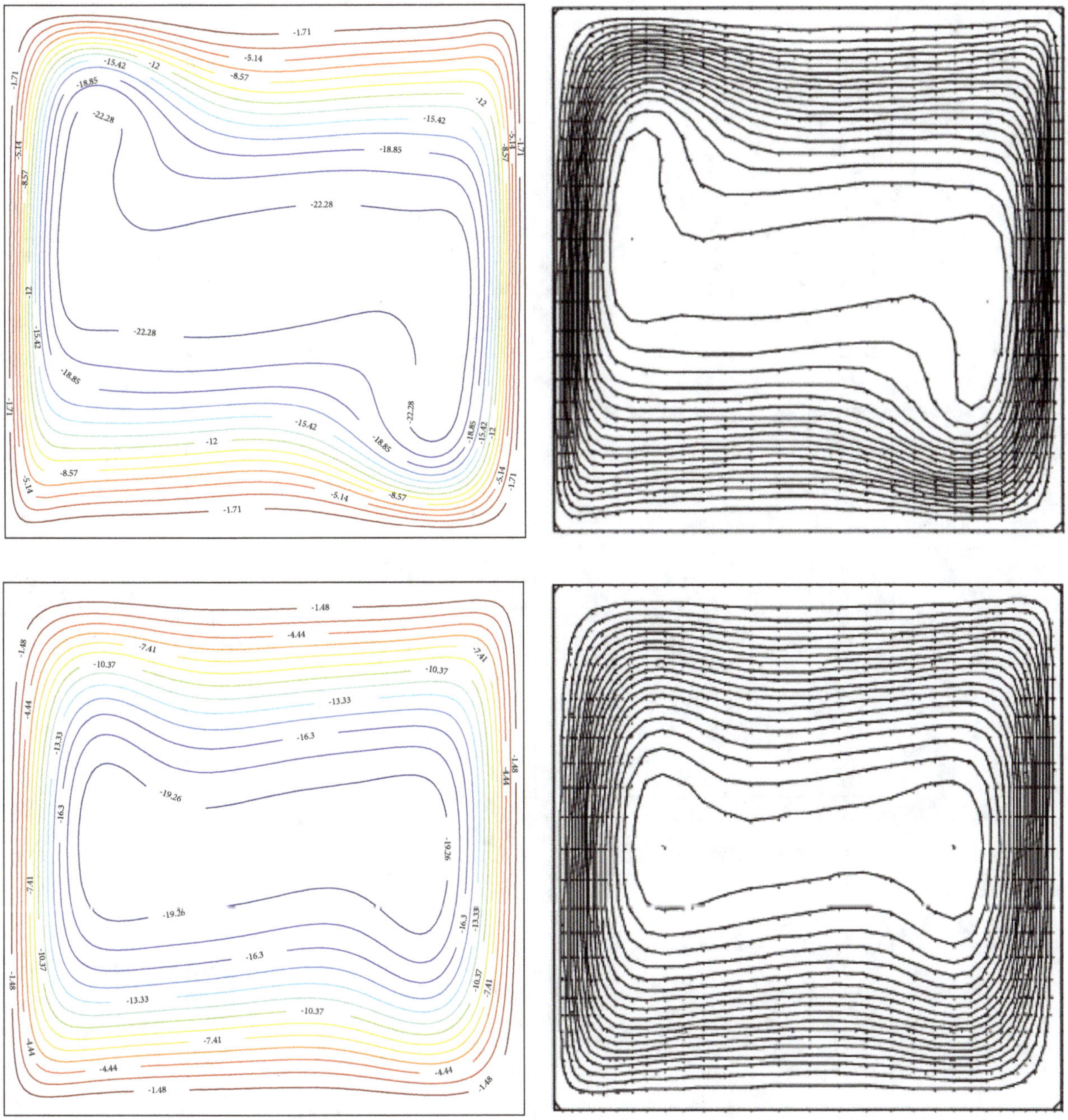


FIGURE 3: Comparison of computed streamlines with Aydin and Pop [10] results for continuous heating left wall at $\phi = 0$, $Ra = 10^6$, $Pr = 0.71$, $K = 0, 2$.

TABLE 2: $|\Psi_{\min}^{\max}|$ against internal heating parameter with different $K$ for discontinuous case at $\phi = 0.1$.

| $G$ | $K = 0$ | $K = 2$ | $K = 4$ | $K = 6$ |
|---|---|---|---|---|
| 0.0 | 2.1126 | 1.1531 | 0.8002 | 0.6003 |
| 0.5 | 2.1130 | 1.1631 | 0.8154 | 0.6186 |
| 1.0 | 2.1112 | 1.1721 | 0.8304 | 0.6373 |
| 1.5 | 2.1064 | 1.1796 | 0.8448 | 0.6566 |
| 2.0 | 2.0981 | 1.1848 | 0.8581 | 0.6786 |

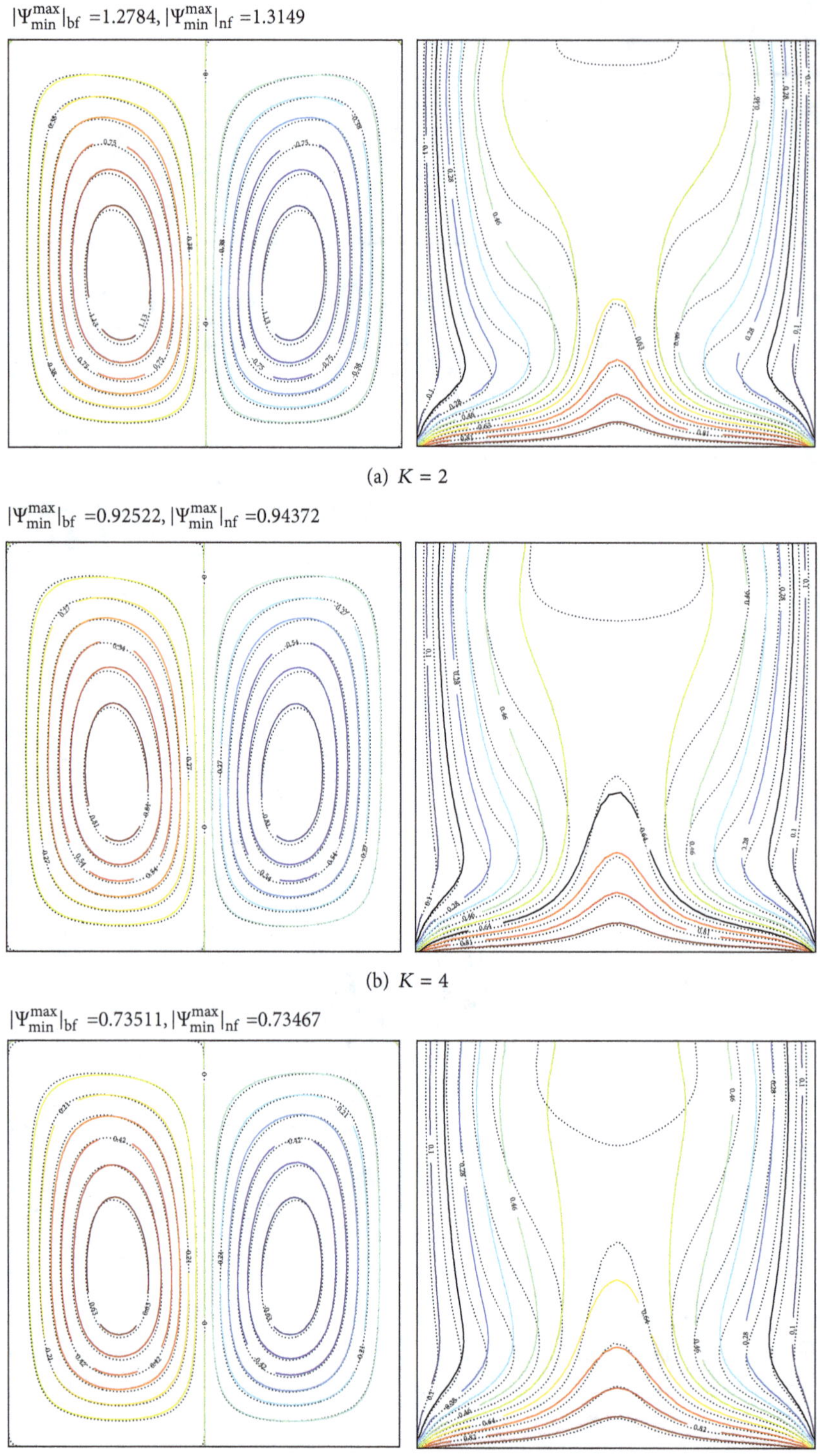


(a) $K = 2$

(b) $K = 4$

(c) $K = 6$

FIGURE 4: Streamlines and isotherms evolutions of nanoliquid (solid lines) and base liquid (dashed lines) by varying the material parameter for continuous heating corresponding to case 1 at $G = 1.0$.

## 5. Conclusions

The main purpose of the present research is to investigate the influence of continuous and discontinuous heating bottom surface and parabolic internal reaction profile on the liquid circulation, thermal distribution, and heat transfer characteristics due to convective heat transfer of micropolar nanoliquid. The dimensionless forms of the model equations

$|\Psi^{max}_{min}|_{bf} = 1.1474$, $|\Psi^{max}_{min}|_{nf} = 1.171$

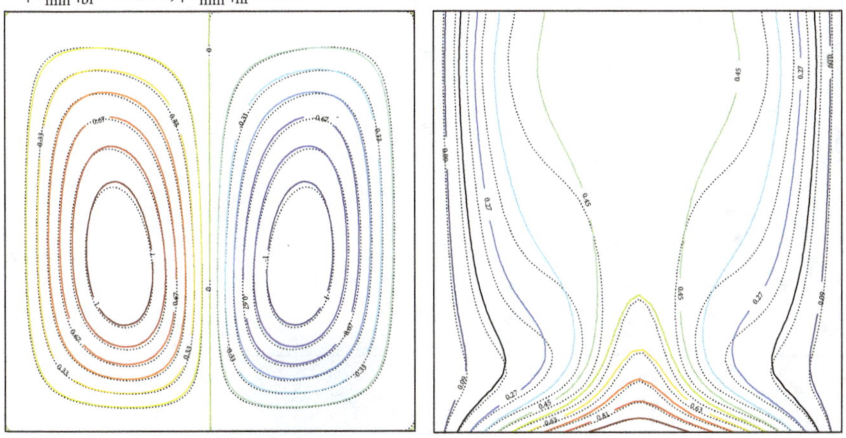

(a) $K = 2$

$|\Psi^{max}_{min}|_{bf} = 0.82352$, $|\Psi^{max}_{min}|_{nf} = 0.82961$

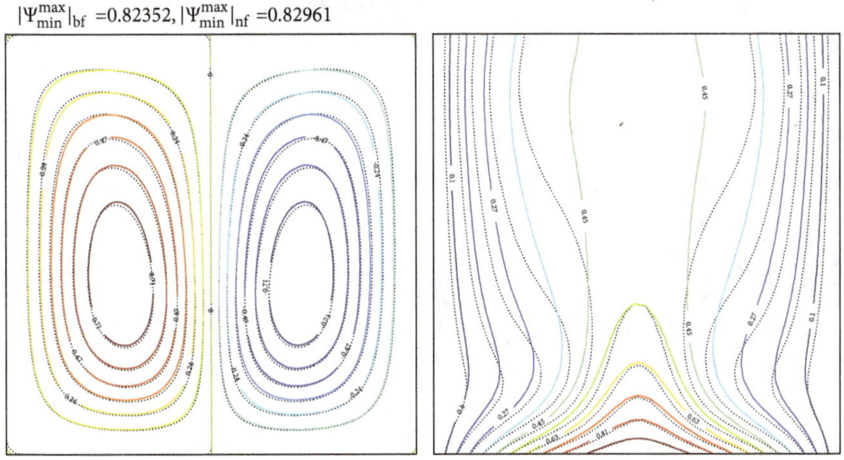

(b) $K = 4$

$|\Psi^{max}_{min}|_{bf} = 0.64788$, $|\Psi^{max}_{min}|_{nf} = 0.63668$

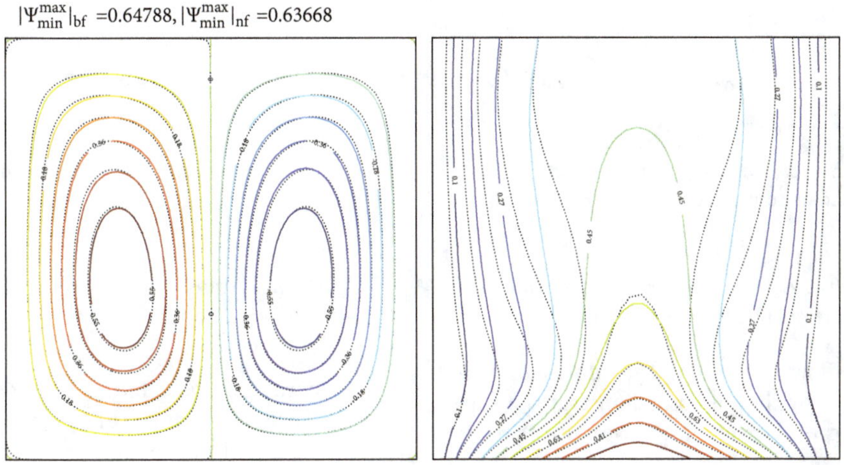

(c) $K = 6$

FIGURE 5: Streamlines and isotherms evolutions of nanoliquid (solid lines) and base liquid (dashed lines) by varying the material parameter for discontinuous heating corresponding to case 2 at $G = 1.0$.

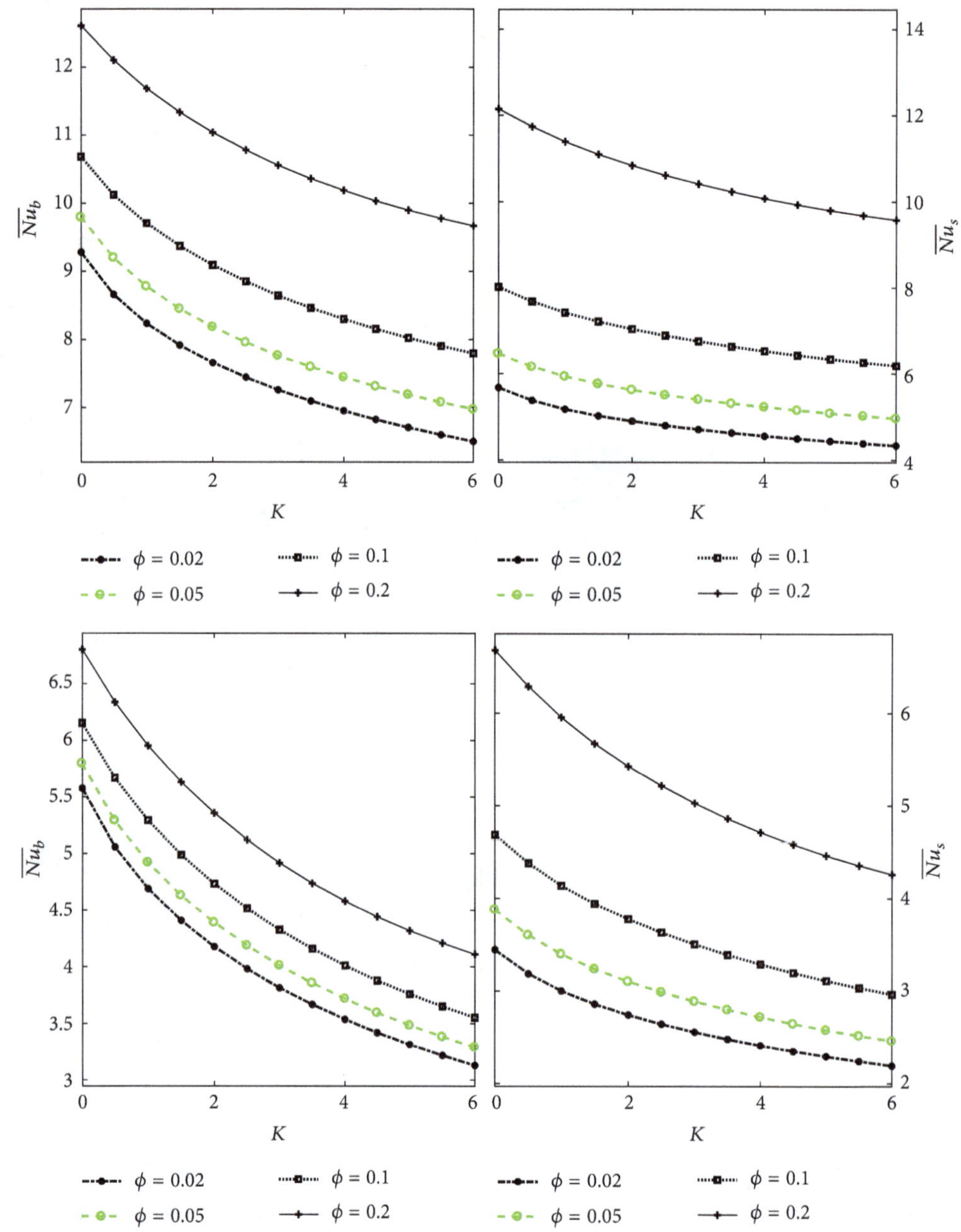


FIGURE 6: Variations of average Nusselt number at bottom (left) and average Nusselt number at side (right) against K by varying $\phi$ for continuous case 1 (top) and discontinuous heating case 2 (bottom) at $G = 1.0$.

are solved using the iterative finite difference technique. The important findings of the current analysis are as follows:

(1) The convective flow acceleration by adding nanoparticles is retarded by the microrotation and the suppression has a great impact on the weak exothermic reaction for both cases.

(2) In general, the heat transfers at hot and cool walls of micropolar nanoliquid are less than that of the Newtonian nanoliquid for both cases. The heat transfer at cool wall tends to be identical between Newtonian and micropolar nanoliquid for the weak exothermic reaction at discontinuous heating case.

(3) The heat transfer at hot and cool walls increases as the solid volume fraction of the nanoparticles increases for both cases.

(4) Increasing the internal reaction decreases the heat transfer at the hot wall but increases the heat transfer at the cool wall for both cases, Newtonian or micropolar nanoliquid.

## References


[1] S. U. S. Choi, "Enhancing thermal conductivity of fluids with nanoparticles," in *Proceedings of the ASME International Mechanical Engineering Congress and Exposition*, pp. 99–105, November 1995.

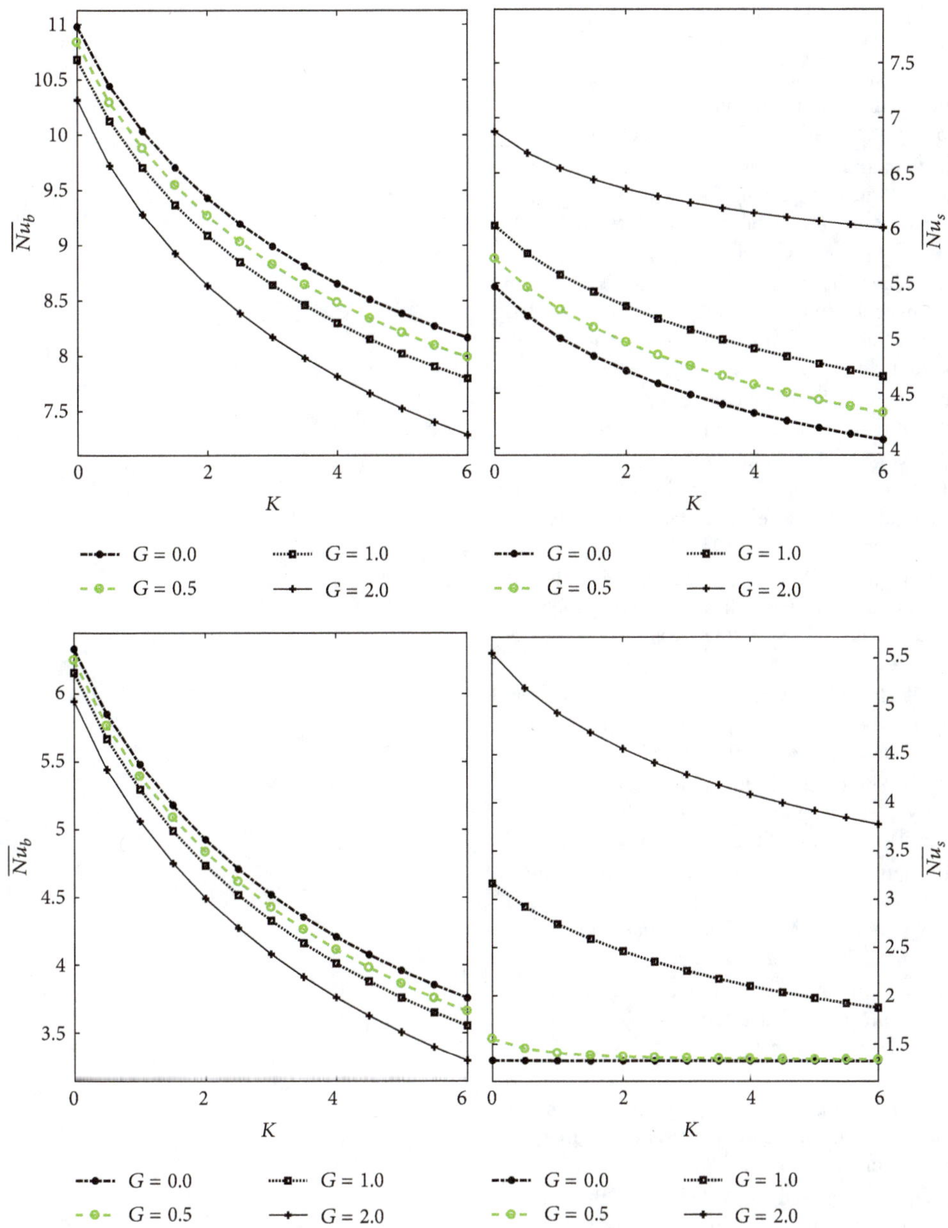


FIGURE 7: Variations of average Nusselt number at bottom (left) and average Nusselt number at side (right) against K by varying $\phi$ for continuous case 1 (top) and discontinuous heating case 2 (bottom) at $G = 1.0$.


[2] K. Khanafer, K. Vafai, and M. Lightstone, "Buoyancy-driven heat transfer enhancement in a two-dimensional enclosure utilizing nanofluids," *International Journal of Heat and Mass Transfer*, vol. 46, no. 19, pp. 3639–3653, 2003.

[3] R.-Y. Jou and S.-C. Tzeng, "Numerical research of nature convective heat transfer enhancement filled with nanofluids in rectangular enclosures," *International Communications in Heat and Mass Transfer*, vol. 33, no. 6, pp. 727–736, 2006.

[4] E. Büyük Öğüt, "Natural convection of water-based nanofluids in an inclined enclosure with a heat source," *International Journal of Thermal Sciences*, vol. 48, no. 11, pp. 2063–2073, 2009.

[5] W. Rashmi, A. F. Ismail, M. Khalid, and Y. Faridah, "CFD studies on natural convection heat transfer of $Al_2O_3$-water nanofluids," *Heat and Mass Transfer*, vol. 47, no. 10, pp. 1301–1310, 2011.

[6] G. A. Sheikhzadeh, M. Dastmalchi, and H. Khorasanizadeh, "Effects of nanoparticles transport mechanisms on Al2O 3-water nanofluid natural convection in a square enclosure," *International Journal of Thermal Sciences*, vol. 66, pp. 51–62, 2013.

[7] Z. Boulahia, A. Wakif, and R. Sehaqui, "Numerical study of mixed convection of the nanofluids in two-sided lid-driven square cavity with a pair of triangular heating cylinders," *Journal of Engineering (United States)*, vol. 2016, 2016.

[8] S. Y. Motlagh and H. Soltanipour, "Natural convection of Al2O3-water nanofluid in an inclined cavity using Buongiorno's two-phase model," *International Journal of Thermal Sciences*, vol. 111, pp. 310–320, 2017.

[9] C.-C. Liao, "Heat transfer transitions of natural convection flows in a differentially heated square enclosure filled with nanoflu-

ids," *International Journal of Heat and Mass Transfer*, vol. 115, pp. 625–634, 2017.

[10] O. Aydin and I. Pop, "Natural convection in a differentially heated enclosure filled with a micropolar fluid," *International Journal of Thermal Sciences*, vol. 46, no. 10, pp. 963–969, 2007.

[11] M. Saleem, S. Asghar, and M. A. Hossain, "Natural convection flow of micropolar fluid in a rectangular cavity heated from below with cold sidewalls," *Mathematical and Computer Modelling*, vol. 54, no. 1-2, pp. 508–518, 2011.

[12] S. K. Jena and S. P. Bhattacharyya, "The effect of microstructure on the thermal convection in a rectangular box of fluid heated from below," *International Journal of Engineering Science*, vol. 24, no. 1, pp. 69–78, 1986.

[13] S.-G. Wang and T.-H. Hsu, "Natural convection of micropolar fluids in an inclined rectangular enclosure," *Mathematical and Computer Modelling*, vol. 17, no. 10, pp. 73–80, 1993.

[14] T.-H. Tsu and C.-K. Chen, "Natural convection of micropolar fluids in a rectangular enclosure," *International Journal of Engineering Science*, vol. 34, no. 4, pp. 407–415, 1996.

[15] T.-H. Hsu, P.-T. Hsu, and S. O.-Y. Tsai, "Natural convection of micropolar fluids in an enclosure with heat sources," *International Journal of Heat and Mass Transfer*, vol. 40, no. 17, pp. 4239–4249, 1997.

[16] T. H. Hsu and K. Y. Hong, "Natural Convection of Micropolar Fluids in an Open Cavity," *Numerical Heat Transfer, Part A: Applications*, vol. 50, no. 3, pp. 281–300, 2006.

[17] N. S. Gibanov, M. A. Sheremet, and I. Pop, "Free convection in a trapezoidal cavity filled with a micropolar fluid," *International Journal of Heat and Mass Transfer*, vol. 99, pp. 831–838, 2016.

[18] I. V. Miroshnichenko, M. A. Sheremet, and I. Pop, "Natural convection in a trapezoidal cavity filled with a micropolar fluid under the effect of a local heat source," *International Journal of Mechanical Sciences*, vol. 120, pp. 182–189, 2017.

[19] M. A. Sheremet, I. Pop, and A. Ishak, "Time-dependent natural convection of micropolar fluid in a wavy triangular cavity," *International Journal of Heat and Mass Transfer*, vol. 105, pp. 610–622, 2017.

[20] O. Turk and M. Tezer-Sezgin, "FEM solution to natural convection flow of a micropolar nanofluid in the presence of a magnetic field," *Meccanica*, vol. 52, no. 4-5, pp. 889–901, 2017.

[21] N. Ali, M. Nazeer, T. Javed, and M. A. Siddiqui, "Buoyancy-driven cavity flow of a micropolar fluid with variably heated bottom wall," *Heat Transfer Research*, vol. 49, no. 5, pp. 457–481, 2018.

[22] G. C. Bourantas and V. C. Loukopoulos, "Modeling the natural convective flow of micropolar nanofluids," *International Journal of Heat and Mass Transfer*, vol. 68, pp. 35–41, 2014.

[23] H. Hashemi, Z. Namazian, and S. A. M. Mehryan, "Cu-water micropolar nanofluid natural convection within a porous enclosure with heat generation," *Journal of Molecular Liquids*, vol. 236, pp. 48–60, 2017.

[24] M. Izadi, S. Mehryan, and M. A. Sheremet, "Natural convection of CuO-water micropolar nanofluids inside a porous enclosure using local thermal non-equilibrium condition," *Journal of the Taiwan Institute of Chemical Engineers*, vol. 88, pp. 89–103, 2018.

[25] H. C. Brinkman, "The viscosity of concentrated suspensions and solutions," *The Journal of Chemical Physics*, vol. 20, no. 4, p. 571, 1952.

3

# Experimental Study on RFID Antenna Reading Areas in a Tunnel System

**Kai Kordelin,[1] Jaana Kordelin,[1] Markku Johansson,[2] Johanna Virkki,[1] Leena Ukkonen,[1] and Lauri Sydänheimo[1]**

[1]*BioMediTech Institute and Faculty of Biomedical Sciences and Engineering, Tampere University of Technology, Tampere, Finland*
[2]*Aitosolutions Oyj, Kangasala, Finland*

Correspondence should be addressed to Kai Kordelin; kai@kordelin.eu

Academic Editor: Yuh-Shyan Hwang

We study optimized antenna reading area mappings for a radiofrequency identification- (RFID-) based access monitoring system, used in an underground nuclear waste storage facility. We shortly introduce the access monitoring system developed for the ONKALO tunnel in Finland and describe the antenna mounting points as well as the research area. Finally, we study the measurement results of the antenna reading areas and factors that affect the reading area size. Based on our results, in addition to antenna location and direction, absorption to obstacles, reflections, diffraction, scattering, and refraction affect the antenna reading area.

## 1. Introduction

The spent nuclear fuel accumulated from the Finnish nuclear power plants will be disposed in Olkiluoto, Finland, in a tunnel system known as "ONKALO." The purpose of ONKALO is to ensure that the bed rock in Olkiluoto is suitable for the final storage of the used nuclear fuel. Versatile research has been conducted there since the beginning of its construction. The nuclear waste facilities consist of an encapsulation plant, constructed to encapsulate the spent nuclear fuel; a disposal facility, consisting of an underground repository and other underground rooms; and the above ground service spaces. Access routes to the disposal facility are an inclined access tunnel and vertical shafts. As underground tunnels are usually structurally nonuniform, with a network of interconnected tunnels, crosscuts, shafts, escape ways, first-aid stations, alcoves, and tunnel block-ages, monitoring the access to the tunnel plays a critical part in the work safety [1]. In the deposition area of ONKALO, two parallel central tunnels connect all the deposition tunnels and these central tunnels are interconnected at regular intervals [2].

The developed radiofrequency identification- (RFID-) based access monitoring system was created to increase and upkeep work safety. Its purpose is to know how many persons are located inside ONKALO in a case of an accident. In emergency situations, underground personnel and equipment identification and localization are critical [3–5]. RFID technology is a wireless, contactless technology that provides efficient solutions for identification, warehousing, and logistics and also for reliable and easy access monitoring [6–8]. A comprehensive introduction to today's systems and standardization is provided in [9–12].

There are no specific instructions for the placement and the direction of RFID antennas in an underground tunnel. With good design it is possible to create optimal antenna geometry for an underground tunnel but this kind of optimized antenna works only in the designed point [13], and there is no systematically studied information available on how radio waves propagate in a tunnel. A radio signal is reflected from the walls of the tunnels creating many different propagation paths [14]. The main problems with tunnels are the uneven wall surfaces that cause reflections and disturb radio waves. The final antenna locations and direction can be done only by trial and error. This is also how it was done in ONKALO and the purpose of this study is to introduce how

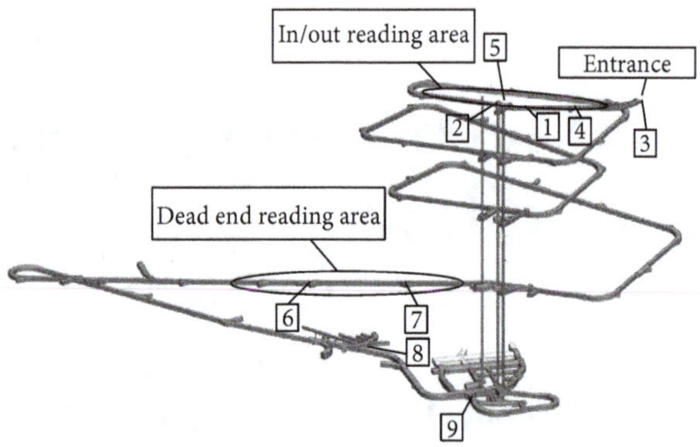


Figure 1: The reading points and reading areas in ONKALO.

the optimized antenna reading area mappings were done in this underground nuclear waste storage facility.

This paper is organized as follows. After Introduction, the ONKALO RFID system and equipment are shortly presented in Section 2. Section 3 introduces the research area and the research plan. In Section 4, the antenna installation points are given and in Section 5 the measurement plans and implementation of the measurements are described. Section 6 presents the antennas measurement results and describes the features that affect the antenna reading area size. Finally, in Section 7, the conclusions of this study are gathered together.

## 2. ONKALO RFID System and Equipment

There are two different information reading areas at ONKALO (See Figure 1). A reading area consists of reading points that give information from both directions of traffic. The first reading area is at the entry of ONKALO. It monitors the traffic going in and out of ONKALO. This reading area has four reading points. The second reading area, with two reading points, is located at the "dead end" and it controls the traffic going in and out of that area. The use of the RFID system and its reliability in ONKALO has been studied for many years [15].

In addition to the given reading areas, there are three additional reading points. One is located in the demonstration area and another at the technical level. The third reading point is at the beginning of the drive ramp, before the entrance to ONKALO.

A reading point consists of a reader and antenna. The reader is only used for reading information; it does not transmit anything. The active RFID tag, that is, ID card, is the only transponder in the system. The process of identification is the same at ONKALOs both reading areas and all reading points. In practice, when a vehicle, as well as people inside it, comes to a reading point area, the reader identifies the ID numbers of the vehicle and the people and records them to the system database.

The readers are Wavetrend L-RX201 RFID readers (Figure 2). The readers have been linked to the system with integrated computers. The readers' sensitivity, that is, threshold value, has been adjusted by program to 65. According to the RX201 manual, the readers RSSI (received signal strength

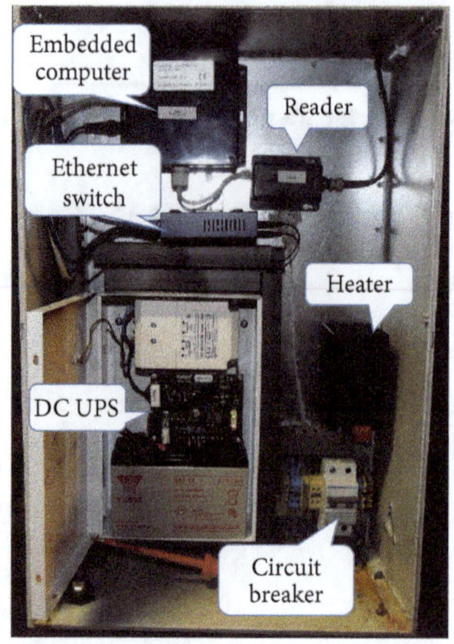


Figure 2: Reader with integrated computer.

indicator) can be adjusted between 60 and 120, where 60 is the RSSI maximum sensitivity. The L-RX201 maximum sensitivity is −103 dBm, which means that RSSI 60 would be equal to −103 dBm. No other dBm values are given and thus it is not possible to make a conversion chart between RSSI and received power.

The used antennas are TwinGain™ 434 directional antennas (Figure 3). The antenna polarization has been informed to be circular, but its reading area is more elliptical. The protection class of the antenna is IP 65. The antenna cable determines the maximum distance between the reader and antenna, which cannot exceed ten meters. The used antenna was initially measured in an anechoic chamber in laboratory environment. In Figure 4, the radiation pattern of the antenna is shown in horizontal and vertical directions. The systems integrated computer is a Raspberry PI Model B, a one-circuit board computer that has a 700 MHz ARM11 processor and 512 MB memory.

The Wavetrend manufactured RFID identificators with active chips using a 434 MHz frequency are used in the ONKALO as ID cards (Figure 3). Each card sends its own ID number once every 1.5 seconds. The ID card is a beacon type tag that can be read only when it sends its own ID number. The card is inside a 85 mm long, 55 mm wide, and 5 mm thick plastic holder.

The tunnel identification program is HorusWindow applications logic program by Elcoplast (Figure 5). The application logic fulfills all tasks given by the program and saves the data to the database. The application logic does not have its own interface, so a web application has been created. The web application and the application logic are not directly connected, as all data goes through the database service (SQL Server).

## 3. Research Area and Research Plan

The object of this paper is to study how radio waves propagate in a tunnel. The ONKALO upper end, that is, so-called in/out

(a) (b)

FIGURE 3: Directional antenna (a) and active ID card (b).

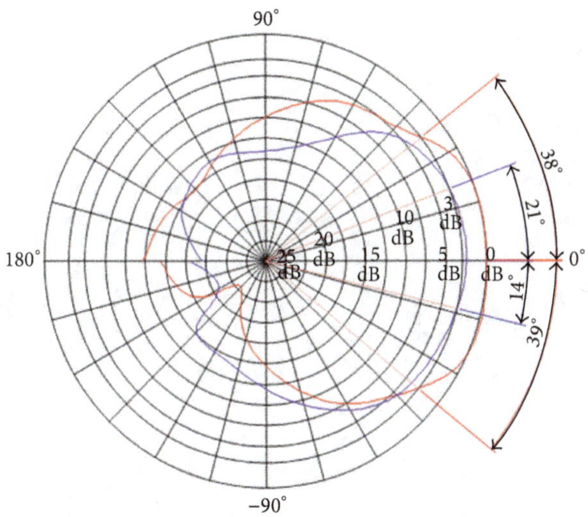


FIGURE 4: Horizontal and vertical radiation patterns of the antenna. The −3 dB ring is marked with the opening angles of both measurements.

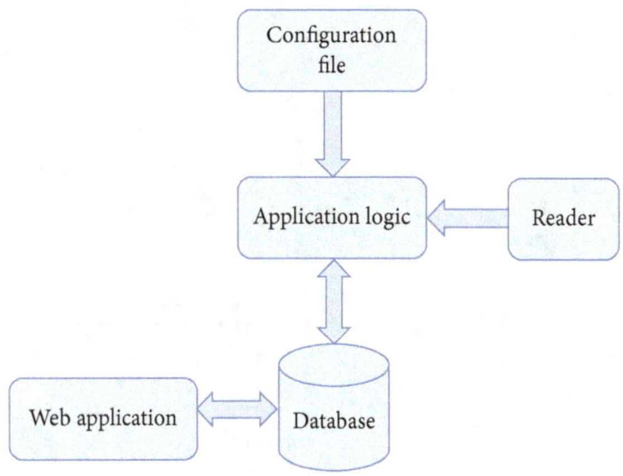


FIGURE 5: Program description.

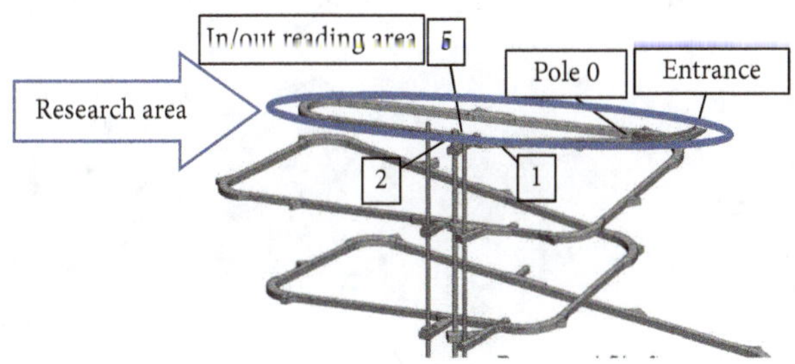


FIGURE 6: Research area, reading points, and reading areas in ONKALO.

reading area, was selected as the research area. This area is marked in Figure 6. The reading area consists of three different readers and antennas.

The ID tags were installed to the safety cones while doing the measurements (See Figure 7). By this method the ID tag could be set to a height of 1 meter. Firstly, safety cones with ID cards attached to them were brought into the ONKALO tunnel system, and they were placed to the reading area for reading tests. Screenshots of the readings were taken of the HorusWindows identification program. By examining these measurement records, it was possible to get a good idea on how the antennas were working in each place.

## 4. Reader Antenna Installation Points

As there initially was no official installation guide for the antenna placement, the antennas were installed by considering the tunnel decline and vehicle size. The installation location information has been given by pole number, which means the current points distance (in meters) from the tunnel opening. Pole 0 is located at the entrance. A pole number with a + in front of it means that it is located before the actual tunnel opening.

*4.1. Ingoing Traffic Reading Antenna Installation Point.* Reader antenna 1, that is, the antenna that reads traffic going into ONKALO, is installed on a firewall at pole number 95 (See Figure 14). The firewall is built from PAROC elements. PAROC elements are called sandwich elements because of their structure (See Figure 8) and the elements consist of a PAROC structural stone fiber core. Thin metal sheets have been glued on to the stone fiber core, which creates a very

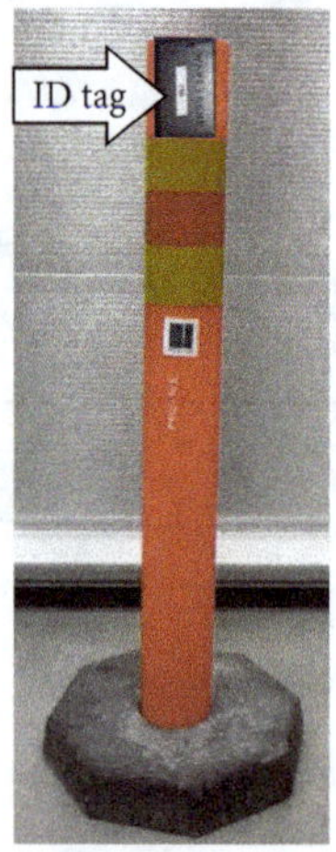


FIGURE 7: Safety cone.

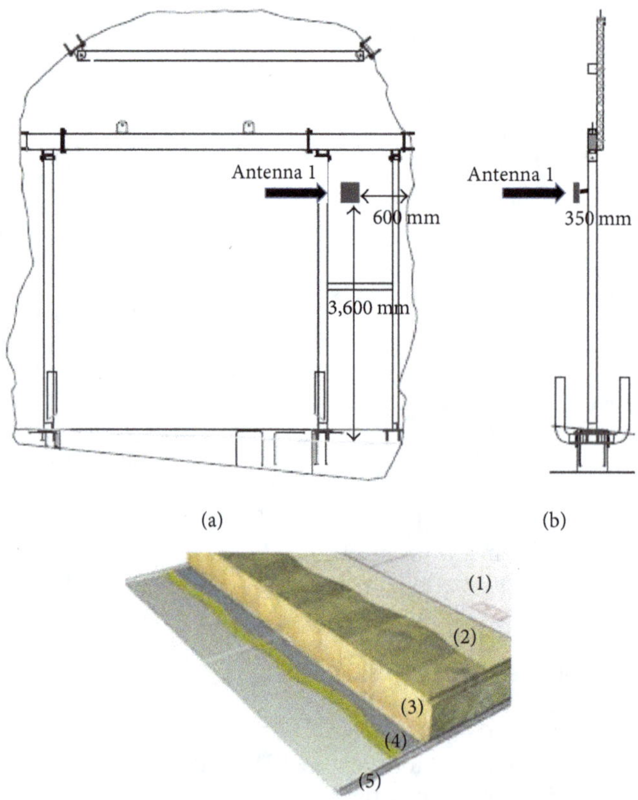


(1) Substrate-coated steel sheets
(2) Strong and durable adhesive
(3) Noncombustible core of PAROC structural stone wool lamellas
(4) Multilayer primer to ensure the bonding between the adhesive and the steel sheet
(5) Fire safe and tight joint design

(c)

FIGURE 8: Firewall construction at pole 95 from the front (a) and from the side (b) and the firewall PAROC-element structure (c).

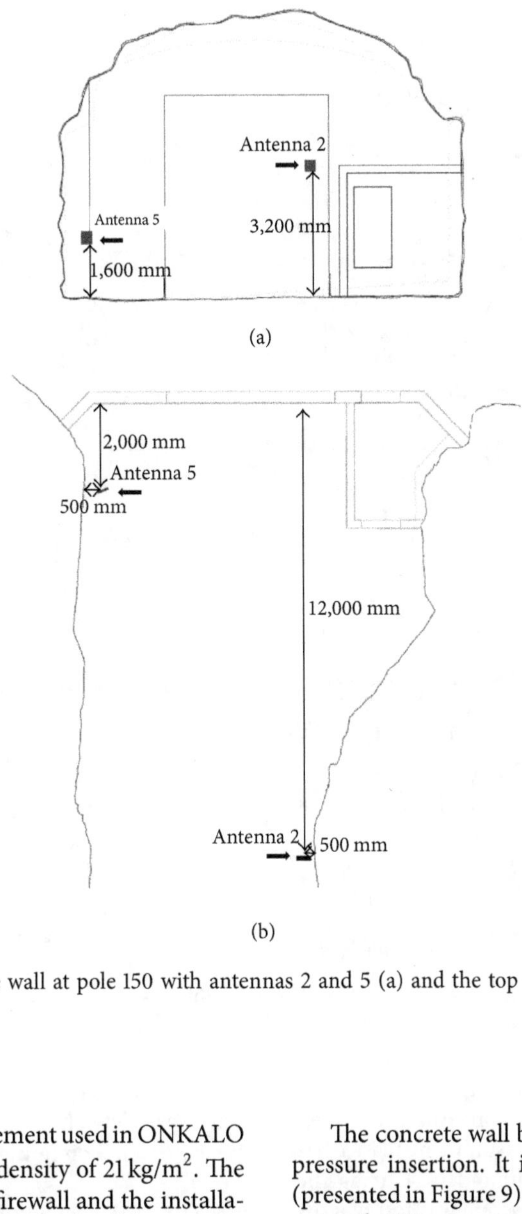


FIGURE 9: The front side of the concrete wall at pole 150 with antennas 2 and 5 (a) and the top side of the concrete wall at pole 150 with antennas 2 and 5 (b).

strong composite structure [16]. The element used in ONKALO has a $U$ value of 0.36 W/m$^2$K and a density of 21 kg/m$^2$. The antenna is placed 350 mm from the firewall and the installation height is 3600 mm. The antenna is 600 mm away from the tunnel wall, as presented in Figure 8.

*4.2. Outgoing Traffic Reading Antenna Installation Points.* Both antennas reading outgoing traffic are attached to the sidewall of the ONKALO tunnel. It was not possible to install the antennas to the concrete structures (shown in Figure 8) as at that point they were not built yet, but the wall was ready when these measurements were conducted.

Reader antenna 2 is attached to the left wall at pole 162. It is mounted 12,000 mm from the concrete wall and at height of 3200 mm from the ground. The side of the antenna is 500 mm from the tunnel rock wall. There is no wall structure behind the antenna, only the driveway. Reader antenna 5 is installed to the right hand side wall at pole 152. It is mounted 2,000 mm from the concrete wall and its height from the ground is 1600 mm. The side of the antenna is 500 mm from the tunnel rock wall (Figure 9).

The concrete wall behind the antennas has been done by pressure insertion. It is a 300 mm thick sectioning firewall (presented in Figure 9) and has a double rebar (stainless steel) net with 200 mm gaps. It has a 4200 mm wide and 5100 mm high section without rebar for the driveway.

## 5. Measurements

The idea of these measurements was to make a preliminary study towards a model for the best antenna location in tunnel conditions. Four measurement configurations were made for the research as the first configurations turned out to be insufficient and results indicated that more detailed measurements were necessary.

*5.1. Measurement Configurations.* At its widest the measurements covered the area from Pole 320 to the drive ramp upper end at pole +70. The measurement points have been marked to the ONKALO layout in Figure 14.

*Measurement Configuration 1.* Measurement configuration 1 consisted of antennas 1, 2, and 5. Safety cones with ID tags

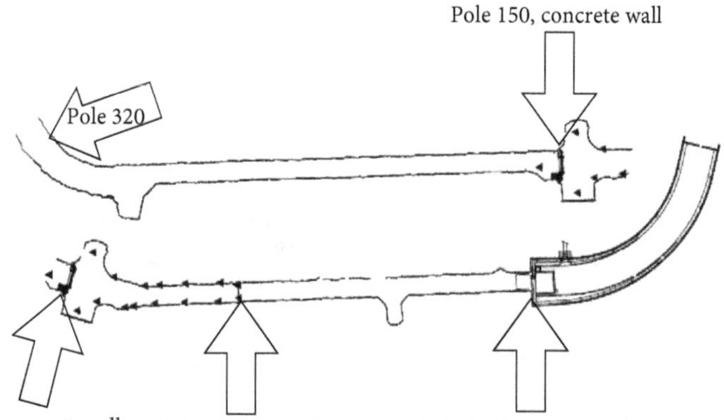


FIGURE 10: Measurement configuration 1.

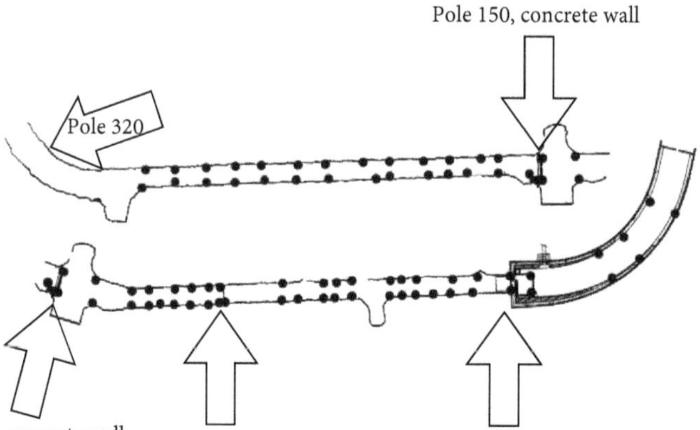


FIGURE 11: Measurement configuration 2.

were set on both sides of the ONKALO driveway between poles 95 and 155. The purpose of this measurement plan was to map the reader's rear area. Measurement plan 1 has been marked by triangles in Figures 10 and 14. During the measurements it was noticed that the scope was too narrow to confirm if the antennas worked properly.

*Measurement Configuration 2.* Measurement configuration 2 consisted of antennas 1, 2, and 5 full reading range, including both frontal and rear area. Safety cones with ID tags were set on both sides of the ONKALO driveway between poles +55 and 280. Plan 2 has been marked by circles in Figures 11 and 14. The reason why measurement plan 3 was considered necessary was the fact that the frontal reading area of the antennas was larger than anticipated.

*Measurement Configuration 3.* This measurement configuration was done deeper in the ONKALO tunnel system. The furthest measuring points were located at the left side of the driveway at pole 320 and on the right side of pole 310. The whole measurement area consisted of the poles from 0 to 320. From these measurements it was possible to get the driveways' both sides' maximum reading area. Plan 3 has been marked by squares in Figures 12 and 14. During the measurements, it was noticed that the results from the driveways' left and right sides differed, which resulted in measurement configuration 4.

*Measurement Configuration 4.* Measurement configuration 4 consisted of measuring the ONKALO driveways' middle section. The measurement started from pole +70 and ended at pole 320. Plan 4 has been marked by stars in Figures 13 and 14. These measurements gave sufficient data for the antenna middle section reading areas.

*5.2. Measurement Implementation.* Authenticating the measurements was difficult as readings from various antennas came almost constantly. The HorusWindows program does not save the readings but shows them only in real time (see Figure 15). For this reason, the only way to get real time data was taking screenshots of the display. These screenshots were saved in the measurement program to their specific places with given names. The measurement plan also had all ID numbers of the tags. By combining these two factors it was possible to get the tag location in the ONKALO tunnel. If the print screen file did not contain the ID card number, it meant the antennas had not recognized that specific tag.

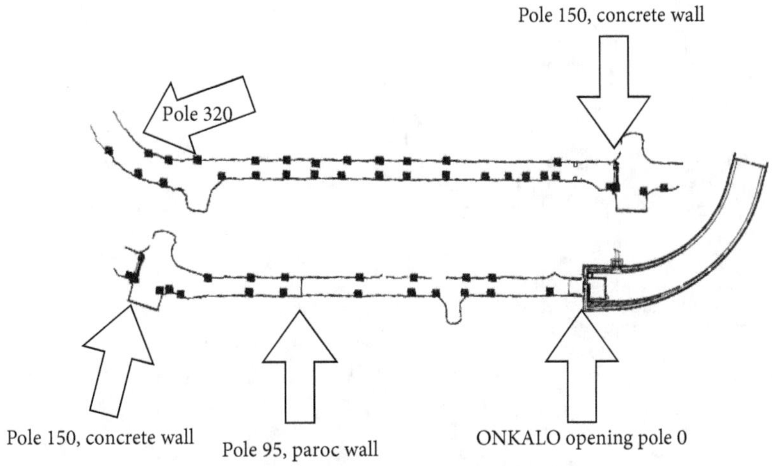


FIGURE 12: Measurement configuration 3.

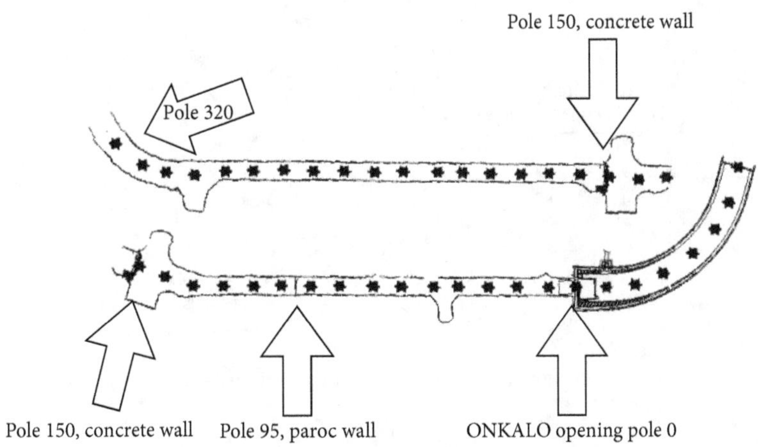


FIGURE 13: Measurement configuration 4.

A chart was made from identified tags where RSSI was marked. Also the pole number where the tag was and the tags distance to the antenna was recorded. From this data it was possible to determine the MRRR (maximum reliable read range) of all antennas. From the produced table all read safety cones were marked on the ONKALO layout. From this, every antenna MRRR range could be marked, that is, the range that the antenna could read all tags at the pole.

## 6. Measurement Results

The measurements for this study were carried out in autumn 2014. Antenna cable connections, possible damage to antennas or wiring, and surrounding conditions affect the antenna reading area measurements. The antennas were installed during 2010 and cabling and connections of Antenna 1 and Antenna 5 were originals. The cable for Antenna 2 has been changed in 2012 when a vehicle hit the antenna and caused it to drop, also damaging the cable. During the measurements, the temperature was 11°C and the relative humidity level was 99%. Condensation gathered on walls and antennas.

*6.1. Antenna 1.* Antenna 1 is the in-reading antenna of the ONKALO tunnel system (Figure 16). This antenna is the most important antenna in the system, due to personnel safety reasons. After the reading of this antenna, the person is logged into the ONKALO system.

The measurement results of this antenna give a clear picture of the antenna MRRR. As the results can be compared to results of antennas 2 and 5, it can be seen that the PAROC walls affect the measurements results of the rear area.

In Figure 16, antenna 1 MRRR area is shown. This indicates the reliable value area starts from pole +5. In the circled area in Figure 16, the RSSI was at its highest at poles 60–70. In Figure 17 you can see how antenna 1 reads from outside the tunnel at the centerline and left side. The left side of the reading area reaches up to 55 mm from the ONKALO opening. Figure 18 shows antenna 1 rear reading area. This figure shows how well the PAROC wall dampens the rear area. The rear area is nearly symmetrical.

Antenna 1 measurement occurrence table values have been summarized to Table 1. On the left side of the table are the measurements of the left side of the tunnel, in the

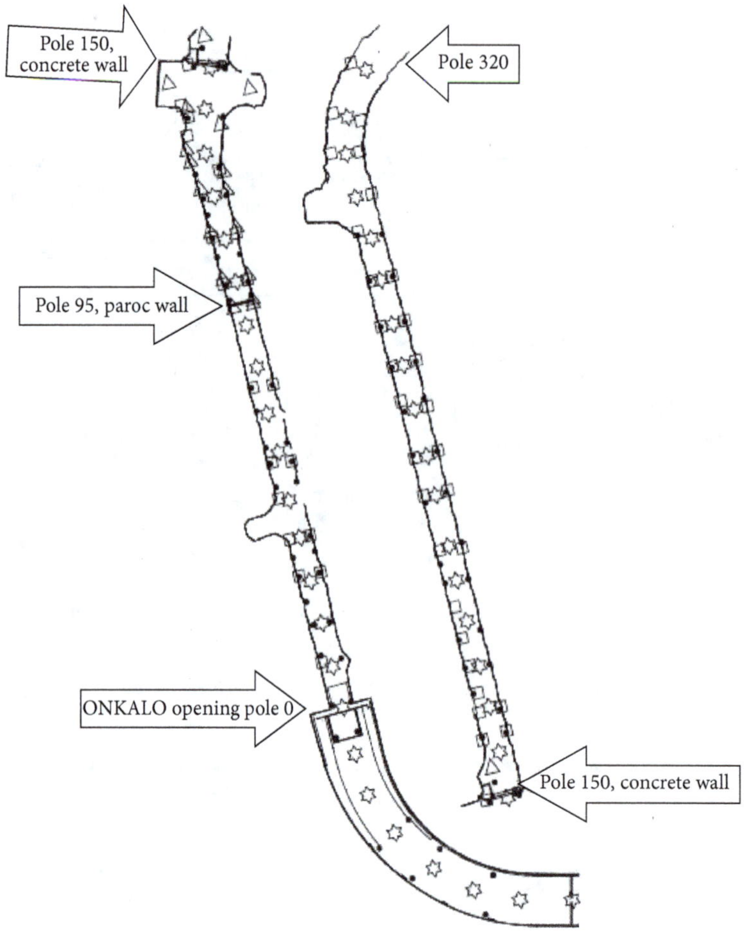


FIGURE 14: Measurement configurations 1–4.

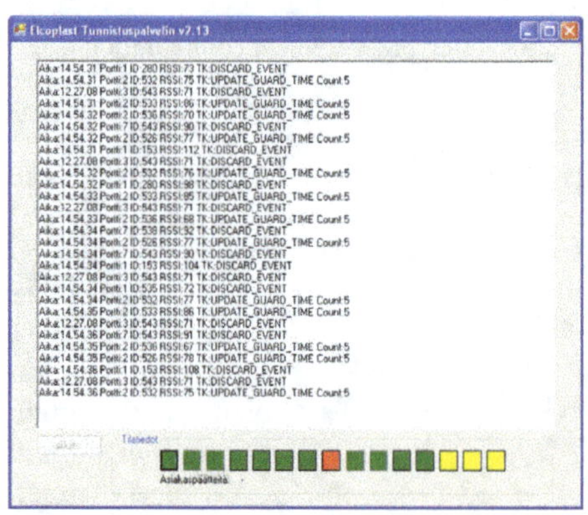

FIGURE 15: An example of a print screen screenshot file.

center are the measurements from the centerline, and on the right are the measurements of the right side. Pole indicates at what pole number the ID tag was when measured. Distance means the direct distance to the antenna. The RSSI value is the average of the measured values. As an example, the line VS135, 529, −40,314, 69,5 means that the ID tag 529 was at the left side at pole 135 and its direct distance to the antenna was −40,314 meters. The "−" here indicates that the ID tag was below the antenna. The number 69,5 is the average RSSI value collected from the documents. In the table, the antenna's MRRR area is limited to Italics.

*6.2. Antenna 2.* Antenna 2 is the second out of ONKALO tunnel reading antenna and it is the "main antenna" for outreading (Figure 19). The measurements of this antenna show how the MRRR area, that is, the reliable reading values

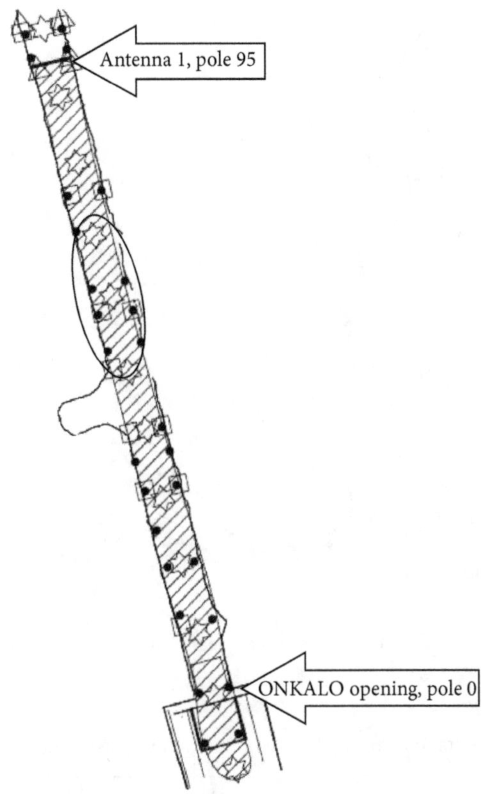


FIGURE 16: Antenna 1 MRRR area.

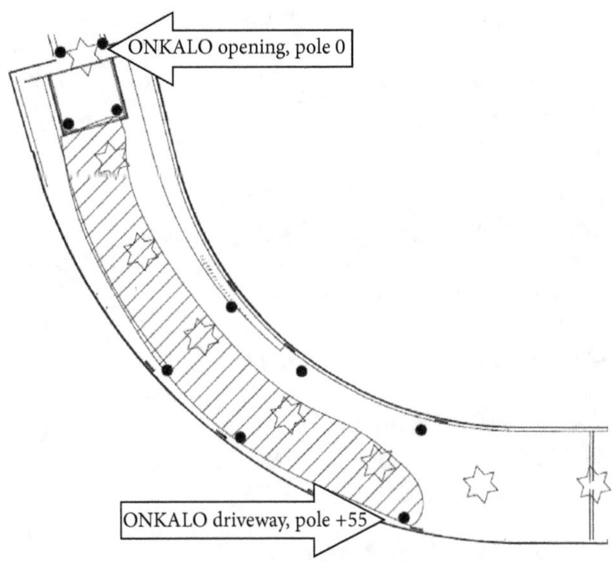


FIGURE 17: Antenna 1 front reading area.

area, changes after equipping the walls. Equipping in this case means installing electricity and instrument housings, water pipes, and other structures. The MRRR area reaches pole 208. There is no wall directly behind antenna 2, which can be seen from the measurement results.

Antenna 2 measurement values have been summarized in Table 2. This table is read similarly to Table 1. Figure 19 shows how antenna 2 MRRR area goes from pole 208 upwards. The ID tag at pole 200 is not read, which can be only explained by measurement failure, as all other tags were readable. ID tags on the sides at the same pole number had an RSSI value over 90.

In Figure 20, it can be seen how antenna 2 reads up to pole 311. It is also evident how equipping the wall causes reflections

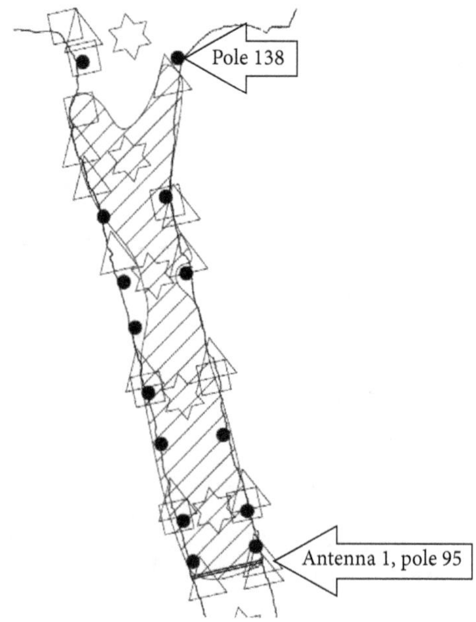


FIGURE 18: Antenna 1 rear reading area.

TABLE 1: Compiled table of antenna 1 measurements.

| Pole | Tag | Distance | RSSI | Pole | Tag | Distance | RSSI | Pole | Tag | Distance | RSSI |
|---|---|---|---|---|---|---|---|---|---|---|---|
| VS 135 | 529 | −40,314 | 69,5 | | | | | OS 138 | 542 | −40,138 | 65,5 |
| VS 130 | 541 | −37,094 | 69 | 130 | 536 | −35,18 | 67,5 | | | | |
| VS 125 | 532 | −31,495 | 67 | | | | 0 | OS 125 | 531 | −30,794 | 71,5 |
| | | | | 120 | 526 | −25,437 | 68,5 | OS 120 | 530 | −25,279 | 80,5 |
| VS 110 | 543 | −17,822 | 83 | 110 | 532 | −15,256 | 74,5 | OS 110 | 542 | −16,964 | 80 |
| VS 100 | 541 | −7,813 | 80,5 | 100 | 533 | −5,942 | 78 | OS 100 | 531 | −5,176 | 89 |
| *VS 95* | *532* | *−4,854* | *106,5* | *90* | *536* | *5,38* | *107* | *OS 95* | *540* | *0* | *104,5* |
| | | | | *80* | *526* | *14,649* | *117,5* | | | | |
| *VS 75* | *529* | *19,426* | *118,5* | | | | *0* | *OS 75* | *531* | *19,231* | *111,5* |
| *VS 70* | *533* | *26,657* | *111,5* | *70* | *532* | *25,337* | *115* | | | | |
| *VS 61* | *530* | *32,988* | *113* | *60* | *533* | *34,431* | *100,5* | *OS 61* | *532* | *32,849* | *107,5* |
| *VS 57* | *529* | *36,584* | *111,5* | | | | | *OS 57* | *531* | *37,061* | *95* |
| *VS 50* | *533* | *44,606* | *94,5* | *50* | *536* | *45,62* | *100* | *OS 53* | *532* | *41,839* | *104* |
| *VS 35* | *530* | *58,361* | *99* | *40* | *526* | *54,287* | *83,5* | *OS 40* | *531* | *54,487* | *79* |
| *VS 30* | *529* | *62,773* | *95,5* | *30* | *532* | *64,344* | *94* | *OS 30* | *531* | *63,063* | *92,5* |
| *VS 20* | *530* | *74,084* | *85* | *20* | *533* | *74,187* | *97,5* | *OS 20* | *532* | *74,285* | *94,5* |
| *VS 15* | *529* | *82,725* | *77* | *10* | *536* | *84,431* | *86,5* | *OS 15* | *531* | *82,845* | *91* |
| *VS 0* | *530* | *93,395* | *98,5* | *0* | *526* | *93,852* | *90,5* | *OS 0* | *532* | *93,17* | *94,5* |
| *VS +5* | *530* | *100,268* | *93* | *+10* | *532* | *104,776* | *72,5* | *OS +5* | *532* | *99,874* | *74* |
| | | | | +20 | 533 | 114,695 | 72,5 | | | | |
| VS +30 | 529 | 126,534 | 65 | +30 | 536 | 124,376 | 73,5 | | | | |
| VS +40 | 530 | 135,669 | 68 | +40 | 526 | 136,067 | 67,5 | | | | |
| VS +55 | 529 | 148,841 | 71,5 | +50 | 532 | 146,694 | 67 | | | | |

that affect the reading. Antenna 2 rear reading point reached all the way to pole 17 (see Figure 21) but this was just one single point. More points near this, where a reading had occurred, could not be found. Thus, it was considered to have happened due to dispersion. The next closest reading point was at pole 58. It can be seen that the fact that there is no signal dampening material behind the antenna affects the rear reading area, as does the PAROC wall door opening at pole 105.

*6.3. Antenna 5.* The main purpose of antenna 5 installation is to verify the readings of antenna 2. This antenna has been

TABLE 2: Compiled table of antenna 2 measurements.

| Pole | tag | Distance | RSSI | Pole | tag | Distance | RSSI | Pole | tag | Distance | RSSI |
|---|---|---|---|---|---|---|---|---|---|---|---|
| VS 311 | 526 | 146,775 | 69,5 | | | | | | | | |
| | | | | 290 | 536 | 127,216 | 67 | OS 290 | 532 | 127,261 | 66,5 |
| | | | | | | | | OS 280 | 531 | 117,867 | 67 |
| | | | | 270 | 532 | 107,788 | 67 | OS 270 | 533 | 107,985 | 71 |
| VS 260 | 526 | 97,523 | 66,5 | | | | | OS 260 | 532 | 97,56 | 74 |
| | | | | | | | | OS 250 | 533 | 88,148 | 70,5 |
| VS 240 | 526 | 77,697 | 73,5 | | | | | OS 240 | 532 | 77,772 | 67,5 |
| VS 230 | 536 | 67,146 | 70 | 230 | 532 | 66,849 | 78,5 | | | | |
| VS 220 | 526 | 58,025 | 75 | 220 | 533 | 57,713 | 72,5 | OS 220 | 532 | 58,133 | 76 |
| *VS 208* | *536* | *45,054* | *72* | 210 | 536 | 45,737 | 80,5 | *OS 208* | *533* | *45,255* | *84* |
| *VS 200* | *529* | *37,812* | *90,5* | | | | | *OS 200* | *530* | *38,12* | *93,5* |
| *VS 190* | *532* | *24,503* | *87,5* | *190* | *532* | *28,434* | *90,5* | *OS 190* | *531* | *26,372* | *79,5* |
| *VS 180* | *529* | *18,088* | *88,5* | *180* | *533* | *18,028* | *94,5* | *OS 180* | *530* | *18,146* | *92,5* |
| *VS 170* | *532* | *7,901* | *95* | *170* | *536* | *8,709* | *114,5* | *OS 170* | *531* | *8,933* | *105* |
| | | | | 160 | 526 | −3,165 | 116,5 | | | | |
| VS 150 | 532 | −12,626 | 85,5 | 150 | 532 | −12,648 | 110 | OS 150 | 531 | −13,372 | 99,5 |
| VS 145 | 500 | −16,819 | 92,5 | | | | | OS 145 | 540 | −16,896 | 78 |
| VS 140 | 529 | −22,091 | 85 | 140 | 533 | −21,993 | 92,5 | OS 140 | 530 | −23,67 | 100 |
| VS 139 | 533 | −24,317 | 77,5 | 130 | 536 | −31,92 | 81 | OS 138 | 542 | −22,825 | 93 |
| VS 120 | 533 | −42,82 | 75 | 120 | 526 | −41,578 | 86,5 | OS 120 | 542 | −37,587 | 82,5 |
| VS 110 | 529 | −50,271 | 77,5 | 110 | 532 | −51,676 | 71 | OS 110 | 531 | −49,846 | 69,5 |
| VS 105 | 533 | −55,585 | 82 | | | | | OS 105 | 530 | −55,428 | 71,5 |
| VS 100 | 529 | −61,236 | 72 | 100 | 533 | −61,157 | 70 | OS 100 | 540 | −57,309 | 84,5 |
| VS 97 | 533 | −65,029 | 74 | 90 | 536 | −71,231 | 70 | OS 97 | 541 | −62,438 | 75,5 |
| | | | | 80 | 526 | −80,893 | 66,5 | | | | |
| VS 57 | 529 | −102,843 | 67 | 70 | 532 | −91,608 | 67,5 | | | | |
| VS 17 | 533 | −147,427 | 70,5 | | | | | | | | |

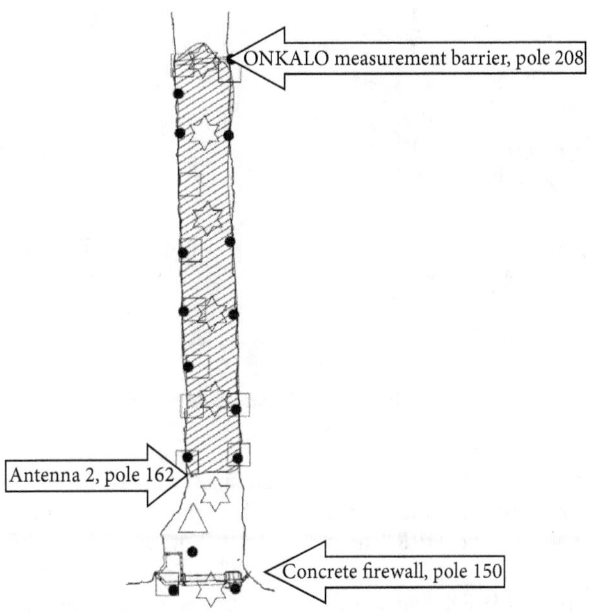


FIGURE 19: Antenna 2 MRRR area.

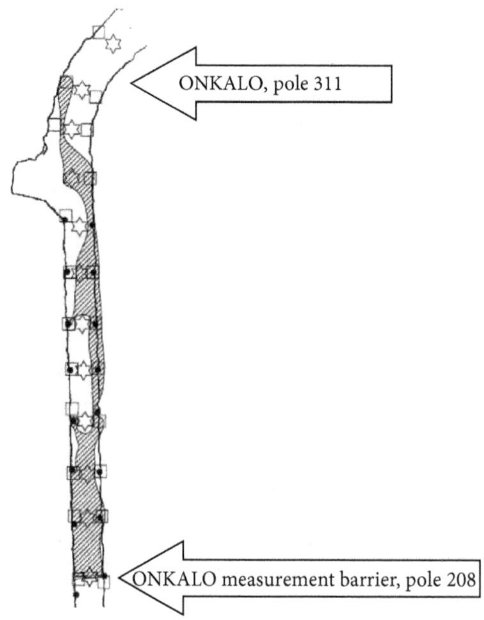


FIGURE 20: Antenna 2 front reading area.

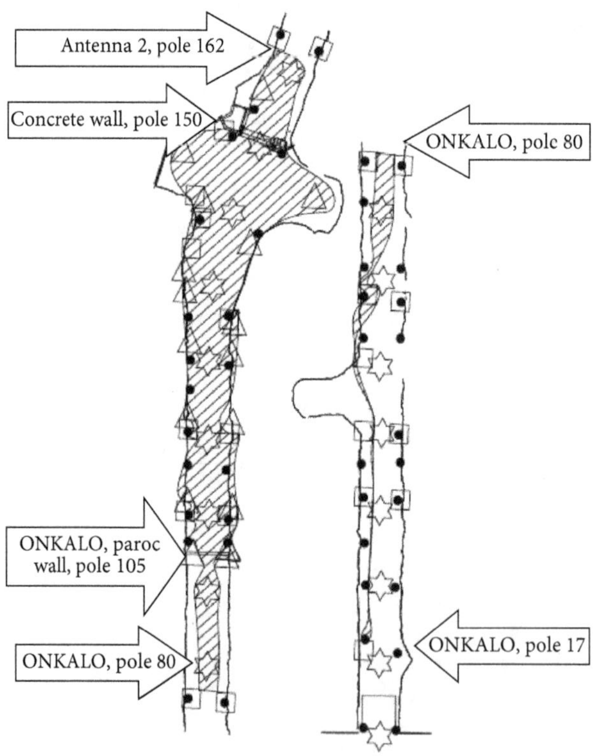


FIGURE 21: Antenna 2 rear reading area.

installed lower than the other antennas. The lower location is meant to ensure reading from delivery vans. The differences due to lower installation can be seen from the MRRR area results.

In Figure 22, it is shown that antenna 5 MRRR area reaches pole number 170 in the ONKALO tunnel. It can also be seen how much the antenna direction and height affect the MRRR area size. Figure 23 shows how antenna 5 reads the left side of the driveway up to pole 208. The antennas strong alignment over the center towards the left wall explains this result. Antenna 5 rear reading area is affected by the concrete wall located at pole 150, as well as the door opening, which can be seen in Figure 23.

Antenna 5 measurement occurrence table values have been summarized in Table 3, which can be read similarly to Tables 1 and 2.

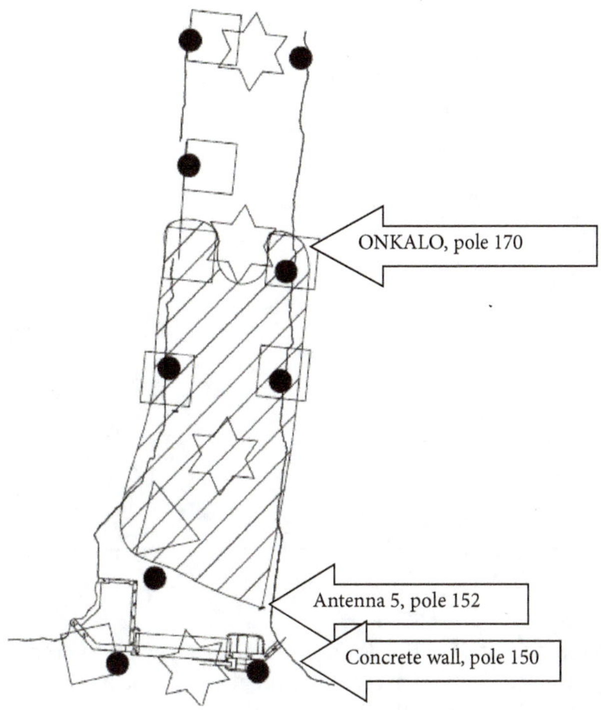


FIGURE 22: Antenna 5 MRRR area.

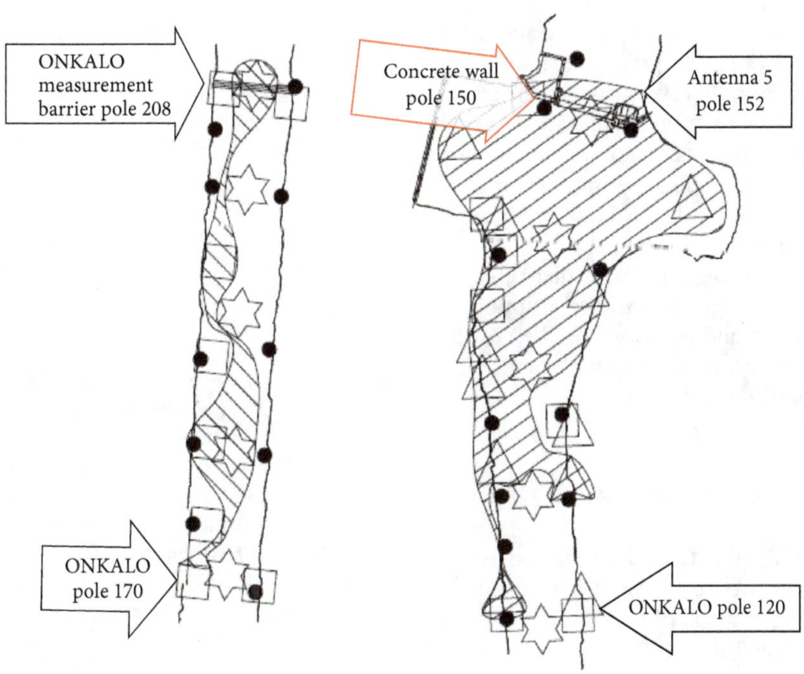


FIGURE 23: Antenna 5 front reading area (a) and rear reading area (b).

*6.4. Further Discussion.* In general, the features that affect the antenna reading area size are antenna direction, dampening from obstacles, reflections, diffraction dampening, dispersion dampening, and fading. Based on our measurement results, obstacles, dispersion, and reflection are the main features affecting the antenna reading area in a tunnel environment.

The effect of dampening because of obstacles can be seen from the antenna front to rear balance. Also the antenna direction can be seen affecting the dampening. From antenna 2

TABLE 3: Compiled table of antenna 5 measurements.

| Pole | Tag | Distance | RSSI | Pole | Tag | Distance | RSSI | Pole | Tag | Distance | RSSI |
|---|---|---|---|---|---|---|---|---|---|---|---|
| VS 195 | 533 | 41,647 | 70 | | | | | | | | |
| VS 180 | 529 | 27,88 | 70,5 | 180 | 533 | 27,371 | 73,5 | | | | |
| *VS 170* | *532* | *17,877* | *83* | | | | | *OS 170* | *531* | *16,879* | *70* |
| *VS 165* | *529* | *12,486* | *84* | *160* | *526* | *7,771* | *77* | *OS 165* | *530* | *11,391* | *86* |
| VS 150 | 541 | −8,966 | 72,5 | 150 | 532 | −4,585 | 94,5 | OS 150 | 531 | −3,268 | 92,5 |
| VS 145 | 500 | −15,276 | 75,5 | | | | | OS 145 | 542 | −9,059 | 76,5 |
| VS 140 | 500 | −15,595 | 91 | 140 | 533 | −13,203 | 86,5 | OS 140 | 530 | −13,803 | 85 |
| VS 139 | 533 | −16,591 | 90 | | | | | OS 138 | 542 | −15,757 | 73,5 |
| VS 135 | 529 | −20,76 | 84,5 | | | | | | | | |
| VS 130 | 543 | −23,257 | 77 | 130 | 536 | −22,728 | 69 | | | | |
| VS 125 | 532 | −27,591 | 70 | | | | | | | | |
| VS 128 | 500 | −25,276 | 79 | | | | | | | | |
| VS 122 | 543 | −30,894 | 77,5 | | | | | OS 120 | 542 | −30,51 | 72 |
| VS 110 | 543 | −40,046 | 67 | | | | | | | | |

measurement results, it can be seen that the effect of the door opening in the rear reading area on the left of the driveway is larger than in the front area (see Figures 20 and 21). The wall behind antenna 1 causes uniform dampening along the driveway of the rear reading area.

The ONKALO walls are coarse due to drilling and blasting. The coarse wall surface causes surprising reflections and dispersion especially at the so-called breaks, which means the junction of two blasting areas. An example of dispersion can be seen at the rear reading area of antenna 2. The antenna managed to read a tag at its rear reading area from 147 meters.

Reflection can be seen in ONKALO in a way that in some areas reading of tags does not follow common order (Figures 20, 21, and 23), and this definitely requires further research. Diffraction dampening or fading do not have a noticeable effect in this implementation and it was not studied in this case. The configurations were built so that the antennas and tags had a clear visual contact to the readers.

## 7. Conclusion

In this study, the effects of RFID antenna location and direction on the antenna reading area size were studied in an underground access monitoring system. In addition to antenna location and direction, also absorption to obstacles, reflections, diffraction, scattering, and refraction affect the antenna reading area.

The irregularities of the reading areas showed the need for further research, which means that the tunnel rock surfaces need to be studied and possible water leakages need to be located, in order to examine their effects on the reading area size.

Before future measurements, also the collection of data will be improved. All measurement results will be saved to a database. In addition, measurements will be made so that there are tags at different heights. Based on these measurements, it will be possible to define the best antenna direction and location.

## References


[1] P. Misra, S. Kanhere, D. Ostry, and S. Jha, "Safety assurance and rescue communication systems in high-stress environments: a mining case study," *IEEE Communications Magazine*, vol. 48, no. 4, pp. 66–73, 2010.

[2] T. Saanio, A. Ikonen, P. Keto et al., "Design of the Disposal Facility 2012," Working Report 2013-17, Posiva Oy, Eurajoki, Finland, 2012.

[3] Z. Shi, Y. Li, and Q. Gu, "Based on TWA adaptive centroid wireless localization algorithm of area in underground mine," *Advances in Information Sciences and Service Sciences*, vol. 4, no. 19, pp. 191–196, 2012.

[4] L. Bai-Ping, L. Shang, L. Wen-Feng, and L. Chen, "Research on coal mine personnel orientation rescuing system based on RFID," in *Proceedings of the International Conference on MultiMedia and Information Technology (MMIT '08)*, pp. 555–557, Three Gorges, China, December 2008.

[5] A. Wang and Y. Lu, "The design and implementation of mine workers' precise location system," in *Proceedings of the 2nd Conference on Environmental Science and Information Application Technology (ESIAT '10)*, pp. 303–305, Wuhan, China, July 2010.

[6] U. Farooq, K. M. Hasan, F. Rafiq, and M. U. Asad, "An integrated approach towards designing an embedded wireless monitoring and access control system using RFID and MMS technologies," in *Proceedings of the 16th International Multi Topic Conference (INMIC '13)*, pp. 182–188, Lahore, Pakistan, December 2013.

[7] C. Hurjui, C. Turcu, and A. Graur, "Management system of the products on warranty based on RFID technologies," in *Proceedings of the 11th International Conference on Optimization of Electrical and Electronic Equipment (OPTIM '08)*, pp. 231–236, IEEE, Braşov, Romania, May 2008.

[8] A. Mai, Z. Wei, and M. Gao, "An access control and positioning security management system based on RFID," in *Proceedings of the 7th International Conference on Intelligent Human-Machine*

*Systems and Cybernetics (IHMSC '15)*, pp. 537–540, Hangzhou, China, August 2015.

[9] J. Landt, "The history of RFID," *IEEE Potentials*, vol. 24, no. 4, pp. 8–11, 2005.

[10] P. Nikitin, "Leon Theremin (Lev Termen)," *IEEE Antennas and Propagation Magazine*, vol. 54, no. 5, pp. 252–257, 2012.

[11] R. Want, "An introduction to RFID technology," *IEEE Pervasive Computing*, vol. 5, no. 1, pp. 25–33, 2006.

[12] D. Dobkin, *The RF in RFID: Passive UHF RFID in Practice, Newnes-Elsevier*, Netherland, Amsterdam, 2008.

[13] Z. Sun and I. F. Akyildiz, "Optimal MIMO antenna geometry analysis for wireless networks in underground tunnels," in *Proceedings of the IEEE Global Telecommunications Conference (GLOBECOM '09)*, pp. 1–6, Honolulu, Hawaii, USA, December 2009.

[14] L. Sydanheimo, M. Keskilammi, and M. Kivikoski, "Reliable mobile computing to underground mine," in *Proceedings of the IEEE International Conference on Communications (ICC '00)*, pp. 882–888, June 2000.

[15] K. Kordelin, J. Kordelin, M. Johansson, J. Virkki, L. Ukkonen, and L. Sydänheimo, "Development and implementation of an RFID-based tunnel access monitoring system," *Science and Technology of Nuclear Installations*, vol. 2016, Article ID 9897675, 10 pages, 2016.

[16] Paroc Web-pages, 2016, http://www.paroc.com.

# 4

# Energy and Exergy Analysis of Ocean Compressed Air Energy Storage Concepts

**Vikram C. Patil and Paul I. Ro**

*North Carolina State University, Raleigh, NC, USA*

Correspondence should be addressed to Paul I. Ro; ro@ncsu.edu

Academic Editor: Kiyan Parham

Optimal utilization of renewable energy resources needs energy storage capability in integration with the electric grid. Ocean compressed air energy storage (OCAES) can provide promising large-scale energy storage. In OCAES, energy is stored in the form of compressed air under the ocean. Underwater energy storage results in a constant-pressure storage system which has potential to show high efficiency compared to constant-volume energy storage. Various OCAES concepts, namely, diabatic, adiabatic, and isothermal OCAES, are possible based on the handling of heat in the system. These OCAES concepts are assessed using energy and exergy analysis in this paper. Roundtrip efficiency of liquid piston based OCAES is also investigated using an experimental liquid piston compressor. Further, the potential of improved efficiency of liquid piston based OCAES with use of various heat transfer enhancement techniques is investigated. Results show that adiabatic OCAES shows improved efficiency over diabatic OCAES by storing thermal exergy in thermal energy storage and isothermal OCAES shows significantly higher efficiency over adiabatic and diabatic OCAES. Liquid piston based OCAES is estimated to show roundtrip efficiency of about 45% and use of heat transfer enhancement in liquid piston has potential to improve roundtrip efficiency of liquid piston based OCAES up to 62%.

## 1. Introduction

Electricity generation from renewable energy sources plays important role in reducing dependence on fossil fuels and in curtailing greenhouse gas emissions. However, renewable energy sources such as wind, solar, tidal, and wave are sporadic in nature. The variability of power from the renewable energy sources makes it hard to integrate with the electric grid [1]. Utility-scale energy storage systems are needed to improve the utilization of renewable energy resources in electric grid [2]. Compressed air energy storage (CAES) system is a reliable large-scale energy storage method with relatively low specific investment cost [3]. In CAES system, intermittent energy is used to compress the atmospheric air to the high-pressure and compressed air is stored in the high-pressure reservoir. When electricity demand is high, the stored high pressurized air is expanded through the turbine to generate electricity. Two large-scale CAES plants are in operation, one is in Huntorf, Germany [4], and the other is in McIntosh, Alabama, USA [5]. In both the plants, compressed air is stored in an underground cavern. However, these underground caverns are constant-volume air storage reservoirs. In constant-volume air storage systems, charging and discharging processes result in pressure variation. These varying conditions can result in low efficiencies of compression and expansion due to deviation from the designed points [6]. This can be avoided by utilizing ocean depth for storage of the compressed air in which high-pressure environment under the water can be effectively used for creating constant-pressure storage system [7]. Such an ocean compressed air storage (OCAES) system can effectively integrate multiple energy sources located offshore with high system efficiency.

Various types of OCAES configurations are possibly based on the idealized thermodynamic process of compression and expansion of air. These thermodynamic processes are identified based on how heat is handled during compression and expansion processes. The three major types are diabatic OCAES, adiabatic OCAES, and isothermal OCAES. These types are discussed in the next section with more details. Roundtrip efficiency (also called end-to-end efficiency) is an important parameter to assess the efficacy of an energy storage system. As processes in the OCAES system are accompanied by heat and work transfer, it is very difficult to understand the efficiency of an OCAES system using first law of thermodynamics. This is because the first law of thermodynamics deals with heat and work equally. Therefore, roundtrip efficiency based on exergy analysis which is based on the second law of thermodynamics would be beneficial in better understanding the characteristics of the different types of OCAES systems.

Various studies have investigated different types of CAES configurations based on energy and exergy analysis. Kim et al. reviewed the main drawbacks of the existing CAES systems and presented energy and exergy analysis of various innovative CAES concepts [8]. They investigated concepts like adiabatic CAES, isothermal CAES, micro-CAES combined with air-cycle heating and cooling, and constant-pressure CAES combined with pumped hydrostorage. Their analysis illustrated that drawbacks of existing CAES systems can be addressed by employing innovative CAES concepts. Bagdanavicius and Jenkins investigated the potential for using heat generated during compression stage of CAES with a district energy system [9]. Exergy and exergoeconomic analysis of CAES and CAES with thermal storage were performed by them. Their analysis showed that utilization of waste heat increases energy efficiency from 48% for the CAES to almost 86% for CAES with thermal storage. They also observed that highest exergy destruction occurs in the heat exchangers during compression stage.

Different adiabatic CAES configurations were simulated and analyzed by Hartmann et al. [10] using energy balance. Polytropic CAES with one, two, and three stages and an isentropic CAES were considered. They observed that high value of 70% efficiency is only achieved for isentropic configuration and efficiency of the polytropic configuration is about 60%. Their analysis also suggested that developing high-temperature thermal storage (>600°C) and temperature resistant materials for compressors are key elements in achieving higher efficiency. Grazzini and Milazzo presented a thermodynamic analysis of multistage adiabatic CAES [11]. They proposed a comprehensive set of criteria for the design of adiabatic CAES based on a detailed thermodynamic analysis of the design parameters and influence on system efficiency, with attention to heat transfer devices. An exergy analysis was presented by Tessier et al. on adiabatic CAES system utilizing a cascade of phase change materials for waste heat storage and recovery [12]. Incorporation of phase change materials predicted to show 15% increased efficiencies of storage and recovery over the current design. An experimental study of CAES system with thermal energy storage by Wang et al. has shown a mere 22.6% roundtrip efficiency [13]. A recent detailed review on CAES has been presented in [14].

The existing CAES plants in Huntorf and McIntosh are of a diabatic type and show roundtrip efficiency of 42% and 54%, respectively. The Huntorf plant was primarily designed to provide reserve power and blackstart capability where high efficiency is of minor importance. The McIntosh plant was designed to perform load shifting on a weekly basis, which requires the cycle efficiency to be as high as possible. The McIntosh plant could achieve considerably higher efficiency over Huntorf plant using a recuperator to reduce exergy loss. The concept of adiabatic CAES using thermal energy storage (TES) was also considered during development of these plants; however, diabatic CAES is preferred at that time due to technical and economic advantages [14].

Adiabatic CAES needs a high-temperature TES withstanding the combination of thermal and mechanical stresses which requires special material and complex system engineering. Also, considerable engineering effort is needed to design an electrically driven compressor that operates at the high outlet temperature essential for adiabatic operation. Recent developments in TES have shown good prospects in achieving adiabatic CAES in practice [15]. An adiabatic CAES is under development under the project name "ADELE-ING" which could show roundtrip efficiency up to 70% [16]. Another approach in achieving high efficiency is through isothermal CAES. Technically, it was very difficult to achieve isothermal operation at high power density. However, recent developments in liquid piston compressor [17] and heat transfer enhancement techniques in the liquid piston [18–20] could result in near-isothermal compression and expansion at high power density. Thermodynamic and economic review of CAES by Rogers et al. [21] indicates that efficiencies of advanced adiabatic CAES and isothermal CAES have been increased by over 30% and energy storage densities have been improved by a factor of 5 using near-surface piping.

A multilevel underwater CAES system integrated with battery pack is proposed by Wang et al. [22]. Their thermodynamic analysis shows that roundtrip exergy efficiency of the multilevel underwater CAES varies from 62% to 81% in different working mode. Advanced exergy analysis of an underwater CAES by Wang et al. indicates exergy efficiency of 53.6% under real conditions with theoretical maximum exergy efficiency of 84.3% [23]. Clearly, there is a great potential for performance improvement of underwater CAES. A study by Cheung et al. indicates that pipe diameter, turbine, air compressor, and air storage depth have the greatest influence on system performance of underwater CAES [24]. Multiobjective optimization of an underwater CAES with objectives of maximizing roundtrip efficiency and operating profit and minimizing cost rate is performed by Cheung et al. using genetic algorithm [25]. Their analysis indicated roundtrip efficiency of 68.5% and operating profit of $53.5 per cycle for the preferred system designs.

Based on earlier studies on CAES, it can be observed that there could be a significant variation in efficiencies of various OCAES configurations. Comparative analysis of different OCAES systems would help in understanding these better. Preliminary studies of exergy analysis for various OCAES configurations and end-to-end efficiency of liquid piston based OCAES have been presented earlier in [26, 27]. In this paper, detailed analysis of various OCAES concepts using energy and exergy analysis is presented. This would help in assessing improvement areas in achieving higher roundtrip efficiency with OCAES.

Investment costs associated with different types of OCAES would also vary as the technology used in these configurations differs considerably. The compressor and expander in the diabatic OCAES are mature technologies whereas the same in the adiabatic and isothermal OCAES are still in the development stages. Also, the cost of thermal energy storage (TES) used in adiabatic OCAES differs significantly based on the kind of TES system used. Broadly, there are three types of TES systems, sensible heat TES, latent heat TES, and thermochemical TES. In general, latent heat TES and thermochemical TES are more expensive than sensible heat TES; however, former could be economically viable with a high number of operating cycles [15]. Although the cost of different OCAES configurations might differ significantly, this paper only focuses on efficiency aspect without accounting any cost difference. However, an economic assessment of these OCAES configurations would be necessary before investment decisions. This study could provide a framework for an economic assessment of different OCAES by monetizing efficiencies incorporating cost difference.

## 2. OCAES Configurations

OCAES configurations can be broadly distinguished depending on the targeted idealized process of compression and expansion of air. It is decided based on how heat is handled during compression and prior to expansion of the air. Three major OCAES configurations would be diabatic OCAES, adiabatic OCAES, and isothermal OCAES. These system configurations are discussed in this section.

*2.1. Diabatic OCAES.* In the diabatic OCAES system, in energy storage mode, air is compressed using conventional compressors and cooled to the surrounding temperature before sending it to the storage system. Compression process increases the temperature of air due to the heat of compression which is dissipated before sending air to the storage device. This results in loss of thermal energy of compression due to cooling. In energy recovery mode, the air from the storage is heated using fuel and then passed through the expander to generate electricity.

The schematic of diabatic OCAES is shown in Figure 1. Processes 1-2 represent compression of atmospheric air using air compressor run using electric motor operating on excess electric energy. The motor efficiency and losses in the compressor would result in loss of energy/exergy during this process. Compressed air from the compressor is then passed through the cooler (processes 2-3) before sending it to the underwater air storage system. A significant amount of heat energy (thermal exergy) is lost in the cooler. Processes 3-4 indicate charging and discharging from the air storage system. Mechanical exergy (energy in pressure form) in the high-pressure air is stored in the underground storage. There can be a small amount of loss of energy/exergy due to leakage and pressure drop in the storage system. In process 4-5, high-pressure air is heated using external heat input (thermal exergy) to increase expansion work. The processes 5-6 are the expansion process in which mechanical exergy in the form of electrical energy is delivered. However, processes 5-6 involve loss of energy/exergy due to inefficiencies in expansion process and also loss of energy/exergy from the exhaust gas.

*2.2. Adiabatic OCAES.* The construction of adiabatic system would require a compressor design delivering higher temperature at the outlet of compression, a reliable TES system to store thermal energy at high temperature and an expander design operating at high inlet air temperature with broad operation range. These components installed and connected with pipelines carrying air with the layout shown in Figure 2 would result in the adiabatic OCAES system.

It can be seen that cooler and heater in the diabatic OCAES are replaced with thermal energy storage (TES) in the adiabatic OCAES configuration. The diabatic OCAES uses fuel in the heater; therefore, it cannot be considered as a pure storage system and is actually a combination of storage and

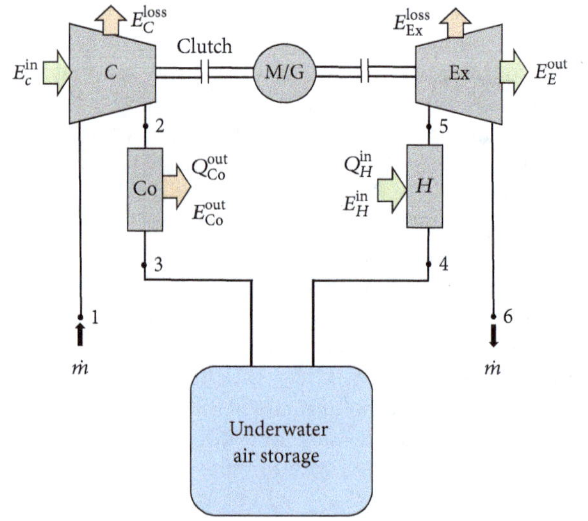


FIGURE 1: Schematic of diabatic OCAES.

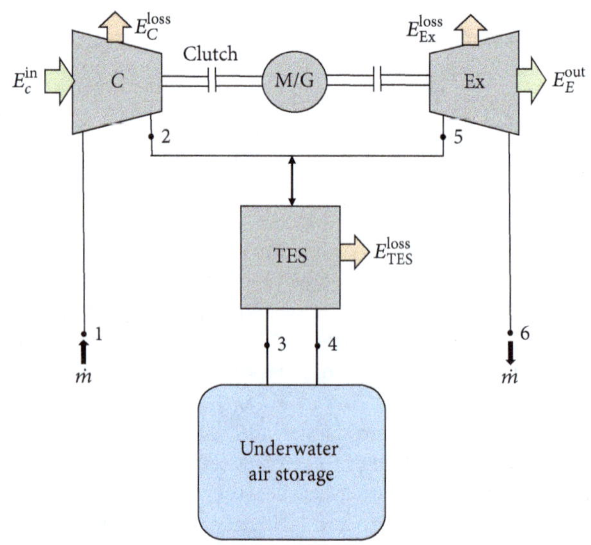


FIGURE 2: Schematic of adiabatic OCAES.

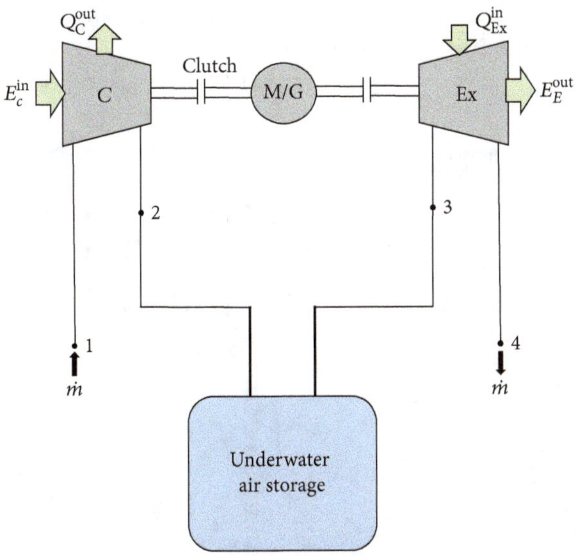


FIGURE 3: Schematic of isothermal OCAES.

power plant. This can be overcome in adiabatic OCAES which uses TES to store heat from the compressed air before sending it to the air storage. The stored heat in TES is used to heat the air before passing it to the expander. This eliminates the need for cooler and heater in the adiabatic system as TES works as both. Therefore, thermal energy and thermal exergy losses in the cooler and thermal heat input in the heater of diabatic OCAES are completely eliminated in the adiabatic OCAES. However, TES involves some energy/exergy losses which result in added inefficiencies in adiabatic OCAES.

*2.3. Isothermal OCAES.* Isothermal OCAES would eliminate the need for fuel and high-temperature thermal energy storage. This can be done by isothermal compression and expansion process which minimizes compression work and maximizes expansion work. Figure 3 shows a schematic of isothermal OCAES system. The compressor in the isothermal OCAES dissipates heat energy during compression process resulting in the conversion of electrical energy into mechanical exergy form in the compressed air. Ideal isothermal compression does not add any thermal exergy in the compressed air; therefore, loss of exergy is totally avoided. Similarly, in the ideal isothermal expansion, mechanical exergy from the compressed air is completely converted into electrical energy.

It is very difficult to achieve isothermal compression and expansion using conventional compressors and expansion devices as conventional compressors and expanders work at high speed with the nearly adiabatic process. Special types of compressors and expanders are required to achieve isothermal OCAES in reality. Liquid piston compressor can be used to achieve near-isothermal compression and expansion operation. In the liquid piston compressor, a column of liquid (usually water) is utilized to compress a gas in the fixed volume chamber. A hydraulic pump is used to generate the flow of the liquid for the liquid pistons. The liquid flow in and out of compression chamber is controlled with valves. As a liquid can conform to an irregular chamber volume, the surface area to volume ratio in the gas chamber can be maximized using a liquid piston. This results in increasing the heat transfer during the gas compression/expansion which facilitates near-isothermal operation [17]. Various heat transfer enhancement techniques like the use of porous media inserts [19], spray cooling [18], and use of hollow spheres [20] can be effectively used in the liquid piston to achieve near-isothermal operation.

A typical liquid piston based isothermal OCAES would have electric motor/generator, hydraulic pump/motor, liquid piston compressor/expander, air cooler/heater, pipelines connecting various components, control valves, and underwater air storage. Figure 4 shows a schematic of a liquid piston based OCAES system. Although liquid piston compression is efficient compared to the existing compressors technologies, the added components like hydraulic pump/motor and hydraulic lines have some inefficiencies which would affect the overall efficiency of OCAES system.

## 3. Energy and Exergy Analysis

Inefficiencies in various components of the OCAES contribute to the loss of energy in the storage system. Energy efficiency of a component is the ratio of energy out from the component to energy into the component. In energy analysis, energy efficiencies of various components in the OCAES are modeled and used to evaluate the energy efficiency of overall OCAES system.

The exergy transfer to the system can happen by work, heat, and mass transfer. The exergy transfer by heat is given by

$$\dot{E}_q = \int 1 - \frac{T_0}{T}\delta\dot{Q}, \tag{1}$$

where $T$ is temperature, $\dot{Q}$ is heat transfer rate, and subscript 0 indicates properties at environmental conditions [28].

Exergy transfer by mass flow ($\dot{m}$) is given by

$$\dot{E}_m = \dot{m}e, \tag{2}$$

where specific exergy ($e$) of an ideal gas is given by

$$e = C_p\left(T - T_0\right) - T_0\left[C_p \ln\left(\frac{T}{T_0}\right) - R\ln\left(\frac{P}{P_0}\right)\right], \tag{3}$$

where $C_p$ is specific heat at constant pressure, $R$ is the specific gas constant, and $P$ is pressure.

The above specific exergy consists of two parts, mechanical exergy and thermal exergy. The mechanical exergy is associated with the system pressure and is the exergy change when the system is brought to the state $[T_0, P_0]$ from the state $[T_0, P]$. The thermal exergy is associated with the system temperature and is the exergy change when the system is

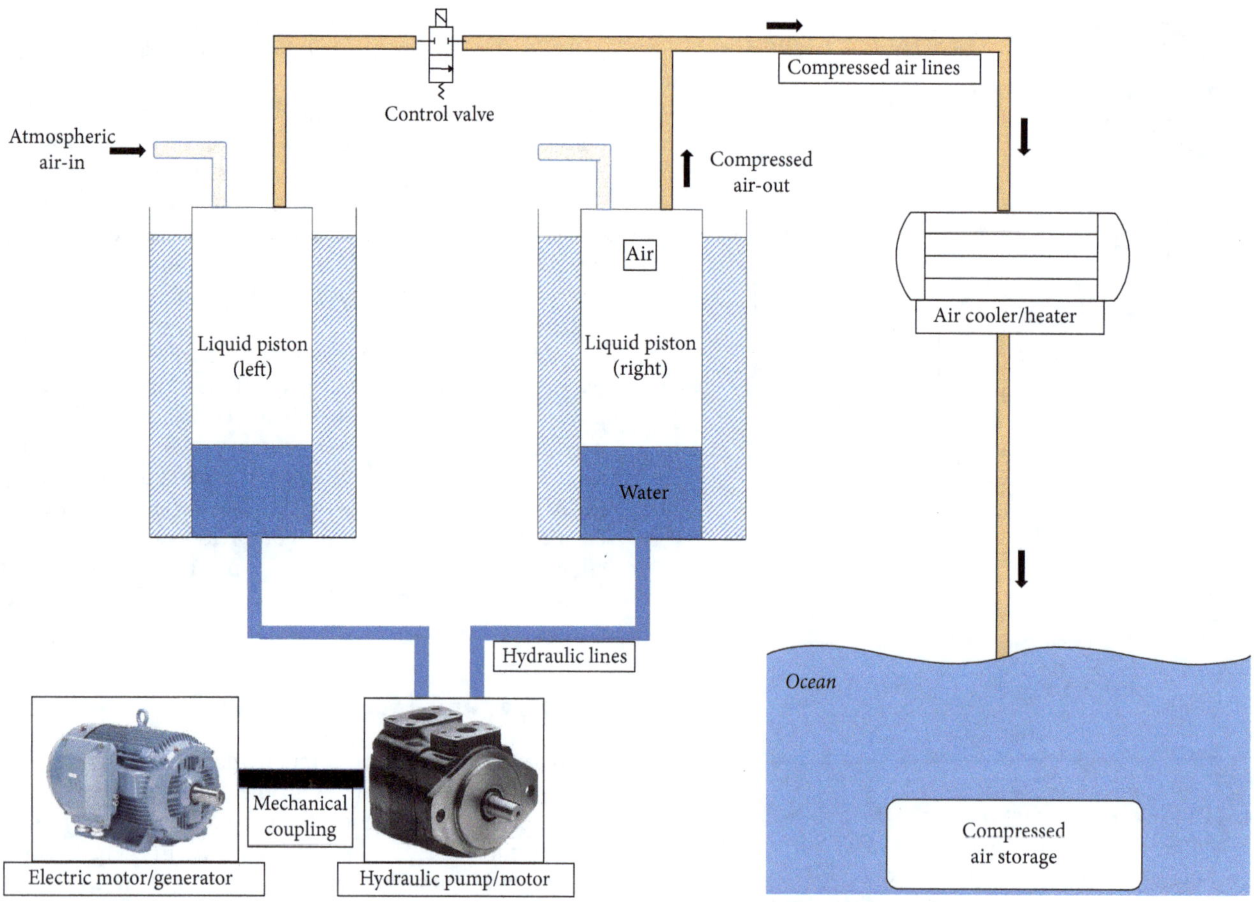


FIGURE 4: Schematic of a liquid piston based isothermal OCAES system.

brought to the state $[T_0, P]$ from the state $[T, P]$. Mechanical and thermal exergies are given by (4) and (5), respectively.

$$e^M = RT_0 \ln\left(\frac{P}{P_0}\right), \tag{4}$$

$$e^T = C_p\left(T - T_0 - T_0 \ln\left(\frac{T}{T_0}\right)\right), \tag{5}$$

where superscripts $M$ and $T$ indicate mechanical and thermal parts, respectively.

Individual components in the OCAES system can be analyzed based on exergy analysis. Exergy efficiency of a component is given by

$$\varepsilon_{\text{com}} = \frac{\dot{E}^-_{\text{com}}}{\dot{E}^+_{\text{com}}}, \tag{6}$$

where $\varepsilon$ denotes exergy efficiency and subscripts indicates component.

Energy efficiency and exergy efficiency of individual components in OCAES are discussed below.

*3.1. Electric Motor and Generator.* Electric motor/generator has mechanical and electrical losses during its operation. Energy efficiency of electric motor/generator (M/G) is the ratio of power output from M/G to the power input to M/G as calculated using

$$\eta_{\text{M/G}} = \frac{P_n \times \text{Load}}{P_i}. \tag{7}$$

As electric motor/generator deals with work transfer (electrical energy and shaft work) only, exergy efficiency of electric motor/generator is its energy efficiency.

*3.2. Hydraulic Pump/Motor.* Energy efficiency of hydraulic pump/motor depends on its displacement, speed of operation, and the pressure differential between inlet and outlet. The overall efficiency of a hydraulic pump for a particular power input is given by

$$\eta_{\text{HP}} = \frac{D \times N \times \Delta p}{P_{\text{in}}}. \tag{8}$$

Exergy efficiency of hydraulic pump/motor is also same as its energy efficiency because it deals only with work transfer.

*3.3. Compressor.* The exergy transfer to the compressor ($\dot{E}^+_C$) is in the form of shaft work from the motor whereas exergy transfer from the compressor ($\dot{E}_C^-$) is due to air mass transfer from the compressor at high pressure and temperature. There are exergy losses in the compressor due to mechanical losses. Isentropic efficiency [29] and mechanical efficiency of the compressor can be used to evaluate exergy efficiency of compressor.

In case of isothermal OCAES, liquid piston compressor efficiency is defined as the ratio of stored energy to work input. Storage energy is the amount of work extracted from the isothermal expansion of compressed air to the atmospheric pressure. Work input consists of compression work, cooling work and friction work. Atmospheric air compressed to high-pressure results in increasing temperature. This compressed air is cooled to initial temperature to maintain storage pressure [30]. This adds cooling work. In case of liquid pistons, friction work is comparatively small (unless diameter is too small) and can be neglected [17]. Liquid piston compressor efficiency ($\eta_C$) after neglecting viscous friction is given by

$$\eta_C = \frac{\overbrace{\ln(P_r) + 1/P_r - 1}^{E_{\text{Storage}}}}{\underbrace{\left(P_r^{(n-1)/n} - 1\right)/(n-1) + P_r^{-1/n} - 1}_{W_{\text{compression}}} + \underbrace{(P_r - 1)\left(P_r^{-1/n} - 1/P_r\right)}_{W_{\text{Cooling}}}}, \quad (9)$$

where $P_r$ is the pressure ratio (ratio of storage pressure to the atmospheric pressure) and $n$ is a polytropic index of compression. Storage pressure (hence $P_r$) depends on the underwater air storage depth and $n$ depends on the magnitude of heat transfer in liquid piston compressor.

*3.4. Cooler.* The compressed air from the compressor contains both mechanical and thermal exergy. In the cooler, the compressed air is cooled to the atmospheric temperature at a constant pressure by dissipating thermal exergy of the compressed air to cooling media. The output compressed air from the cooler would contain only mechanical exergy. Therefore, exergy efficiency of the cooler neglecting pressure losses in the cooler is given by

$$\varepsilon_{\text{co}} = \frac{\dot{m}e^M}{\dot{m}\left(e^M + e^T\right)} = \frac{RT_0 \ln(P_{\text{co}}/P_0)}{RT_0 \ln(P_{\text{co}}/P_0) + C_p\left(T_{\text{co}} - T_0 - T_0 \ln(T_{\text{co}}/T_0)\right)}, \quad (10)$$

where $P_{\text{co}}$ and $T_{\text{co}}$ are pressure and temperature of compressed air at inlet of the cooler, respectively.

*3.5. Air Pipeline Connecting Various Components.* Energy/exergy loss inside the pipeline carrying air is equal to product of pressure drop and flow rate. Pressure drop for steady, fully developed, incompressible flow in the pipe can be calculated using Darcy–Weisbach equation [31].

*3.6. Thermal Energy Storage (TES).* In adiabatic OCAES, high-temperature high-pressure compressed air is passed through the TES to store thermal exergy of compressed air in the TES. This stored thermal energy is used to increase thermal exergy of compressed air before sending it through the expander. The thermal and pressure losses in the TES result in the loss of energy/exergy. The exergy efficiency of TES is given by

$$\varepsilon_{\text{TES}} = \frac{RT_0 \ln\left(P_{\text{TES}}^{\text{out}}/P_0\right)}{RT_0 \ln\left(P_{\text{TES}}^{\text{in}}/P_0\right) + C_p\left(T_{\text{TES}}^{\text{in}} - T_0 - T_0 \ln\left(T_{\text{TES}}^{\text{in}}/T_0\right)\right)}, \quad (11)$$

where $P_{\text{TES}}^{\text{in}}$ and $T_{\text{TES}}^{\text{in}}$ are pressure and temperature of compressed air at inlet of the TES. $P_{\text{TES}}^{\text{out}}$ and $T_{\text{TES}}^{\text{out}}$ are pressure and temperature of compressed air at outlet of the TES, respectively.

*3.7. Air Storage.* Leakage and pressure losses in the underwater air storage system result in energy/exergy losses. Leakage per unit volume per unit time can be calculated by measuring pressure drop in an isolated air storage system given by equation [32]

$$\dot{L}_S = \frac{\rho_s}{t_e}\frac{(\Delta P)_S}{P_S}, \quad (12)$$

where $\dot{L}_S$ is the leakage rate (kg/hr·m$^3$), $\rho_s$ is the density of air at storage pressure and temperature, $(\Delta P)S$ is the pressure drop in the isolated air storage system in time $t_e$, and $P_S$ is the storage pressure.

The energy/exergy efficiency of an air storage system is given by [27]

$$\varepsilon_S = 1 - \frac{t_0 \times \dot{L}_S}{\rho_s}, \quad (13)$$

where $t_o$ is the operation time of the OCAES.

*3.8. Heater.* In the diabatic OCAES, external fuel is used to heat the air. The heat (thermal exergy) added in the heater increases exergy of the air. Exergy efficiency of a heater neglecting pressure losses in the heater and considering the constant rate of heat transfer is given by

$$\varepsilon_H = \frac{C_p\left(T_H - T_0 - T_0 \ln(T_H/T_0)\right) + RT_0 \ln\left(P_{H,\text{out}}/P_0\right)}{Q_s\left(1 - T_0/T_s\right) + RT_0 \ln\left(P_{H,\text{in}}/P_0\right)}, \quad (14)$$

where $P_{H,\text{in}}$ and $P_{H,\text{out}}$ are pressures at inlet and outlet of the heater, respectively, $T_H$ is the temperature of air at heater output, $T_S$ is the temperature of heat source, and $Q_S$ is the heat transfer per unit mass of air.

*3.9. Expander.* The exergy transfer to the expander ($\dot{E}_{\text{Ex}}^+$) is by compressed air inlet whereas exergy transfer from the expander ($\dot{E}_{\text{Ex}}^-$) is in the form of shaft work delivered to the generator. Similar to compressor, exergy efficiency of

Table 1: Stochastic assignments in Monte Carlo simulation.

| Variable | Mean/max value [$\mu$] (%) | Standard deviation[a] or max/min value (%) | Distribution |
|---|---|---|---|
| $\varepsilon_{M/G}$ [34] | 96 | 0.5 | Normal |
| $\varepsilon_{HP/HM}$ [8] | 93 | 1 | Normal |
| $\eta_{C,isen}$ | 85 | 2 | Normal |
| $\eta_{C,mech}$ | 95 | 1 | Normal |
| $\eta_P$ | $(P_{in} - \Delta P)/P_{in}$ | Max = $\mu$,<br>Min = $\mu - 0.5$ | Triangular |
| $\varepsilon_{TES}$ | 80 | 2 | Normal |
| $\varepsilon_S$ | Using (13) | Max = $\mu + 0.5$<br>Min = $\mu - 0.5$ | Triangular |
| $\varepsilon_H$ | 95 | 1 | Normal |
| $\eta_{Ex,isen}$ | 85 | 1 | Normal |
| $\eta_{Ex,mech}$ | 95 | 1 | Normal |

[a]For normal distribution.

expander can be calculated using isentropic efficiency [29] and mechanical efficiency of the expander.

In case of isothermal OCAES using a liquid piston, liquid piston expansion efficiency ($\eta_E$) for polytropic expansion index $n$ is given by

$$\eta_E = \frac{\overbrace{\left(1-(1/P_r)^{(n-1)/n}\right)/(n-1)-(1/P_r)^{(n-1)/n}+1/P_r}^{W_{\text{Expansion}}}}{\underbrace{\ln(P_r)+1/P_r-1}_{E_{\text{Storage}}}}. \quad (15)$$

## 4. Numerical Simulations

Different types of OCAES systems are modeled based on energy and exergy analysis of individual components in the OCAES system. Storage pressure of 10 bar gauge (100 m ocean depth) is considered for analysis. Various components specifications designed for maximum power capacity of 0.5 MW with 2 MWh energy storage were used [33]. Efficiencies of motor/generator and hydraulic pump/motor are considered from the industry standards [8, 34]. Pipelines connecting cooler to air storage and air storage to the heater are considered with 1000 m in length, 0.2 m in diameter, and of 15 $\mu$m surface roughness [35]. For air storage, the leakage rate of 0.01 kg/hr·m$^3$ [36] and operation time of 8 hours were assumed [37].

Uncertainty analysis is performed using Monte-Carlo Simulations (100000 runs) to estimate mean and confidence interval values of energy efficiency and exergy efficiency. Stochastic assignments considered are given in Table 1.

All the simulations were performed considering 1 atm and 20°C environmental conditions. In all the configurations, single stage compression and single stage expansion were considered. Adiabatic and isothermal OCAES systems are considered without the use of fuel. In the diabatic configuration, heat source temperature of 1500°C is considered. The amount of heat transfer from the heat source is evaluated

Table 2: Thermodynamic properties and mass flow rates of air at different state points for diabatic OCAES system (see Figure 1 for state points).

| State point | Pressure (kPa) | Temperature (K) | Mass flow rate (kg/s) |
|---|---|---|---|
| 1 | 101.3 | 293 | 1.33 |
| 2 | 1119.6 | 633 | 1.33 |
| 3 | 1114.6 | 293 | 1.33 |
| 4 | 1109.6 | 293 | 1.32 |
| 5 | 4200.0 | 823 | 1.32 |
| 6 | 101.3 | 364 | 1.32 |

to achieve inlet conditions to expander with 42 bar and 550°C. These values are referred from HP turbine operating conditions of Huntorf plant [38]. In adiabatic OCAES configuration, TES storage temperature of 327°C is considered [39] and inlet air conditions to the expander of 10 bar and 327°C are considered. In isothermal OCAES configuration, liquid piston based compression and expansion are considered.

To simplify the thermodynamic model, all the analysis is performed for steady state operation of the system. Also, the kinetic and potential energy of the fluids are assumed to be negligible and the air is considered as an ideal gas [23]. For these system considerations, the thermodynamic properties and mass flow rate of air at different state points of the system are evaluated. Those are listed in Tables 2, 3, and 4 for diabatic, adiabatic, and isothermal OCAES, respectively. Finally, different types of OCAES systems are compared using energy/exergy flow, energy efficiency, and exergy efficiency evaluations.

## 5. Results and Discussion

Figures 5, 6, and 7 show exergy flow in various OCAES configurations. In figures, each box represents a component in the OCAES system. The height of the box represents qualitative exergy flow to the component. The arrows indicate the exergy flow direction and the amount of exergy flow is

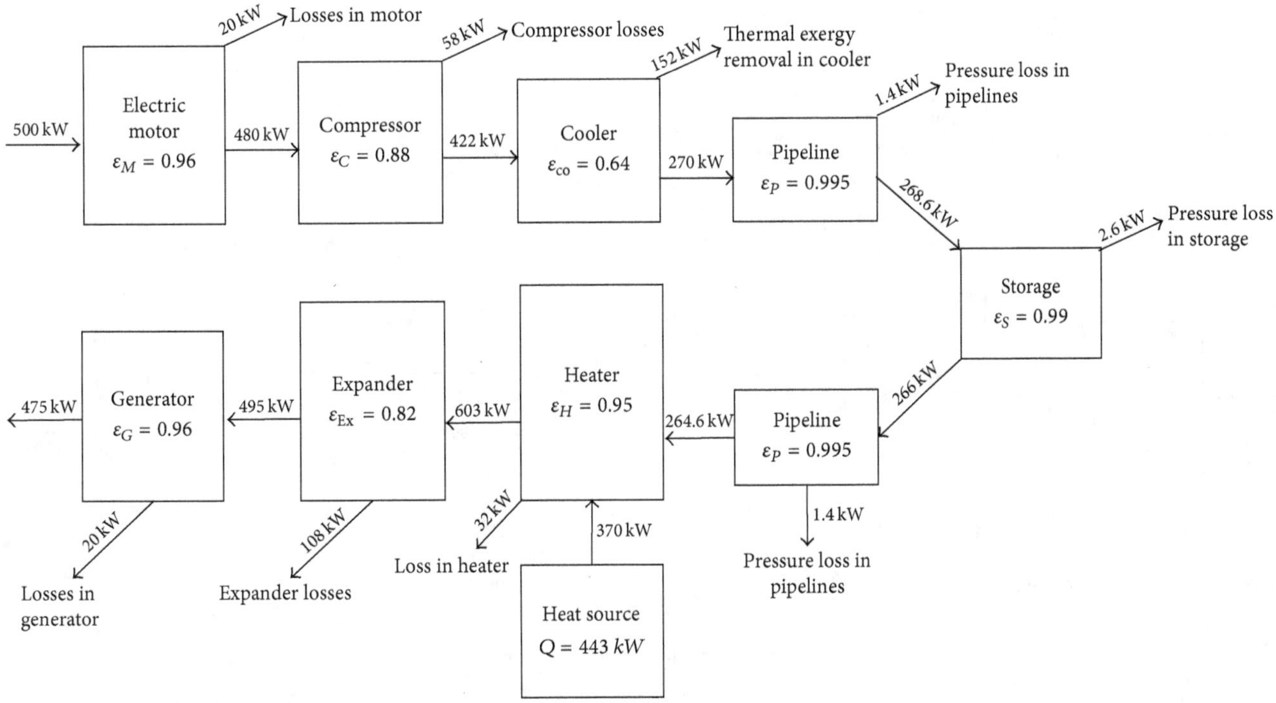


FIGURE 5: Exergy flow in diabatic OCAES.

TABLE 3: Thermodynamic properties and mass flow rates of air at different state points for adiabatic OCAES system (see Figure 2 for state points).

| State point | Pressure (kPa) | Temperature (K) | Mass flow rate (kg/s) |
|---|---|---|---|
| 1 | 101.3 | 293 | 1.33 |
| 2 | 1119.6 | 633 | 1.33 |
| 3 | 1114.6 | 293 | 1.33 |
| 4 | 1109.6 | 293 | 1.32 |
| 5 | 1104.6 | 600 | 1.32 |
| 6 | 101.3 | 354 | 1.32 |

TABLE 4: Thermodynamic properties and mass flow rates of air at different state points for isothermal OCAES system (see Figure 3 for state points).

| State point | Pressure (kPa) | Temperature (K) | Mass flow rate (kg/s) |
|---|---|---|---|
| 1 | 101.3 | 293 | 2.10 |
| 2 | 1119.6 | 300 | 2.10 |
| 3 | 1114.6 | 293 | 2.08 |
| 4 | 101.3 | 286 | 2.08 |

mentioned on each arrow. The arrows pointing away from the boxes indicate exergy loss to the environment. The exergy efficiencies of the individual components are mentioned in the boxes. The input power of 500 kW is considered for the analysis.

*5.1. Diabatic OCAES.* Exergy flow in the diabatic OCAES is presented in Figure 5. Inefficiencies in electric motor result in exergy loss of 20 kW. It is observed that a significant amount of exergy is lost from the cooler to the surrounding. Losses in the pipelines and storage are very small compared to other losses. The addition of heat energy to the heater from the heat source adds exergy to the system. Although energy input from the heat source to the heater is 443 kW, the exergy value of this heat is only about 370 kW. This addition of exergy in heater compensates the loss of exergy in the cooler to achieve the same level of power output. The expander is the next exergy inefficient component after the cooler in the diabatic system. Inlet air to the expander contains exergy in both thermal and mechanical forms. Irreversibility in expansion process results in exhaust air with a significant amount of thermal exergy, which results in higher exergy loss in the expander. The recuperator is not considered in the current configuration. In recuperative process waste heat from the exhaust of the expander can be used to heat expander inlet air. This would reduce exergy burden from the heat source and hence improve overall exergy efficiency of OCAES system.

The overall exergy efficiency of diabatic OCAES is 55% whereas energy efficiency is 50%. As diabatic OCAES has fuel input in the heater system, it is not a pure energy storage system but the combination of energy storage and power plant. Therefore, exergy efficiency is a good measure of diabatic OCAES for comparison with other systems.

*5.2. Adiabatic OCAES.* The thermal exergy lost from the cooler in the diabatic OCAES is stored in the adiabatic OCAES using TES which can improve exergy efficiency of

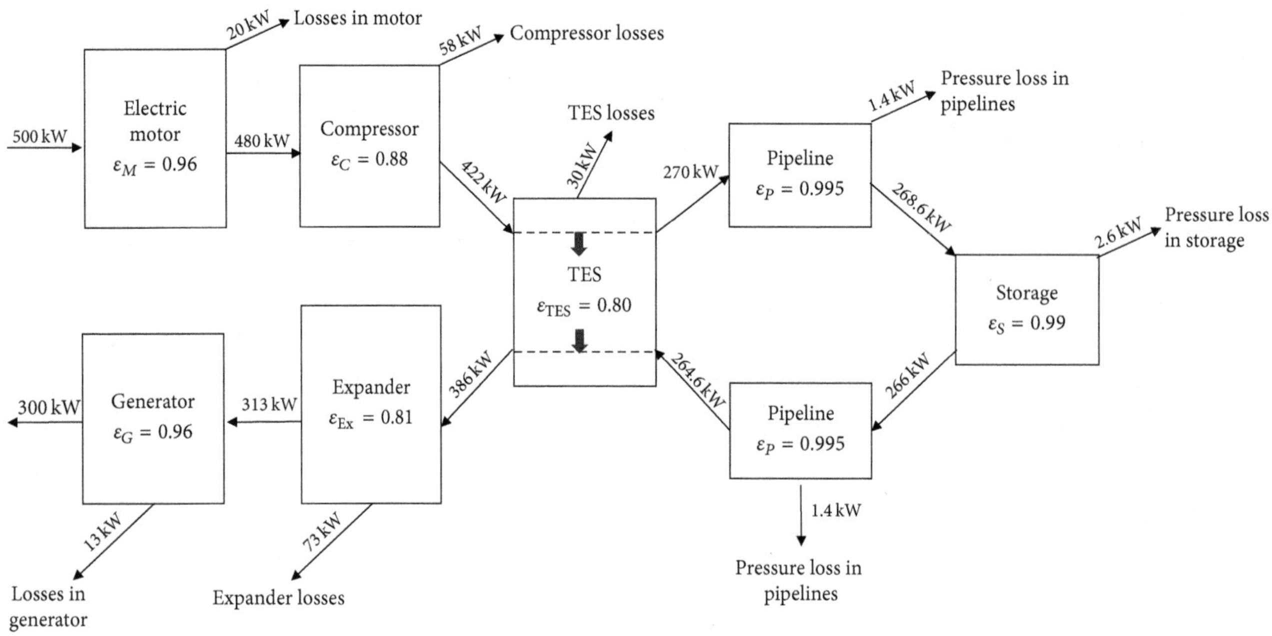


FIGURE 6: Exergy flow in adiabatic OCAES.

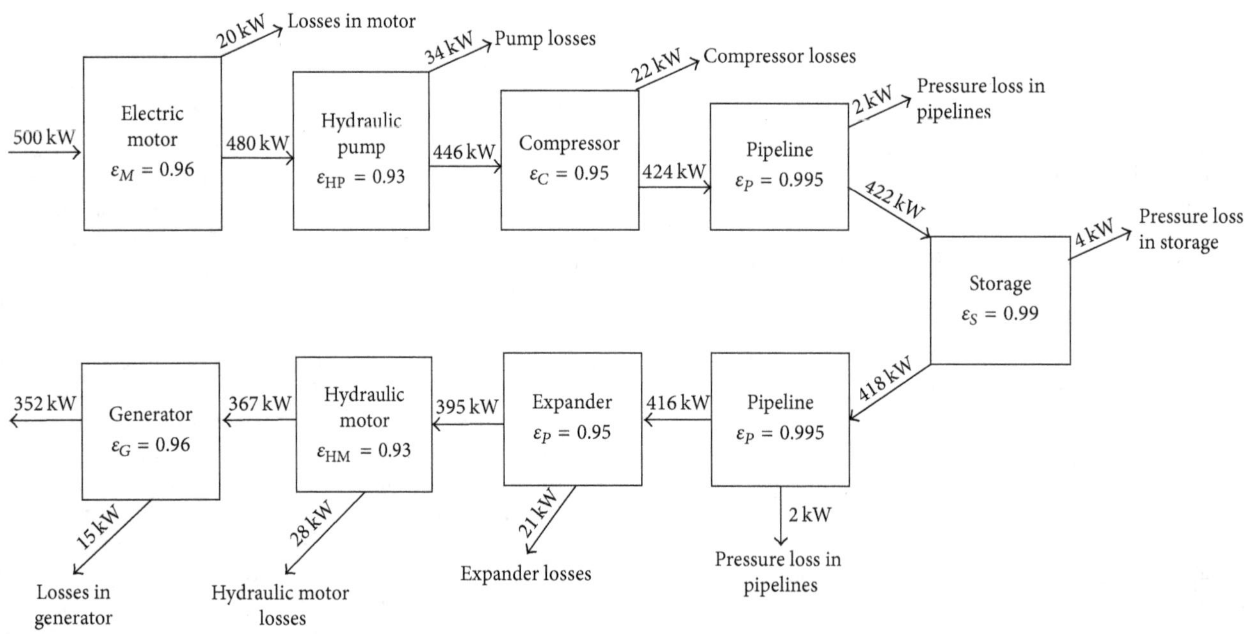


FIGURE 7: Exergy flow in isothermal OCAES.

adiabatic OCAES significantly over diabatic OCAES. Figure 6 shows exergy flow in adiabatic OCAES. Exergy loss in the cooler of diabatic OCAES can be avoided by use of TES which stores a significant amount of thermal energy. TES supplies heat to the air before the expander thus increasing the exergy potential of air. With the use of TES, the external heat source is removed in the adiabatic OCAES configuration. However, inefficiencies in TES account for exergy loss which contribute to reduction energy output. Comparison of diabatic and adiabatic exergy flow reveals that adiabatic CAES gives less exergy output (electric energy) from the generator. This is because the fuel source used in the diabatic OCAES allows the expander to be operated with higher power capacity.

Overall exergy efficiency of adiabatic OCAES is 60% which is 5% higher than that of diabatic OCAES. This improvement is due to reuse of thermal exergy of the air using TES. Improvement in TES efficiency from current consideration of 80% would further improve the efficiency of adiabatic OCAES. As external fuel is not used in adiabatic OCAES, overall energy efficiency of adiabatic OCAES is same

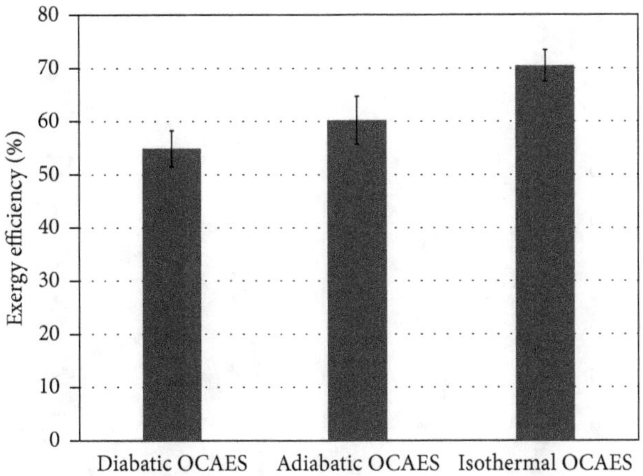


FIGURE 8: Overall exergy efficiency of different types of OCAES.

as overall exergy efficiency. Careful observation of exergy losses in adiabatic OCAES shows that losses in compressor and expander are major contributors of inefficiencies in adiabatic OCAES.

5.3. *Isothermal OCAES.* The need of TES and fuel source is eliminated in the isothermal OCAES. Exergy flow in the isothermal OCAES is shown in Figure 7. Liquid piston based compressor and expander requires a hydraulic pump and a hydraulic motor as added components in the isothermal OCAES. Inefficiencies of these components would contribute to the exergy loss in the system. It can be observed in Figure 7 that hydraulic pump/motor show high exergy losses in comparison with losses in other components in isothermal OCAES. However, the use of liquid piston in conjunction with hydraulic pump and motor has the potential to improve efficiencies of the compressor and expander significantly. This results in a reduction of exergy losses in compressor and expander. Also, the absence of TES and heater eliminates losses associated with those which helps in improving the overall efficiency of the system. Overall exergy efficiency of the isothermal OCAES is about 70% which is significantly higher than diabatic and adiabatic OCAES.

The overall exergy efficiencies of all the three configurations with 95% confidence interval bounds are shown in Figure 8 for comparison. The uncertainties in various assumptions show about 3-4% variation in exergy efficiencies. Clearly, isothermal OCAES is the most efficient and diabatic OCAES is the least efficient among three configurations considered based on exergy analysis. Energy efficiency of diabatic OCAES is about 50% whereas that of adiabatic and isothermal OCAES is same as their exergy efficiencies. Energy efficiency might not be a reliable comparative parameter as it would undervalue the efficiency of the diabatic system.

5.4. *Liquid Piston Based OCAES.* Although isothermal OCAES shows high exergy efficiency, such a high level of efficiency is contingent upon near-isothermal compression and expansion. Liquid piston compressor is experimentally tested to investigate its effectiveness in achieving near-isothermal

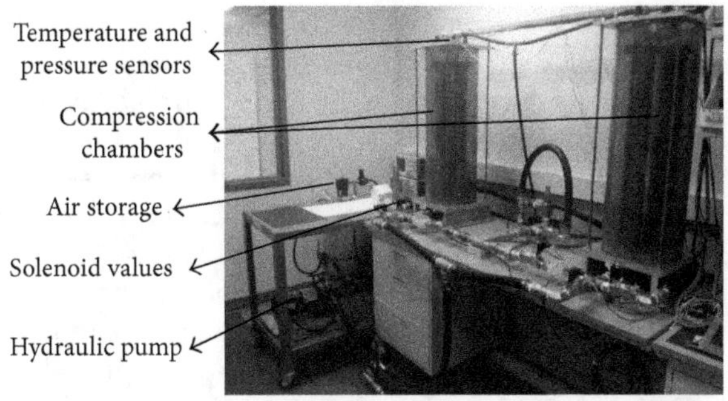


FIGURE 9: Experimental setup of liquid piston compressor.

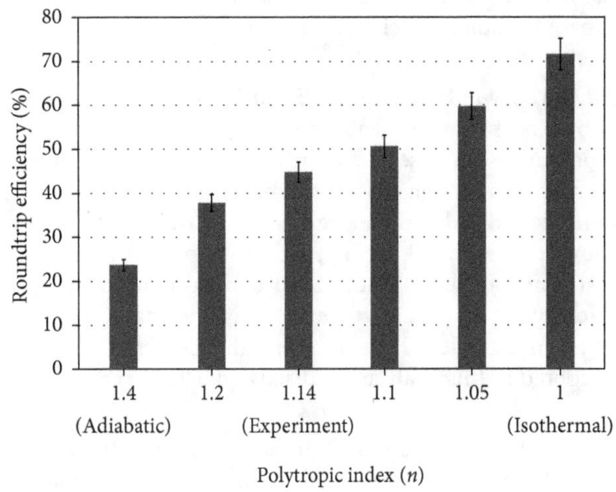


FIGURE 10: Roundtrip efficiency of liquid piston based OCAES for different polytropic indices.

compression. Figure 9 shows the experimental setup of a liquid piston compressor. The compression chamber was divided into four parallel copper pipes, each having an inner diameter of 76 mm and length of 760 mm. Two compression chambers were used in the experimental setup to ensure continuous production of compressed air. The compression chambers were enclosed in a plastic cylinder filled with water to maintain the temperature of the outer copper wall a constant. A hydraulic pump was used to alternatively drive water from one compression chamber to the other. The pump delivers a constant flow rate of 10 gpm at a maximum pressure of 13.1 bar gauge (190 psi). A K-type thermocouple of diameter 0.0508 mm (0.002 in) and a pressure sensor were installed at the top of each compression chamber as indicated in Figure 9. Experiments were performed with a pressure ratio of 6 and the stroke time of 10 s.

The liquid piston compressor and expansion efficiency can be calculated using the polytropic index of compression and expansion in (9) and (15), respectively. The polytropic index is calculated from the $P$-$T$ curve (pressure-temperature curve) considering $P^{(1-n)}T^{n}$ = constant relation.

Roundtrip efficiencies of liquid piston based OCAES system with various polytropic indexes of compression/expansion are shown in Figure 10. Uncertainty bars represent

95% confidence interval valves. It can be observed that estimated mean value of end-to-end efficiency increases from 24% to 72% with a decrease in the polytropic index of compression from 1.4 (adiabatic process) to 1 (isothermal process). This clearly indicates liquid piston compression and expansion efficiency has a major influence on the end-to-end efficiency. For a polytropic index of 1.14 observed with an experimental liquid piston, a roundtrip efficiency of 45% was shown. Noticeably, this efficiency value is way below efficiency level of isothermal OCAES. However, the liquid piston setup used in the experimental investigation did not involve any heat transfer enhancement mechanism to abate temperature rise. Various designs of liquid pistons leading to a lower polytropic index of compression/expansion would increase roundtrip efficiency for the OCAES system.

*5.5. OCAES with Heat Transfer Enhancement in Liquid Piston.* Various heat transfer enhancement techniques in liquid piston have been observed to be effective in improving efficiency and power density. Various heat transfer enhancement techniques considered are use of optimal trajectories, use of hollow spheres, spray cooling, and use of porous media. The effectiveness of these heat transfer enhancement techniques in achieving an efficiency level of isothermal OCAES is further investigated. Heat transfer enhancement techniques considered in this analysis are discussed below.

*5.5.1. Optimal Trajectories.* The Pareto optimal trajectories for the liquid piston compressor/expander that maximizes efficiency for a given power have been found by Saadat et al. [40]. They considered general heat transfer models, the viscous friction, and system constraints in the optimization process. These optimal trajectories were experimentally tested by Shirazi et al. [41]. It was observed that optimal profiles show up to 4% higher efficiency for the same power density or 30% higher power density for the same the efficiency compared to ad hoc constant flow rate profiles.

*5.5.2. Use of Hollow Spheres.* Hollow spheres floating at the liquid-air interface in the liquid piston have been observed to be effective in reducing the temperature of the compressed air in the liquid piston compressor. Hollow spheres made of Silicon Carbide (SiC), High-Density Polyethylene (HDPE), and Polypropylene (PP) were tested by Ramakrishnan et al. in a liquid piston compressor [20]. It was observed that those hollow spheres are effective in bringing down the temperature of the compressed gas and hence enhance heat transfer in the system. The polytropic index of compression with use of hollow spheres reduces to 1.08 from 1.15 which corresponds to increase in compression efficiency from 79% to 87% for a compression ratio of 10.

*5.5.3. Spray Cooling.* In the spray cooling concept, small water droplets and high mass loading create a large interfacial surface area for heat transfer. The droplet spray heat transfer in the liquid piston compressor has been investigated by Qin and Loth [42]. They developed a detailed multiphase thermodynamic model and validated with experimental data.

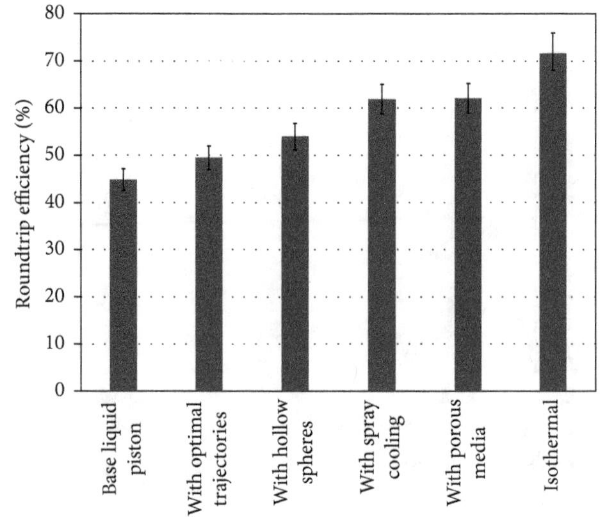


Figure 11: Roundtrip efficiency of liquid piston based OCAES with various heat transfer enhancement techniques.

It was observed that the total surface area of aloft droplets is critical to achieving high performance in a liquid piston and the best can be achieved with small droplets and high mass loading combined with direct injection. It is shown that compression efficiency could be increased from 71% for adiabatic compression to as high as 98% with spray injection for a compression ratio of 10. The mass loading, droplet diameters, and direct versus premixed injection influence effectiveness of spray cooling and hence compression efficiency. For a compression ratio of 10 and mass loading of 1, the compression efficiency of 93% is shown with spray cooling in the liquid piston.

*5.5.4. Porous Media.* The porous media inserts increase heat transfer surface area significantly; hence their addition to the liquid piston compression/expander increases the compression efficiency at a fixed power density. Experimental investigations on heat transfer with porous media in a liquid piston during compression and expansion have been carried out by Yan et al. [19]. A baseline case without inserts and five cases with different porous inserts were tested in a compression experiment. It was found that, in compression, porous inserts increase power density by 39-fold at 95% efficiency and increase efficiency by 18% at 100 $kW/m^3$. Similarly, in the expansion, porous media inserts increase power density threefold at 89% efficiency and increase efficiency by 7% at 150 $kW/m^3$.

The end-to-end efficiency of liquid piston based OCAES system with various heat transfer enhancement techniques in liquid piston is shown in Figure 11. The base liquid piston without any heat transfer enhancement technique considers experimental observed liquid piston compressor/expander efficiency. The isothermal OCAES is one in which compression and expansion happen isothermally which indicates 100% compression and expansion efficiency. Uncertainty bars represent 95% confidence interval valves.

It can be observed that estimated mean value of end-to-end efficiency is consistently higher with the use of heat transfer enhancement technique over the base liquid piston. The optimal trajectories can improve OCAES efficiency by about 5%. Although this improvement is comparatively a small value, this improvement happens without introducing any other media in the liquid piston. Therefore, optimal trajectory technique can be considered as an efficiency improvement technique without affecting other performance of the liquid piston like volumetric efficiency, power density, and long-term reliability. The use of hollow spheres has potential to improve the efficiency of OCAES system by about 9%. The addition of a layer of floating spheres in liquid piston increases heat transfer area and helps in temperature abatement in the liquid piston during compression. This results in improved efficiency of compression and similarly for expansion.

Further, spray cooling and porous media inserts can show about 17% improvement in the OCAES efficiency achieving end-to-end efficiency of about 62%. Both of these methods increase heat transfer medium inside the liquid piston and hence help to curtail the temperature rise in the liquid piston during compression and temperature fall during expansion. The high surface area and high specific heat of the water spray help in heat transfer enhancement and hence improving compression/expansion efficiency of liquid piston. In porous media inserts, higher heat transfer surface area in porous media helps in absorbing the thermal energy from the air during compression and dissipates it to the liquid. In expansion mode, the same porous media transfers thermal energy from the liquid to the air. Although spray cooling and porous media improves the efficiency of OCAES significantly, these heat transfer enhancement techniques require other media to be introduced in the liquid piston. This results in a reduced volumetric efficiency of compression/expansion. Also, spray cooling and porous media inserts might require special care for continuous reliable heat transfer performance. This is because of the change in spray characteristics and porous media characteristics over time due to their degradation.

An isothermal liquid piston compressor would define an upper limit for roundtrip efficiency of OCAES, which is about 72% for the given system considerations. This indicates that inefficiencies in the motor/generator, hydraulic pump/motor, pipelines, control valves, and storage result in about 28% of energy loss. End-to-end efficiencies of existing compressed air energy storage (CAES) plants in Huntorf (Germany) and McIntosh AL (USA) are 42% and 54%, respectively [14]. Clearly, liquid piston based OCAES with the use of heat transfer enhancement technique such as spray cooling or porous media inserts in liquid piston can show significantly higher end-to-end efficiency over existing CAES plants.

## 6. Conclusions

In pursuit of developing efficient economical large-scale energy storage, ocean compressed air energy storage can play an important role. Various OCAES concepts are possible, namely, diabatic, adiabatic, and isothermal OCAES. Energy and exergy analysis of these concepts is performed for OCAES system of the maximum power capacity of 0.5 MW and 2 MWh energy storage with storage pressure of 10 bar (100 m of ocean depth). Analytical models for energy and exergy analysis of various components in OCAES are presented. The exergy flow, energy efficiency, and exergy efficiency of various OCAES concepts are analyzed for comparative assessment.

The analysis shows that energy efficiency of diabatic OCAES is about 50% whereas its exergy efficiency is about 55%. Clearly, energy efficiency undervalues efficiency of diabatic OCAES; therefore, exergy efficiency would be a good measure of efficiency for comparison with other storage concepts. Adiabatic OCAES shows about 5% improvement in exergy efficiency over diabatic OCAES. Isothermal OCAES shows significantly higher efficiency over diabatic and adiabatic OCAES. Analysis of liquid piston based OCAES with the use of experimental liquid piston compressor indicated roundtrip efficiency of 45%. With heat transfer enhancement in the liquid piston, roundtrip efficiency of about 62% is possible with the use of either spray cooling or porous media inserts. Overall, liquid piston based OCAES with use of heat transfer enhancement has potential to show significantly higher efficiency than existing compressed air energy storage plants.

## Acknowledgments

The authors gratefully acknowledge support from the UNC Coastal Studies Institute of North Carolina, USA.

## References

[1] P. Denholm, E. Ela, B. Kirby, and M. Milligan, "The role of energy storage with renewable electricity generation," *Energy Storage: Issues and Applications*, pp. 1–58, 2011.

[2] H. Ibrahim, A. Ilinca, and J. Perron, "Energy storage systems-Characteristics and comparisons," *Renewable & Sustainable Energy Reviews*, vol. 12, no. 5, pp. 1221–1250, 2008.

[3] D. O. Akinyele and R. K. Rayudu, "Review of energy storage technologies for sustainable power networks," *Sustainable Energy Technologies and Assessments*, vol. 8, pp. 74–91, 2014.

[4] F. Crotogino, K. Mohmeyer, and R. Scharf, "Huntorf CAES: More than 20 Years of Successful Operation," Orlando, Fla, USA, 2001.

[5] R. Pollak, *History of First US Compressed-Air Energy Storage (CAES) Plant (110 MW 26 h) Volume 2: Construction*, Electric Power Research Institute (EPRI), 1994.

[6] Y. M. Kim and D. Favrat, "Energy and exergy analysis of a micro-compressed air energy storage and air cycle heating and cooling system," *Energy*, vol. 35, no. 1, pp. 213–220, 2009.

[7] S. D. Lim, A. P. Mazzoleni, J.-K. Park, P. I. Ro, and B. Quinlan, "Conceptual design of ocean compressed air energy storage system," *Marine Technology Society Journal*, vol. 47, no. 2, pp. 70–81, 2013.

[8] Y.-M. Kim, J.-H. Lee, S.-J. Kim, and D. Favrat, "Potential and evolution of compressed air energy storage: Energy and exergy analyses," *Entropy*, vol. 14, no. 8, pp. 1501–1521, 2012.

[9] A. Bagdanavicius and N. Jenkins, "Exergy and exergoeconomic analysis of a Compressed Air Energy Storage combined with a district energy system," *Energy Conversion and Management*, vol. 77, pp. 432–440, 2014.

[10] N. Hartmann, O. Vöhringer, C. Kruck, and L. Eltrop, "Simulation and analysis of different adiabatic Compressed Air Energy Storage plant configurations," *Applied Energy*, vol. 93, pp. 541–548, 2012.

[11] G. Grazzini and A. Milazzo, "A thermodynamic analysis of multistage adiabatic CAES," *Proceedings of the IEEE*, vol. 100, no. 2, pp. 461–472, 2012.

[12] M. J. Tessier, M. C. Floros, L. Bouzidi, and S. S. Narine, "Exergy analysis of an adiabatic compressed air energy storage system using a cascade of phase change materials," *Energy*, vol. 106, pp. 528–534, 2016.

[13] S. Wang, X. Zhang, L. Yang, Y. Zhou, and J. Wang, "Experimental study of compressed air energy storage system with thermal energy storage," *Energy*, vol. 103, pp. 182–191, 2016.

[14] M. Budt, D. Wolf, R. Span, and J. Yan, "A review on compressed air energy storage: Basic principles, past milestones and recent developments," *Applied Energy*, vol. 170, pp. 250–268, 2016.

[15] A. Hauer, "Thermal Energy Storage," IEA-ETSAP and IRENA Technology Brief E, 2013.

[16] S. Zunft, "Adiabatic CAES: The ADELE-ING Project," 2015.

[17] J. D. Van de Ven and P. Y. Li, "Liquid piston gas compression," *Applied Energy*, vol. 86, no. 10, pp. 2183–2191, 2009.

[18] C. Qin, E. Loth, P. Li, T. Simon, and J. Van De Ven, "Spray-cooling concept for wind-based compressed air energy storage," *Journal of Renewable and Sustainable Energy*, vol. 6, no. 4, Article ID 043125, 2014.

[19] B. Yan, J. Wieberdink, F. Shirazi, P. Y. Li, T. W. Simon, and J. D. Van de Ven, "Experimental study of heat transfer enhancement in a liquid piston compressor/expander using porous media inserts," *Applied Energy*, vol. 154, pp. 40–50, 2015.

[20] K. R. Ramakrishnan, P. I. Ro, and V. C. Patil, "Temperature abatement using hollow spheres in liquid piston compressor for Ocean Compressed Air Energy Storage system," in *Proceedings of the 2016 OCEANS MTS/IEEE Monterey, OCE 2016*, USA, September 2016.

[21] A. Rogers, A. Henderson, X. Wang, and M. Negnevitsky, "Compressed air energy storage: Thermodynamic and economic review," in *Proceedings of the 2014 IEEE Power and Energy Society General Meeting*, USA, July 2014.

[22] Z. Wang, D. S.-K. Ting, R. Carriveau, W. Xiong, and Z. Wang, "Design and thermodynamic analysis of a multi-level underwater compressed air energy storage system," *Journal of Energy Storage*, vol. 5, pp. 203–211, 2016.

[23] Z. Wang, W. Xiong, D. S.-K. Ting, R. Carriveau, and Z. Wang, "Conventional and advanced exergy analyses of an underwater compressed air energy storage system," *Applied Energy*, vol. 180, pp. 810–822, 2016.

[24] B. C. Cheung, R. Carriveau, and D. S.-K. Ting, "Parameters affecting scalable underwater compressed air energy storage," *Applied Energy*, vol. 134, pp. 239–247, 2014.

[25] B. C. Cheung, R. Carriveau, and D. S. K. Ting, "Multi-objective optimization of an underwater compressed air energy storage system using genetic algorithm," *Energy*, vol. 74, pp. 396–404, 2014.

[26] V. C. Patil and P. I. Ro, "Comparative Assessment of Different Types of Ocean Compressed Air Energy Storage Systems Based on Exergy Analysis," in *ASME 2017 Power Conference Joint With ICOPE-17 collocated with the ASME 2017 11th International Conference on Energy Sustainability, the ASME 2017 15th International Conference on Fuel Cell Science, Engineering and Technology, and the ASME, 2017, Nuclear Forum*, American Society of Mechanical Engineers, 2017.

[27] V. C. Patil, P. I. Ro, and R. Kishore Ranganath, "End-to-end efficiency of liquid piston based ocean compressed air energy storage," in *Proceedings of the 2016 OCEANS MTS/IEEE Monterey, OCE 2016*, USA, September 2016.

[28] G. Tsatsaronis, "Definitions and nomenclature in exergy analysis and exergoeconomics," *Energy*, vol. 32, no. 4, pp. 249–253, 2007.

[29] Y. A. Çengel and M. A. Boles, *Thermodynamics*, McGraw-Hill Education, New York, NY, USA, 2015.

[30] C. Zhang, B. Yan, J. Wieberdink et al., "Thermal analysis of a compressor for application to Compressed Air Energy Storage," *Applied Thermal Engineering*, vol. 73, no. 2, pp. 1402–1411, 2014.

[31] B. R. Munson, D. F. Young, and T. H. Okiishi, "Fundamentals of Fluid Mechanics," *New York*, vol. 3, no. 4, 1990.

[32] N. Benton, "Compressed Air Evaluation Protocol," 2014.

[33] G. Huff, "DOE Global Energy Storage Database," 2015.

[34] S.-M. Lu, "A review of high-efficiency motors: Specification, policy, and technology," *Renewable & Sustainable Energy Reviews*, vol. 59, pp. 1–12, 2016.

[35] E. W. McAllister, *Pipeline Rules of Thumb Handbook: A Manual of Quick, Accurate Solutions to Everyday Pipeline Engineering Problems*, Gulf Professional Publishing, 2013.

[36] A. J. Pimm, S. D. Garvey, and M. de Jong, "Design and testing of Energy Bags for underwater compressed air energy storage," *Energy*, vol. 66, pp. 496–508, 2014.

[37] H. Hoffeins, "Huntorf air storage gas turbine power plant," *Energy Supply, Brown Boveri Publication DGK*, vol. 90, 202 pages, 1994.

[38] M. Raju and S. Kumar Khaitan, "Modelling and simulation of compressed air storage in caverns: a case study of the Huntorf plant," *Applied Energy*, vol. 89, no. 1, pp. 474–481, 2012.

[39] M. M. Kenisarin, "High-temperature phase change materials for thermal energy storage," *Renewable & Sustainable Energy Reviews*, vol. 14, no. 3, pp. 955–970, 2010.

[40] M. Saadat, P. Y. Li, and T. W. Simon, "Optimal trajectories for a liquid piston compressor/expander in a Compressed Air Energy Storage system with consideration of heat transfer and friction," in *Proceedings of the 2012 American Control Conference, ACC 2012*, pp. 1800–1805, can, June 2012.

[41] F. A. Shirazi, M. Saadat, and B. Yan, "Optimal Control Experimentation of Compression Trajectories for a Liquid Piston Air Compressor," in *ASME 2013 Heat Transfer Summer Conference collocated with the ASME 2013 7th International Conference on Energy Sustainability and the ASME 11th International Conference on Fuel Cell Science, Engineering and Technology*, American Society of Mechanical Engineers, 2013.

[42] C. Qin and E. Loth, "Liquid piston compression efficiency with droplet heat transfer," *Applied Energy*, vol. 114, pp. 539–550, 2014.

# 5

# Thermodynamic Analysis of a Solar Combined Ejector Absorption Cooling System

**Doniazed Sioud, Raoudha Garma, and Ahmed Bellagi**

*Energy Engineering Department, Ecole Nationale d'Ingénieurs de Monastir (ENIM), University of Monastir, Tunisia*

Correspondence should be addressed to Doniazed Sioud; siouddoniazed@gmail.com

Academic Editor: Angelo Maiorino

The objective of this paper is to investigate theoretically a solar driven 60 kW absorption cooling system. The system is constituted of a combined ejector single-effect absorption cycle coupled with a linear Fresnel solar concentrator and using water/lithium bromide as working fluid. The combined ejector single-effect absorption cycle exhibits high performances, almost equal to that of double-effect absorption device. However, higher driving heat temperatures are required than in the case of conventional single-effect machines. A mathematical model is set up to analyze the optical performance of the linear Fresnel concentrator. Simulations are carried out to study the overall system performance $COP_{\text{system}}$ and the performances of the combined absorption machine $COP_{\text{cycle}}$ for generator driving temperatures and pressures in the ranges 180°C – 210°C and 198 kPa – 270 kPa, respectively. Further, the effect of operating parameters such as the cooling medium and chilled water temperatures is investigated. A maximum cycle performance of 1.03 is found for a generator pressure of 272 kPa and chilled and cooling water temperatures of 7°C and 25°C, respectively. A case study is investigated for a typical summer Tunisian day, from 8:00 to 18:00. The effect of ambient temperature and solar radiation on cycle and system performances is simulated. The optical performances of the concentrator are also analyzed. Simulation results show that between 11:00 and 14:00 the collector efficiency is 0.61 and that the $COP_{cycle}$ reaches values always higher than 0.9 and the $COP_{system}$ is larger than 0.55. Globally the performances of the investigated cycle are similar to those of double-effect conventional absorption system.

## 1. Introduction

In summer, the demand of electrical energy for air-conditioning and refrigeration becomes important, particularly in hot climate zones like the Mediterranean region. In many countries in this climate zone, the electric energy consumption reaches, and sometimes exceeds, in hot summer days the limit of production capacity. Consequently, the overall energy consumption and costs and the environmental impact become important. To reduce these impacts, solar cooling technologies may constitute a promising and attractive solution.

Solar energy applications and practices are investigated for different domains [1] such as pyrolysis [2], refrigeration and cooling [3], power generation [4], drying and various other industrial uses [5, 6]. In view of the increased interest in high temperature solar applications, many studies focus on the amelioration of collector configurations and concentrator techniques in order to improve their optical efficiency and the heat transfer characteristics of the absorber by using for instance nanofluids as a heat transfer fluids [7].

In the domain of refrigeration and air-conditioning, absorption techniques for production of cold are considered particularly suitable for solar applications. Single-effect absorption systems can be activated by flat plate collectors and/or evacuated tube collectors [8, 9] since they require relatively low temperatures of the driving heat sources. They have however low performances. Therefore, investigators focus attention on double-effect absorption cycles that need higher temperature heat sources provided by parabolic through collectors or linear Fresnel concentrators. Several investigations consider the technologies of double-effect absorption systems coupled with parabolic through collectors. An 11 kW solar absorption cooling system was studied by Florides *et al.* [10] in Nicosia and optimized using TRNSYS. The required parabolic collector area was $15 m^2$ and the capacity of hot

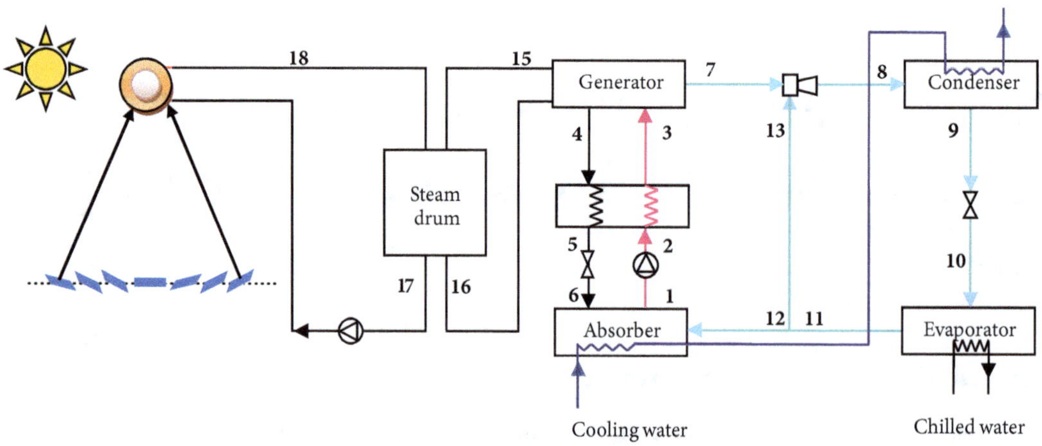


FIGURE 1: Solar combined ejector single-effect chiller.

water storage, 600 L. Shirazi *et al.* [11] investigated the feasibility of producing cooling via the connection of LiBr-$H_2O$ single, double, and triple-effect absorption cycles to solar collectors under various weather conditions, namely, 1023 kW single-effect absorption chiller coupled with evacuated tube collectors (ETCs), parabolic trough collectors (PTC) coupled with 1163 kW double-effect and triple-effect units. The authors concluded that the energetic and economic performance of multieffect absorption machines coupled with concentrating collectors in low direct normal irradiation (DNI) fraction (less than 50%) climate zones is poor due to large solar field sizes required and high payback time compared to a single-effect chiller system driven by heat from evacuated tube collector array. On the other hand, in climatic regions with very high DNI faction (more than 65%), the combination of triple-effect chiller and parabolic trough collectors results in the most energy-efficient and cost-effective plant configuration, achieving the smallest solar field and the lowest payback period. Chemisana *et al.* [12] compared cooling systems connected and not connected to solar collectors. The solar cooling devices were basic single-effect absorption system that used heat from evacuated tube collectors, double-effect absorption system coupled with a Fresnel reflective collector incorporated on a building façade. These authors found that the solar absorber area of the Fresnel collector system was lower than that of standard solar thermal machine. Beltagy *et al.* [13] investigated theoretically and experimentally a solar cooling prototype with Fresnel collectors. The setup was designed basing on developed theoretical model, and experimental data was compared to numerical results. It was found that the measured and estimated produced thermal power from collectors and the efficiency per day were comparable. In addition, they found that the efficiency exceeds 40% per day for the 250 kW thermal prototype.

A 174 kW solar/gas double-effect LiBr/water absorption plant was tested at Seville in Spain [14]. The system was driven by a pressurized hot water flow using a 352 $m^2$ heated by a linear Fresnel collector. A direct-fired natural gas burner was used as auxiliary heat source. For 75% of generator's total heat input covered by solar energy, the average collector efficiency was found to vary between 0.35 and 0.4, and the mean coefficient of performance of the chiller, between 1.1 and 1.25.

Major drawbacks of the solar cooling systems are low efficiency in case of use of solar collectors with single-effect absorption systems and huge areas of thermal collectors required in case of double-effect absorption cycles. New machine configurations with high efficiency are needed. In literature, it is reported that some modified designs of single-effect chillers might exhibit coefficients of performance similar to that machines applying double-effect cycles [15, 16]. In these configurations, ejectors are incorporated in between generator and condenser.

The aim of this paper is to investigate a solar combined single-effect cycle using LiBr-$H_2O$ as working fluid and coupled to solar collector. Because high temperature levels are needed to active the installation, a linear Fresnel collector is considered for the production of high temperature steam required to regenerate the refrigerant in the desorber. Driving temperature and chilled water temperature as well as cooling water temperature affect the performance of the chiller, $COP_{cycle}$, and the overall system performance, $COP_{system}$. Following simulations analyze thoroughly the effect of these three parameters. Also a case study of a typical Tunisian summer day is performed.

## 2. Description of the System

The proposed solar 60 kW cooling system is coupled with a 164 $m^2$ linear Fresnel collector, as schematically illustrated in Figure 1. The adopted collector (type IS-LF11 [17]) is used to produce steam at temperatures between 180°C and 230°C (pressures of 10 bar and 28 bar, respectively) for driving the generator of the combined cooling cycle. Cold is produced at temperatures between 12°C and 4°C.

The combined ejector single-effect cycle looks like an ordinary absorption single-effect cycle where an ejector is inserted between the generator and the condenser. The basic operational principle of the system is as follows: In the

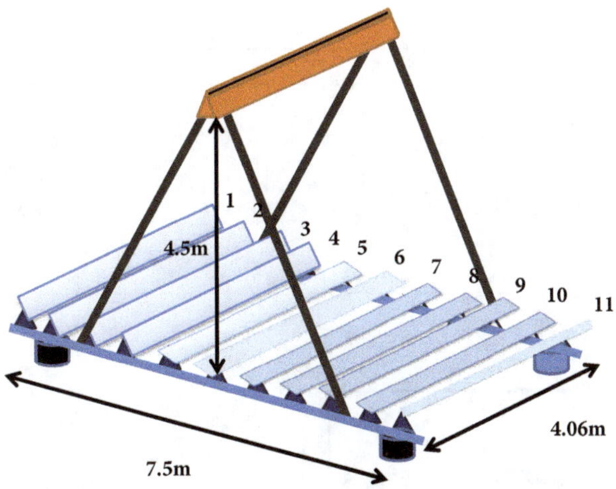

FIGURE 2: A module of industrial linear Fresnel collector.

TABLE 1: Technical data of basic IS-LF11 module [17]

| | |
|---|---|
| Width (m) | 7.5 |
| Length (m) | 4.06 |
| Primary mirror area ($m^2$) | 22 |
| Receiver height (m) | 4.5 |
| Mirror rows number | 11 |
| Thermal loss of primary reflector, $u_1$ ($W/m^2K^2$) | 0.00043 |
| Clearance between mirror rows (m) | 0-0.5 |
| Specific weight (kg/$m^2$ installation surface area) | 27 |
| Maximum operational temperature (°C) | 400 |
| Life expectancy (years) | +20 |
| Heat transfer flow | Pressurized Water or thermal oil |

generator, the aqueous salt solution ($H_2O$/LiBr) is brought to boiling by heat provided by the solar source, in order to vaporize the refrigerant (water) and separate it from the solution. These high-pressure refrigerant vapors (7) are introduced in the ejector nozzle as primary flow where they are accelerated and strongly depressurized. A fraction of the vapor exiting the evaporator (13) is entrained by the primary flow into the ejector. The two gases are then well mixed in the mixing chamber and compressed in the diffuser (8) before feeding the condenser (9) where they liquefy by rejecting condensation heat to the cooling water (cw). The condensate expands through an adiabatic expansion valve (10) and proceeds to the evaporator where it evaporates (11) by absorbing heat from the chilled water. While in the basic cycle, all the refrigerant vapor leaving the evaporator proceeds to the absorber, only a part of it flows into the absorber in the combined cycle where it dissolves in the salt rich solution. The rest of the vapor is sucked in the ejector as secondary flow. The salt solution that leaves the absorber (1) is pumped to the generator (3), after preheating in the solution heat exchanger.

The generator is activated by the heat $\dot{Q}_g$ liberated by condensation of hot saturated steam (15) which is then pumped back as liquid (16) to the steam drum. A primary thermal transfer fluid loop (17-18) vehicles heat from solar collector to steam drum.

*2.1. Solar Linear Fresnel Collector.* IS-LF11 solar linear Fresnel collectors considered here are combined with boiler and can be used for industrial processes. Process heat in the range of 100 kW to 10 MW at standard pressure of 40 bar is generated, but up to 120 bar and temperatures up to 400°C are possible [17]. A "module" is one block of collectors associated with 4.06 m long vacuum absorber tubes. Figure 2 presents two modules of the adopted industrial collector. It is possible to use different kinds of heat transfer fluids (HTF): pressurized water, thermal oil, superheated steam, etc. Table 1 gives the manufacturer data of the commercial industrial Fresnel collector used in the simulations.

*2.2. LiBr-$H_2O$ Combined Ejector Single-Effect Absorption Chiller.* In the investigated combined ejector single-effect absorption cycle the ejector is located between condenser and generator. It uses the vapor produced in the generator to entrain and compress the refrigerant vapor coming from the evaporator before flowing in the condenser. Consequently, the pressure in generator and condenser—equal in a conventional single-effect cycle—are now distinct: intermediate pressure in condenser, equal to the high pressure in the original cycle, and a higher pressure in the generator. Hence, higher driving source temperatures are now required, between 180°C and 210°C. Table 2 gives the assumed operating conditions in the simulations. In particular, the nominal cooling power of the installation is $\dot{Q}_e$ = 60 kW, available in form of chilled water at temperatures between 4°C and 12°C for driving temperatures and pressures in the range 180°C – 210°C and 198 kPa – 270 kPa, respectively.

*2.3. Ejector.* A typical ejector configuration is shown in Figure 3. A steam ejector is composed of four zones: primary nozzle, suction chamber, mixing chamber (convergent duct followed by constant section area), and diffuser. Table 3 gives the ejector geometrical data [16].

## 3. Theoretical Model

*3.1. Linear Fresnel Collector Model.* The optical efficiency $\eta_{optical}$ of the collector depends on the incidence angle of the direct irradiation. As reference of $\eta_{optical}$ is its value $\eta_0$ under ideal conditions: clean reflectors and absorber tubes, sun at zenith. For the adopted collector $\eta_0$ = 0.635 and $\eta_{max}$ = 0.663 (sun at 5° transversal zenith angle, and 0° longitudinal zenith angle). For arbitrary incidence angle $\eta_0$ is multiplied by a correction factor, the so-called incidence angle modifier ($IAM$). $IAM$ has two components: a field independent transversal component, $IAM_t$, and a longitudinal component, $IAM_l$, that depends on the number of in string assembled modules (one in Figure 2).

TABLE 2: Operating conditions assumed for the simulations.

| Input parameter | Fixed value | Ranges |
|---|---|---|
| Capacity (kW) | 60 | |
| Generator temperature (°C) | | 180-210 |
| Steam pressure (kPa) | | 198-270 |
| Inlet temperature cooling medium (°C) | 32 | 25-35 |
| Chilled water outlet (°C) | 7 | 7- 12 |
| Heat exchanger efficiency | 0.8 | |

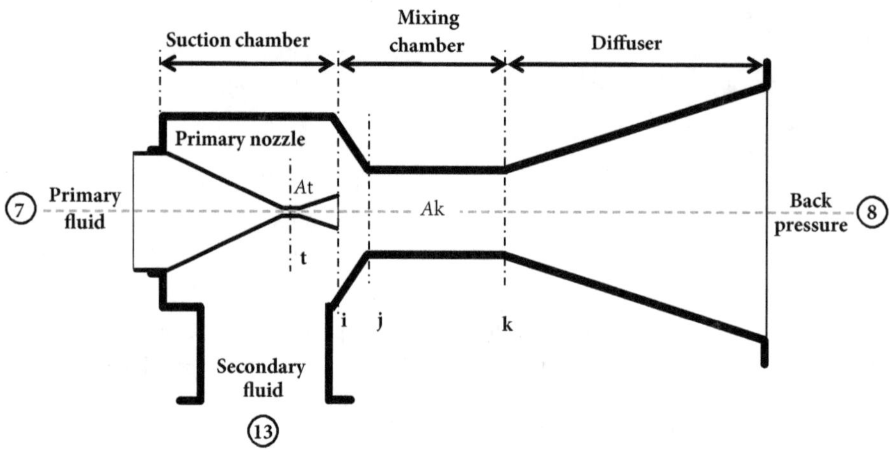


FIGURE 3: Schematic diagram of ejector.

TABLE 3: Ejector characteristics [16].

| Parameter | Value |
|---|---|
| Nozzle throat diameter, $d_t$ (mm) | 1.6 |
| Nozzle exit diameter, $d_i$ (mm) | 6 |
| Mixing chamber diameter, $d_j$ (mm) | 18 |
| Mixing chamber length (mm) | 140 |
| Diffuser divergent length (mm) | 210 |
| Diffuser exit diameter (mm) | 40 |

The thermal losses of the reflector, $\dot{q}_{loss}$, are dominated by radiative losses and approximately estimated by

$$\dot{q}_{loss} = u_0 A_{aperture}\Delta T + u_1 A_{aperture}\Delta T^2 \quad (1)$$

$A_{aperture}$ stands for the active aperture area of the collector and $\Delta T$ for the temperature difference between that of the heat transfer flow HTF and the ambient, as expressed in

$$\Delta T = T_{HTF} - T_{amb} \quad (2)$$

The thermal loss coefficients of the IS-LF11 are $u_0 = 0$ and $u_1 = 43.\ 10^{-5}\ \mathrm{W/m^2K^2}$ [17]. Equation (1) reduces then to

$$\dot{q}_{loss} = 43.10^{-5} A_{aperture}\Delta T^2 \quad (3)$$

The collected thermal power $\dot{q}$ writes as

$$\begin{aligned}\dot{q} &= \eta_{optical} AG - \dot{q}_{loss} \\ &= \eta_0 IAM_l(\theta_{l,\delta}) IAM_t(\theta_{t,\delta}) A_{aperture} G \\ &\quad - u_1 A_{aperture}\Delta T^2\end{aligned} \quad (4)$$

where $\theta$ is the zenith angle, $\delta$, the angle between absorber pipe and the north-south direction, and $G$, the direct normal irradiance (DNI).

The thermal efficiency of the collector $\eta$ is the ratio of the usable thermal power $\dot{q}$ and the solar irradiance collected ($A_{aperture}G$), detailed in

$$\begin{aligned}\eta &= \frac{\dot{q}}{A_{aperture}G} = \eta_{optical} - \frac{\dot{q}_{loss}}{A_{aperture}G} \\ &= \eta_0 IAM_l(\theta_{l,\delta}) IAM_t(\theta_{t,\delta}) - u_1 A_{aperture}\frac{\Delta T^2}{G}\end{aligned} \quad (5)$$

Figure 4 depicts this correlation, $\eta$ *versus*$(\Delta T/G)$, for the adopted collector and various values of $G$ [17]. It is worthy noticing that knowing thermal efficiency, irradiation, and actual collector size, the expected gross heat produced by the collector can be predicted.

*3.2. Combined Ejector Single-Effect Absorption Cycle Model.* The mathematical model of the cycle is based on mass and energy balances written for each machine element. For a steady state process and neglecting kinetic and potential

TABLE 4: Comparison of results from [16] with those of present work.

| | Ref. [16] | Present work |
|---|---|---|
| Condenser pressure (kPa) | < 3 | 3.78 |
| Generator temperature (°C) | 192 | 192 |
| Generator Pressure (°C) | 198 | 198 |
| Concentration at absorber exit (mass. %) | 62.3 | 62.3 |
| Refrigerant mass flow rate, generator (kg/min) | 0.0393 | 0.0393 |
| Refrigerant mass flow rate, evaporator (kg/min) | 0.059 | 0.063 |
| $\dot{Q}_g$(kW) | 2.8 | 2.8 |
| $\dot{Q}_e$(kW) | 2.4 | 2.5 |
| *COP* | 0.87 | 0.91 |

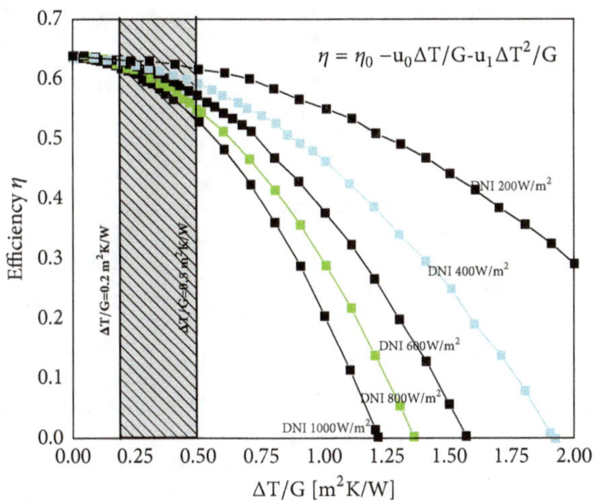


FIGURE 4: Collector thermal efficiency $\eta$ *versus* $(\Delta T/G)$ for various values of G.

energy variations, balance equations for global mass (equation (6)), salt mass (equation (7)), and energy (equation (8)) write, respectively,

$$\sum \dot{m}_{in} = \sum \dot{m}_{out} \tag{6}$$

$$\sum \dot{m}_{in} X_{in} = \sum \dot{m}_{out} X_{out} \tag{7}$$

$$\sum \dot{Q} - \sum \dot{W} = \sum \dot{m}_{out} h_{out} - \sum \dot{m}_{in} h_{in} \tag{8}$$

$\dot{Q}$ and $\dot{W}$ are heat and transfer rates between machine element and its surroundings. The $COP_{\text{cycle}}$ of the absorption system is expressed as

$$COP_{\text{cycle}} = \frac{\dot{Q}_e}{\dot{Q}_g + \sum \dot{W}_P} \tag{9}$$

*3.3. Ejector Model.* The adopted 1D model of the ejector is detailed in reference [15]. It is based on quadratic equations of each zone and enables the prediction of the ejector performance, i.e., the entrainment ratio (ratio of mass flow rate of the primary flow $\dot{m}_7$ and secondary flow rate $\dot{m}_{13}$) for given ejector geometry and inlet and outlet pressures and temperatures.

## 4. Validations

A computer programs are developed using the software Engineering Equation Solver (*EES*) [20] in order to validate thermodynamically the models of combined ejector single-effect absorption unit and steam ejector.

*4.1. Validation of Combined Ejector Single-Effect Model.* The simulations are run under following conditions, according to the data of [16]:

(i) Evaporator temperature: 5∘C

(ii) Condenser temperature: 28∘C

(iii) Concentrations of salt solution leaving absorber: 62.3%

(iv) Heat exchanger effectiveness: 0.8

(v) Solution flow rate leaving generator: 0.393 kg/min

(vi) Generator pressure: 198 kPa and temperature: 192∘C

Table 4 compares the obtained simulations results with those reported by Aphornratana *et al.* [16]. As can be noted, deviations are generally small, less than 1%, with a maximum of 6% for the mass flow rate of refrigerant leaving the evaporator. This can be attributed to the simplifying assumption of perfect gas behavior for the steam in the ejector model. Deviations of driving power and cooling load are respectively 0.3% and 6%. Thus, the proposed model for the simulation of the chiller is fairly validated.

*4.2. Validation of Ejector Model.* In the experimental work of Sriveerakul *et al.* [19], various combinations of ejector geometry were tested for different operating temperatures and pressures. Primary and secondary flows were vapors. The 1D model used for the machine simulations is tested against the data of this experimental work for validation. Calculated results are compared with data reported in [19] in Table 5.

Deviations ε% for backpressure, entrainment ratio, and ejector area ratio are determined as

$$\varepsilon\% = \frac{\text{calculated value} - \text{experimental value}}{\text{experimental value}} \tag{10}$$

Minimum absolute percent deviation of backpressure is 3% and maximum, 23%. For entrainment ratio the corresponding values are 0.3% and 18%, and for ejector area ratio, 0% and

TABLE 5: Ejector model validation ($T_p$, primary flow temperature;$T_s$, entrained flow temperature; $(A_k/A_t)$, ejector area ratio, *i.e.*, ratio of constant area section, $A_k$, and nozzle throat area, $A_t$).

| Backpressure (mbar) | | | Entrainment ratio | | | $(A_k/A_t)$ | | |
|---|---|---|---|---|---|---|---|---|
| Exp. [19] | Cal. | ε (%) | Exp. [19] | Cal. | ε (%) | Exp. [19] | Cal. | ε (%) |
| | | | $T_p$ = 120°C, $T_s$ = 10°C | | | | | |
| 0.53 | 0.48 | -10. | 90.25 | 103 | 14. | 37 | 37 | 0 |
| | | | $T_p$ = 130°C, $T_s$ = 10°C | | | | | |
| 0.4 | 0.39 | -3. | 90.25 | 97.71 | 18. | 50 | 50 | 0 |
| | | | $T_p$ = 140°C, $T_s$ = 15°C | | | | | |
| 0.28 | 0.33 | 18. | 90.25 | 96.02 | 6. | 65 | 65 | 0 |
| | | | $T_p$ = 130°C, $T_s$ = 5°C | | | | | |
| 0.31 | 0.32 | 5. | 90.25 | 96.55 | 7. | 48 | 48 | 0 |
| 0.39 | 0.43 | 13. | 117.9 | 117.5 | -0.3 | 41 | 37 | 11. |
| 0.47 | 0. 57 | 23. | 160.5 | 144.2 | -10. | 35 | 31 | 13. |
| 0.31 | 0.36 | 19. | 90.25 | 104.6 | 16. | 45 | 43 | 5. |
| 0.3 | 0.35 | 18. | 90.25 | 101.7 | 13. | 46 | 46 | 0 |

13%. Based on the simulation results and percent deviations reported in Table 5, it can be concluded that the proposed 1D ejector model describes fairly well the experimental data.

## 5. Solar Chiller Simulation Results and Discussion

Simulations of a 60 kW solar chiller are performed in order to analyze thermodynamically and predict the performances of a combined ejector single-effect absorption refrigeration cycle coupled with linear Fresnel collector modules. The computer program was developed using the software Engineering Equation Solver (*EES*) [20]. Thermophysical properties of the refrigerant system LiBr-$H_2O$ are determined by applying internal procedures of the software. Tunisian climate and meteorological data are taken into consideration in the program.

A solar absorption chiller is characterized by an overall coefficient of performance, $COP_{system}$, a key factor to describe the efficiency of the entire system, expressed in (11), and is defined as the product of the efficiency of the solar collector $\eta$ and the thermal coefficient of performance of the absorption refrigerant cycle $COP_{cycle}$ [21]:

$$COP_{system} = \eta COP_{cycle} \tag{11}$$

*5.1. Effect of Generator Temperature and Pressure on System Performance (for Constant Chilled Water Temperature).* Simulations are performed by varying gene rotor temperature and pressure as well as the cooling medium temperature, for fixed chilled outlet-temperature, $T_c$ = 7°C. The results are represented graphically in Figures 5(a), 5(b) and 5(c), for, respectively, three values of the generator pressure 198 kPa, 232 kPa, and 272 kPa. The coefficient of performance of the chiller $COP_{cycle}$ together with the performance of the whole installation $COP_{system}$, including the solar facility, is depicted *versus* generator temperature for three values of the generator pressure and cooling medium temperature.

TABLE 6: Maximum *COP* reached for solution generator temperature set at 210°C (maximum considered) and $T_c$ = 7°C.

| $P_{generator}$ (kPa) | $COP_{system}$ | $COP_{cycle}$ | $T_{cw}$ (°C) |
|---|---|---|---|
| | 0.57 | 0.99 | 25 |
| 198 | 0.52 | 0.85 | 30 |
| | 0.46 | 0.75 | 35 |
| | 0.59 | 1.02 | 25 |
| 232 | 0.53 | 0.88 | 30 |
| | 0.47 | 0.77 | 35 |
| | 0.6 | 1.03 | 25 |
| 270 | 0.54 | 0.89 | 30 |
| | 0.48 | 0.79 | 35 |

Given these findings, we note that

(i) The *COP* increases with increasing generator temperature, $T_g$. For fixed $T_g$, it decreases with decreasing cooling water temperature.

(ii) The $COP_{cycle}$ reaches values ranging from 0.4 to 1.03, reaching the performances of double- effect absorption cycles, thus larger than what can be obtained in case conventional single-effect without ejector [22].

(iii) The $COP_{system}$ is almost half $COP_{cycle}$ due to the limited efficiency of the solar collector installation. Even in the case of favorable conditions, in gray high-lightened zone in Figure 4 and representing a common situation in Tunisian summer, this efficiency is ranging between 0.5 and 0.6.

(iv) It is observed that the maximum $COP_{cycle}$ (and consequently the $COP_{system}$) attains higher values when the generator pressure is larger. The highest performances are reached for the maximum generator temperature considered in this study, $T_{g,max}$ = 210∘C. Table 6 summarizes these results. In fact, for minimal cooling medium temperature (25∘C) and highest generator temperature (210∘C), the $COP_{cycle}$ = 0.99, 1.02, 1.03

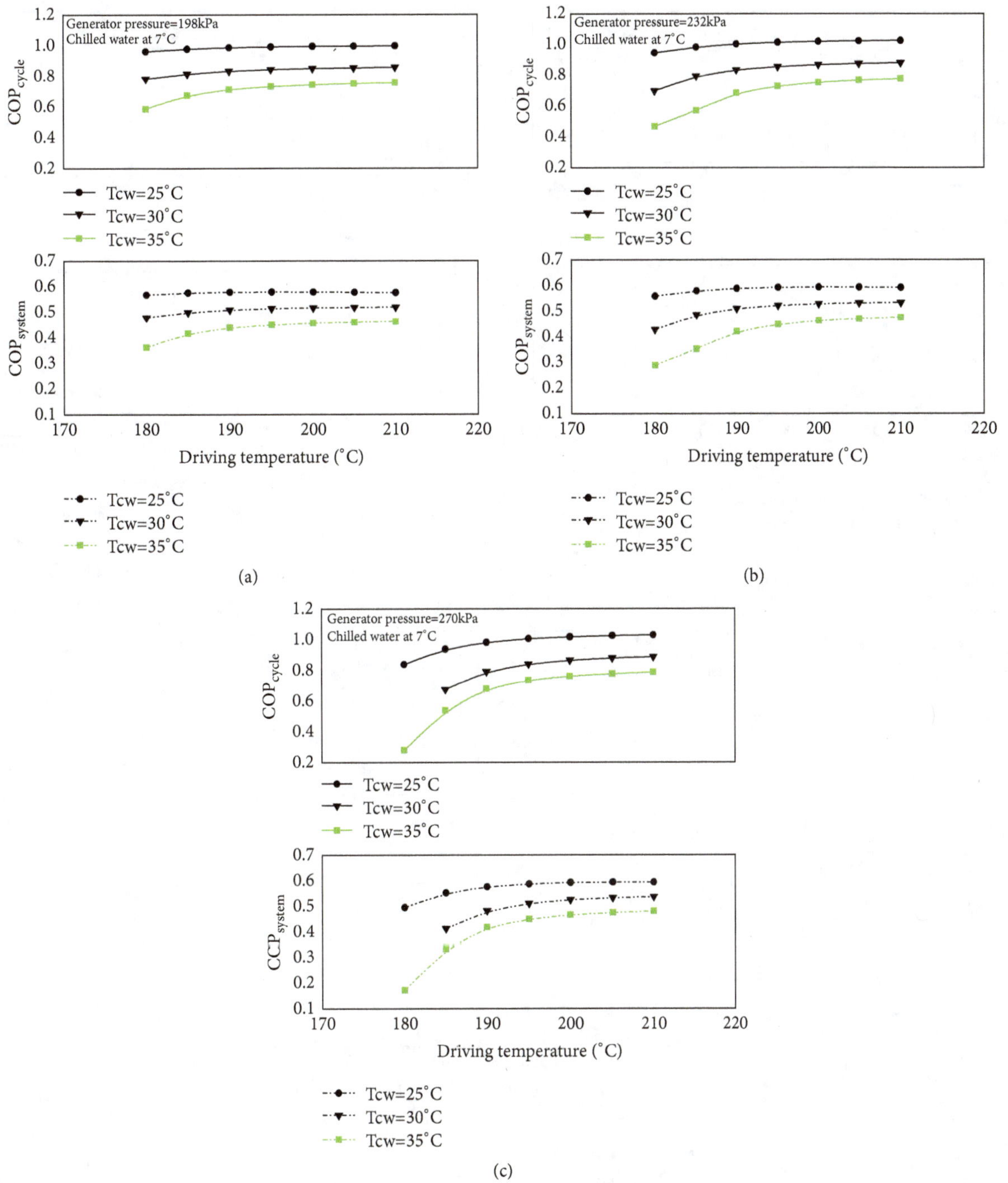


FIGURE 5: Effect of generator temperature on system performance for different generator pressures and cooling medium temperatures.

and the $COP_{system}$ = 0.57, 0.59, 0.6 for generator pressures at 198 kPa, 232 kPa, 272 kPa, respectively. It is noticed further that

(v) For a fixed generator pressure, the maximum $COP$ decreases when the temperature of the cooling medium rises. So when the pressure in the generator is at 198 kPa, the maximum $COP_{cycle}$ = 0.99, 0.85, 0.75 for cooling medium temperature, respectively, set at 25∘C, 30∘C, 35∘C, with the corresponding values for $COP_{system}$ = 0.57, 0.52, 0.46.

(vi) The performances of cycle and system are optimal for the highest generator pressure (270 kPa), lowest cooling medium temperature (25∘C), and largest chilled water temperature (12∘C).

*5.2. Effect of Generator Temperature and Pressure on System Performance (for Constant Cooling Medium Temperature).* Similar simulations are run by maintaining constant the cooling medium temperature $T_{cw}$ = 32°C and varying the chilled water temperature $T_c$. Figures 6(a), 6(b) and 6(c) present the evolution of $COP_{cycle}$ and $COP_{system}$ of the solar

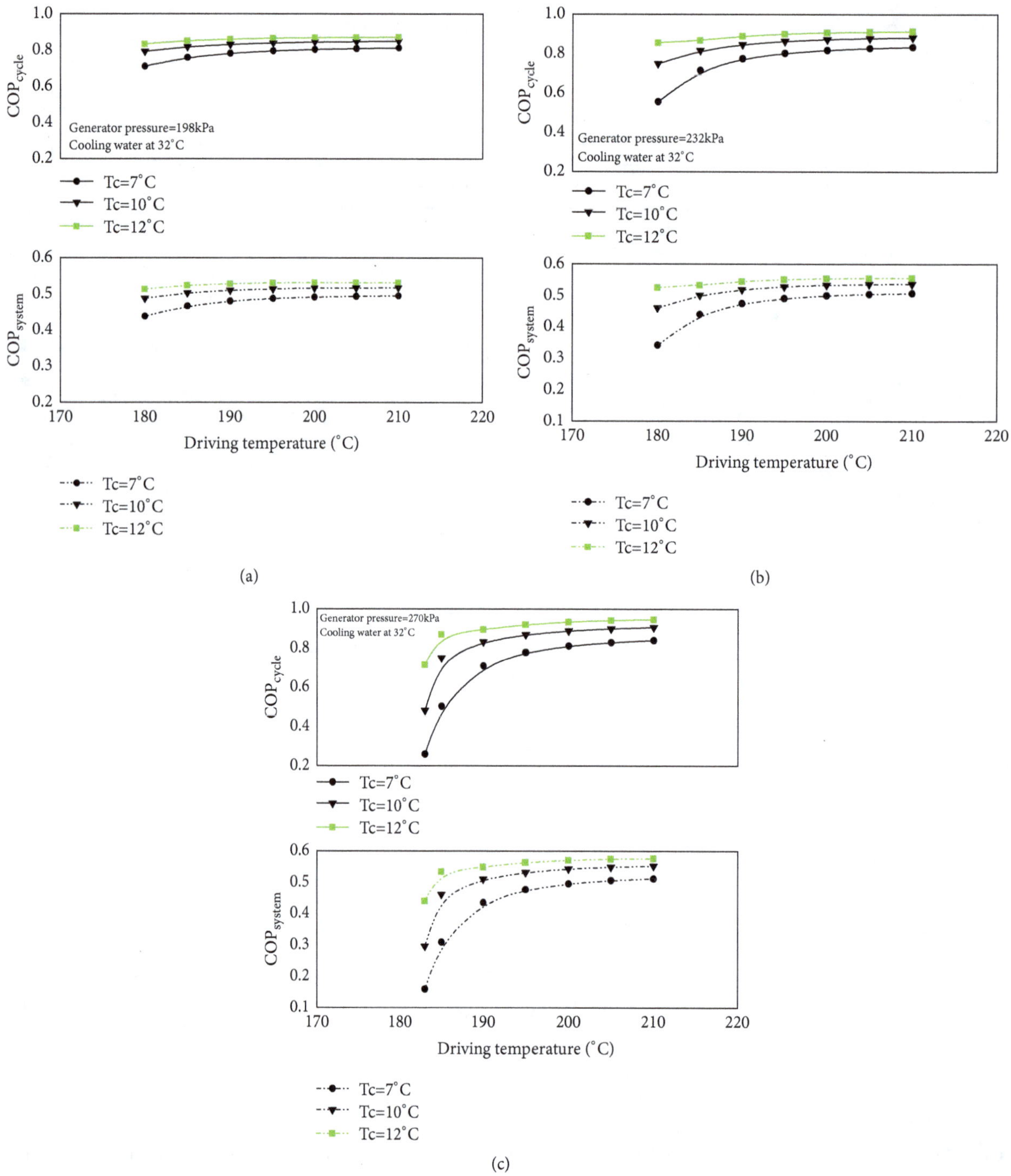


FIGURE 6: Effect of generator temperature on system performance for different generator pressures and chilled water temperatures.

combined single-effect absorption refrigeration system for three values of the generator pressure 198 kPa, 232 kPa, and 272 kPa, respectively. It can be noted that

(i) Again, both $COP_{cycle}$ and $COP_{system}$ increase when the generator temperature is increased.

(ii) $COP_{cycle}$ and $COP_{system}$ increase with the driving temperature, similarly to the behavior already observed in Figure 5. Table 7 gives the maximum of cycle and system performances, when the driving temperature reaches its highest value, 210∘C. For $T_c$ = 12∘C $COP_{cycle}$ reaches 0.87, 0.91, and 0.95 for generator pressure 198 kPa, 232 kPa, and 270 kPa, respectively.

(iii) The performances of cycle and system ($COP_{cycle}$ and the $COP_{system}$) increase with increasing chilled water temperature. For example, as shown in Table 7, for generator pressure equal to 270 kPa, the maximum

TABLE 7: Maximum *COP* reached for solution generator temperature set at 210°C (maximum considered) and $T_{cw}$ = 32°C.

| $P_{generator}$ (kPa) | $COP_{system}$ | $COP_{cycle}$ | $T_c$ (°C) |
|---|---|---|---|
| | 0.50 | 0.81 | 7 |
| 198 | 0.52 | 0.85 | 10 |
| | 0.53 | 0.87 | 12 |
| | 0.51 | 0.83 | 7 |
| 232 | 0.54 | 0.88 | 10 |
| | 0.56 | 0.91 | 12 |
| | 0.51 | 0.84 | 7 |
| 272 | 0.55 | 0.91 | 10 |
| | 0.58 | 0.95 | 12 |

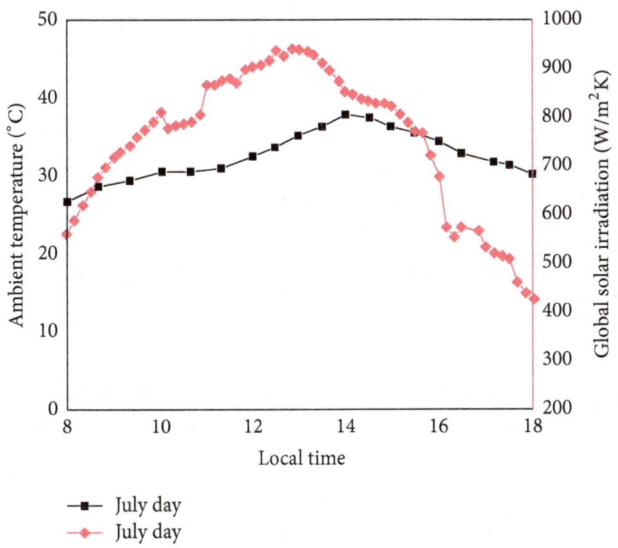


FIGURE 7: Evolution of solar radiation and ambient temperature in a typical summer day (July 16) in Tunis-Tunisia [18].

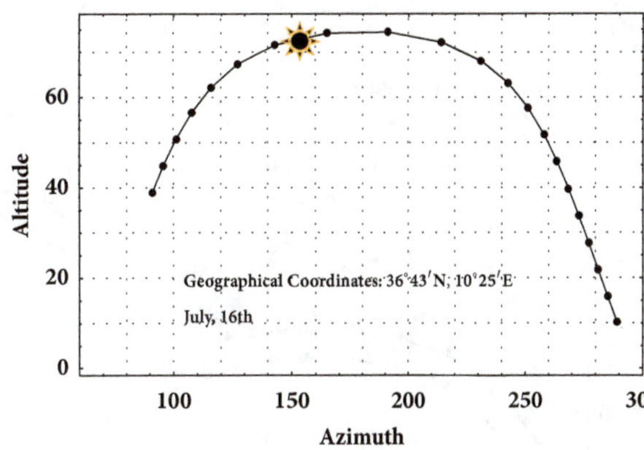


FIGURE 8: Solar chart for a typical summer day in Tunis-Tunisia (8:00–18:00 in 0:30 intervals).

$COP_{cycle}$ = 0.84, 0.91, 0.95 for chilled water temperature, respectively, set at 7∘C, 10∘C, 12∘C, with corresponding values for $COP_{system}$ = 0.51, 0.55, 0.58.

(iv) $COP_{system}$ is 60% lower than $COP_{cycle}$. It ranges between 0.1 and 0.6, comparable with the results other authors [22, 23].

(v) When the cooling medium temperature is set at 32∘C, the performances of the cycle and the system reach their maximum at the highest pressure in the generator (270 kPa) and for a chilled water temperature $T_c$ = 12∘C.

*5.3. Case Study: Hourly Local Data.* Belghouthi *et al.* [18] performed a thermal test of weather and solar conditions during July 2008 at a research centre in Borj Cedria, near Tunis. The research centre is located at 36°43' latitude and 10°25' longitude. Figure 7 shows the data for a particular day (16th of July 2008), of this Mediterranean climate zone characterized by a high level of solar resources. It receives an average of 17 kJ/m²/day with a total insolation period of 3500 h/year and 350 sunny days per year. Ambient temperature varies between 39°C and 27°C, and solar radiation in the range 250 W/m² 900 W/m². Maximum solar radiation of 958 W/m² is observed at 12:00. Maximum ambient temperature of 37°C is found at 14:00. Figure 8 reproduces the solar chart for the test day.

The efficiency of the Fresnel collector as expressed in (5) depends on the quadratic temperature difference $\Delta T = T_{HTF} - T_{amb}$ (equation (2)) between that of heat transfer flow and ambient. In the high-lightened gray interval in Figure 4, the temperature $T_{HTF}$ affects more the collector efficiency than the irradiance G. Considering the model equations (1)-(5) and Figure 4, the lower is $T_{HTF}$ the higher is the collector efficiency. Due to the compensating effects of solar radiation and ambient temperature during the time interval 11:00-15:00, the efficiency of the collector is at its maximum at this period as shown in Figures 9 and 10.

Figure 11 shows the evolution $COP_{system}$ and $COP_{cycle}$ during the test summer day. The $COP_{cycle}$ reaches always values higher than 0.9 and the $COP_{system}$ is higher than 0.55. The cycle performances are independent on ambient temperature, set at 25°C; this is the reason of the constant $COP_{cycle}$. The maximum $COP_{system}$ is equal to 0.99, 1.02, 1.03 for generator pressure set, respectively, at 198 kPa, 232 kPa, and 270 kPa. The evolution of $COP_{system}$ is similar to that of the collector efficiency; it reaches its maximum in the middle of the day from 12:00 to 14 pm, namely, 0.61, 0.63, and 0.63 for

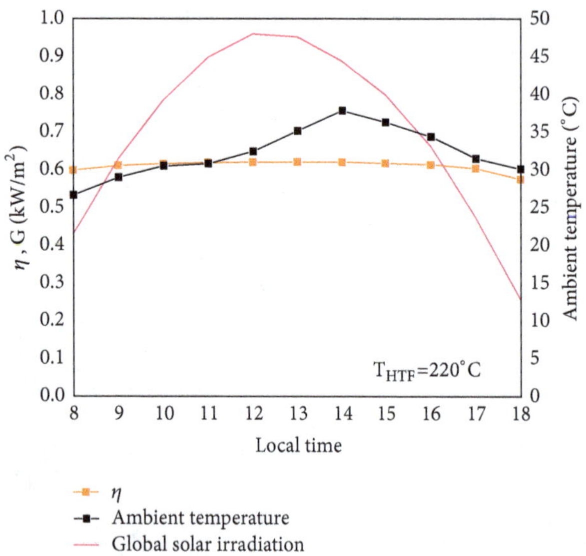


FIGURE 9: Collector efficiency *versus* ambient temperature and solar radiation for a fixed temperature $T_{HTF}$ = 220°C and climate conditions of Figure 7.

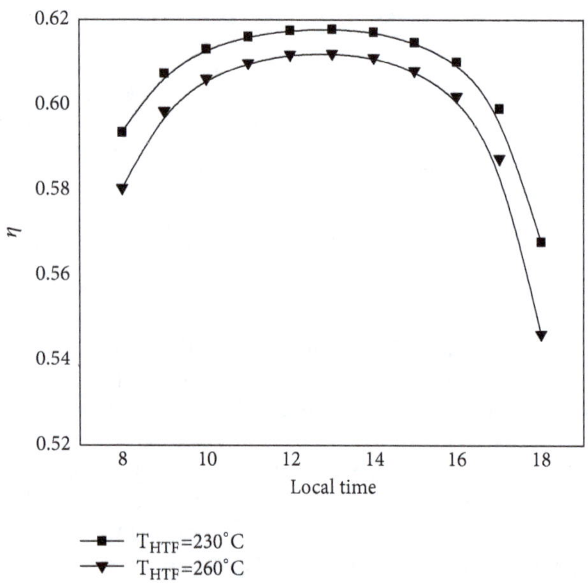


FIGURE 10: Collector efficiency during typical Tunisian summer day for different HTF temperatures.

generator pressure set, respectively, at 198 kPa, 232 kPa, and 270 kPa.

## 6. Conclusion

This paper investigated solar combined ejector single-effect absorption refrigeration cycle. The considered chiller is driven by condensation of steam at temperatures between 180°C and 210°C, using pressurized water heated by a linear Fresnel collector.

The model of the collector is detailed and simulations of the overall system and the cooling machine are performed.

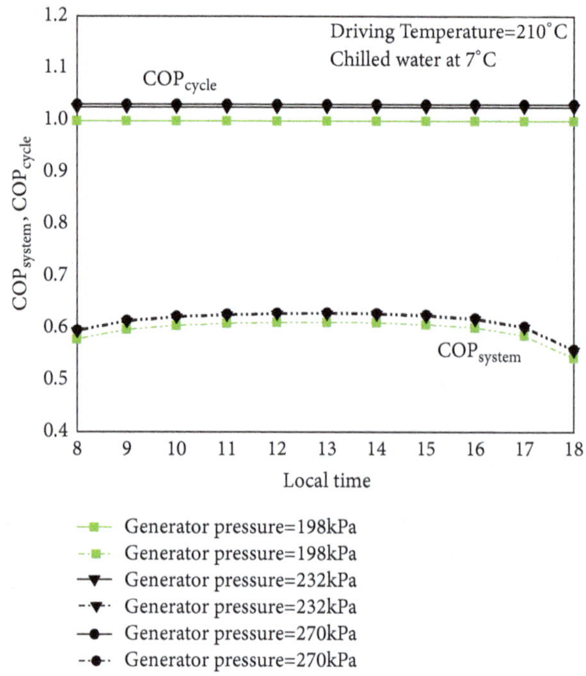


FIGURE 11: Hourly evolution of $COP_{system}$ and $COP_{cycle}$ during a typical Tunisian summer day for different generator pressure.

The produced water chilled water is varied between 7°C and 12°C. In regard to the simulation results, it can be concluded that

(i) The *COP* increases with the driving the heat temperature increases, decreases when cooling medium temperature gets higher, and increases with the chilled water temperature.

(ii) The $COP_{cycle}$ of this particular single-effect cycle can reach 1.03, a value comparable to that of a conventional double-effect cycle.

(iii) For fixed cooling medium temperature, the maximum *COP* reaches higher values when the generator pressure is larger, *e.g.*, for a cooling medium set at 25∘C and solution generator temperature at 210∘C, the $COP_{cycle}$ = 0.99, 1.02, 1.03 and the $COP_{system}$ = 0.57, 0.59, 0.6, respectively, for generator pressure is at 198 kPa, 232 kPa, and 272 kPa.

(iv) $COP_{system}$ is almost half of $COP_{cycle}$ due to the limited efficiency of the solar collector compartment, between 0.5 and 0.6.

(v) For a fixed generator pressure, the maximum *COP* decreases when the temperature of the cooling medium rises. For a generator pressure of 198 kPa and chilled water temperature at 7∘C, the maximum $COP_{cycle}$ is 0.99, 0.85, and 0.75 for cooling medium temperature of 25∘C, 30∘C, and 35∘C, respectively. The corresponding values of $COP_{system}$ are 0.57, 0.52, and 0.46.

(vi) For a cooling medium temperature of 32∘C and by fixed chilled water temperature, the $COP_{cycle}$ and the

$COP_{system}$ increase with the driving heat temperature, reaching a maximum for $T_g$ = 210∘C: 0.87, 0.91, and 0.95 for, respectively, 198 kPa, 232 kPa, and 270 kPa generator pressure and at 12∘C chilled water temperature.

(vii) The performances of cycle and system are optimal for the highest generator pressure (270 kPa in this study) and chilled water temperature (12∘C).

A case study of a typical summer day is simulated; results showed that

(i) The average efficiency of linear Fresnel collector considered in the simulations varies between 0.6 and 0.62.

(ii) In the test period, $COP_{cycle}$ reaches values always higher than 0.9 and $COP_{system}$ larger than 0.55. The cycle performances are independent from ambient temperature with a maximum equal to 0.99, 1.02, and 1.03 for generator pressure, respectively, at 198 kPa, 232 kPa, and 270 kPa.

(iii) The system performance evolves parallel to that of the collector efficiency, reaching its maximum in the middle of the day from 12:00 to 14 pm.

## Data Availability

Previously reported solar collector data used to support this study are available at [W. Christine, B. Michael, M. Florian, H. Alexander, N. Tomas, "Solar cooling with water-ammonia absorption chillers and concentrating solar collector-operational experience", International Journal of refrigeration, vol. 39, pp. 57-76, 2014.]. These prior studies and datasets are cited at relevant places within the text as [17]. Also the local metrological data were used to support this study and are available at [M. Balghouthi, M. H. Chahbani, and A. Guizani, "Investigation of a solar cooling installation in Tunisia," Applied Energy, vol. 98, pp. 138–148, 2012.]. These prior studies and datasets are cited at relevant places within the text as [18].

## References

[1] K. Kaygusuz, "Prospect of concentrating solar power in Turkey: the sustainable future," *Renewable & Sustainable Energy Reviews*, vol. 15, no. 1, pp. 808–814, 2011.

[2] M. U. H. Joardder, P. K. Halder, A. Rahim, and N. Paul, "Solar Assisted Fast Pyrolysis: A Novel Approach of Renewable Energy Production," *Journal of Engineering*, vol. 2014, Article ID 252848, 10 pages, 2014.

[3] F. J. Cabrera, A. Fernández-García, R. M. P. Silva, and M. Pérez-García, "Use of parabolic trough solar collectors for solar refrigeration and air-conditioning applications," *Renewable & Sustainable Energy Reviews*, vol. 20, pp. 103–118, 2013.

[4] S. Beerbaum and G. Weinrebe, "Solar thermal power generation in India - A techno-economic analysis," *Journal of Renewable Energy*, vol. 21, no. 2, pp. 153–174, 2000.

[5] G. Manzolini, M. Bellarmino, E. Macchi, and P. Silva, "Solar thermodynamic plants for cogenerative industrial applications in southern Europe," *Journal of Renewable Energy*, vol. 36, no. 1, pp. 235–243, 2011.

[6] A. Maiorino and M. Valentini, "On the technical and economic feasibility of a solar thermodynamic power plant in an area of medium-high direct sunlight intensity: an actual case study," *Journal of the Brazilian Society of Mechanical Sciences and Engineering*, vol. 39, no. 1, pp. 245–257, 2016.

[7] K. Ajay and L. Kundan, "Combined Experimental and CFD Investigation of the Parabolic Shaped Solar Collector Utilizing Nanofluid (CuO-H2O and SiO2-H2O) as a Working Fluid," *Journal of Engineering*, vol. 2016, Article ID 5729576, 10 pages, 2016.

[8] O. Marc, G. Anies, F. Lucas, and J. Castaing-Lasvignottes, "Assessing performance and controlling operating conditions of a solar driven absorption chiller using simplified numerical models," *Solar Energy*, vol. 86, no. 9, pp. 2231–2239, 2012.

[9] G. A. Florides, S. A. Kalogirou, S. A. Tassou, and L. C. Wrobel, "Modelling and simulation of an absorption solar cooling system for Cyprus," *Solar Energy*, vol. 72, no. 1, pp. 43–51, 2002.

[10] G. A. Florides, S. A. Kalogirou, S. A. Tassou, and L. C. Wrobel, "Modelling, simulation and warming impact assessment of a domestic-size absorption solar cooling system," *Applied Thermal Engineering*, vol. 22, no. 12, pp. 1313–1325, 2002.

[11] A. Shirazi, R. A. Taylor, S. D. White, and G. L. Morrison, "Multi-effect absorption chillers powered by the sun: reality or a reverie," *Energy Procedia*, vol. 91, pp. 844–856, 2016.

[12] D. Chemisana, J. López-Villada, A. Coronas, J. I. Rosell, and C. Lodi, "Building integration of concentrating systems for solar cooling applications," *Applied Thermal Engineering*, vol. 50, no. 2, pp. 1472–1479, 2013.

[13] H. Beltagy, D. Semmar, C. Lehaut, and N. Said, "Theoretical and experimental performance analysis of a Fresnel type solar concentrator," *Journal of Renewable Energy*, vol. 101, pp. 782–793, 2017.

[14] P. Bermejo, F. J. Pino, and F. Rosa, "Solar absorption cooling plant in Seville," *Solar Energy*, vol. 84, no. 8, pp. 1503–1512, 2010.

[15] D.-W. Sun, I. W. Eames, and S. Aphornratana, "Evaluation of a novel combined ejector - Absorption refrigeration cycle - I: Computer simulation," *International Journal of Refrigeration*, vol. 19, no. 3, pp. 172–180, 1996.

[16] S. Aphornratana and I. W. Eames, "Experimental investigation of a combined ejector-absorption refrigerator," *International Journal of Energy Research*, vol. 22, pp. 195–207, 1998.

[17] C. Weber, M. Berger, F. Mehling, A. Heinrich, and T. Núñez, "Solar cooling with water-ammonia absorption chillers and concentrating solar collector - Operational experience," *International Journal of Refrigeration*, vol. 39, pp. 57–76, 2014.

[18] M. Balghouthi, M. H. Chahbani, and A. Guizani, "Investigation of a solar cooling installation in Tunisia," *Applied Energy*, vol. 98, pp. 138–148, 2012.

[19] T. Sriveerakul, S. Aphornratana, and K. Chunnanond, "Performance prediction of steam ejector using computational fluid dynamics: Part 1. Validation of the CFD results," *International Journal of Thermal Sciences*, vol. 46, no. 8, pp. 812–822, 2007.

[20] S. A. Klein and F. Alvarado, *Engineering equation solver*, F-chart software, Middletonm, Wis, USA.

[21] R. Fathi, C. Guemimi, and S. Ouaskit, "An irreversible thermodynamic model for solar absorption refrigerator," *Journal of Renewable Energy*, vol. 29, no. 8, pp. 1349–1365, 2004.

6

# A Condensed Introduction to the Doubly Fed Induction Generator Wind Energy Conversion Systems

**Julius Mwaniki,[1] Hui Lin,[1] and Zhiyong Dai[2]**

[1]*Department of Electrical Engineering, School of Automation, Northwestern Polytechnical University, Xi'an 710129, China*
[2]*Automatic Control Department, School of Mechano-Electronic Engineering, Xidian University, Xi'an 710071, China*

Correspondence should be addressed to Julius Mwaniki; jukamwa@gmail.com

Academic Editor: Yuh-Shyan Hwang

The increase in wind power penetration, at 456 GW as of June 2016, has resulted in more stringent grid codes which specify that the wind energy conversion systems (WECS) must remain connected to the system during and after a grid fault and, furthermore, must offer grid support by providing reactive currents. The doubly fed induction generator (DFIG) WECS is a well-proven technology, having been in use in wind power generation for many years and having a large world market share due to its many merits. Newer technologies such as the direct drive gearless permanent magnet synchronous generator have come up to challenge its market share, but the large number of installed machines ensures that it remains of interest in the wind industry. This paper presents a concise introduction of the DFIG WECS covering its construction, operation, merits, demerits, modelling, control types, levels and strategies, faults and their proposed solutions, and, finally, simulation. Qualities for the optimal control strategy are then proposed. The paper is intended to cover major issues related to the DFIG WECS that are a must for an overview of the system and hence serve as an introduction especially for new entrants into this area of study.

## 1. Introduction

Wind power was expected to supply 5% of electrical power worldwide when it hit 500 GW at the end of 2016 up from 456 GW at end of June 2016. By the end of 2020, the total installed capacity is projected to be 1900 GW [1]. Among the applied wind energy conversion systems (WECS) configurations, the four most common ones are designated types A to D. Type A utilizes a fixed speed turbine connected to a squirrel cage induction generator (SCIG) through a gearbox. Type B is a variable speed turbine connected through a gearbox to a wound rotor induction generator (WRIG) whose rotor is connected to a variable resistance. Both types A and B utilize a soft starter to smoothen connection to the grid during start-up operations and a capacitor bank for reactive power supply. Type C is also a variable speed turbine connected through a gearbox to a WRIG. However, the rotor output is fed to the grid through a partially rated frequency converter. This is also referred to as the doubly fed induction generator (DFIG). Type D involves a variable speed turbine connected either directly or through a gearbox to a WRIG, a wound rotor synchronous generator (WRSG), or a permanent magnet synchronous generator (PMSG) which connects to the grid through a fully rated frequency converter. The gearless type D WECS are direct driven large diameter multipole generators. Their large size and associated weight, in addition to the higher cost of the fully rated converter, erode the advantage of reduced cost and maintenance afforded by eliminating the gearbox [2]. The merits of the DFIG make it one of the most widely used generators for WECS. The DFIG WECS have been in use in wind power generation for many years now and have a large share of the world market. Whereas newer technologies such as the direct drive gearless PMSG have come up to challenge its market share, the large number of installed machines ensures that it remains of interest in the wind industry.

The DFIG WECS is a mix of aerodynamic, mechanical, electromagnetic, and electronic systems, and as such, control of the various subsystems in both steady and transient states is quite complex, especially since the increase in wind power penetration has resulted in more stringent grid codes which

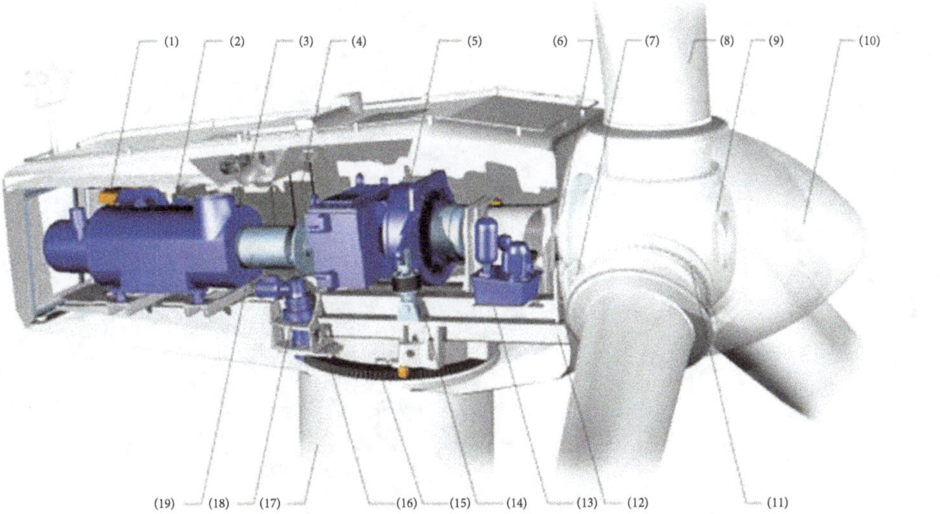


FIGURE 1: Components of a wind turbine [5]. (1) Service crane. (2) Generator. (3) Cooling system. (4) Top control unit. (5) Gearbox. (6) Main shaft with two bearings. (7) Rotor lock system. (8) Blade. (9) Hub. (10) Hub cover. (11) Blade bearing. (12) Bed frame. (13) Hydraulic unit. (14) Shock absorbers. (15) Yaw ring. (16) Brake. (17) Tower. (18) Yaw gears. (19) Transmission high speed shaft.

specify that the WECS must remain connected to the system even when a fault occurs and, furthermore, must provide reactive currents to support the grid voltages. This is made more difficult by the intermittent and uncertain nature of the prime mover, the wind [3, 4]. Many works have been done on various aspects of the DFIG WECS, such as the aerodynamic energy conversion, the electromechanical energy conversion, control underbalanced and imbalanced grid voltages, power electronic converters and their modulation schemes, and power evacuation to the grid. These lay emphasize on a single aspect at a time, but none has, to the authors' knowledge, tackled the complete operation of the system. This paper seeks to bridge these works for a complete picture into this multifaceted system.

This paper is divided into ten sections. Starting with an introduction in Section 1, Section 2 covers construction while Section 3 gives insight into the operation of the system. Section 4 presents the merits and demerits of the DFIG WECS. Section 5 gives modelling of the wind turbine and the DFIG while Section 6 covers grid connection requirements, control types, levels, and strategies. Section 7 covers possible faults and proposed solutions while simulation is handled in Section 8. Finally, discussion and conclusion are carried out in Sections 9 and 10, respectively.

## 2. Construction

The DFIG WECS consists of a wind turbine (WT) system connected to a DFIG through a gearbox for the purpose of extracting mechanical power from the air flows and converting it into electrical power. The wind turbine power output increases as to the cube of wind speed. Wind speed increases with increase in height; hence, horizontal axis wind turbines are mounted on a tower. The generator, gearbox, and associated control equipment are housed in a nacelle at the top of the tower. Some designs have some control equipment housed within the base of the tower. Figure 1 shows the components of a wind turbine.

The DFIG WECS utilizes three bladed horizontal axis wind turbines with active blade pitch capability. The gearbox consists of two helical and one planetary gear stages with the low speed shaft from the turbine rotor connected to the planetary stage and the high speed generator rotor shaft to the last helical stage as shown in Figure 2 [5, 17, 18].

The DFIG has a wound stator which consists of three copper windings with a 120° spatial offset in generators with two poles while those with four poles have six windings at 60° offset [19]. The stator is connected, either directly or through a transformer depending on the level of its output voltage in relation to the grid voltage, to the grid bus. It also has a wound rotor which is brought out through slip rings and carbon brushes for connection to a power electronic converter (PEC) and then either directly or through a transformer, to the stator terminals.

The components in the rotor circuit include the following [14, 20–22].

*2.1. dv/dt Filter.* It is a resistance in parallel with an inductance that is connected between the rotor windings terminals and the rotor side converter (RSC) terminals to protect the rotor windings against harmful effects from the RSC which include bearing currents, capacitive leakage currents, and degradation of insulation due to polarization effect by the resultant capacitive displacement current from sudden rise in voltage.

*2.2. Power Electronic Converter.* It includes the following:

*(a) RSC.* This is a voltage source converter (VSC) type, chosen over current source type due to its faster response to system

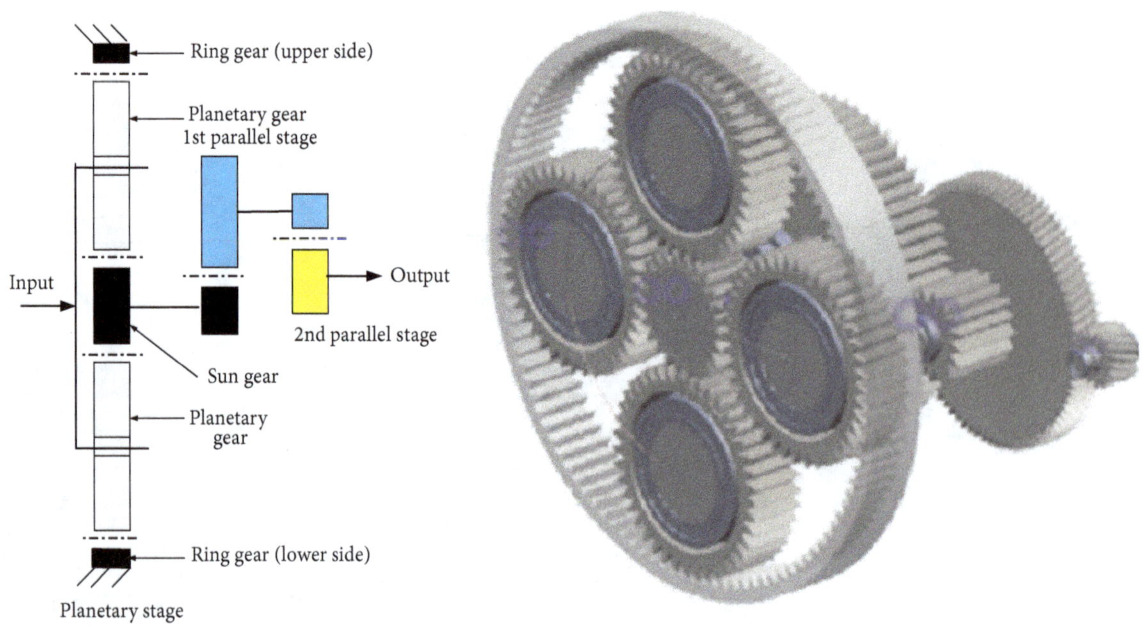


FIGURE 2: One-stage planetary and two-stage helical/parallel gear arrangement [6].

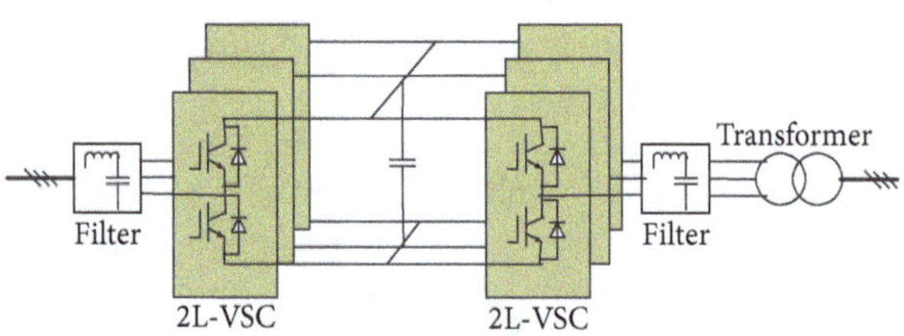


FIGURE 3: Two-level back-to-back voltage source converter [7].

changes and lower losses in the DC-link. It is composed of a rectifier bridge utilizing insulated gate bipolar transistors (IGBT) and diodes. It supplies magnetization current to the rotor. It also controls active and reactive generator power output by injecting currents of varying amplitude and frequency into the rotor windings. Slip power is also sent to the DC-link. The IGBTs switching is controlled by pulses which are generated by either sinusoidal pulse width modulation (SPWM) or space vector modulation (SVM) techniques.

*(b) DC-Link.* It links the RSC to the GSC as shown in Figure 3. It is composed of an energy storage capacitor or combination of several capacitors that try to maintain constant voltage at their terminals. A VSC is characterized by a high value of capacitance in the DC-link.

*(c) Grid Side Converter (GSC).* The GSC is a bidirectional rectifier bridge, utilizing IGBTs as the switching device, responsible for maintaining the DC bus voltage within certain limits by transferring the power from the rotor, which is stored in the DC-link capacitor, to the grid. It also exchanges reactive power with the grid by either absorbing reactive power from the grid or exporting reactive power to the grid as per the set value. Usually, its power factor is set at unity to minimize current flowing in it.

As DFIG WECS increase in individual size, the high power ratings have resulted in different converter configurations to accommodate the high currents and voltages with existing semiconductor device ratings. These include the following [7, 8, 14, 23].

*(a) Two-Level Back-to-Back VSC.* This is the most established solution with a simple structure and few components. Bulky filters are required to limit the high dv/dt stresses on generator and transformer due to the only two voltage levels and to reduce the total harmonic distortion (THD) of grid voltages and currents.

*(b) Parallel Connection of Two-Level Back-to-Back VSCs.* This is as shown in Figure 4. It has the advantage of redundancy and hence increases reliability of the converter.

*(c) Multilevel Topology.* This is classified into the following:

(i) Three-level neutral point diode clamped (NPC) structure

(ii) Four-level flying capacitor clamped structure

(iii) Five-level cascade H-bridge (CHB) connection which is bulky and complex as it requires many isolated DC sources

*(d) Matrix Converters.* A matrix converter, also called a cycloconverter, transfers AC supply directly into an AC load of a different frequency. It does not utilize a DC-link capacitor. It has lower harmonics and lower switching losses.

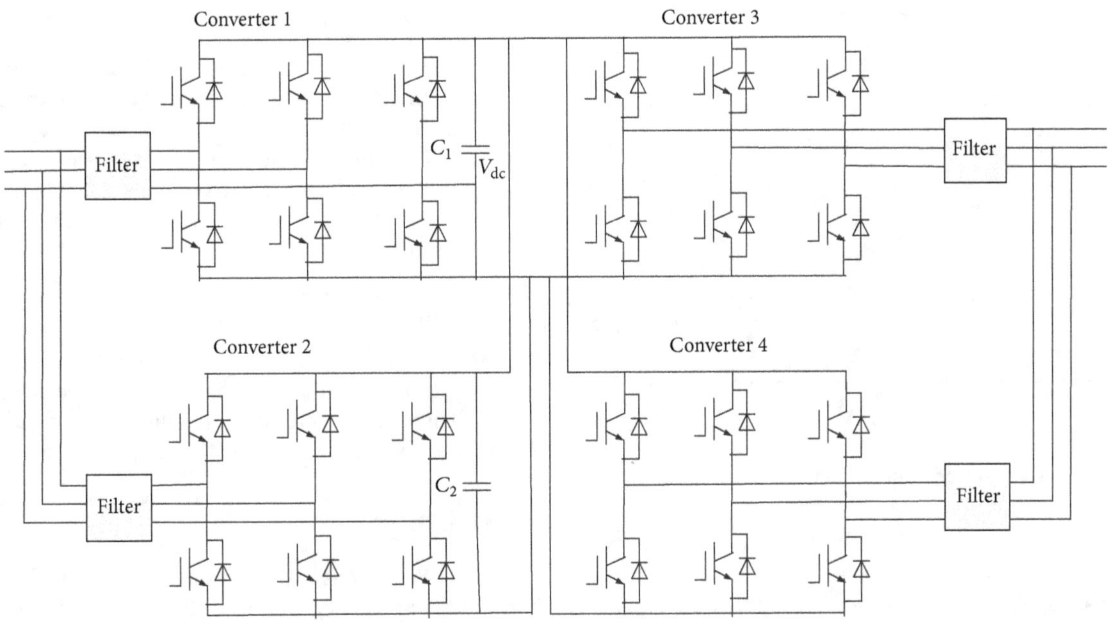


FIGURE 4: Two-level back-to-back VSC in parallel connection [8].

It however has higher conduction losses, limitation on output voltage, and poor protection during faults [2].

*(e) Resonant Converters.* The main topology, the natural clamped converter, consists of two two-level back-to-back VSCs with a circuit to provide resonance in between them.

*(f) Tandem Converters.* These consist of a two-level back-to-back current source converter (CSC), designated the primary converter, in parallel with a two-level back-to-back VSC, and designated the secondary converter.

*2.3. Harmonic Filter.* The VSC produces harmonics, in multiples of the switching frequency, which cause distortion of the grid voltages and currents. The harmonics can be reduced by use of filters, either active, passive, or hybrid, and by control that utilizes RSC or GSC capacity [21, 24].

*2.4. Electromagnetic Compatibility (EMC) Filter.* The switching on and off of IGBTs causes interference over a wide signal spectrum. The interference signals can be transmitted or received by electrical connections through impedances between the emitting source and the susceptible equipment, through magnetic and electric fields causing capacitive and inductive connections, and by radiation. The EMC filter reduces these emissions to a level where other electrical devices can be operated in the vicinity without problems. The PEC itself must also not be affected by low levels of external interference to avoid giving rise to cases of dangerous operating states [19].

*2.5. Converter Control Unit (CCU).* It carries out the control of the various elements in the rotor circuit for coordinated, reliable, and accurate operation. The CCU is a programmable electronic device with various inputs from its own measurements and the WECS controller, and it sends out output signals to the elements to be controlled. It also has communication channels for remote sensing and control. Controls for the RSC and GSC under both normal conditions and grid disturbance conditions must ensure that predetermined setpoints are met.

## 3. Operation

The DFIG WECS is a variable speed wind turbine (VSWT) and is operated at varying speeds corresponding to the varying wind speeds from the cut-in speed (Point A) through the rated wind speed (Point C) to the cut-out speed (Point E) as shown on the turbine torque characteristic curves in Figure 5.

Below cut-in speed (Point A), the machine is at standstill as the wind speed is too low for efficient and economical operation. Once sufficient wind is available, the turbine starts rotating and the machine starts generating power. The power coefficient is maintained constant at its maximum between points B and C to maximize power output as the wind speed increases. This is done by applying a maximum power point tracking (MPPT) strategy by which the turbine rotor speed is adjusted relative to the prevailing wind speed. During this period, the blade pitch angle is fixed at 0°. Beyond point D, power regulation is introduced to ensure output power does not exceed rated power as wind speed continues rising. The power coefficient is lowered by pitching the rotor blades, hence reducing the speed of turbine rotation. At point E, the cut-out speed, the machine is stopped by pitching the turbine blades out of the wind and applying the emergency brake to avoid structural damage due to high wind speeds. Running at high speeds can also result in high noise emission [4, 14].

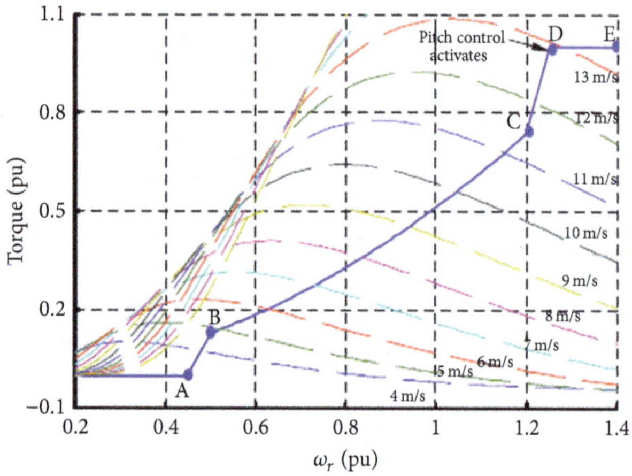


FIGURE 5: Turbine torque characteristics [9].

Whereas DFIG WECS have been utilized in stand-alone applications, the most common use is to supply power to the grid. The stator of the DFIG draws a reactive magnetizing current which is supplied either by the grid or the GSC. Since the DFIG is an asynchronous generator, its electromagnetic torque is developed when currents induced in the rotor windings set up a magnetic field which interacts with the stator field. These currents in the rotor are induced by both the voltage induced by the rotating stator field and the voltage supplied by the RSC. The stator field is of constant frequency set by the grid frequency. The rotor field, on the other hand, is of varying frequency which is dependent on the slip [2, 14].

The variable speed operation of the DFIG is possible due to use of the PEC. The bidirectional power transfer in the PEC enables two modes of operation [14]:

(i) Subsynchronous mode or partial load mode at below rated speeds: the stator of the DFIG supplies power to the grid and slip power to the rotor via the converters and slip rings.

(ii) Supersynchronous mode or nominal load range at above rated speed: both the stator output power and the rotor slip power are fed to the grid.

## 4. Merits and Demerits

The DFIG WECS has the following advantages [25–27]:

(i) Higher efficiency even at low wind speeds

(ii) Decoupled control of active ($P$) and reactive ($Q$) powers, hence being more responsive to system needs

(iii) Reduced acoustic noise making the system more environmentally acceptable

(iv) Reduced mechanical stresses which reduces maintenance costs and prolongs system lifetime

(v) Contributing to system power quality improvement without using external reactive compensation devices

(vi) Less cost as PEC is rated for lower power levels, usually 30%

And it has the following disadvantages [3, 14, 28]:

(i) The direct connection of the stator to the grid leaves the machine more vulnerable to system disturbances such as voltage dips and swells as the partially rated VSC is limited in its ability to withstand over currents and over voltages induced in it by stator fault currents due to the magnetic coupling between the stator and the rotor.

(ii) The VSC is a source of harmonics due to switching actions.

(iii) Use of PEC, gearbox, and slip rings introduces increased costs and maintenance needs, increased losses, and decreased reliability as they increase the number of components that can fail.

## 5. Modelling

The DFIG WECS has two main subsystems. The first handles harnessing of energy in the airflows and the subsequent conversion into rotational mechanical power. This subsystem includes the turbine rotors, the hub, the low speed shaft, the gearbox, and the high speed shaft that is connected to the generator rotor. The second subsystem carries out the conversion of the mechanical power into electrical power. This is handled by the generator and the PEC.

*5.1. The Wind Turbine Model.* The mechanical power extracted from the wind is given by [11, 29]

$$P_m = \frac{1}{2} A \rho V^3 C_p (\lambda, \beta), \tag{1}$$

where $P_m$ is mechanical power (W), $A$ is area swept by turbine rotor (m$^2$), $V$ is wind speed (m/s), $C_p$ is the power coefficient, $\rho$ is air density (kg/m$^3$), $\lambda$ is tip speed ratio (TSR) given as $(\omega_m * R/V)$, $\beta$ is blade pitch angle (degrees), and $\omega_m$ is angular speed of wind turbine (rad/s) given as $(p\Omega_m/2)$, where $p$ is number of poles and $\Omega_m$ is rotor mechanical rotational speed (rad/s).

And the mechanical torque developed by the wind turbine shaft is given as

$$T_m = \frac{P_m}{\omega_m} = \frac{1}{2} A \rho V^2 C_p (\lambda, \beta) \frac{R}{\lambda}, \tag{2}$$

where $T_m$ is mechanical torque (N·m) and $R$ is radius of turbine rotor (m). The power coefficient is given by

$$C_p (\lambda, \beta) = 0.5 \left( \Gamma - 0.02\beta^2 - 5.6 \right) e^{-0.17\Gamma}, \tag{3}$$

where

$$\Gamma = \frac{R\,(3600)}{\lambda\,(1609)}. \tag{4}$$

*5.2. The DFIG Model.* The DFIG park model in the $d$-$q$ reference frame, assuming sinusoidal flux, constant winding resistance, and neglecting magnetic saturation, is given in Figure 6.

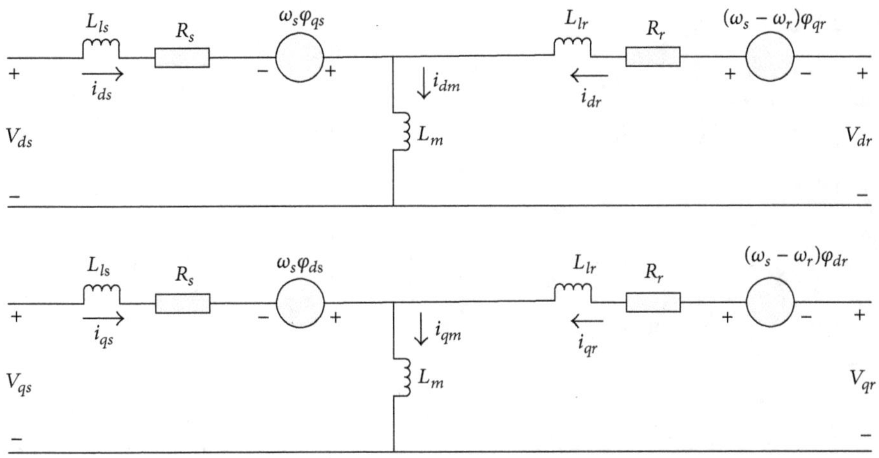


FIGURE 6: DFIG equivalent circuit in $d$-$q$ reference frame [10].

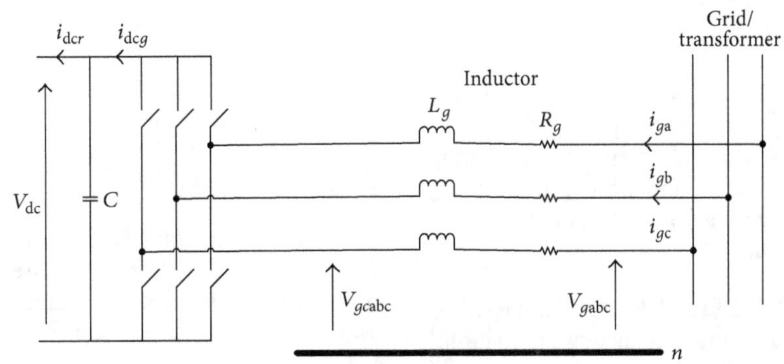


FIGURE 7: Schematic of GSC connection to grid [11].

The equations relating the DFIG park model stator and rotor voltages and fluxes are

$$
\begin{aligned}
v_{ds} &= R_s i_{ds} - \omega_s \varphi_{qs} + \frac{d\varphi_{ds}}{dt} \\
v_{qs} &= R_s i_{qs} + \omega_s \varphi_{ds} + \frac{d\varphi_{qs}}{dt} \\
v_{dr} &= R_r i_{dr} - \omega \varphi_{qr} + \frac{d\varphi_{dr}}{dt} \\
v_{qr} &= R_r i_{qr} + \omega \varphi_{dr} + \frac{d\varphi_{qr}}{dt},
\end{aligned}
\tag{5}
$$

where $v$ are voltages (V), $i$ are currents (A), $R$ are resistances ($\Omega$), and $\varphi$ are flux linkages (V·s). Indices $d$ and $q$ indicate direct and quadrature axis components, respectively, while $s$ and $r$ indicate stator and rotor quantities, respectively, and $\omega_s$ is grid electrical angular speed, $\omega_r$ is rotor electrical angular speed and slip speed, and $\omega$ is given by $\omega_s - \omega_r$.

The direct and quadrature stator and rotor flux components are decoupled magnetically and are given by

$$
\begin{aligned}
\varphi_{ds} &= L_s i_{ds} + L_m i_{dr} \\
\varphi_{qs} &= L_s i_{qs} + L_m i_{qr} \\
\varphi_{dr} &= L_r i_{dr} + L_m i_{ds} \\
\varphi_{qr} &= L_r i_{qr} + L_m i_{qs},
\end{aligned}
\tag{6}
$$

where $L_s = L_{ls} + L_m$ and $L_r = L_{lr} + L_m$, $L_{ls}$ and $L_{lr}$ are stator leakage and rotor self-inductances, and $L_m$ is mutual inductance (H).

The equivalent circuit for the grid side converter connection to the grid is given in Figure 7.

The voltage across the inductor is given, in the $d$-$q$ reference frame, by

$$
\begin{aligned}
v_{gd} &= R_g i_{gd} + L_g \frac{di_{gd}}{dt} - \omega_s L_g i_{gd} + v_{gcd} \\
v_{gq} &= R_g i_{gq} + L_g \frac{di_{gq}}{dt} + \omega_s L_g i_{gq} + v_{gcq},
\end{aligned}
\tag{7}
$$

where $V_{gcd}$ and $V_{gcq}$ are voltages at GSC output terminals while $V_{gd}$ and $V_{gq}$ are voltages at the grid.

## 6. Control

*6.1. Grid Utility Operator's Specifications.* Control for DFIG WECS is aimed at power production that meets the grid

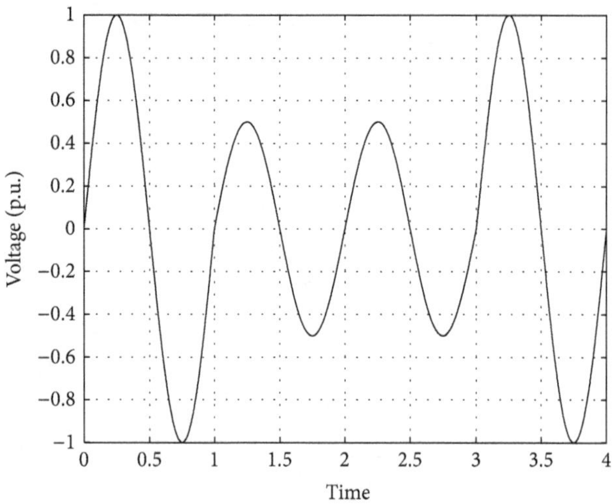


FIGURE 8: Voltage sag (dip) [12].

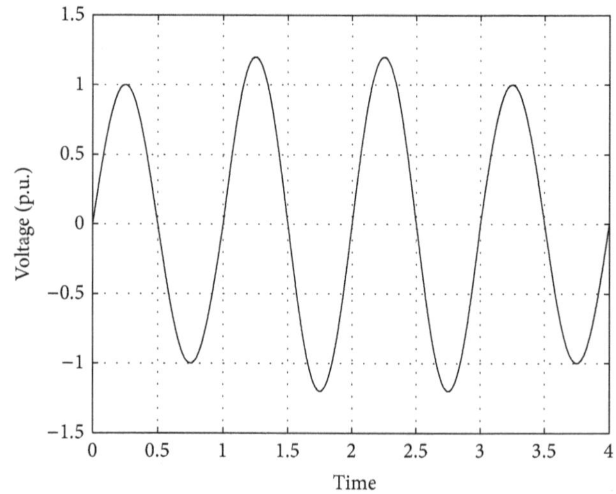


FIGURE 9: Voltage swell [13].

utility operator's specifications. These usually vary as to the operator, based on the complexity of the grid system. IEEE standard 1549 defines the basic requirements for integrating distributed generation units including wind power to the utility grids. Common grid connection requirements include [2, 19, 27, 30–34] the following.

*(a) Fault Ride-Through (FRT).* DFIG WECS are impacted by faults from the grid. These include frequency fluctuations, voltage sags, and voltage swells. FRT capability is critical system requirements for wind farms as disconnection of the wind farm on occurrence of a fault is no longer an option since the sudden loss of a large source of active power can result in cascade trip of other plants due to undervoltage and/or underfrequency due to power-load mismatch in the system. Thus, the main aim of FRT requirement is the instantaneous resumption of active power as soon as the fault is cleared [3]. IEEE std. 1159/95 defines voltage sag as a reduction of between 0.1 to 0.9 p.u. of the voltage and current (on an rms basis) at the power frequency for any duration ranging from 0.5 cycles to 1 minute. The amplitude is the remaining voltage value during the sag. On the other hand, IEC defines a voltage dip as a sudden voltage reduction at any point in the electrical system lasting from anywhere between half a cycle to a few seconds [35]. Figure 8 shows the waveform of a voltage sag.

Voltage swell as defined by IEEE 1159/95 is the increase in the RMS voltage level to 1.1 p.u–1.8 p.u. of nominal, at the power frequency, for durations of 0.5 cycles to 1 minute. Figure 9 shows the waveform of a voltage swell.

There are two subtypes of FRT.

*(i) Low-Voltage Ride-Through (LVRT).* Some grid codes also include zero-voltage ride-through (ZVRT), where the nominal voltage drops to zero, under LVRT. Voltage sags result from sudden loss of large generating units, switching in of large loads such as induction motors, and energizing of transformers and system faults such as short circuits and faults to ground [23, 36, 37]. On occurrence of voltage sag, the WECS must remain connected for a given set time before disconnecting. In some cases, the WECS must help support grid voltage during the fault. LVRT curves for four countries are given in Figure 10.

*(ii) High-Voltage Ride-Through (HVRT).* Voltage swells result from single-phase short-term interruptions, reactive power overcompensation from capacitor banks, lightning strikes, or switching off of large loads such as in response to voltage sag [23, 38, 39]. When voltage goes above its upper limit, usually at 1.1% though being set at 1.3% for Spain and Australia [15], the WECS must stay connected for a given period of time before disconnecting.

Since voltage sags are more frequent than swells, LVRT is given greater focus than HVRT.

*(b) Active Power (P) Control.* The WECS must provide active power control to ensure stable system frequency and avoid overloading of transmission lines. The ramp rate should also be within the set range.

*(c) Reactive Power (Q) Control.* Dynamic reactive power control capability is required as the power factor and reactive power balance must be maintained in the desired range.

*(d) Voltage Regulation.* The voltage at each WECS terminal must be maintained at a constant value.

*(e) Voltage Operating Ranges.* The WECS must operate within typical grid voltage limits.

*(f) Frequency Control.* This entails frequency regulation capability so as to maintain desired grid frequency.

*(g) Frequency Operating Ranges.* The WECS must operate within typical grid frequency limits.

*(h) Power Quality.* The WECS outputs in terms of voltage fluctuations which could result in flicker and voltage/current harmonics are limited to within set ranges.

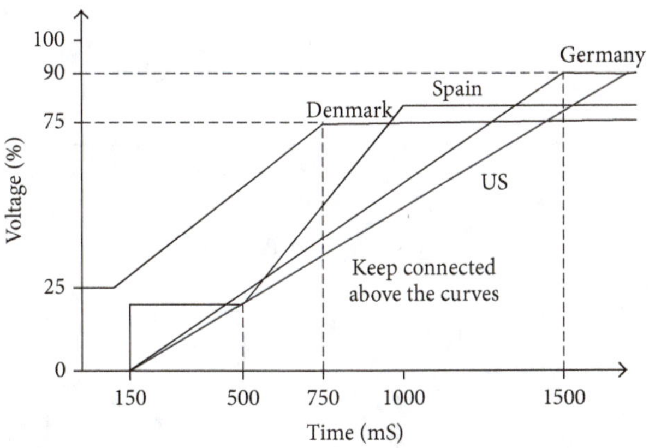


Figure 10: Curves for low-voltage ride-through requirements for different countries [7].

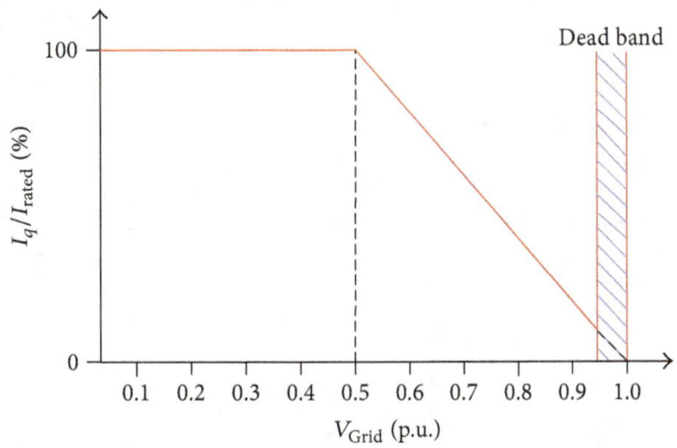


Figure 11: German and Danish grid code reactive current requirements during sags [10].

*(i) Stability.* The WECS must be able to ride out transient disturbances that result from short circuits and faults causing momentary power imbalances. This is because if WECS disconnect at such times, they worsen the situation with the loss of power production causing system instability. However, the generator response must go beyond just securing machine operation during the fault. Instead, the response must assist in maintaining and even improving the system stability. In Germany, the transmission operator E.oN has an LVRT requirement that, for voltages below 0.5 p.u, the wind turbine must supply, within 20 mS, 100% of its rated current as reactive current to help support the grid [40, 41]. This is shown in Figure 11.

*(j) Wind Farm Modelling and Verification.* Models and system data of the WECS are used to carry out simulations to investigate interactions between the wind farm and the grid. Monitoring equipment must also be set up to verify actual behaviour during faults.

*(k) Communication and External Control.* The WECS must provide certain parameters to the grid operator for proper operation of the power system. Remote connection and disconnection must also be possible.

*6.2. DFIG WECS Control Types and Levels.* The control and supervision of a wind turbine are automated so as to realize benefits such as reduced cost of operation, increased staff safety, and increased energy yield. It is handled by three different systems: supervision, control, and safety [4, 14, 19, 42].

*6.2.1. The Supervisory System.* The management or supervisory system provides up-to-date desired system values and reacts to medium and long-term variations in ranges from minutes to years. It carries out tasks such as monitoring wind and other loading conditions, sending information to the control system, monitoring operation for safe conditions, monitoring grid conditions, actuating emergency systems, and switching between operating states that facilitate connection and disconnection to the grid.

The supervisory system operates in two modes:

(i) Continuous fault monitoring: to facilitate preventive maintenance and planned system service

(ii) Emergency shutdown: for cases where there are high winds, catastrophic failure, and other emergencies

It also incorporates a failsafe system so that in case of power failure to either the controls, the control logic, or the sensors, the turbine can be safely brought to a stop or shutdown state.

Except in emergency situations, management system decisions are always transferred to the final component under control through the control system. The management system is thus hierarchically above the control system, acting only when necessary to direct the turbine operations both collectively in case of a wind farm and individually. For wind farms, the management system monitors the overall operation of the farm as well as that of specific machines and is called a supervisory control and data acquisition system (SCADA).

*6.2.2. The Control System.* The control system acts under the supervision of a management system. It deals with direct intervention measures on the turbine, the gearbox, the drive shaft, and the generator, collectively referred to as the power drive train, to create adjustments that result in the output meeting the set limits on a point by point basis. The control system operates with regard to functionality and integration of the turbine, taking into consideration the characteristics of each component and its reaction to external effects as well as adjustment efforts. It thus handles the internal conditions of the system by maintaining values specified for the turbine by the management system. It acts continuously as it has to react to fast changes in wind speed and generator load for as long as the system dynamics are within limits set by the management system. The control system operation is aimed at regulating the turbine power in high winds and optimizing it in low speed winds.

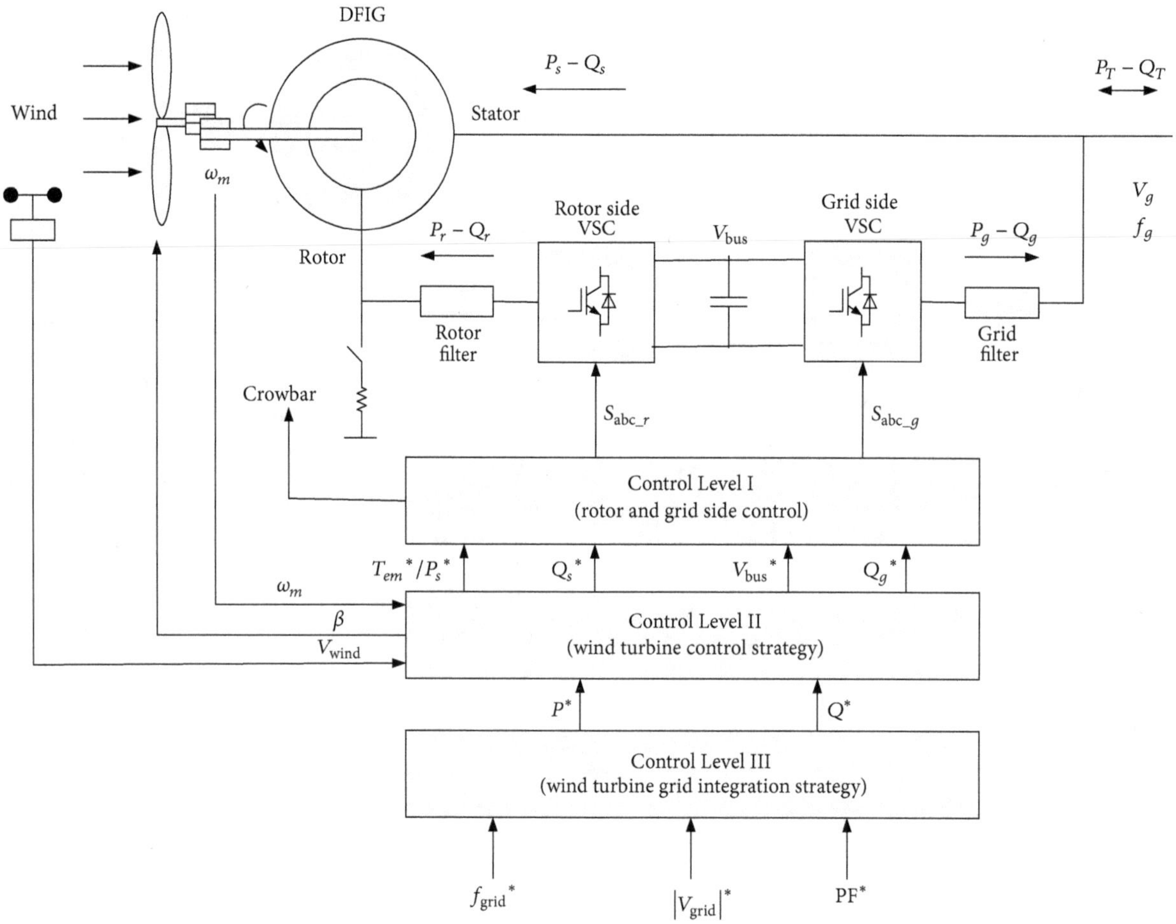


FIGURE 12: Conceptual diagram for DFIG wind turbine with partial scale power converter showing control strategy levels [14].

The five major processes to be controlled in a wind turbine are as follows:

(i) Development of aerodynamic torque: which is at the rotor blades and hence includes control of blade pitch and rotor orientation

(ii) Development of generator torque: by controlling mechanical inputs to the generator such as by change of shaft speed and controlling the air-gap rotating magnetic field

(iii) Conversion of electrical current into motion: which involves use of electrical signals to effect actions such as blade pitch angle change, rotor/yaw orientation control, or applying brakes (brake torque control)

(iv) Conversion of electrical power: the conversion of power generated to desired voltage, frequency, and power factor by use of PEC

(v) Conversion of wind energy to electrical power: which is the overall process, incorporating all the above 4 processes but, in addition, involving the management system inputs

*6.2.3. The Safety System.* Due to the high cost of WECS installations, as well as for grid system and personnel safety, a safety system that is independent of the main control system is essential. It functions to bring the turbine to a safe condition in the event of a serious problem by bringing the turbine to a stop or slow idling speed with the blades feathered and the generator switched off. It acts as a backup system when the main control system fails or when activated by an operator-controlled emergency stop button.

Control strategy for DFIG WECS is on three interlinked levels as shown in Figure 12 [25].

*(a) Level I.* This controls power flow from the DFIG, both from stator and from rotor, to the grid using the PEC. The converter control is divided into RSC control and GSC control [43].

(i) RSC control: RSC has decoupled control of stator active power (electromechanical torque) and the stator reactive power. It has a two-stage controller, oriented on either stator flux or grid/stator voltage reference frame. It has a faster inner current control loop and a slower outer power control loop.

(ii) GSC control: it aims at maintaining the DC-link capacitor voltage at a set value in addition to maintaining the desired converter power factor. It has a two-stage controller on grid/stator voltage oriented reference frame where the $d$-axis current component controls the active power and hence the DC-link voltage on the outer slower loop while the $q$-axis current component controls the reactive power on the faster inner loop.

*(b) Level II.* This handles control of conversion of wind energy in the airflows into mechanical energy by the turbine rotor, always aiming at maximum power production except when a reserve capacity is ordered by Level III control. It receives its set-points from Level III control and sends control set-points to Level I control.

*(c) Level III.* This is the highest level of control as it controls integration with the grid and provides control related to the grid voltage, power factor, and frequency as well as responding to grid operator requirements in terms of active and reactive power demands. It is similar to the wind farm control system but deals only with individual WECS values.

*6.3. DFIG WECS Control Strategies.* Many control strategies are proposed and they include the following.

*6.3.1. Vector Oriented Control (VOC).* This has the VSC being either voltage controlled (VC-VSC) or current controlled (CC-VSC) and being based on either of the following reference frames.

(i) Grid/stator voltage oriented frame [44]: in stator voltage oriented (SVO) control, the $d$-axis of the synchronous frame is fixed on the stator voltage vector. Thus, $V_s = V_{ds}$ and $V_{qs} = 0$. Stator active and reactive powers are derived as

$$P_s = v_{ds} i_{ds} = \frac{v_{ds}(\varphi_{ds} - L_m i_{dr})}{L_s}$$
$$Q_s = -v_{ds} i_{qs} = -\frac{v_{ds}(\varphi_{qs} - L_m i_{qr})}{L_s}. \quad (8)$$

Hence, the active power is controlled by adjusting $i_{dr}$ and reactive power by adjusting $i_{qr}$, respectively. This is possible through control of $V_{dr}$ and $V_{qr}$. In the GSC, the $d$-axis current controls the active power and so indirectly the DC-link voltage, whereas the $q$-axis current controls the reactive power.

(ii) Stator flux oriented frame [11, 45]: in stator flux orientation, the reference frame rotates synchronously with respect to the stator flux such that the reference frame has its $d$-axis simultaneously overlapping the axis of the stator winding flux. Hence, $\omega = \omega_s$ and $\varphi_{qs} = 0$. The dynamic equations then change to

$$v_{ds} = R_s i_{ds} + \frac{d\varphi_{ds}}{dt}$$
$$v_{qs} = R_s i_{qs} + \omega_s \varphi_{ds} \quad (9)$$
$$\varphi_{ds} = L_s i_{ds} + L_m i_{dr}$$
$$\varphi_{qs} = 0 = L_s i_{qs} + L_m i_{qr}.$$

The electromagnetic torque and stator active power are obtained as below where $p$ is number of poles:

$$T_e = \frac{3}{2}\frac{p}{2}\varphi_{ds} i_{qr}$$
$$P_s = \frac{3}{2}\frac{p}{2}\omega_s \varphi_{ds} i_{qr}. \quad (10)$$

But DFIG stator flux is restricted by the constant magnitude and frequency of the stator voltage; hence, torque control and thus active power can be achieved by controlling $i_{qr}$. Also, with stator flux remaining unchanged, reactive power is derived as

$$Q_s = \frac{3}{2}\frac{p}{2}\omega_s \varphi_{ds} i_{ds}. \quad (11)$$

Regardless of magnitude and direction of rotor power, the GSC keeps the DC-link voltage constant. The DC-link voltage balance equation is thus

$$C V_{dc} \frac{dV_{dc}}{dt} = P_r - P_g, \quad (12)$$

where $C$ is the DC-link capacitance (F), $P_r$ is active power from rotor into RSC (W), and $P_g$ is active power flow into the grid from the GSC (W). The capacitor current is

$$i_{dc} = C\frac{dV_{dc}}{dt} = \frac{3}{4} m i_{gcd} - i_{dcr}, \quad (13)$$

where $i_{gcd}$ is $d$-axis current flowing between the grid and the GSC (A), $i_{dcr}$ is the rotor side DC current (A), and $m$ is the pulse width modulation index of GSC. The reactive power flow into the grid from the GSC is given by

$$Q_q = \frac{3}{2} V_q i_{gcq}, \quad (14)$$

where $V_g$ is magnitude of grid phase voltage (V) and $i_{gcq}$ is $q$-axis current of GSC (A). As such, DC-link voltage can be controlled by adjusting $i_{gcd}$ while reactive power from GSC is controlled by adjusting $i_{gcq}$.

(iii) Combination where the RSC is under stator flux orientation while the GSC is under stator voltage oriented frame [46]

(iv) Rotor flux oriented reference frame [47, 48]

*6.3.2. Direct Torque Control (DTC) [49].* Since the electromagnetic torque is related to the stator flux, the converter switch state is obtained from a look-up table based on the position of stator flux vector. It has a simple control structure and good dynamic response and greatly reduces machine torque oscillations.

*6.3.3. Direct Power Control (DPC).* The error between calculated instantaneous active and reactive power and their estimated values is used to select, from a switching table, the converter switching state. This has a simple control structure, good dynamic response, and improved robustness to machine parameter variations. It however requires a high frequency sampling for satisfactory performance and has variable switching frequency. The variable switching frequency makes design of appropriate filters difficult. It can be grid/stator voltage oriented or stator flux oriented [31, 50, 51].

It is noted that VOC is achieved using proportional-integral (PI) controllers while proportional-integral plus resonant (PI + R) controllers give improved performance. VOC relies heavily on machine parameters and hence is not robust to machine variations. VOC introduces lower THD with satisfactory steady state performance but the direct methods are faster in dynamic response. DPC however has high power losses. Also, estimation of stator voltage space vector is easier than that of stator flux space vector [25, 52].

*6.3.4. Sliding Mode Control (SMC) [53].* SMC is a nonlinear control technique derived from variable structure control. It includes several different continuous functions for mapping plant state to a control surface. The switching between the different functions is determined by plant states represented by a switching function and so can change at any time during system operation unlike in feedback control.

The steps of SMC design are as follows:

(i) Selecting a sliding surface that models the desired closed loop performance

(ii) Designing a control law such that the system state trajectory is forced towards the sliding surface [54]

(iii) Designing a boundary layer of the switching function such that the control of the state variables is as follows:

  (a) Outside of the boundary layer: being constrained to keep discontinuous switching

  (b) Inside of boundary layer: the locus of state variables being restricted by a linear constraint

The thicker the boundary layer, the better the control. This however increases the control gain and reduces the inhibitory effect on chattering. It could also result in steady state error [55].

For a nonlinear system [37],

$$\dot{X}(t) = f(x,t) + B(x,t) * U(x,t), \quad (15)$$

where $X$ is the state vector and $U$ is the control vector.

The control law can then be defined as

$$U_{dq} = U_{\text{eq}} + U_n, \quad (16)$$

where the Signum function is defined by $U_n = -K_f * \text{sign}(S(x,t))$ given that $U_{dq}$ is the control vector, $U_{\text{eq}}$ is the equivalent control vector which is valid only in the sliding surface, $U_n$ is the switching part of control (the correction vector), and $K_f$ is the controller gain.

SMC is superior to PI controllers used in VOC in that it [53, 54, 56, 57]

(i) improves performance of the system against unmodeled dynamics,

(ii) reduces overshoot and transient time,

(iii) is robust to external disturbances and machine parameter changes due to generator temperature rise, magnetic saturation, and skin effect, among others; this is because the control signals only depend on the selected switching variables states,

(iv) has simple implementation as it does not require the decomposition of the voltage and current sequences in the VSC or use of a phase locked loop (PLL) to obtain the grid voltage positive sequence; this avoids time consuming computations as well as the challenges of the PLL losing synchronism with system frequency following a fault [40, 58, 59],

(v) has better reference tracking.

SMC has a problem with chattering, the high frequency oscillation of system state trajectory around the sliding surface. Chattering results in undesirable mechanical stresses [60]. This is caused by idealized modelling of the system, time delay of actual switching, and the discretization of digital control [55]. It can be reduced by using a finite gain $K_f$ with a small boundary [54].

There are many different subtypes of SMC which aim at reducing the shortcomings of SMC while amplifying its advantages to various degrees of success. These include discrete time SMC, terminal SMC, back stepping terminal SMC, integral terminal SMC, nonsingular SMC, fast terminal SMC, and super twisting SMC.

Various combinations of the four control strategies above are also utilized with the aim of taking advantage of the strong points of each of the control strategies in the combination while minimizing their shortcomings. Other control strategies for the whole converter or just either the RSC or GSC are proposed for use in conjunction with the methods stated above. These include the following.

(a) Fuzzy logic control (FLC): it is used for control of nonlinear, uncertain, and adaptive systems with parameter variation. It does not strictly require a mathematical model of the plant. Its control rule is qualitatively expressed on the basis of logic-language variation [56]. A fuzzy logic controller can be used to measure power and rotational speed and then optimally increment the operating speed to attain the required power output change. This avoids use of a wind speed sensor with its associated cost and implementation difficulty. This allows optimal rotor speed reference tracking [60]. FLC can be used in conjunction with the SMC to generate the value of the gain $K_f$

(b) Predictive direct current control (PDCC) [52]

(c) Internal model control (IMC) [61]

(d) Artificial neural networks (ANN) [57]

(e) Genetic algorithm (GA) [62]

(f) Particle swarm optimization (PSO) [63]

TABLE 1: DFIG WECS internal faults, their effects, and proposed solutions.

| Component affected | Effect | Causes | Risk | Solutions |
|---|---|---|---|---|
| Turbine | Blade fracture | Excessive loading from wind gusts or icing | Blade pieces and dislodged ice flying off risking damage to other turbines and personnel | Emergency shutdown in high gust or ice situations<br>Deicing systems incorporated in rotor blades |
| | Lightning strike | Strikes by lightning | Fire | Lightning earthing system incorporated in rotor blade |
| | Hydraulic failure | Control system malfunction, environmental degradation | Loss of timely control<br>Fire from leaking oil | Timely preventive maintenance |
| Gearbox | Gearbox failure | Insufficient lubrication | Worn or fractured gears,<br>fire | Use of sufficient and correct high quality lubricants<br>Sufficient lubricant cooling system |
| Generator | Windings open circuit,<br>winding insulation failure | High operating temperatures,<br>transient voltages and currents | Reduced or stoppage of power production | Regulation of voltages and currents to ensure stable operating temperatures |
| PEC | Semiconductor device failure<br>Failure of bond wire connection | Thermal cycling | Output power fluctuations | Control that reduces thermal variations |
| | DC-link capacitor failures | Voltage fluctuations, harmonic currents | Reduced output power,<br>increased fluctuations in output currents | Control that reduces voltage variations across the capacitor |

## 7. Faults

The main challenge in operation of the DFIG WECS is continued operation under fault conditions. Faults could originate internally from the machine or externally from the grid. Table 1 gives a summary of the internal faults, their effects, and solutions proposed [6, 7, 19, 64–73].

Over 50% of internal faults are attributed to the PEC with 30% being from the DC-link capacitor. While the gearbox accounts for only 5% of failures, they are usually more costly, at times requiring complete gearbox replacements, and hence result in more downtime.

For external faults, voltage swells cause high temperatures in the generator windings. This results in insulation degradation and eventual failure leading to inter-turn short circuits. High voltages can also interfere with the control system [13]. Voltage sag results in high stator currents which in turn induce high rotor currents and reduced capability of the WECS to export power as to the square of the voltage drop. The latter leads to generator speed increase and increase in DC-link voltage. The two approaches to deal with effects of system faults are given in Table 2 [3, 15, 21, 34, 46, 50, 74–90].

Some approaches utilize combinations of the solutions outlined above in a bid to cover the effects of severe faults. Examples include use of braking chopper with SRC [3] or with antiparallel thyristors [28]. Figure 13 shows application of a SMES for compensation.

## 8. Simulation

The DFIG WECS performance under different operating conditions is explored using simulations. Different computer tools are utilized and include power system computer aided design and electromagnetic transient including DC (PSCAD/EMTDC) [3], MATLAB/Simulink [91], power system simulation for engineering (PSS/E) [25], and electromagnetic transient program (EMTP-RV) [92]. Use of MATLAB-PSIM cosimulation [93] and DigSILENT power factory [94, 95] has also been reported in earlier works. Programs such as PSS/E are particularly suited for large power system analysis while PSCAD/EMTDC and EMTP-RV are applicable in large power system transient studies. The modelling is done to different levels of detail, with more detailed models taking more time to run. While high order models can be used, the 5th-order model is sufficient for detailed dynamic representation of the DFIG WECS while 3rd-order model can be obtained by neglecting stator flux transients. This limits its accuracy in modelling rotor current transient behaviour; however, the reduced complexity makes it best suited for large-scale wind farm simulations. The electromagnetic subsystem has very short time constants in comparison with the electromechanical subsystems which are modelled with the turbine and generator shaft masses, inertia, and damping either lumped (single-mass) or apart (two-mass). However, two-mass models are required for power system transient

TABLE 2: DFIG WECS proposed solutions to faults from grid system.

| Approach | Aim | Location of application | Method of application | Examples | Methodology |
|---|---|---|---|---|---|
| Control | Keep WECS within operation constraints without additional components or devices | Control system | Software | Virtual resistance/inductance | Increased resistance and inductance in virtual rotor circuit results in decreased rotor currents and their decay times |
| | | | | Supplementary rotor current (SRC) controller | Introduces a multiplication factor $k$ to the measured rotor currents effectively reducing the control variable to within acceptable limits |
| | | | | Flux damping/demagnetization | Rotor current controlled to counteract the stator flux negative sequence and DC components of flux |
| Hardware | Maintain WECS within operational limits by addition of components | Rotor | Shunt | Active crowbar, passive crowbar | High currents shunted through dissipating resistors |
| | | | Series | Series dynamic braking resistor (SDBR) | High currents passed through dissipating resistors |
| | | DC-link | Shunt | Braking chopper | Charging currents passed through dissipating resistors |
| | | | | Energy storage systems (ESS), for example, battery (BESS), energy capacitor storage (ECS), superconducting magnetic energy storage (SMES) | Excess energy stored for later release during system deficit |
| | | Stator | Series | SDBR | Added resistance reduces fault currents to within limits |
| | | | | Series antiparallel thyristors | Provide quick disconnection of stator circuit to interrupt stator flux oscillations |
| | | | | Dynamic voltage restorer (DVR), series grid side converter (SGSC) | Maintain stator terminal voltage at nominal values by injecting voltage |
| | | PCC | Series | Flexible AC transmissions systems (FACTS) devices such as static synchronous series compensator (SSSC), DVR | Maintain voltage at PCC nominal values by injecting voltage |
| | | | Shunt | FACTS devices such as static synchronous compensator (STATCOM), static Var compensator (SVC), SMES | Maintain voltage at PCC nominal values by injecting reactive power during sags. SMES also draws reactive power during swells |
| | | | Hybrid | FACTS devices such as unified power flow compensator (UPFC) | Combined action of series and shunt connected FACTs devices as given above |

studies [96]. While individual machine performance is of interest, grid operators are more interested in the overall performance of the wind farm and make this a prerequisite for connection.

Figure 14 shows the MATLAB/Simulink model of a 9 MW wind farm composed of 6 number 1.5 MW DFIG machines, without FRT capability, connected through a 25 kV feeder to a grid at 120 kV [16]. Each machine has a turbine rated at 1.5 MW mechanical power with turbine power characteristics as depicted in Figure 5. The generators each have a nominal power rating of 1.5 MVA at 60 Hz and utilize power converters rated at 0.5 p.u. with a DC-link composed of a 10,000 pF capacitor at 1200 kV. Figure 15 presents simulation results for normal operation with a step change increase in wind speed at 5 s where the blades are pitched at 20 s once rated power is achieved. In Figure 16, waveforms for case with a three-phase to ground fault applied at 25 s and removed after 150 ms are given. It is seen that the machines are disconnected due to overcurrents which go up to 1.25 p.u. In addition, the speed increases to 1.52 p.u., forcing further pitching of

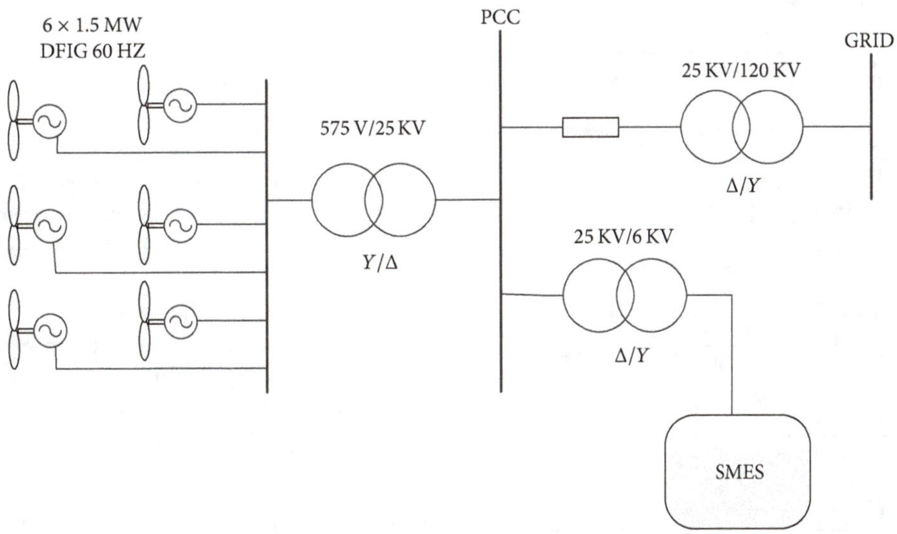


FIGURE 13: Single-line diagram showing SMES connection for reactive power compensation [15].

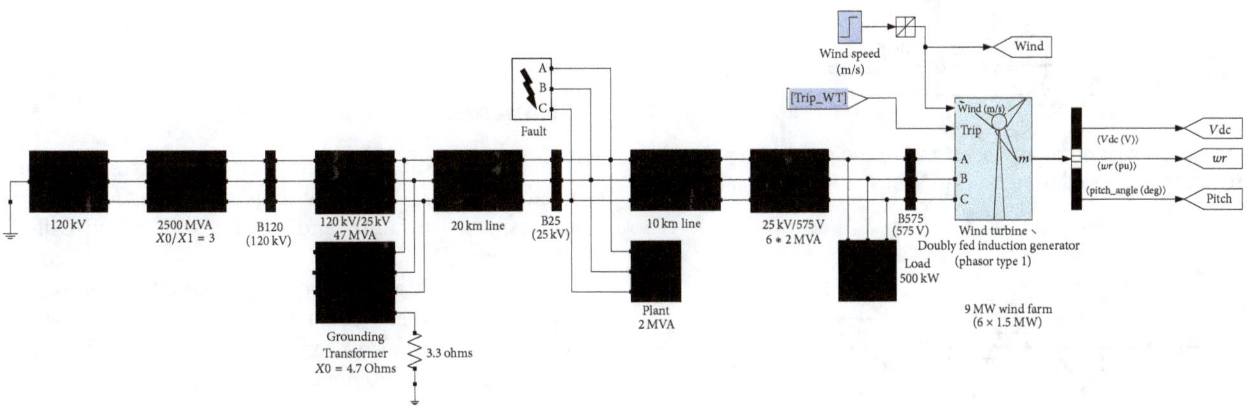


FIGURE 14: MATLAB/Simulink model of a 9 MW DFIG WECS wind farm [16].

the blades to reduce it. Thus, the need for FRT capability is clearly demonstrated. The simulations were done on a desktop computer running MATLAB 2010b.

## 9. Discussion

The DFIG WECS must meet the grid connection requirements which are only expected to become stricter as grid systems evolve further. For cost effectiveness, the DFIG WECS must operate at full efficiency during both steady and transient states while producing as much power as possible under the prevailing wind and grid conditions. As such, the choice of an appropriate control strategy is critical.

*9.1. The Ideal Control Strategy.* With the many proposed control strategies, and taking into consideration other aspects of the DFIG WECS such as operation with both MPPT and curtailed production, as well as operation during fault conditions, the qualities sought in deciding on an optimal solution include the following:

(i) Independent of machine parameters: machine parameter variations cannot be completely avoided and occur from environmental changes such as seasonal air density changes, from component aging, or from change in operating conditions such as loading, steady state, or transient operation. A control system that does not rely on preset values such as a look-up table but instead relies on instantaneous values or has a learning ability can self-adjust for these variations giving a more accurate response.

(ii) Low component count: additional hardware and components introduce higher costs, increased space requirements, increased weight, increased control complexity and design, reduced system reliability as there are more components that can fail, increased maintenance, and challenges and limitations inherent in such components such as requirements for heat dissipation and introduction of delays and measurement errors [51].

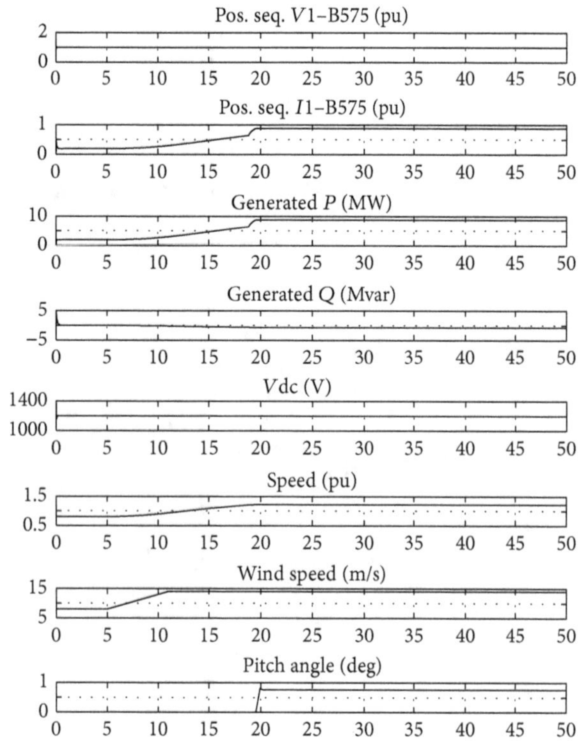


Figure 15: Simulation results for normal operation with step change increase in wind speed from 8 m/s to 14 m/s at 5-second mark.

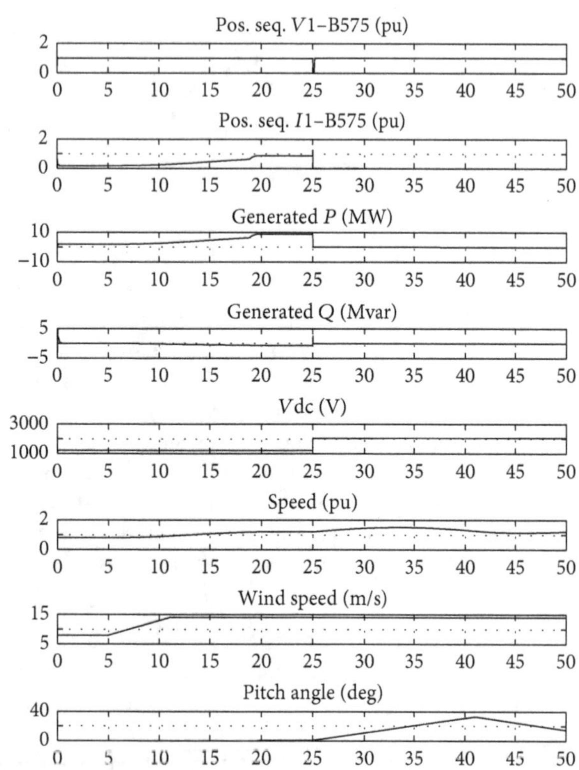


Figure 16: Simulation results for fault condition with a step change increase in wind speed from 8 m/s to 14 m/s at 5-second mark followed by a 150 ms duration three phases to ground fault at 25-second mark.

(iii) Good performance under both steady state and transient/fault states: this is in terms of accuracy of reference tracking, stability, and speed of response.

(iv) Reduced computation burden: this includes reduction or complete elimination of sequence decomposition and reference frame transformations. This reduces computing power and memory requirements and as such costs. It also reduces time delays and errors [97].

(v) Simplicity of implementation: the control system values such as resistor and gain values should be easily and quickly obtained. The system should also not introduce other problems that will require suppression or elimination.

Among the reviewed proposed control systems, the SMC strategy, especially in combination with FLC as proposed in [56], seems the most attractive choice as it best meets the qualities given above.

## 10. Conclusion

The DFIG WECS is presented from its construction, operation, advantages and shortcomings, modelling, control solutions, response to faults, and, finally, simulation. Different control approaches used to date have been explored with various control strategies being presented. Possible faults expected in the operation of the DFIG WECS and ways of tackling them have also been presented. Sample simulation results have been presented underlining the need for FRT capability. The paper has brought out issues to consider for an optimal control strategy which must integrate FRT capabilities so as to satisfy the ever stringent grid connection codes as well as taking into consideration the cost implications of its implementation.

## References

[1] World Wind Energy Association, 2016, http://www.wwindea.org/wwea-half-year-report-worldwind-wind-capacity-reached-456-gw.

[2] T. Ackerman, *Wind Power in Power Systems*, John Wiley and Sons Ltd, England, UK, 2005.

[3] K. E. Okedu and U. Roland, "Low voltage ride through methods for DFIG VSWT," *Journal of Emerging Trends in Engineering and Applied Sciences*, vol. 5, no. 8, pp. 221–226, 2014.

[4] D. A. Rivkin, L. D. Anderson, and L. Silk, *Wind Turbine Control Systems*, Jones and Bartlett Learning, Massachusetts, Mass, USA, 2013.

[5] Vestas Wind Systems A/S, 2014, https://www.vestas.com/en/products/turbines.

[6] A. Bonanomi, "Powerful analysis of wind turbine gearboxes," *Power Transmission World*, 2014, http://www.powertransmission-world.com/powerful-analysis-of-wind-turbine-gearboxes/.

[7] F. Blaabjerg, M. Liserre, and K. Ma, "Power electronics converters for wind turbine systems," *IEEE Transactions on Industry Applications*, vol. 48, no. 2, pp. 708–719, 2012.

[8] J. Zhang, G. Chen, and X. Cai, "Thermal smooth control for Multi-MW parallel wind power converter," in *Proceedings of the 2013 IEEE International Conference of IEEE Region 10, IEEE TENCON 2013*, pp. 1–4, October 2013.

[9] T. Lei, M. Barnes, and A. C. Smith, "Thermal cycling evaluation for DFIG wind turbine power converter based on joint modelling," in *Proceedings of the 5th Annual IEEE Energy Conversion Congress and Exhibition, ECCE 2013*, pp. 3845–3851, usa, September 2013.

[10] Y. M. Alsmadi, L. Xu, F. Blaabjerg, A. P. Ortega, and A. Wang, "Comprehensive analysis of the dynamic behavior of grid-connected DFIG-based wind turbines under LVRT conditions," in *Proceedings of the 7th Annual IEEE Energy Conversion Congress and Exposition, ECCE 2015*, pp. 4178–4187, can, September 2015.

[11] S. Sharma, J. P. Mishra, and S. Datta, "Sliding mode power control of a DFIG based variable speed wind energy conversion system," in *Proceedings of the 12th IEEE International Conference Electronics, Energy, Environment, Communication, Computer, Control, INDICON '15*, December 2015.

[12] "Power quality basics: voltage sags or dips," Power Quality in Electrical Systems, 2011, http://www.powerqualityworld.com/2011/03/voltage-sags-dips-power-quality-basics.html.

[13] "Power quality basics: voltage swell," *Power Quality in Electrical Systems*, 2011, http://www.powerqualityworld.com/2011/04/voltage-swell-power-quality-basics.html.

[14] G. Abad, J. Lopez, M. Rodriguez, L. Marroyo, and G. Iwanski, *Doubly Fed Induction Machine: Modelling and Control for Wind Energy Generation*, John Wiley and Sons, New Jersey, NJ, USA, 2011.

[15] A. M. S. Yunus, M. A. S. Masoum, and A. Abu-Siada, "Application of SMES to enhance the dynamic performance of DFIG during voltage sag and swell," *IEEE Transactions on Applied Superconductivity*, vol. 22, no. 4, 2012.

[16] The MathWorks, Inc., MATLAB© 1984-2010, http://www.mathworks.com.

[17] Gamesa Corporacion Tecnologica, S.A., 2014, http://www.gamesacorp.com/en/products-and-services/wind-turbines/.

[18] Siemens Wind Power GmbH, 2015, http://www.siemens.com/global/en/home/markets/wind/turbines/technology/features.html.

[19] S. Heir, *Grid Integration of Wind Energy Conversion Systems*, John Wiley and Sons Ltd, England, UK, 2nd edition, 2006.

[20] M. Rahimi and M. Parniani, "Low voltage ride-through capability improvement of DFIG-based wind turbines under unbalanced voltage dips," *International Journal of Electrical Power and Energy Systems*, vol. 60, pp. 82–95, 2014.

[21] Ingeteam, "Power converter training manual," 2014, http://www.ingeteam.com.

[22] D. Zhou, F. Blaabjerg, M. Lau, and M. Tonnes, "Thermal behavior optimization in multi-MW wind power converter by reactive power circulation," *IEEE Transactions on Industry Applications*, vol. 50, no. 1, pp. 433–440, 2014.

[23] H. L. Hansen, L. Helle, F. Blaabjerg, E. Ritchie, S. Munk-Nielsen, H. Bindner et al., *Conceptual survey of generators and power electronics for wind turbines*, RISO National Laboratory, Roskilde, Denmark, 2001.

[24] M. Boutoubat, L. Mokrani, and M. Machmoum, "Control of a wind energy conversion system equipped by a DFIG for active power generation and power quality improvement," *Renewable Energy*, vol. 50, pp. 378–386, 2013.

[25] A. Gupta, S. N. Singh, and D. K. Khatod, "Modeling and simulation of doubly fed induction generator coupled with wind turbine-An overview," *Journal of Engineering, Computers & Applied Sciences*, vol. 2, no. 8, 2013.

[26] K. E. Okedu, "Applications of wind power generation in grid connected power system," *The Pacific Journal of Science and Technology*, vol. 13, no. 2, 2012.

[27] T. Burton, N. Jenkins, D. Sharpe, and E. Bossanyi, *Wind Energy Handbook*, John Wiley and Sons Ltd, England, UK, 2nd edition, 2011.

[28] A. Petersson, *Analysis, modeling and control of doubly-fed induction generators for wind turbines [Ph.D. thesis]*, Chalmers University of Technology, Göteborg, Sweden, 2005.

[29] K. E. Okedu, R. Takahashi, J. Tamura, and S. M. Muyeen, "Comparative study on current and voltage controlled voltage source converter based variable speed wind generator," in *Proceedings of the 2011 2nd International Conference on Electric Power and Energy Conversion Systems, EPECS '11*, November 2011.

[30] J. J. Justo, F. Mwasilu, and J.-W. Jung, "Doubly-fed induction generator based wind turbines: a comprehensive review of fault ride-through strategies," *Renewable and Sustainable Energy Reviews*, vol. 45, pp. 447–467, 2015.

[31] R. Teodorescu, M. Liserre, and P. Rodriquez, *Grid Converters for Photovoltaic and Wind Power systems*, John Wiley and Sons Ltd, England, UK, 2011.

[32] D. A. Rivkin, M. Randall, and L. Silk, *Wind Power Generation and Distribution*, Jones and Bartlett Learning, Massachusetts, Mass, USA, 2013.

[33] M. Ali, *Wind Energy Systems Solutions for Power Quality and Stabilization*, CRC Press, Boca Raton, Fla, USA, 2012.

[34] O. P. Mahela and A. G. Shaik, "Comprehensive overview of grid interfaced wind energy generation systems," *Renewable and Sustainable Energy Reviews*, vol. 57, pp. 260–281, 2016.

[35] R. Prasad, *Low voltage ride-through capability for matrix converters fed adjustable-speed induction machine drives for industrial and wind appliances [Ph.D. thesis]*, University of Minnesota, 2011.

[36] M. A. Abdullah, A. H. M. Yatim, C. W. Tan, and R. Saidur, "A review of maximum power point tracking algorithms for wind energy systems," *Renewable and Sustainable Energy Reviews*, vol. 16, no. 5, pp. 3220–3227, 2012.

[37] N. Cherfia, D. Kerdoun, and A. Boumassata, "Sliding mode control indirect strategy of the active and reactive power for the wind turbine DFIG," in *Proceedings of the Conference Internationale des Energies Renouvelables (CIER '13)*, Sousse, Tunisie, 2013.

[38] Z. Zheng, G. Yang, and H. Geng, "High voltage ride-through control strategy of grid-side converter for DFIG-based WECS," in *Proceedings of the 39th Annual Conference of the IEEE Industrial Electronics Society (IECON '13)*, pp. 5282–5287, IEEE, Vienna, Austria, November 2013.

[39] C. Feltes, S. Engelhardt, J. Kretschmann, J. Fortmann, F. Koch, and I. Erlich, "High voltage ride-through of DFIG-based wind turbines," in *Proceedings of the IEEE Power and Energy Society General Meeting—Conversion and Delivery of Electrical Energy in the 21st Century*, pp. 1–8, IEEE, Pittsburgh, Pa, USA, July 2008.

[40] J. Morneau, *A comparative evaluation of low voltage ride through solutions in wind turbines using doubly fed induction generators [M. S. thesis]*, McGill University, Montreal, Canada, 2008.

[41] M. M. Baggu, *Advanced control techniques for DFIG-based wind turbine converters to improve low voltage ride-through during system imbalances [Ph.D. thesis]*, Missouri University of Science and Technology, Missouri, Mo, USA, 2009.

[42] D. A. Rivkin, A. Liddell, and L. Silk, *Wind Turbine Systems*, Jones and Bartlett Learning, Massachusetts, Mass, USA, 2013.

[43] K. E. Okedu, S. M. Muyeen, R. Takahashi, and J. Tamura, "Comparative study of wind farm stabilization using variable speed generator and FACTS device," in *Proceedings of the 2011 IEEE GCC Conference and Exhibition, GCC '11*, pp. 569–572, February 2011.

[44] L. Shuhui and T. A. Haskew, "Analysis of decoupled d-q vector control in DFIG back-to-back PWM converter," in *Proceedings of the 2007 IEEE Power Engineering Society General Meeting*, pp. 1–7, June 2007.

[45] M. Mohseni, S. M. Islam, and M. A. S. Masoum, "Impacts of symmetrical and asymmetrical voltage sags on DFIG-based wind turbines considering phase-angle jump, voltage recovery, and sag parameters," *IEEE Transactions on Power Electronics*, vol. 26, no. 5, pp. 1587–1598, 2011.

[46] W. Qiao, G. K. Venayagamoorthy, and R. G. Harley, "Real-time implementation of a STATCOM on a wind farm equipped with doubly fed induction generators," *IEEE Transactions on Industry Applications*, vol. 45, no. 1, pp. 98–107, 2009.

[47] T. M. Masaud and P. K. Sen, "Modeling and control of doubly fed induction generator for wind power," in *Proceedings of the 43rd North American Power Symposium, NAPS '11*, 8, 1 pages, August 2011.

[48] J. Fletcher and J. Yang, "Introduction to the doubly-fed induction generator for wind power applications," in *Paths to Sustainable Energy*, A. Ng, Ed., InTech, 2010, http://www.intechopen.com/books/paths-to-sustainable-energy/introduction-to-the-doubly-fed-inductiongenerator-for-wind-power-applications.

[49] S. Y. Rao and J. A. Laxmi, "Direct torque control of DFIG based wind turbine under voltage dips," *International Journal of Advances in Engineering and Technology (IJAET)*, vol. 3, no. 2, pp. 711–720, 2012.

[50] Q. Liu, L. Yu, and G. Wu, "Comparison of control strategy for double-fed induction generator (DFIG)," in *Proceedings of the 3rd International Conference on Measuring Technology and Mechatronics Automation, ICMTMA'11*, pp. 741–744, January 2011.

[51] B. Rückert and W. Hofmann, "Commutation strategy for minimized common mode voltage using direct power control," in *Proceedings of the 27th Annual IEEE Applied Power Electronics Conference and Exposition, APEC '12*, pp. 106–112, February 2012.

[52] M. E. Zarei and B. Asaei, "A simple predictive direct current control for grid side converter of DFIG under ideal and non-ideal supply voltage," in *Proceedings of the 2013 13th International Conference on Environment and Electrical Engineering, EEEIC '13*, pp. 278–283, November 2013.

[53] N. H. Saad, A. A. Sattar, and A. E.-A. M. Mansour, "Low voltage ride through of doubly-fed induction generator connected to the grid using sliding mode control strategy," *Renewable Energy*, vol. 80, pp. 583–594, 2015.

[54] A. M. Kassem, K. M. Hasaneen, and A. M. Yousef, "Dynamic modeling and robust power control of DFIG driven by wind turbine at infinite grid," *International Journal of Electrical Power and Energy Systems*, vol. 44, no. 1, pp. 375–382, 2013.

[55] Y. Wang, D. Zhao, B. Zhao, and H. Xu, "A new type of control strategy of DFIG in wind power system based on SMC," in *Proceedings of the 11th International Conference on Electrical Machines and Systems, ICEMS '08*, pp. 2428–2431, October 2008.

[56] D. Kairous and R. Wamkeue, "DFIG-based fuzzy sliding-mode control of WECS with a flywheel energy storage," *Electric Power Systems Research*, vol. 93, pp. 16–23, 2012.

[57] K. Belgacem, A. Mezouar, and A. Massoum, "Sliding mode control of a doubly-fed induction generator for wind energy conversion," *International Journal of Energy Engineering*, vol. 3, no. 1, pp. 30–36, 2013.

[58] M. I. Martinez, G. Tapia, A. Susperregui, and H. Camblong, "Sliding-mode control for DFIG rotor- and grid-side converters under unbalanced and harmonically distorted grid voltage," *IEEE Transactions on Energy Conversion*, vol. 27, no. 2, pp. 328–339, 2012.

[59] Y. Zou, *Modeling, control and maximum power point tracking (MPPT) of doubly-fed induction generator (DFIG) wind power system [Ph.D. thesis]*, The University of Akron, Ohio, USA, 2012.

[60] S. Abdeddaim and A. Betka, "Optimal tracking and robust power control of the DFIG wind turbine," *International Journal of Electrical Power and Energy Systems*, vol. 49, no. 1, pp. 234–242, 2013.

[61] D. Campos-Gaona, E. L. Moreno-Goytia, and O. Anaya-Lara, "Fault ride-through improvement of DFIG-WT by integrating a two-degrees-of-freedom internal model control," *IEEE Transactions on Industrial Electronics*, vol. 60, no. 3, pp. 1133–1145, 2013.

[62] J. P. A. Vieira, M. V. A. Nunes, and U. H. Bezerra, "Using genetic algorithm to obtain optimal controllers for the DFIG converters to enhance power system operational security," in *Proceedings of the 2010 IREP Symposium - Bulk Power System Dynamics and Control—VIII, IREP '10*, August 2010.

[63] M. T. Hagh, S. Roozbehani, F. Najaty, S. Ghaemi, Y. Tan, and K. M. Muttaqi, "Direct power control of DFIG based wind turbine based on wind speed estimation and particle swarm optimization," in *Proceedings of the 25th Australasian Universities Power Engineering Conference, AUPEC '15*, pp. 1–6, September 2015.

[64] J. Puigcorbe and A. De-Beaumont, "Wind turbine gearbox reliability: the impact of rotor support," 2010, http://www.renewableenergyworld.com/articles/print/volume-13/issue-3/wind-power/wind-turbine-gearbox-reliability.html.

[65] A. M. Ragheb and M. Ragheb, *Wind Turbine Gearbox Technologies*, 2011, http://www.intechopen.com/books/fundamental-and-advancedtopics-in-wind-power.

[66] "Wind Power Engineering," http://www.windpowerengineering.com/design/mechanical/gearboxes/gears-gearboxes-101/.

[67] W. Qiao and R. G. Harley, "Improved control of DFIG wind turbines for operation with unbalanced network voltages," in *Proceedings of the 2008 IEEE Industry Applications Society Annual Meeting, IAS '08*, pp. 1–7, October 2008.

[68] S. Karimi, A. Gaillard, P. Poure, and S. Saadate, "FPGA-based real-time power converter failure diagnosis for wind energy conversion systems," *IEEE Transactions on Industrial Electronics*, vol. 55, no. 12, pp. 4299–4308, 2008.

[69] M. Shahbazi, M. R. Zolghadri, P. Poure, and S. Saadate, "Wind energy conversion system based on DFIG with open switch fault tolerant six-legs AC_DC_AC converter," in *Proceedings of the 2013 IEEE International Conference on Industrial Technology, ICIT '13*, pp. 1656–1661, February 2013.

[70] A. F. Abdou, A. Abu-Siada, and H. R. Pota, "Application of STATCOM to improve the LVRT of DFIG during RSC fire-through fault," in *Proceedings of the 2012 22nd Australasian Universities Power Engineering Conference: "Green Smart Grid Systems", AUPEC '12*, pp. 1-6, September 2012.

[71] E. Giraldo and A. Garces, "An adaptive control strategy for a wind energy conversion system based on PWM-CSC and PMSG," *IEEE Transactions on Power Systems*, vol. 29, no. 3, pp. 1446–1453, 2014.

[72] C. Liu, D. Xu, N. Zhu, F. Blaabjerg, and M. Chen, "DC-voltage fluctuation elimination through a DC-capacitor current control for DFIG converters under unbalanced grid voltage conditions," *IEEE Transactions on Power Electronics*, vol. 28, no. 7, pp. 3206–3218, 2013.

[73] A. M. S. Yunus, A. Abu-Siada, and M. A. S. Masoum, "Impact of DC-link fault on the dynamic performance of DFIG," in *Proceedings of the 2012 IEEE Power and Energy Society General Meeting, PES '12*, pp. 1–6, July 2012.

[74] S. Hu, X. Lin, Y. Kang, and X. Zou, "An improved low-voltage ride-through control strategy of doubly fed induction generator during grid faults," *IEEE Transactions on Power Electronics*, vol. 26, no. 12, pp. 3653–3665, 2011.

[75] D. Zhu, X. Zou, Y. Kang, L. Deng, and Q. Huang, "Inductance-simulating control for DFIG-based wind turbine to ride-through grid faults," in *Proceedings of the 31st Annual IEEE Applied Power Electronics Conference and Exposition, APEC '16*, pp. 3521–3525, March 2016.

[76] K. E. Okedu, S. M. Muyeen, R. Takahashi, and J. Tamura, "Wind farms fault ride through using DFIG with new protection scheme," *IEEE Transactions on Sustainable Energy*, vol. 3, no. 2, pp. 242–254, 2012.

[77] D. Xiang, L. Ran, P. J. Tavner, and S. Yang, "Control of a doubly fed induction generator in a wind turbine during grid fault ride-through," *IEEE Transactions on Energy Conversion*, vol. 21, no. 3, pp. 652–662, 2006.

[78] K.E. Okedu, *Low Voltage Ride through Enhancement of Grid Connected Wind Farms: Augmentation of Variable Speed Wind Turbines Fault Ride through (FRT) Capability*, LAP Lambert Academic Publishing, Saarbrücken, Germany, 2012.

[79] V. Akhmatov, "Variable-speed wind turbines with doubly-fed induction generators part IV: uninterrupted operation features at grid faults with converter control coordination," *Wind Engineering*, vol. 27, no. 6, pp. 519–529, 2003.

[80] V. Akhmatov, "Variable-speed wind turbines with doubly-fed induction generators part III: model with the back-to-back converters," *Wind Engineering*, vol. 27, no. 2, pp. 79–91, 2003.

[81] K. E. Okedu, S. M. Muyeen, R. Takahashi, and J. Tamura, "Application of SDBR with DFIG to augment wind farm fault ride through," in *Proceedings of the 2011 International Conference on Electrical Machines and Systems, ICEMS '11*, August 2011.

[82] K. E. Okedu, S. M. Muyeen, R. Takahashi, and J. Tamura, "Effectiveness of current-controlled voltage source converter excited doubly fed induction generator for wind farm stabilization," *Electric Power Components and Systems*, vol. 40, no. 5, pp. 556–574, 2012.

[83] M. Nasiri, J. Milimonfared, and S. H. Fathi, "A review of low-voltage ride-through enhancement methods for permanent magnet synchronous generator based wind turbines," *Renewable and Sustainable Energy Reviews*, vol. 47, pp. 399–415, 2015.

[84] C. Laxmi, K. S. Latha, and Himani, "Improving the low voltage ride through capability of wind generator system using crowbar and battery energy storage system," *International Journal of Engineering Science Invention*, vol. 1, no. 7, pp. 14–19, 2013.

[85] A. O. Ibrahim, T. H. Nguyen, D.-C. Lee, and S.-C. Kim, "A fault ride-through technique of DFIG wind turbine systems using dynamic voltage restorers," *IEEE Transactions on Energy Conversion*, vol. 26, no. 3, pp. 871–882, 2011.

[86] A. Moghadasi, A. Sarwat, and J. M. Guerrero, "A comprehensive review of low-voltage-ride-through methods for fixed-speed wind power generators," *Renewable and Sustainable Energy Reviews*, vol. 55, pp. 823–839, 2016.

[87] J. Yao, H. Li, Z. Chen et al., "Enhanced control of a DFIG-based wind-power generation system with series grid-side converter under unbalanced grid voltage conditions," *IEEE Transactions on Power Electronics*, vol. 28, no. 7, pp. 3167–3181, 2013.

[88] K. E. Okedu, "A Study of wind farm stabilization using DFIG or STATCOM considering grid requirements," *Journal of Engineering Science and Technology Review*, vol. 3, no. 1, pp. 200–209, 2010.

[89] A. J. Prabhakar, *Application of STATCOM for improved dynamic performance of wind farms in a power grid [M. S. thesis]*, Missouri University of Science and Technology, Missouri, Mo, USA, 2008.

[90] K. R. Padiyar, *FACTS Controllers in Power Transmission and Distribution*, New Age International Publishers, New Delhi, India, 2007.

[91] M. A. Abomahdi and A. K. Bharadwaj, "Modeling and simulation of DFIG to grid connected wind power generation using MatLab," *International Journal of Electrical Engineering & Technology (IJEET)*, pp. 29–40, 2015.

[92] S. O. Faried, I. Unal, D. Rai, and J. Mahseredjian, "Utilizing DFIG-based wind farms for damping subsynchronous resonance in nearby turbine-generators," *IEEE Transactions on Power Systems*, vol. 28, no. 1, pp. 452–459, 2013.

[93] C. Alvarez, H. Amarís, O. Samuelsson, D. Flórez, and L. González, "Custom power systems and software platforms for wind farms under voltage dips situations," in *Proceedings of the 43rd International Universities Power Engineering Conference, UPEC '08*, pp. 1–4, September 2008.

[94] M. Kayikci and J. Milanovic, "DFIG modelling and the relevance of model simplification," in *Proceedings of the 18th International Conference and Exhibition on Electricity Distribution (CIRED '05)*, 2005.

[95] C. Hamon, K. Elkington, and M. Ghandhari, "Doubly-fed induction generator modeling and control in DigSilent PowerFactory," in *Proceedings of the 2010 International Conference on Power System Technology: Technological Innovations Making Power Grid Smarter, POWERCON '10*, October 2010.

[96] J. He, Q. Li, S. Qin, and R. Wang, "DFIG wind turbine modeling and validation for LVRT behavior," in *Proceedings of the 2012 IEEE Innovative Smart Grid Technologies—Asia, ISGT Asia*, pp. 1–5, May 2012.

[97] J. Kearney, M. F. Conlon, and E. Coyle, "The application of multi frequency resonant controllers in in a DFIG to improve performance by reducing unwanted power and torque pulsations and reducing current harmonics," in *Proceedings of the 2010 45th International Universities' Power Engineering Conference, UPEC '10*, September 2010.

# 7

# Experimental Study on Strain Reliability of Embroidered Passive UHF RFID Textile Tag Antennas and Interconnections

**Xiaochen Chen,[1] Aruhan Liu,[2] Zhigang Wei,[2] Leena Ukkonen,[1] and Johanna Virkki[1]**

[1]*Department of Electronics and Communications Engineering, Tampere University of Technology, P.O. Box 692, 33101 Tampere, Finland*
[2]*Department of Electronic Engineering, City University of Hong Kong, 83 Tat Chee Avenue, Kowloon, Hong Kong*

Correspondence should be addressed to Xiaochen Chen; xiaochen.chen@tut.fi

Academic Editor: Guy A. E. Vandenbosch

We present embroidered antennas and interconnections in passive UHF RFID textile tags and test their strain reliability. Firstly, we fabricate tag antennas on two different stretchable fabric substrates by five different embroidery patterns and choose the most stretchable ones for testing. Next, the tag ICs are attached by sewing and gluing, and the tag reliability during repeated stretching cycles is evaluated through wireless measurements. Initially, the chosen tags achieve read ranges of 6–8 meters and can strain up to 140–150% of their original length. After 100 stretching cycles to 80% of their maximum strain, the read ranges of the tags with glued interconnections are similar to the initial values. In addition, also the read ranges of the tags with sewed interconnections are still more than 70%–85% of their initial values. However, some challenges with the reproducibility need to be solved next.

## 1. Introduction

Passive UHF (ultra-high frequency) RFID (radio frequency identification) inspired technology has been recognized as a compelling approach to achieve versatile energy efficient wireless technologies for future WBANs (wireless body area networks) [1–4]. The development of WBAN technologies has gained a lot of research attention during the recent years as they can offer remarkable benefits for healthcare and welfare sectors [1–5], as well as novel sports-related applications [5, 6].

Light-weight, flexible antennas and interconnections are needed for these versatile wireless systems and wearable applications also require the antennas and interconnections to be an integral part of clothing. In addition, antennas and interconnections in wearable applications need to endure different mechanical stresses, such as bending and stretching. Additive manufacturing methods have been found to provide the foundation for these novel applications, as they have the capacity to produce conductors on soft and stretchable materials, such as textiles [7–9]. One highly potential additive textile antenna fabrication method is embroidery with conductive yarn. Sewing is a simple manufacturing method with great possibilities due to its compatibility with various textile materials [9–13]. In embroidery, we have a full control of the conductive pattern: shape, stitch density, and stitch type [9]. In addition to conductor and antenna fabrication, sewing has also been found to be a highly potential method for embedding electrical interconnections into textile materials [9–13].

The effects of the antenna sewing pattern on the tag performance have been previously studied in [9, 14]. Also the strain reliability of sewed interconnections in general has been recently studied [15]. However, to the best of our knowledge, this is the first strain reliability study of embroidered RFID tag antennas and interconnections. During normal movement, the human skin may stretch up to 15%–20%, which means we need to implement antennas and interconnections that function well at least under such elongations [16].

In this study, dipole antennas for passive UHF RFID tags were embroidered on two different stretchable textile materials. After choosing the most suitable textile substrate and sewing patters, stretchable tag antennas were fabricated

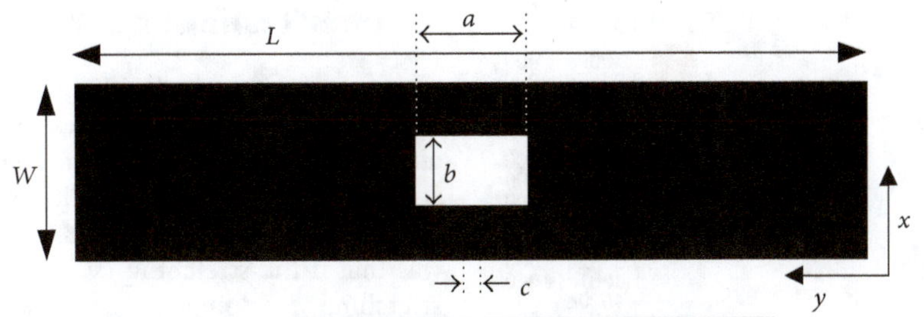


| Geometrical parameters in millimeters | | | | |
|---|---|---|---|---|
| $L$ | $W$ | $a$ | $b$ | $c$ |
| 100 | 20 | 14.3 | 8.125 | 2 |

FIGURE 1: The tag antenna geometry.

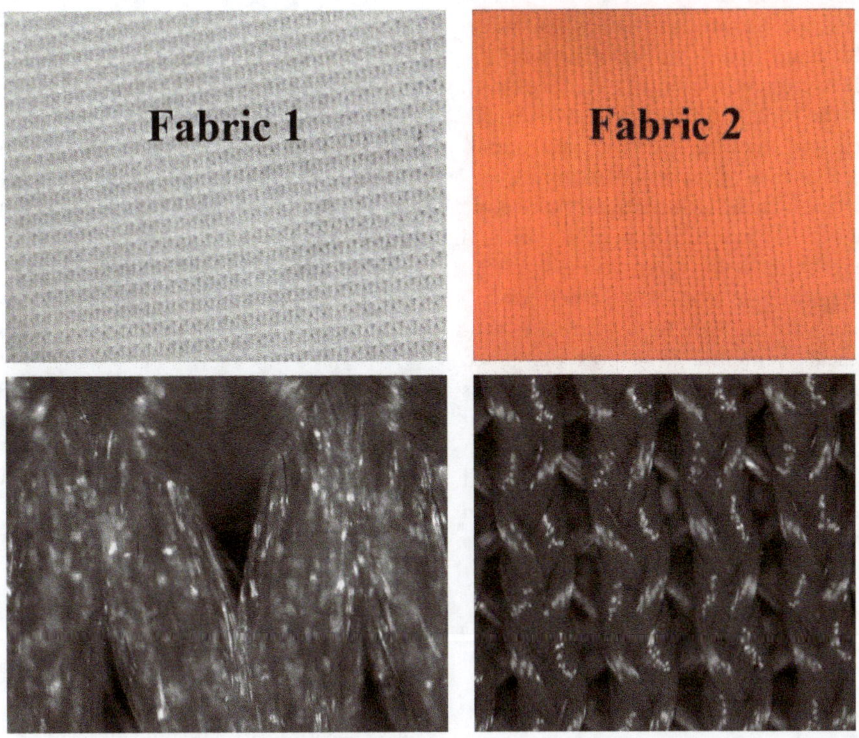


(a) (b)

FIGURE 2: Photos and microscope pictures of Fabric 1 (a) and Fabric 2 (b), magnification ×5.

by embroidery, tag ICs were attached by sewing and by gluing, and the tag reliability was tested under cyclic strain. The performance of the tags before and after testing was evaluated by wireless measurements.

## 2. Tag Fabrication

The studied tag antenna (shown in Figure 1) is a straight dipole, which is a widely used antenna in UHF RFID tags. This antenna has been successfully used in several UHF RFID studies [7, 8] and it originated from a strain reliability study of stretchable electrotextile and screen-printed tags [17].

The tag antennas were fabricated on two different stretchable fabrics (Fabric 1 is an elastic band and Fabric 2 is a mixture of viscose and polyester), by using five different embroidery patterns. The two fabric materials are shown in Figure 2. The used thread was multifilament silver plated thread (Shieldex multifilament thread 110f34 dtex 2-ply HC). The DC linear resistivity of the thread is 500 ± 100 Ω/m, and the diameter is approximately 0.16 mm.

The simulated amplitude of the surface current [A/m] on the dipole antenna at 915 MHz is shown in Figure 3. In the simulation, 0.62 W was delivered to the antenna. As can be seen, the current density at 915 MHz was high around the embedded inductive matching loop and the nearby antenna edges. Two of the studied embroidery patterns were fully embroidered antennas (numbers 1 and 2 in Figure 4), sewed vertically ($x$-axis in Figure 1) and horizontally ($y$-axis in Figure 1). We also fabricated two dipoles with optimized patterns (numbers 2 and 3 in Figure 4), where full vertical and

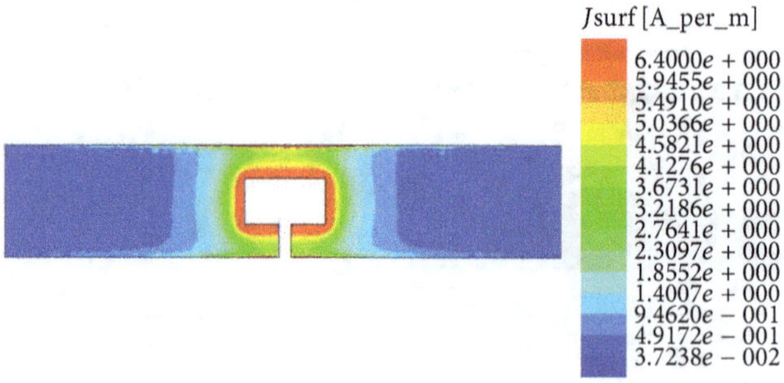


FIGURE 3: Current density on the dipole antenna.

horizontal patterns were used in places with highest surface current density and the rest of the antenna was only partially embroidered. In the fifth antenna, only the borders, that is, the contour of the antenna, were sewed (number 5 in Figure 4). It should be noted that the focus of this study was to test the strain reliability of the sewed antennas and interconnections, not to optimize the performance of the sewed antennas.

During sewing, it was noticed that embroidery on Fabric 2 is a major challenge because the cloth is soft, is thin, and has small tension. The ready-made tag antennas are shown in Figure 4, and for each tag type, we fabricated 5 antennas.

A vector network analyzer was used to measure the input impedance of the fabricated antennas. The port was glued with conductive silver epoxy on the position where the ICs were to be attached (the measurements were done without ICs to reduce the uncertainty). The input impedance of embroidered antennas with different patterns was measured at 915 MHz and the results are shown in Table 1. The reflection coefficient Γ and return loss RL can be calculated by

$$\Gamma = \frac{Z_A - Z_{\text{ic}}}{Z_A + Z_{\text{ic}}}; \qquad \text{RL} = -20 \log |\Gamma|, \tag{1}$$

where $Z_A$ is the impedance of the antenna and $Z_{\text{ic}}$ is the impedance of the IC, which is $12.76 - 190.28j$ at 915 MHz. Γ and RL of the embroidered antennas are shown in Table 1.

Next, the strain ability of the different antennas was tested (see Table 2). Based on the stretching results and the experience gathered during sewing, Fabric 2 was omitted from further research and two sewing ways of making the antenna on Fabric 1 were chosen: contour pattern (Pattern 5) and partial vertical pattern (Pattern 3), as shown in Figure 4, marked with the red rectangles.

The used tag IC (integrated circuit) is NXP UCODE G2iL series RFID IC, provided by the manufacturer in a fixture patterned from copper on a plastic film (see Figure 5, where the IC strap structure is presented).

We attached the fixture pads to the antennas by embroidering over them with the conductive yarn during antenna fabrication, as shown in Figure 6. Finally, for comparison, we created another set of identical tags where the IC was attached using conductive epoxy (Circuit Works CW2400). See Figure 6 for the ready tags with ICs attached.

## 3. Strain Testing and Wireless Measurements

The tags were stretched a 100 times to 80% of their maximum strain and back, and the wireless performance was measured after 1, 10, 20, 30, 40, 50, and 100 stretching cycles. The stretching was done by hand with each stretching cycle lasting about 2 seconds. The reason for the 2 second cycle time was that if the stretching cycle was any shorter, that is, the stretching was done any faster, the glued antenna-IC attachments were easily broken (see Figure 7). This kind of reliability problem was not found in case of the sewed antenna-IC interconnections.

The wireless performance of the tags was evaluated using Voyantic Tagformance RFID measurement system. It contains an RFID reader with an adjustable transmission frequency (800–1000 MHz) and output power (up to 30 dBm) and provides the recording of the backscattered signal strength (down to −80 dBm) from the tag under test.

We wanted to focus on the performance of the sewed tags in absence of environmental stress factors and the effects of the proximity of the human body, so that the source of any observed performance variation is limited to the antenna fabrication parameters and effects of cyclic strain. We conducted all the measurements with the tag suspended on a foam fixture in an anechoic chamber.

During the test, we recorded the lowest continuous-wave transmission power (threshold power: $P_{\text{th}}$) at which the tag remained responsive. The wireless channel from the reader antenna to the location of the tag under test was characterized using a system reference tag with known properties. As explained in [17], this enabled us to estimate the attainable read range of the tag ($d_{\text{tag}}$) versus frequency from

$$d_{\text{tag}} = \frac{\lambda}{4\pi} \sqrt{\frac{\text{EIRP}}{\Lambda} \frac{P_{\text{th}^*}}{P_{\text{th}}}}, \tag{2}$$

where $\lambda$ is the wavelength transmitted from the reader antenna, $P_{\text{th}}$ is the measured threshold power of the sensor tag, $\Lambda$ is a known constant describing the sensitivity of the system reference tag, $P_{\text{th}^*}$ is the measured threshold power of the system reference tag, and EIRP is the emission limit of an RFID reader given as equivalent isotropic radiated power. We present all the results corresponding to EIRP = 3.28 W, which is the emission limit for instance in European countries.

## 4. Results and Discussion

In Figure 8, the read range results for two tags of each tag type are shown in a frequency range of 800–1000 MHz. As shown in Figure 8, the peak read ranges of the contour pattern tags with the sewed antenna-IC attachment are initially around 5.5–6.5 meters, whereas with the glued antenna-IC attachment, the read ranges can be slightly more than 7 meters. Thus, the glued antenna-IC interconnection seems to provide a better electrical connection. The peak read ranges of the Contour pattern tags are all around 900–920 MHz.

For the partial vertical pattern tags, the peak read ranges are around 7 meters with the glued IC and around 7-8 meters with the sewed IC. In this case the performance of both

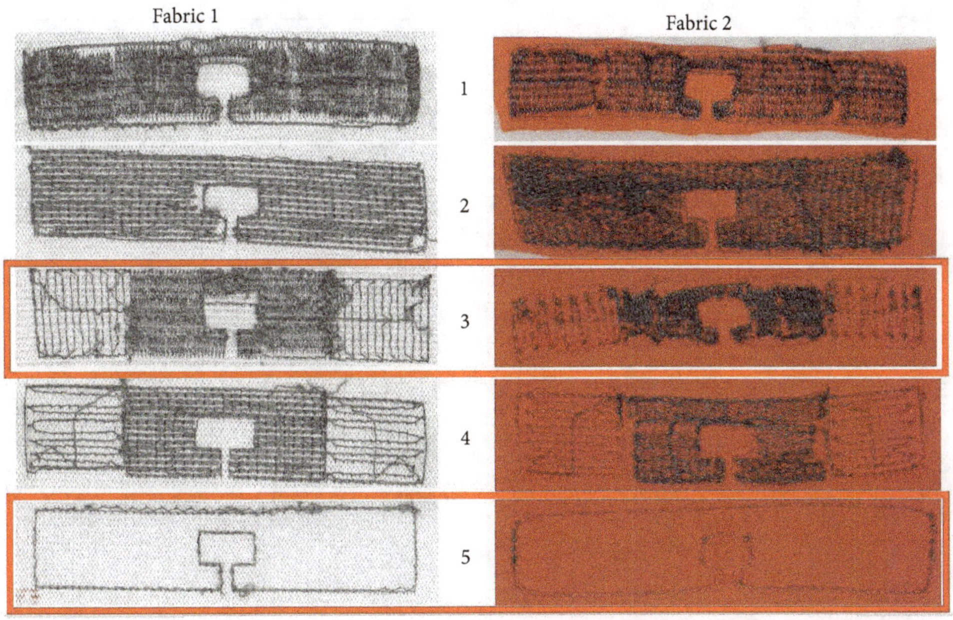


FIGURE 4: The 5 different antenna sewing patterns on both fabrics, the selected two sewing patters on Fabric 1 are marked with a red rectangular.

TABLE 1: Reflection coefficient and return loss of the antennas.

| | Input impedance | Reflection coefficient (Γ) | Return loss (RL) |
|---|---|---|---|
| Contour pattern | $241.83 - 132.8j$ | $0.05 + 0.3j$ | 10.34 dB |
| Partial vertical pattern | $152.35 - 9.99j$ | $-0.19 - 0.88j$ | 1.1 dB |

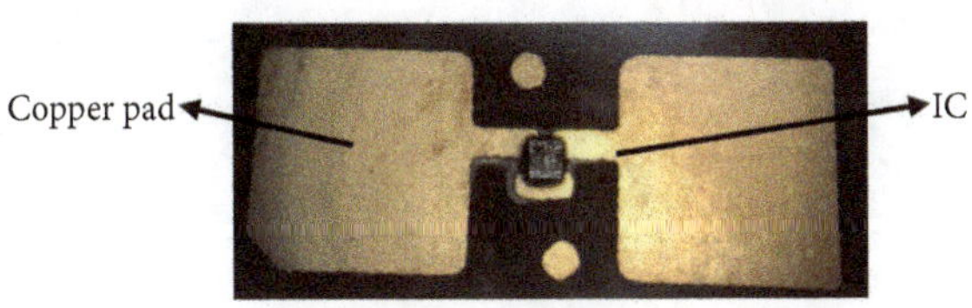


FIGURE 5: The used IC.

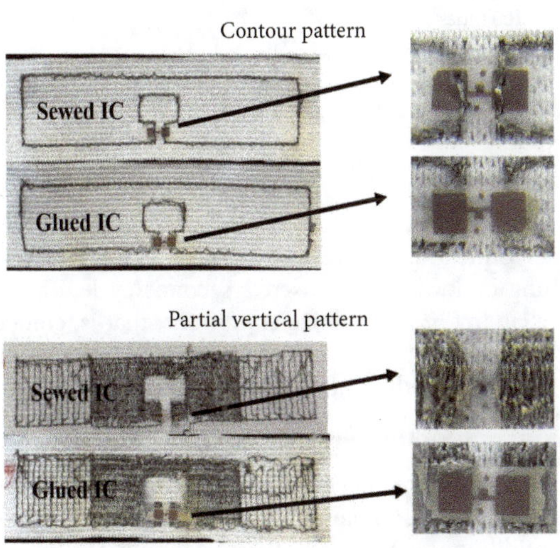


FIGURE 6: Ready-made tags and antenna-IC interconnections.

FIGURE 7: A broken glued antenna-IC interconnection.

interconnection types is similar. This can be explained by the fact that, in the contour pattern antenna, the IC was only attached by sewing a single line over the pads. In the partial vertical pattern, there are several stiches of the conductive yarn over the IC pads (see Figure 6), which makes the embroidered interconnection more conductive than in the case of the contour pattern. The peak read ranges of the tags with glued ICs are around 950 MHz, whereas the peak read ranges of the tags with the sewed ICs are around 910–920 MHz.

It can be noticed from Figure 8 that the reproducibility of the tags with the sewed interconnections is a challenge, whereas the read range figures of the tags with the glued interconnections are very similar. Thus, optimizing the antenna-IC sewing patter for good reproducibility will definitely be the next step of this research.

Next, the tags were stretched 100 times to 80% of their maximum strain and the wireless performance was measured after 1, 10, 20, 30, 40, 50, and 100 stretching cycles. As can be seen from Figures 9 and 10, after 50 harsh stretching

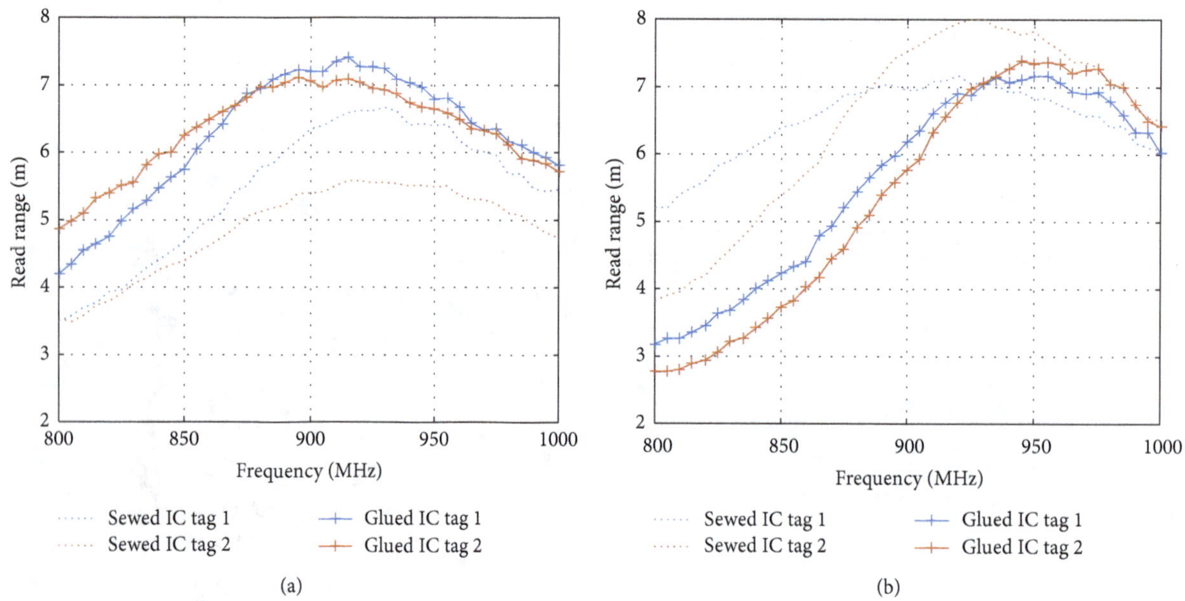


FIGURE 8: The initial attainable read ranges of the contour pattern (a) tag and partial vertical pattern (b) tag.

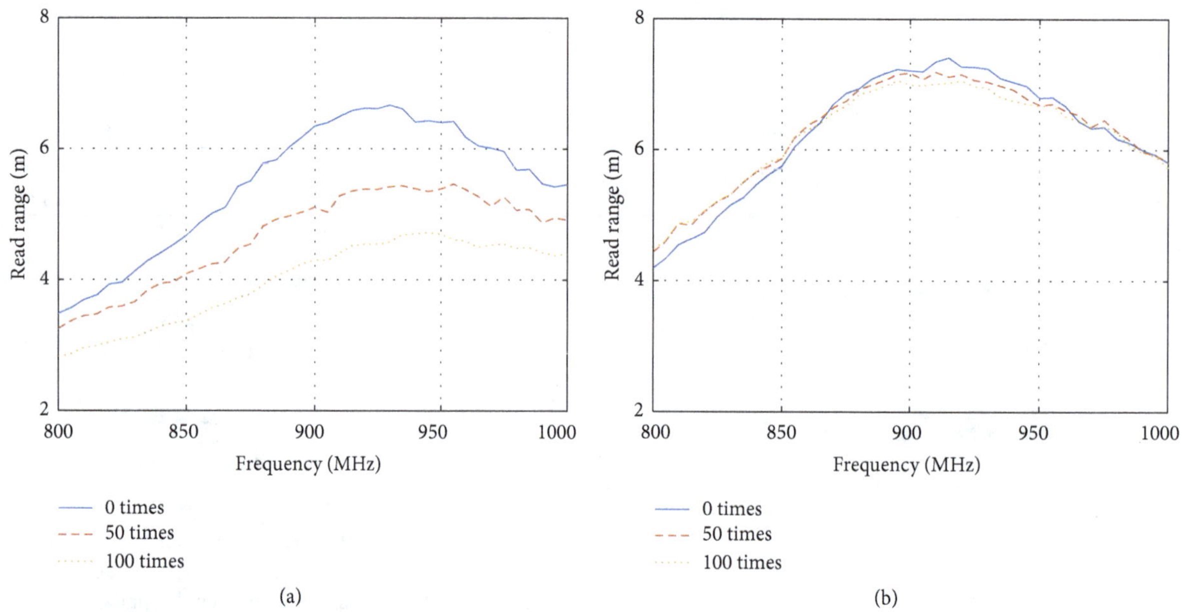


FIGURE 9: Repeated stretching test of contour pattern with sewed IC (a) and glued IC (b).

cycles, the peak read ranges of the tags with the sewed interconnections have decreased about 1 meter compared to the initial read ranges. After 100 stretching cycles, the read ranges of the contour pattern tags had still slightly decreased, to 4.7 meters, but the read ranges of the partial vertical pattern tags remained about the same as after 50 stretching cycles.

All the achieved read ranges are very promising. Based on our results, the sewed antenna-IC interconnections could be useful in future textile tag applications. However, the fabrication process needs to be further studied, for example, by trying different types of stiches around and through the IC pads. Also the challenges with the reproducibility need to be solved. As mentioned earlier, the tags with the glued antenna-IC interconnections were not able to endure any faster stretching cycles than the used 2 seconds/cycle. Thus, also their reliability in normal use conditions, that is, embedded into clothing, can introduce challenges and requires further studying. On the other hand, the used type of stretching did not seem to have an effect on the read ranges of the tags with the glued ICs.

It should be noted here that this cycled stretching test was relatively harsh in comparison with the commonly expected field conditions of wearable antennas, with the exception of the special scenario where the tag on textile would be

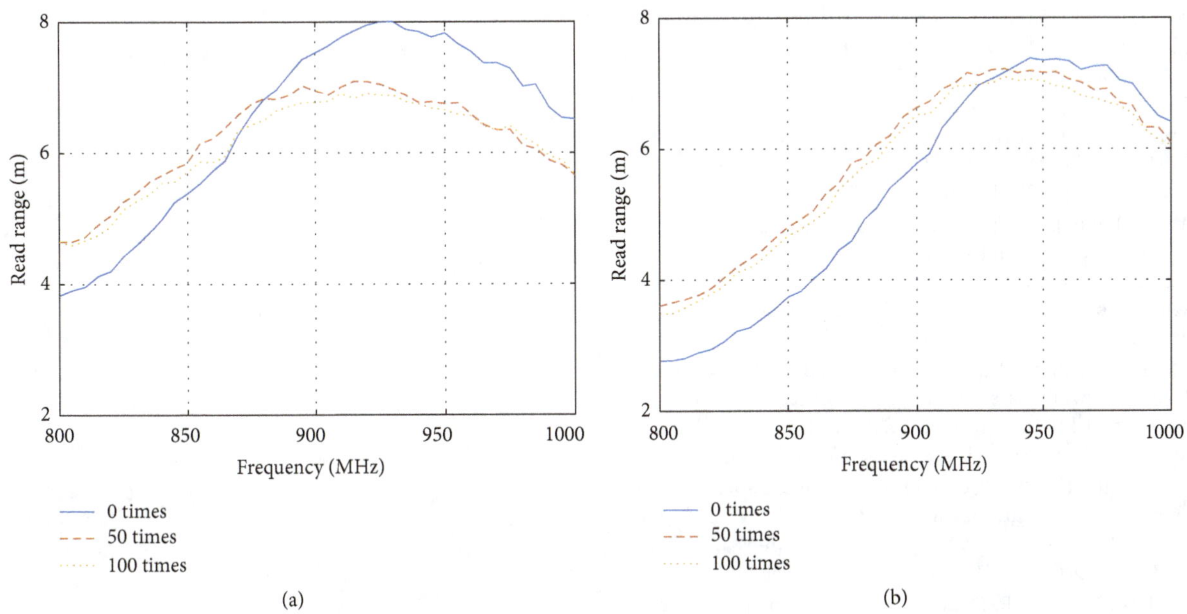


FIGURE 10: Repeated stretching test of partial vertical pattern with sewed IC (a) and glued IC (b).

TABLE 2: The maximum stretching lengths (% of the initial length) of tag patters on Fabric 1 and Fabric 2.

| Tag pattern | Maximum length (%) | |
|---|---|---|
| | Fabric 1 | Fabric 2 |
| N/A | 180% | 180% |
| 1 | 150% | 125 % |
| 2 | 125% | 120 % |
| 3 | 140% | 120 % |
| 4 | 110% | 115 % |
| 5 | 120% | 120 % |

placed at a continually strained location, for example, for the purpose of strain monitoring. In such harsh conditions, one possibility would be to use a suitable coating to protect the antenna-IC interconnection from any mechanical stress, as suggested in a recent washing reliability study of electrotextile RFID tags [18]. Next, we will also study the reliability of these tags in less severe strain conditions.

## 5. Conclusions

We studied the fabrication and impact of strain on the performance of passive UHF RFID textile tags with embroidered antennas on two structurally dissimilar substrates. Two fabrication methods, sewing and gluing, were also tested for the antenna-IC interconnections. According to our results, the characteristics of the fabric substrate have a huge effect on the fabrication of the tags. Our measurement results showed that a simple contour pattern and a partial vertically sewed pattern can be utilized to fabricate stretchable tag antennas on an elastic band-based substrate. However, both the sewed and glued interconnections showed some reliability/reproducibility issues that need to be solved. Also, the degree of stretching on different parts of clothing will be examined and further reliability testing will be done. Future work also includes electromagnetic optimization of the antennas and their measurements near the human body.

## Competing Interests

The authors declare that there is no conflict of interests regarding the publication of this paper.

## Acknowledgments

This research work was supported by the European Commission AdvIOT IRSES project, Academy of Finland, TEKES, and Jane and Aatos Erkko Foundation.

## References

[1] A. Sani, M. Rajab, R. Foster, and Y. Hao, "Antennas and propagation of implanted RFIDs for pervasive healthcare applications," *Proceedings of the IEEE*, vol. 98, no. 9, pp. 1648–1655, 2010.

[2] A. Yakovlev, S. Kim, and A. Poon, "Implantable biomedical devices: wireless powering and communication," *IEEE Communications Magazine*, vol. 50, no. 4, pp. 152–159, 2012.

[3] J. Grosinger, "Feasibility of backscatter RFID systems on the human body," *EURASIP Journal on Embedded Systems*, vol. 2013, article 2, 2013.

[4] E. Moradi, K. Koski, T. Björninen et al., "Miniature implantable and wearable on-body antennas: towards the new era of wireless body-centric systems," *IEEE Antennas and Propagation Magazine*, vol. 56, no. 1, pp. 271–291, 2014.

[5] S. Merilampi, H. He, L. Sydänheimo, L. Ukkonen, and J. Virkki, "The possibilities of passive UHF RFID textile tags as comfortable wearable sweat rate sensors," in *Proceedings of the Progress in Electromagnetics Research Symposium (PIERS '16)*, Shanghai, China, August 2016.

[6] P. Salvo, F. Di Francesco, D. Costanzo, C. Ferrari, M. G. Trivella, and D. De Rossi, "A wearable sensor for measuring sweat rate," *IEEE Sensors Journal*, vol. 10, no. 10, pp. 1557–1558, 2010.

[7] M. Akbari, J. Virkki, L. Sydanheimo, and L. Ukkonen, "Toward graphene-based passive UHF RFID textile tags: A Reliability Study," *IEEE Transactions on Device and Materials Reliability*, vol. 16, no. 3, pp. 429–431, 2016.

[8] J. Wang, J. Liu, J. Virkki et al., "Brush-painting and photonical sintering of copper and silver inks on cotton fabric to form antennas for wearable ultra-high-frequency radio-frequency identification tags," *Textile Research Journal*, vol. 86, no. 15, pp. 1616–1624, 2016.

[9] E. Moradi, T. Björninen, L. Ukkonen, and Y. Rahmat-Samii, "Effects of sewing pattern on the performance of embroidered dipole-type RFID tag antennas," *IEEE Antennas and Wireless Propagation Letters*, vol. 11, pp. 1482–1485, 2012.

[10] M. E. Berglund, J. Duvall, C. Simon, and L. E. Dunne, "Surface-mount component attachment for e-textiles," in *Proceedings of the 19th ACM International Symposium on Wearable Computers (ISWC '15)*, pp. 65–66, ACM, Osaka, Japan, September 2015.

[11] E. R. Post, M. Orth, R. R. Russo, and N. Gershenfeld, "E-broidery: design and fabrication of textile-based computing," *IBM Systems Journal*, vol. 39, no. 3-4, pp. 840–860, 2000.

[12] T. Linz, C. Kallmayer, R. Aschenbrenner, and H. Reichl, "Embroidering electrical interconnects with conductive yarn for the integration of flexible electronic modules into fabric," in *Proceedings of the 9th IEEE International Symposium on Wearable Computers (ISWC '05)*, pp. 86–89, Osaka, Japan, October 2005.

[13] T. Linz, R. Vieroth, C. Dils et al., "Embroidered interconnections and encapsulation for electronics in textiles for wearable electronics applications," *Advances in Science and Technology*, vol. 60, pp. 85–94, 2008.

[14] B. Ivsic, D. Bonefacic, and J. Bartolic, "Considerations on embroidered textile antennas for wearable applications," *IEEE Antennas and Wireless Propagation Letters*, vol. 12, pp. 1708–1711, 2013.

[15] R. Rahimi, W. Yu, T. Parupudi, M. Ochoa, and B. Ziaie, "A low-cost fabrication technique for direct sewing stretchable interconnetions for wearable electronics," in *Proceedings of the 18th International Conference on Solid-State Sensors, Actuators and Microsystems (TRANSDUCERS '15)*, pp. 1350–1353, Anchorage, Alaska, USA, June 2015.

[16] B. Huyghe, H. Rogier, J. Vanfleteren, and F. Axisa, "Design and manufacturing of stretchable high-frequency interconnects," *IEEE Transactions on Advanced Packaging*, vol. 31, no. 4, pp. 802–808, 2008.

[17] J. Virkki, T. Björninen, S. Merilampi, L. Sydänheimo, and L. Ukkonen, "The effects of recurrent stretching on the performance of electro-textile and screen-printed ultra-high-frequency radio-frequency identification tags," *Textile Research Journal*, vol. 85, no. 3, pp. 294–301, 2015.

[18] S. Wang, N. L. Chong, J. Virkki, T. Björninen, L. Sydänheimo, and L. Ukkonen, "Towards washable electrotextile UHF RFID tags: reliability study of epoxy-coated copper fabric antennas," *International Journal of Antennas and Propagation*, vol. 2015, Article ID 424150, 8 pages, 2015.

8

# Bidding Strategy of Virtual Power Plant with Energy Storage Power Station and Photovoltaic and Wind Power

**Zhongfu Tan, Qingkun Tan, and Yuwei Wang**

*School of Economics and Management, North China Electric Power University, Beijing 102206, China*

Correspondence should be addressed to Zhongfu Tan; tanzhongfubeijing@126.com

Academic Editor: Yuh-Shyan Hwang

For the virtual power plants containing energy storage power stations and photovoltaic and wind power, the output of PV and wind power is uncertain and virtual power plants must consider this uncertainty when they participate in the auction in the electricity market. In this context, this paper studies the bidding strategy of the virtual power plant with photovoltaic and wind power. Assuming that the upper and lower limits of the combined output of photovoltaic and wind power are stochastically variable, the fluctuation range of the day-ahead energy market and capacity price is stochastically variable. If the capacity of the storage station is large enough to stabilize the fluctuation of the output of the wind and photovoltaic power, virtual power plants can participate in the electricity market bidding. This paper constructs a robust optimization model of virtual power plant bidding strategy in the electricity market, which considers the cost of charge and discharge of energy storage power station and transmission congestion. The model proposed in this paper is solved by CPLEX; the example results show that the model is reasonable and the method is valid.

## 1. Introduction

The randomness and fluctuation of the wind and photovoltaic power output restrain its grid connected capability [1, 2]. In 2016, China's wind and photovoltaic power installed capacity was 182.15 GW, which ranked first in the world, but the amount of wind power curtailment electricity reached 20 TW·h. The average number of hours of wind power utilization was 1728 h, especially in the "Three-North" area of China, where the wind and solar energy resource is relatively abundant and the equivalent utilization time of wind power is only about 1445 h. In order to achieve large-scale wind power integration, the system needs to arrange reserve services for wind and photovoltaic power, mainly including conventional thermal power units, pumped storage power stations, and energy storage power stations. As the environmental pressure reduction and transmission capacities caused storage power station reserve services inapplicability, this will limit the energy storage station providing the reserve service for wind and photovoltaic power. This paper constructs a virtual power plant with energy storage power station and photovoltaic and wind power which bids in the electricity market, maximizes the benefit of virtual power plant, and promotes the grid-connected generation of photovoltaic and wind power.

In recent years, many scholars have carried out research on the influencing factors of new energy utilization. Nick et al. [3] and Al Kaabi et al. [4] proposed the concept of generating units-power grid-system load interaction, which can effectively reduce the operation cost of the system and promote the grid-connected generation of wind power. Wu et al. [5] used demand side management and storage technology and construct power dispatching model with wind power and energy storage; Georgilakis and Hatziargyriou [6] established a day-ahead market clearing model of wind power based on demand side management, which provided a reasonable operation mechanism for the marketization of wind power. Sadeghi-Barzani et al. [7] established two-stage demand response of stochastic programming model with the wind power to maximize the lower cost of the demand response and enhance grid-connected capacity of wind power. Zhipeng et al. [8] considered price demand response and security constraints, where the system scheduling cost minimization was objective function, and established real-time optimization scheduling model. Huajie et al. [9] proposed regional and provincial two-level optimization scheduling model by adjusting the conventional unit output and interprovincial electricity trade balance wind power output; results show that the model could enhance grid-connected capacity of wind power. Binato et al. [10] proposed a new calculation method to reduce the amount of wind power curtailment in the peak period of the grid. Xiaogang et al. [11] established a dual-objective scheduling model with wind power, based on the second-generation noninferior dominating sorting genetic algorithm. Muñoz-Delgado et al. [12] considered the thermal power, hydropower and wind power, electric and heating load balance, and transmission capacity constraints, with operation cost minimization of the system as the objective function, and set up wind power grid-connected analysis model. The above literatures focus on the multiobjective optimization of wind power system and the improvement of the scheduling model. But the actual situation of China's large-scale wind power centralized access only using these techniques is not enough.

Robust optimization theory is a tool to solve the problem of uncertainty [13, 14]. Haffner et al. [15] introduced the uncertainty set to the large-scale photovoltaic power optimal scheduling systems to make up for the conservative optimization of the box set to partial conservative. Wen and Kumar [16] introduced the game model into the robust optimization problem of the power system and obtained an economic scheduling strategy with strong robustness to short-term uncertainty of wind power generation. Ruozhen et al. [17] introduced the dual theory into the multiperiod inventory robust optimization model; the convex programming problem could be easily solved, which could effectively restrain the influence of the uncertainty of demand distribution on the effectiveness of the strategy. Wen and David (2001) [18] adjusted the boundary of the uncertain set to control the conservativeness of the robust optimization model and realized the economic and security balance of the decision. Pandžić et al. [19] proposed a robust optimization method based on moment uncertain distribution to solve the problem of economic and environmental scheduling with wind power. Kirschen et al. [20] optimized the power flow calculation of distribution network with the strong duality theory; the robust optimization model was transformed into mixed-integer linear programming to compensate for the shortcomings of robust optimization conservatism. Haipeng et al. [21] transformed a two-layer deterministic optimization model into double-layer optimization method, which was optimized by interior point method and had good engineering practicality. Ling et al. [22] introduced the sorted truncation method to improve the efficiency of the solution. Feng et al. [23] proposed an improved robust model for the problem of uncertainty on the right side, which improved the efficiency of the solution. Qiang et al. [24] proposed the scheduling model of the wind power system based on the robust optimization framework, which solved the problem of the objective function deterioration.

In electricity market bidding, Chao et al. [25] proposed a competitive bidding mechanism of the power market with both the supply and demand sides. Peiyi and Xinyan [26] took minimization of purchase cost as the objective function and proposed a bidding method of electricity market based on genetic algorithm. Zheng et al. [27] proposed the formation price mechanism and bidding strategy of renewable energy power generation and analyzed the influence of renewable energy on the electricity market bidding. Jianxue et al. [28] established mixed bidding model of power market, which solved the difficulty of dealing with sub-time constraint and improved the feasibility of the market clearing result. Li et al. [29] analyzed the competition psychology of the power supply market participants with the game theory and established the bidding strategy under different conditions. Yixin et al. [30] analyzed the bidding method of the electricity market in the power supply side with the participation of the microgrid and determined the optimal bidding strategy of the microgrid operators in the alliance and nonalliance. Long et al. [31] analyzed the relationship between the pricing mechanism, the energy substitution, and the bidding strategy and obtained the optimal strategy of the integrated energy sales company to participate in the market bidding purchase. Carrión and Arroyo [32] proposed a transferable bid strategy based on the autoregressive integral moving average mode. Mashhour and Moghaddas-Tafreshi [33] established the profitability model of the electricity sales company under the deviation assessment mechanism and constructed the optimal purchase model of the interruptible load with the minimum cost of the evaluation. Raab et al. [34] introduced the multiagent technology into bidding and optimization scheduling of the virtual power plant, which increased the distributed power generation participating in the electric power transaction.

As a method to deal with the uncertain factors, robust optimization has been paid more attention in many fields such as natural science and engineering technology. On the above background, this paper takes the virtual power plant to maximize the economic benefits as the objective function and the virtual power plant output and market price as uncertainties and constructs the virtual power plant bidding strategy robust optimization model with energy storage station and photovoltaic and wind power.

## 2. Unit Output Model

*2.1. Wind Power Output Model.* Wind turbine power is subject to the wind speed, but if the wind speed is lower than the cut-in speed or higher than the cut-out speed, wind turbines cannot generate electricity; the relationship between wind turbines power and wind speed is as follows:

$$g_{j,w}(t) = \begin{cases} 0, & v(t) \le v_{i,w} \text{ or } v(t) > v_{o,w} \\ g_r \left(v(t) - v_{i,w}\right)\left(v_{r,w} - v_{i,w}\right), & v_{i,w} \le v(t) \le v_{r,w} \\ g_r, & v_{r,w} \le v(t) \le v_{o,w}, \end{cases} \quad (1)$$

where $g_{j,w}(t)$ is the output of $j$ wind turbine at time $t$; $g_r$ is the rated power of wind turbine; $v_{i,w}$, $v_{o,w}$ are the cut-in speed and the cut-out speed of wind turbine; $v_{r,w}$ is the rated speed of wind turbine; $v(t)$ is the speed of wind turbine at time $t$.

*2.2. Photovoltaic Power Model.* The output curve of photovoltaic power system generally satisfies the *Beta* distribution; the specific formula is as follows:

$$f(\theta) = \begin{cases} \dfrac{\Gamma(\alpha)\Gamma(\beta)}{\Gamma(\alpha)+\Gamma(\beta)}\theta^{\alpha-1}(1-\theta)^{\beta-1}, & 0 \le \theta \le 1,\ \alpha \ge 0,\ \beta \ge 0 \\ 0, & \text{else}, \end{cases} \quad (2)$$

where $\alpha, \beta$ are the parameters of the *Beta* distribution; $\theta$ is the radiosity correlation coefficient; the parameters of Beta are calculated by formula (3) as follows:

$$\beta = (1-\mu) \times \left(\frac{\mu \times (1+\mu)}{\delta^2} - 1\right)$$
$$\alpha = \frac{\mu \times \beta}{1-\mu}, \quad (3)$$

where $\mu$ is the mean and normal distribution of solar radiation; $\delta$ is the variance of solar radiation; the probability of solar radiation can be calculated by the following formula:

$$P(\theta) = \int_{\theta_c}^{\theta_d} f(\theta)\, d\theta, \quad (4)$$

where $P(\theta)$ is the radiation intensity; $\theta_c, \theta_d$ are the upper and lower limits of solar radiation $\theta$; the formula for calculating the conversion of solar radiation into electrical energy is as follows:

$$g_{m,\text{pv}}(t) = \varepsilon \times P(\theta) \times S_{\text{pv}} \times \theta_t, \quad (5)$$

where $S_{\text{pv}}$ is area of photoelectricity; $\theta_t$ is photoelectric conversion efficiency of photovoltaic array; $\varepsilon$ is output efficiency of photoelectric inverter.

*2.3. Photovoltaic and Wind Power Combined Output Model.* Both photovoltaic power and wind power generation outputs have volatility and randomness, and their output conditions are closely related to weather conditions. In this paper, the Clay-Copula function is used to simulate the joint probability distribution of wind and photovoltaic power output; the formula is as follows:

$$H(g_w, g_{\text{pv}}) = C\left[F_w(g_w), F_{\text{pv}}(g_{\text{pv}})\right]$$
$$= \max\left\{\left[(F_w(g_w))^{-\varphi} + \left(F_{\text{pv}}(g_{\text{pv}})\right)^{-\varphi} - 1\right]^{1/\varphi}, 0\right\}, \quad (6)$$

where $g_w, g_{\text{pv}}$ are the wind power and photovoltaic power output; $F_w(g_w), F_{\text{pv}}(g_{\text{pv}})$ are the probability distribution functions for wind and photovoltaic power output; $\varphi$ is the connection parameter, for wind and photovoltaic power outputs are not consistent with linear or normal assumptions. Therefore, the Spearman correlation coefficient is introduced in this paper.

In order to convert the edge distribution into corresponding ranks, the uniform probability distribution function is generated by transforming the cumulative probability density function $F(x), G(y)$. $F(x)$ obeys uniform distribution $U(0,1)$, for $r \in (0,1]$; the formula is as follows:

$$P[F(x) \le r] = P\left[x \le F^{-1}(r)\right] = F\left[F^{-1}(r)\right] = r. \quad (7)$$

Similarly, $G(y)$ is transformed into $(0,1]$ uniform distribution $V, U$ and $V$ are uniform distributions on $[0,1]$, $E(U) = E(V) = 1/2$, and $\text{var}(U) = \text{var}(V) = 1/12$. The formula for connection parameter $\varphi$ is as follows:

$$\begin{aligned} \varphi(X,Y) &= \varphi(F(X), G(Y)) = \varphi(U,V) \\ &= \frac{\text{cov}(U,V)}{\sqrt{\text{var}(U)}\sqrt{\text{var}(V)}} \\ &= 12\left[E(UV) - E(U)E(V)\right] \\ &= 12\left[E(UV) - \frac{1}{4}\right] \\ &= 12\int_0^1\int_0^1 uvC(u,v) - 3. \end{aligned} \quad (8)$$

For a given wind and photovoltaic power plant output $(g_w, g_{\text{pv}})$, the paper can construct the joint distribution function $H(g_w, g_{\text{pv}})$; the formula of Spearman correlation coefficient $\varphi$ is shown as follows:

$$\varphi(g_w, g_{\text{pv}}) = 12\int_0^1\int_0^1 F_w(g_w) F_{\text{pv}}(g_{\text{pv}})\, dH(g_w, g_{\text{pv}}) - 3. \quad (9)$$

*2.4. Energy Storage Power Output Model.* Storage power station is charging at night and discharging at daytime, which can effectively reduce the load curve peak valley difference, improve the stability of the grid, and promote wind and photovoltaic power connected grid.

$$g_{k,s}^{\min}(t) \le g_{k,s}(t) \le g_{k,s}^{\max}(t), \quad (10)$$

where $g_{k,s}^{\max}(t)$ and $g_{k,s}^{\min}(t)$ are the upper and lower limits of capacity for energy storage power plants; $g_{k,s}(t)$ is energy storage power plants capacity at time $t$.

When the energy storage power station is in the discharging state,

$$g_{k,s}(t+1) = g_{k,s}(t) - \sum_{s=1}^{S} g_{k,s}^{d}(t)\left(1 + \rho_{k,s}^{d}\right). \quad (11)$$

When the energy storage power station is in the charging state,

$$g_{k,s}(t+1) = g_{k,s}(t) + \sum_{k=1}^{N_s} g_{k,s}^{c}(t)\left(1 + \rho_{k,s}^{c}\right), \quad (12)$$

where $g_{k,s}^{d}(t)$ and $g_{k,s}^{c}(t)$ are the charging and discharging power of the energy storage station $k$ at time $t$; $g_{k,s}(t+1)$ is energy storage station $k$ power capacity at time $t+1$; $\rho_{k,s}^{d}$ and

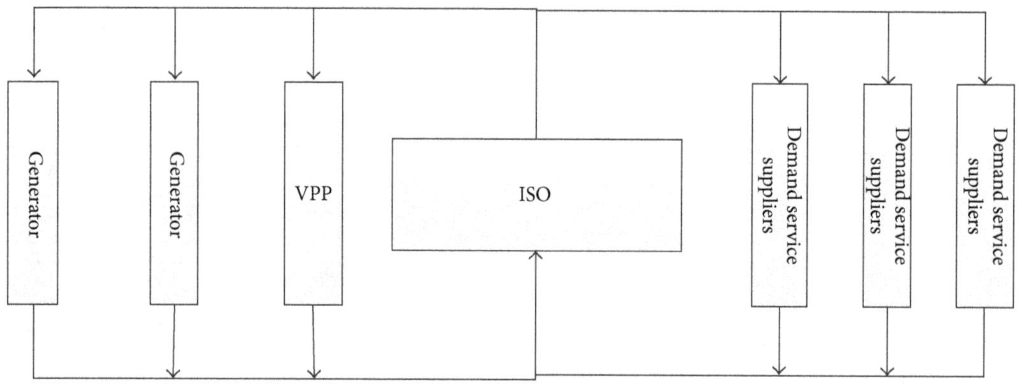


FIGURE 1: Bidding model in electricity market.

$\rho^c_{k,s}$ are discharging and charging loss of energy storage power station; Figure 1 shows bidding process of electricity market.

Figure 1 introduces a virtual power plant including wind, photovoltaic, and energy storage station to compete with traditional energy in the power market. How to realize the maximum benefit of the virtual power plant is the key problem.

## 3. Bidding Strategy of Virtual Power Plant

*3.1. Wind and Photovoltaic Power Jointly Participate in Bidding.* There are differences in the objective function of power market generators due to optimization emphasis, which mainly includes two aspects: technical performance index and economic index. When the virtual power plant wins the bid, the bidding output does not match the actual maximum output; it is necessary to purchase electricity from the real-time market, and VPP should compensate for economic losses.

When the virtual power plant participates in day-ahead power market bidding, $g_{w,\text{pv}}(t)$ is VPP day-ahead bidding output at time $t$, $\Delta t$ is length of a single period, and $\text{MCP}_t$ is the clearing price of electricity market in the period $t$. The revenue of VPP in the market can be expressed by the following formula:

$$R_{w,\text{pv}} = \sum_{t\in T} \text{MCP}_t * g_{w,\text{pv}}(t) * \Delta t. \tag{13}$$

Virtual power plants can use energy storage to provide reserve service for bidding output, which can reduce the deviation between bidding output and actual output of VPP.

*3.2. Objective Function*

*3.2.1. Penalty Cost Minimization.* Assuming that $\Delta g^{\text{up}}(t)$ and $\Delta g^{\text{down}}(t)$ are the deviations between bidding output and actual output of VPP, the corresponding penalty coefficients are $\omega^{\text{up}}(t)$ and $\omega^{\text{down}}(t)$; $E^c$ is the penalty cost resulting from output deviation; the concrete formula is as follows:

$$\begin{aligned} E^c &= \sum_{t\in T} \left[\omega^{\text{up}}(t) * \Delta g^{\text{up}}(t) + \omega^{\text{down}}(t) * \Delta g^{\text{down}}(t)\right] \Delta t \\ \Delta g^{\text{up}}(t) &= g^a(t) - g_s(t) - g(t) \\ \Delta g^{\text{down}}(t) &= g(t) + g_s(t) + g^a(t), \end{aligned} \tag{14}$$

where $g^a(t)$ is the actual output of VPP; $g(t)$ is the bidding output of VPP; $g_s(t)$ is the output of energy storage.

*3.2.2. Congestion Cost Minimization.* Due to the restriction of transmission network, the power system will block during transmission; $E^E$ is congestion cost; the concrete formula is as follows:

$$\begin{aligned} E^E &= P_{\text{loss}} + \lambda_1 \sum_{i=1}^{N_D} \left(V_i - V_{i,\text{lim}}\right)^2 \\ &\quad + \lambda_2 \sum_{i=1}^{N_G+N_C} \left(Q_i - Q_{i,\text{lim}}\right)^2 \\ P_{\text{loss}} &= \sum_{i=1,j\in i}^{N_B} G_{ij}\left[V_i^2 + V_j^2 - 2V_iV_j\cos\left(\delta_i - \delta_j\right)\right], \end{aligned} \tag{15}$$

where $P_{\text{loss}}$ is transmission loss; $N_B$ is the number of branches in the system; $N_D$ is the number of buses in the system; $N_G$ is the number of generators in the system; $N_C$ is the number of Static Var Compensator buses in the system; $\lambda_1$ and $\lambda_2$ are factors of penalty; $G_{ij}$ is conductance between line $i$ and $j$; $V_i$ is the system voltage amplitude; $\delta_i$ is the system voltage phase angle; $V_{i,\text{lim}}$ is the system line voltage; $Q_{i,\text{lim}}$ is the system line capacity.

After introducing the congestion cost in the power system, the objective function can be expressed as follows:

$$\max \quad \pi = \sum_{w=1}^{N} R_{w,\text{pv}} - E^c - \left[P_{\text{loss}} + \lambda_1 \sum_{i=1}^{N_D} \left(V_i - V_{i,\text{lim}}\right)^2 + \lambda_i \sum_{i=1}^{N_G+N_C} \left(Q_i - Q_{i,\text{lim}}\right)^2\right]. \tag{16}$$

*3.3. Constraint Condition.* The constraints of this paper are as follows:

*(1) Load Balance*

$$\sum_{i=1}^{N_G} g_i(t)\,u_i(t)\left(1-\psi_i\right)+\sum_{s=1}^{S} g_s^d(t)\left(1-\rho_s^d\right)+\sum_{w=1}^{W} g_{w,\mathrm{pv}}(t)\left(1-\varphi_w\right)=L^0(t)-\Delta L(t)+\sum_{s=1}^{S} g_s^c(t)\left(1+\rho_s^c\right), \tag{17}$$

where $N_G$ is the number of generators in the system; $S$ is the number of energy storage stations; $W$ is the amount of wind and photovoltaic power; $\psi_i$ is the unit power consumption rate; $\varphi_w$ is the wind and photovoltaic power consumption rate; $L^0(t)$ is system load demand; $\Delta L(t)$ is transmission loss of the system.

*(2) Output Constraint of Energy Storage Station.* In the whole scheduling period $T$, the relationship between the charging capacity and the discharge capacity of the energy storage power station is satisfied as follows:

$$\sum_{t=1}^{T}\left(g_0+\sum_{s=1}^{S} g_{k,s}^c(t)\left(1-\rho_{s,c}\right)-g(t)\right)=\sum_{t=1}^{T}\sum_{s=1}^{S} g_s^d(t)\left(1+\rho_{s,d}\right), \tag{18}$$

where $\rho_{s,c}$ and $\rho_{s,d}$ are charging and discharging efficiency of energy storage power station (0.95); during the same period, the energy storage station cannot charge and discharge simultaneously; the concrete formula is as follows:

$$g_s^d(t)\cdot g_s^c(t)=0. \tag{19}$$

In order to meet the life requirements of energy storage station, the charging and discharging power cannot exceed 20% of the maximum capacity of the energy storage station; the concrete formula is as follows:

$$\begin{gathered} 0\le g_s^d(t)\le g_s^{d,\max}(t) \\ 0\le g_s^c(t)\le g_s^{c,\max}(t), \end{gathered} \tag{20}$$

where $g_s^d(t)$ is the charging capacity of energy storage power station; $g_s^c(t)$ is the discharging capacity of energy storage power station; $g_s^{d,\max}(t)$ and $g_s^{c,\max}(t)$ are the discharging and charging maximum capacity of energy storage station.

Considering the periodicity characteristics of energy storage devices and assuming that the stored energy is consistent at the beginning and end of the dispatch cycle, the concrete formula is as follows:

$$g_{s,0}=g_{s,\mathrm{end}}. \tag{21}$$

*(3) Positive and Negative Rotation Reserve Constraints.* VPP provides positive and negative rotation reserve according to the requirements of power grid; the specific relationship is as follows:

$$\begin{aligned}
&\sum_{i=1}^{N_c} I_i(t)\left[g_i^{\max}(t)-g_i(t)\right]+\sum_{k=1}^{N_s} g_{k,s}^{\max}(t)\ge R^u(t),\\
&g_i^{\max}(t)=\min\left[I_i(t-1)\,g_i^{\max}(t),\,g_i^{\max}(t-1)+D_i^{\max}(t)\,\Delta t\right]I_i(t-1)\\
&R^u(t)=\alpha^c\sum_{i=1}^{N_c} g_i(t)+\alpha^w\sum_{j=1}^{N_w}\left(g_{j,w}(t),g_{j,w_0}(t)\right)\\
&\sum_{i=1}^{N_c} I_i(t)\left[g_i(t)-g_i^{\max}(t)\right]+\sum_{k=1}^{N_s} g_{k,s}^{\min}(t)\ge R^d(t)\\
&g_i^{\max}(t)=\max\left[I_i(t)\,g_i^{\min}(t),\,g_i^{\max}(t)-D_i^{\min}(t)\,\Delta t\right]\cdot I_i(t)\\
&R^d(t)=\beta^w\sum_{i=1}^{N_w}\left(g_{j,w}(t),g_{j,w_0}(t)\right)+\beta^g\sum_{j=1}^{N_s}\left(g_{k,s}(t),g_{k,s_0}(t)\right),
\end{aligned} \tag{22}$$

where $I_i(t)$ is spinning reserve demand coefficient of generating unit $i$ at time $t$; $R^u(t)$ and $R^d(t)$ are up and down spinning reserve requirements of system at time $t$; $g_i^{\max}(t)$ and $g_i^{\min}(t)$ are maximum and minimum output of unit $i$ at time $t$; $g_{k,s}^{\max}(t)$ and $g_{k,s}^{\min}(t)$ are maximum and minimum output of energy storage power station $s$ at time $t$; $D_i^{\max}(t)$ and $D_i^{\min}(t)$ are up and down ramping rate output of unit $i$ at time $t$; $\alpha^c$ and $\alpha^w$ are positive rotation reserve rate and positive rotation reserve rate due to wind power integration; $\beta^g$ and $\beta^w$ are negative rotation reserve rates reduced by VPP integration and negative rotation reserve rate of energy storage station.

*(4) Power Constraint of Transmission Network.* The line power constraints between power systems can be expressed as follows:

$$g_{\mathrm{line}}^{\min}\le g_n(t)\le g_{\mathrm{line}}^{\max}, \tag{23}$$

where $g_n(t)$ is the electric power of line transmission $n$ at time $t$; $g_{\mathrm{line}}^{\min}$ and $g_{\mathrm{line}}^{\max}$ are the maximum and minimum electric power of line transmission $n$ at time $t$.

### 3.4. Construction of Uncertainty Set

*3.4.1. VPP Output Uncertainty Set.* The minimum and maximum output of wind and photovoltaic power at time $t$ can be expressed as follows:

$$\begin{gathered} g_{j,w}^d(t)\le g_{j,w}(t)\le g_{j,w}^u(t) \\ g_{m,\mathrm{pv}}^d(t)\le g_{m,\mathrm{pv}}(t)\le g_{m,\mathrm{pv}}^u(t). \end{gathered} \tag{24}$$

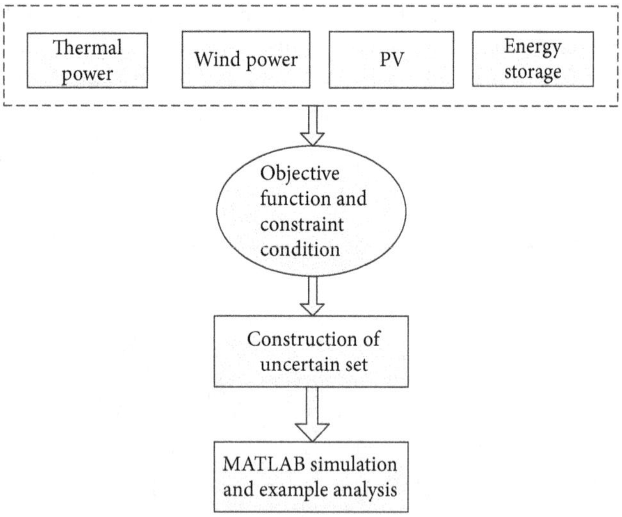


FIGURE 2: VPP bidding flow chart.

In this paper, the absolute predictive error of VPP is constrained by the 1-norm, and the parameter $\Gamma$ is the uncertainty coefficient.

$$\sum_{j=1}^{N_w} \frac{\left| g_{j,w,\mathrm{pv}}^{a}(t) - g_{j,w,\mathrm{pv}}^{f}(t) \right|}{g_{j,w,\mathrm{pv}}^{f}(t)} \leq \Gamma. \tag{25}$$

*3.4.2. Electricity Price Uncertainty Set.* Assume that $\lambda(t)$ is the day-ahead market energy price; $\lambda^{\mathrm{up}}(t)$ is up reserve capacity price; $\lambda^{\mathrm{down}}(t)$ is down reserve capacity price; $\lambda^{R}(t)$ is the spinning reserve capacity price; statistical analysis can be carried out according to historical data:

$$\overline{\lambda}(t) - \widehat{\lambda}(t) \leq \lambda^{R}(t) \leq \overline{\lambda}(t) + \widehat{\lambda}(t), \tag{26}$$

where $\overline{\lambda}(t)$ is mean value of electricity price; $\widehat{\lambda}(t)$ is fluctuation interval radius of electricity price.

## 4. Case Study

*4.1. Solving Method.* As an efficient commercial solver for linear programming, quadratic programming, and mixed-integer linear programming, CPLEX has been widely used in the optimization field. In this paper, AMPL/CPLEX is used to solve the robust optimization model of the energy storage station combined with wind and photovoltaic power. Figure 2 is the algorithm flow of this paper.

In the uncertain constraint programming model constructed by robust constraint, it cannot be directly solved. In terms of nonlinear terms, the computation complexity is high, and the multiobjective optimization problem is solved, but it is difficult to directly solve it. In order to solve the model, it can be transformed into a mixed-integer linear programming model and so on, so that the commercial software such as GAMS/CPLEX can be solved. The core points of the model transformation process include three parts: the deterministic equivalent representation of the stochastic constrained programming model, the normalization of multiple targets, and the linearization of the nonlinear constraints. By setting up different robust control coefficients, the simulation of the cooperative bidding process is carried out, and the optimal allocation of the joint cooperative bidding is obtained. The power system of this paper consists of 6 conventional thermal power units, two wind farms, two energy storage stations, and a photovoltaic power station, as shown in Figure 3.

TABLE 1: Market clearing price for a typical month in PJM electricity market.

| Time | Market clearing price |
|---|---|
| 0 | 20.46 |
| 1 | 16.99 |
| 2 | 14.99 |
| 3 | 15.22 |
| 4 | 18.01 |
| 5 | 25.80 |
| 6 | 30.00 |
| 7 | 31.51 |
| 8 | 31.51 |
| 9 | 31.64 |
| 10 | 32.15 |
| 11 | 31.92 |
| 12 | 32.05 |
| 13 | 32.20 |
| 14 | 31.80 |
| 15 | 31.57 |
| 16 | 30.75 |
| 17 | 30.00 |
| 18 | 32.50 |
| 19 | 33.40 |
| 20 | 32.51 |
| 21 | 32.27 |
| 22 | 31.40 |
| 23 | 28.61 |

*4.2. Data Collection.* In this paper, the cut-in speed is 4 m/s, the cut-out speed is 25 m/s, and the rated wind speed is 15 m/s; the total area of photovoltaic array in photovoltaic power station is 80000 $m^2$, conversion efficiency is 14%, maximum light intensity is 700 w/$m^2$, minimum light intensity is 1 w/$m^2$, and shape parameter and scale parameter of Beta distribution are 0.95; the corresponding penalty coefficient between bidding output and actual output of VPP is 0.95, the rated capacity of two wind farms is 200 MW, the rated capacity of PV is 50 MW, and the rated capacity of energy storage stations is 50 MW; the maximum interrupting capacity of each line is 300 MW. The contract charge is 288 MW·h, $\Delta t$ is 1, and Table 1 shows the mean market clearing price of PJM electricity market.

Assuming that the fluctuation interval radius of the day-ahead market reserve service capacity price is ±15%, Table 2

TABLE 2: Market real-time reserve capacity price.

| Time | Price |
|---|---|
| 0 | 15.57 |
| 1 | 16.18 |
| 2 | 16.44 |
| 3 | 16.49 |
| 4 | 16.52 |
| 5 | 16.56 |
| 6 | 16.56 |
| 7 | 18.93 |
| 8 | 23.26 |
| 9 | 23.28 |
| 10 | 23.65 |
| 11 | 23.65 |
| 12 | 23.76 |
| 13 | 23.84 |
| 14 | 23.87 |
| 15 | 23.92 |
| 16 | 26.45 |
| 17 | 27.95 |
| 18 | 28.02 |
| 19 | 33.40 |
| 20 | 32.51 |
| 21 | 32.27 |
| 22 | 31.40 |
| 23 | 28.61 |

TABLE 3: Optimization results under different constraint violation probabilities.

| Case | Γ | Constraint violation probability | Reserve service income of energy storage power station | Target optimal value |
|---|---|---|---|---|
| 1 | 0 | 53.64 | 5326.88 | 98153.541 |
| 2 | 1 | 50 | 5115.65 | 94096.535 |
| 3 | 3 | 41.36 | 4612.31 | 90249.164 |
| 4 | 5 | 33.13 | 4049.50 | 82971.139 |
| 5 | 12 | 11.5 | 3562.52 | 73302.540 |

shows the mean reserve capacity price of PJM electricity market.

The virtual power plant bidding model needs to predict output of day-ahead wind and photovoltaic power. In order to facilitate the analysis, this paper uses the Clay-Copula function to simulate the joint output of photovoltaic and wind power.

*4.3. Comparison Analysis of Methods.* In Table 3, the optimal values of parameter values and objective functions and the income of the energy storage station are given at different constraint violation probability levels; different constraint violation probabilities correspond to different decision economic risks; from Table 3, the calculation results can be seen; with the decrease of constraint violation probability, the risk of virtual power plant bidding results violating the constraint is also smaller; the auction revenue and storage of the virtual power plant power plant revenues have declined; Figure 4 shows the output curve of units without optimization.

The risk preferences of different scenarios (scenario 1 to scenario 5) in Table 3 correspond to the robust control coefficient, and the constraint violation probability of the uncertainty coefficients of the virtual power plant decreases gradually.

The bidding strategy of virtual power plant under different scenarios is calculated in Figure 5.

It can be seen from the bidding output curve of VPP in Figure 5 that when the decision-maker's risk aversion increases, the VPP bidding output decreases. Accordingly, the reserve power of the energy storage station is reduced to stabilize the fluctuation of the VPP output; Figure 6 shows that the energy storage power station provides reserve power output curve for VPP with optimization.

Energy storage power participates in bidding of the reserve market, which requires coordination between 3 alternate types, so as to maximize the total revenue of the system.

Figures 7 and 8 show the reserve bidding output of the energy storage power station in the market bidding. The bidding strategy of virtual power plant will be affected by the change of energy storage station, wind turbine, and PV output.

In this paper, the robust bidding model of virtual power plant is established, and the robust control coefficient and the chance constrained parameter value are selected according to the risk preference of the decision-maker; Figure 9 is the energy storage power spinning reserve bidding capacity with optimization.

It can be seen from Figures 5–9 that the robust optimization method can improve the reliability of system with considering the uncertainty factors. With the increase of the robust coefficient, the efficiency of the energy storage equipment in VPP will be reduced. The robust optimization algorithm can improve the stability of the system, make the optimization result more in line with the actual situation, and make the decision more reasonable.

*4.4. Sensitivity Analysis of Criteria.* Bidding strategy of virtual power plant will be affected by the change of joint output of energy storage station and VPP.

As can be seen from Figure 10, with the output change of energy storage and wind and photovoltaic power, the sensitivity of the bidding strategy of the virtual power plant to the output of the energy storage station is greater than the sensitivity to the output of the wind and photovoltaic power. In addition to providing reserve for wind and photovoltaic power, the energy storage station can participate in the market bidding with the rich capacity, which can get reserve service fee from the system. When the system wind and photovoltaic power output increased, the wind and photovoltaic power output can participate in the market bidding with correlation of the output of the energy storage station.

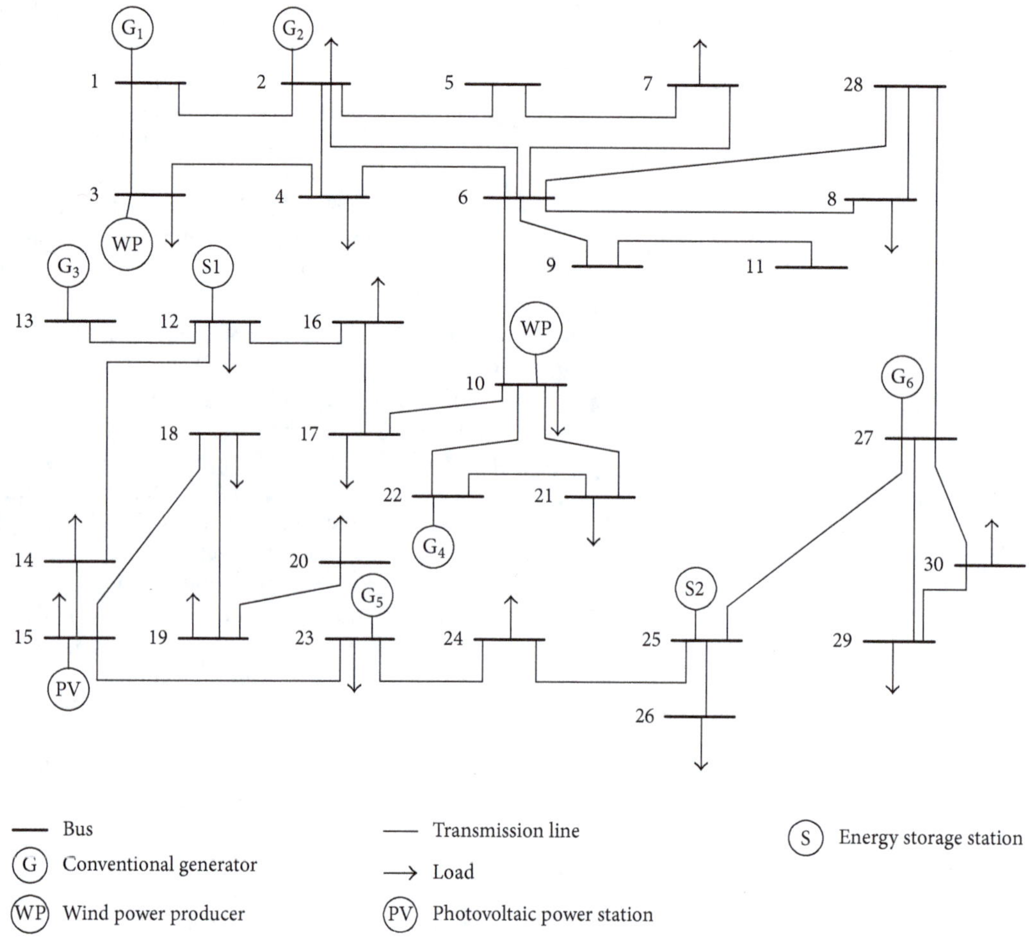


FIGURE 3: Topology distribution of 30 nodes in PJM system.

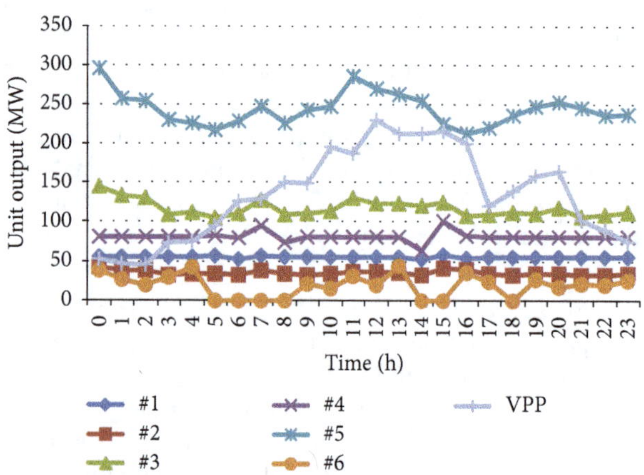


FIGURE 4: The output curve of each unit in the system.

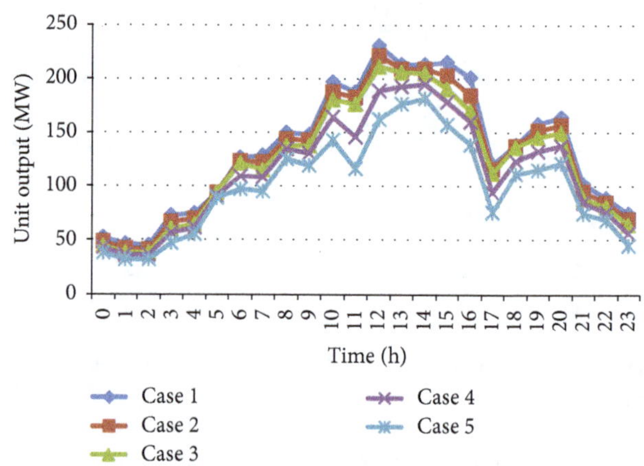


FIGURE 5: Bidding output variation curve of VPP.

## 5. Conclusions

In order to promote large-scale bidding for wind and photovoltaic power, which needs energy storage station providing reserve service, in this paper, a robust optimization bidding model for virtual power plants is proposed, which considers the uncertainty of the wind and photovoltaic power output and the market price. Example results show the following:

(1) Construction of joint probability distribution of wind and photovoltaic power output is based on Clay-Copula function; the combined output model of wind and photovoltaic power is simulated.

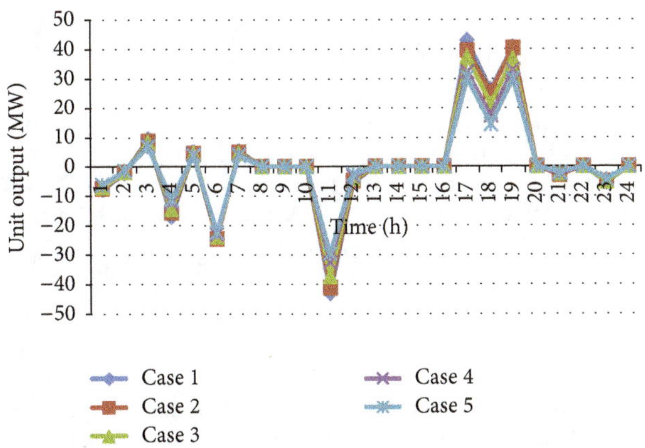


FIGURE 6: The energy storage power station provides reserve power output curve for VPP.

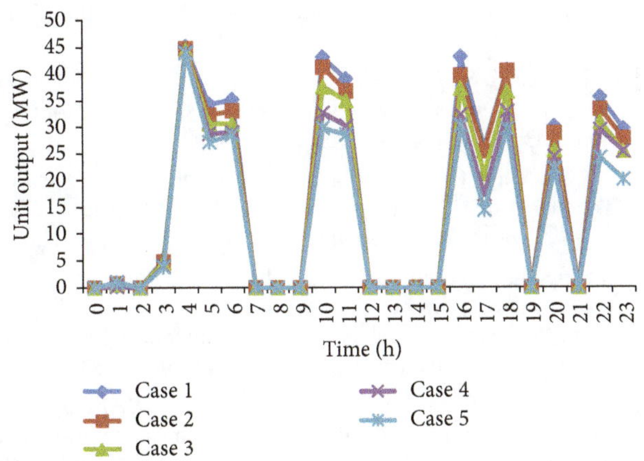


FIGURE 7: Energy storage power upregulation reserve bidding capacity.

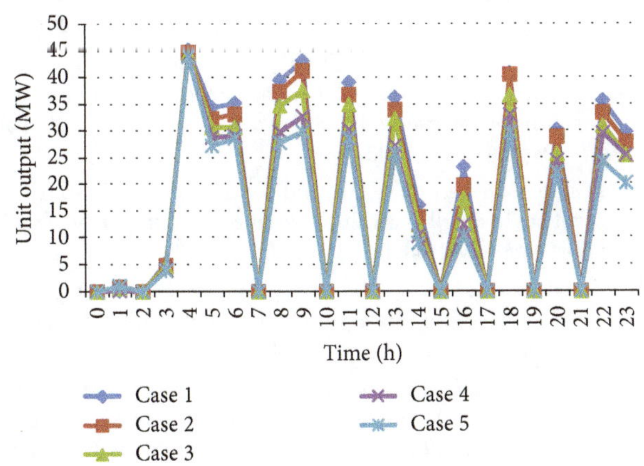


FIGURE 8: Energy storage power downregulation reserve bidding capacity.

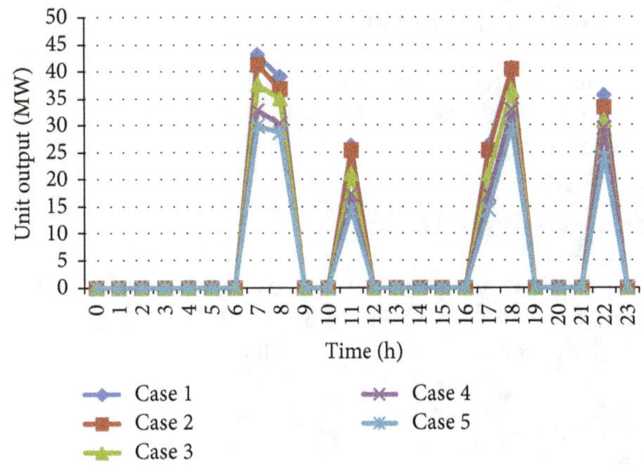


FIGURE 9: Energy storage power spinning reserve bidding capacity.

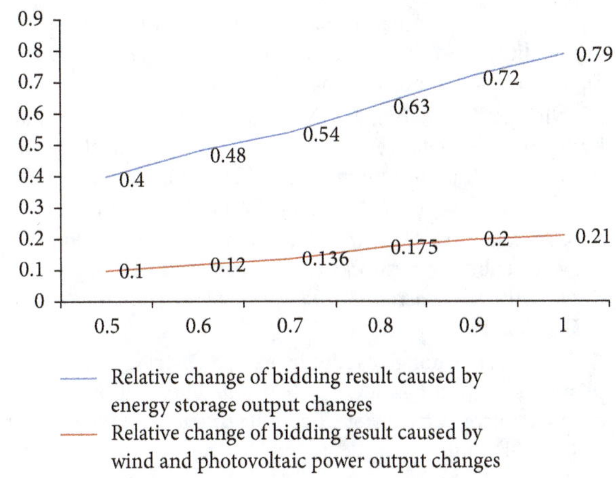


FIGURE 10: Effect on bidding result of energy storage and wind and photovoltaic power output changes.

As a result of the robust optimization, it is seen that the increase of wind and PV power capacity requires storage power station to provide reserve service. In order to promote the enthusiasm of the storage power station and the thermal power unit to participate in the coordination of the wind power, it is necessary to arrange the output of the energy storage power station reasonably, which will realize the maximum income of the virtual power plant.

Intermittence of wind and PV power output is an important issue that cannot be avoided when participating in the bidding of the electricity market. The rapid development of energy storage power station provides a new way to stabilize the fluctuation of wind and PV power output. In this paper, a virtual power station with wind turbines, photovoltaic units, and energy storage power stations is constructed to participate in the collaborative bidding of electricity market; the robust optimization model solves the above problems, and an example is given to illustrate the feasibility and effectiveness of the proposed model and method.

## Authors' Contributions

It should be noted that the whole work was accomplished by the authors collaboratively. All authors read and approved the final manuscript.

## Acknowledgments

This project is supported by the National Natural Science Foundation of China (71573084) and Beijing Municipal Social Science Foundation (16JDYJB044).

## References

[1] B.-H. Zhang, J. Shao, X.-S. Wu, and K. Wang, "Unit commitment with wind farms using scenario tree and chance-constrained programming," *Power System Protection and Control*, vol. 41, no. 1, pp. 127–135, 2013.

[2] X. Liu, "Economic load dispatch constrained by wind power availability: A wait-and-see approach," *IEEE Transactions on Smart Grid*, vol. 1, no. 3, pp. 347–355, 2010.

[3] M. Nick, R. Cherkaoui, and M. Paolone, "Optimal siting and sizing of distributed energy storage systems via alternating direction method of multipliers," *International Journal of Electrical Power & Energy Systems*, vol. 72, pp. 33–39, 2015.

[4] S. S. Al Kaabi, H. H. Zeineldin, and V. Khadkikar, "Planning active distribution networks considering multi-DG configurations," *IEEE Transactions on Power Systems*, vol. 29, no. 2, pp. 785–793, 2014.

[5] O. Wu, C. Haozhong, Z. Xiubin et al., "Distribution network planning method considering distributed generation for peak cutting," *Energy Conversion and Management*, vol. 51, no. 12, pp. 2394–2401, 2010.

[6] P. S. Georgilakis and N. D. Hatziargyriou, "A review of power distribution planning in the modern power systems era: models, methods and future research," *Electric Power Systems Research*, vol. 121, pp. 89–100, 2015.

[7] P. Sadeghi-Barzani, A. Rajabi-Ghahnavieh, and H. Kazemi-Karegar, "Optimal fast charging station placing and sizing," *Applied Energy*, vol. 125, pp. 289–299, 2014.

[8] L. Zhipeng, W. Fushuan, and G. Ledwich, "Optimal planning of electric-vehicle charging stations in distribution systems," *IEEE Transactions on Power Delivery*, vol. 28, no. 1, pp. 102–110, 2013.

[9] D. Huajie, H. Zechun, and S. Yonghua, "Value of the energy storage system in an electric bus fast charging station," *Applied Energy*, vol. 157, pp. 630–639, 2015.

[10] S. Binato, M. V. F. Pereira, and S. Granville, "A new Benders decomposition approach to solve power transmission network design problems," *IEEE Transactions on Power Systems*, vol. 16, no. 2, pp. 235–240, 2001.

[11] X. Xiaogang, W. Biao, L. Xin, and F. Yanjun, "Study on electrical system multi-objective load dispatch optimization with wind farms considering wind power consumption," *Renewable Energy Resources*, vol. 34, no. 1, pp. 49–55, 2016.

[12] G. Muñoz-Delgado, J. Contreras, and J. M. Arroyo, "Joint expansion planning of distributed generation and distribution networks," *IEEE Transactions on Power Systems*, vol. 30, no. 5, pp. 2579–2590, 2015.

[13] A. Ben-Tal and A. Nemirovski, "Robust optimization—methodology and applications," *Mathematical Programming*, vol. 92, no. 3, pp. 453–480, 2002.

[14] A. Ben-Tal and A. Nemirovski, "Robust solutions of linear programming problems contaminated with uncertain data," *Mathematical Programming*, vol. 88, no. 3, pp. 411–424, 2000.

[15] S. Haffner, L. F. A. Pereira, L. A. Pereira, and L. S. Barreto, "Multistage model for distribution expansion planning with distributed generation-part II: numerical results," *IEEE Transactions on Power Delivery*, vol. 23, no. 2, pp. 924-929, 2008.

[16] F. Wen and A. D. Kumar, "Optimal bidding strategies and modeling of imperfect information among competitive generators," *IEEE Transactions on Power Systems*, vol. 16, no. 1, pp. 15–21, 2001.

[17] Q. Ruozhen, H. Xiaoyuan, and Y. Hongtao, "Robust optimization model of multi-period inventory with uncertain demand distribution," *Kongzhi yu Juece/Control and Decision*, vol. 29, no. 9, pp. 1644–1648, 2014.

[18] F. Wen and A. K. David, "Optimal bidding strategies for competitive generators and large consumers," *International Journal of Electrical Power & Energy Systems*, vol. 23, no. 1, pp. 37–43, 2001.

[19] H. Pandžić, J. M. Morales, A. J. Conejo, and I. Kuzle, "Offering model for a virtual power plant based on stochastic programming," *Applied Energy*, vol. 105, pp. 282–292, 2013.

[20] D. S. Kirschen, G. Strbac, P. Cumperayot, and D. P. De Mendes, "Factoring the elasticity of demand in electricity prices," *IEEE Transactions on Power Systems*, vol. 15, no. 2, pp. 612–617, 2000.

[21] Z. Haipeng, L. Hong, and L. Mingbo, "Robust optimal allocation of reactive power compensation in low voltage distribution networks considering uncertainty of photovoltaic generation," *Power System Technology*, vol. 40, no. 12, pp. 3880–3887, 2016.

[22] Q. Ling, L. Yi, D. Sai, W. Hailin, and Z. Weihong, "An improved light robust optimization model and its linear counterpart," *Proceedings of the CSEE*, vol. 36, no. 13, pp. 3365, 3463–3469, 2016.

[23] J. Feng, C. Xingguo, and W. Chaozhu, "Light robustness dispatch model for wind power integrated power system," *Proceedings of the CSEE*, vol. 36, no. 17, pp. 4600–4609, 2016.

[24] W. Qiang, Y. Shiyou, and L. Yuling, "A fast robust optimization methodology based on polynomial chaos and evolutionary algorithm for inverse problems," *Proceedings of the CSEE*, vol. 33, no. 3, pp. 171–175, 2013.

[25] D. Chao, W. Xifan, W. Xiuli, and S. Chengcheng, "Electricity market based on bilateral block bidding," *Automation of Electric Power Systems*, vol. 38, no. 13, pp. 28–32, 2014.

[26] Z. Peiyi and Z. Xinyan, "Research and application of genetic algorithm in bidding strategy for power market," *Shanxi Electric Power*, vol. 38, no. 8, pp. 21–24, 2010.

[27] C. Zheng, Y. Jiajia, J. Xiao-mingl et al., "A literature survey for pricing mechanisms and bidding strategies of renewable energy generation," *Journal of North China Electric Power University*, vol. 41, no. 2, pp. 89–98, 102, 2014.

[28] W. Jianxue, Z. Chenggang, W. Xiuli, and J. Jing, "A mixed bidding mode and its clearing algorithm for electricity market applications," *Automation of Electric Power Systems*, vol. 35, no. 7, pp. 28–33, 2011.

[29] L. Li, H. Yongxiu, and L. Yan, "Research on bidding strategies in electricity market based on gaming theory," *Electric Power*, no. 2, pp. 79–82, 2008.

[30] L. Yixin, G. Li, and W. Chengshan, "Optimal bidding strategy for microgrids in electricity distribution market," *Power System Technology*, vol. 41, no. 8, pp. 2469–2476, 2017.

[31] Y. Long, L. Jichun, G. Hongjun, S. Ning, and L. Yang, "Study on the bidding strategy of comprehensive energy power retailer under the various user-price mechanisms," *Power System Technology*, vol. 38, no. 1, pp. 1–9, 2014.

[32] M. Carrión and J. M. Arroyo, "A computationally efficient mixed-integer linear formulation for the thermal unit commitment problem," *IEEE Transactions on Power Systems*, vol. 21, no. 3, pp. 1371–1378, 2006.

[33] E. Mashhour and S. M. Moghaddas-Tafreshi, "Bidding strategy of virtual power plant for participating in energy and spinning reserve markets—part I: problem formulation," *IEEE Transactions on Power Systems*, vol. 26, no. 2, pp. 949–956, 2011.

[34] A. F. Raab, M. Ferdowsi, E. Karfopoulos et al., "Virtual power plant control concepts with electric vehicles," in *Proceedings of the 16th International Conference on Intelligent System Applications to Power Systems*, pp. 1–6, IEEE, Hersonissos, Greece, September 2011.

9

# Effect of Flow Attack Angle for V-Wavy Plate on Flow and Heat Transfer in a Square Channel Heat Exchanger

**Amnart Boonloi** [1] **and Withada Jedsadaratanachai** [2]

[1] *Department of Mechanical Engineering Technology, College of Industrial Technology, King Mongkut's University of Technology North Bangkok, Bangkok 10800, Thailand*
[2] *Department of Mechanical Engineering, Faculty of Engineering, King Mongkut's Institute of Technology Ladkrabang, Bangkok 10520, Thailand*

Correspondence should be addressed to Withada Jedsadaratanachai; kjwithad@kmitl.ac.th

Academic Editor: Felix Sharipov

Effects of flow attack angles of the V-wavy plate on flow and heat transfer in a square channel heat exchanger are investigated numerically. The V-wavy plates with V-tips pointing downstream and upstream called V-Downstream and V-Upstream, respectively, are examined for the Reynolds number in the range of 3000–10,000. The finite volume method with SIMPLE algorithm is selected to solve the present problem. The numerical results are presented in terms of flow and heat transfer visualization. The thermal performance analysis is also concluded in the form of Nusselt number ratio ($Nu/Nu_0$), friction factor ratio ($f/f_0$), and thermal enhancement factor (TEF). The numerical result shows that the wavy plate can induce the swirling flow through the test section for all cases. The swirling flow disturbs the thermal boundary layer on the channel wall which is the reason for heat transfer enhancement. In range studies, the heat transfer rate increases around 3–6.5 and 2.8–6 times above the smooth channel for V-Downstream and V-Upstream, respectively. The optimum TEF is found at $\alpha = 20^\circ$ and Re = 3000 to be around 2.09 for V-Upstream case.

## 1. Introduction

Heat exchanger is important equipment for various industries. The thermal improvement of heat exchanger can help to save production cost and energy. The thermal improvement in the heat exchanger is separated into two methods: active and passive methods. The active method requires additional power to enhance heat transfer rate of the system. The passive method is done by installing turbulators or vortex generators in the heating/cooling system. The flow created by the vortex generators will disturb the thermal boundary layer on the heat transfer surface which is the reason for heat transfer augmentation. The passive method is worthy way to improve thermal performance in the heat exchanger when compared with active method.

The thermal performance improvement of the heat exchanger with passive method had been performed by many researchers. The V-shaped baffle/rib is a type of vortex generator which gives high effectiveness. The investigations on flow and heat transfer in heat exchanger with V-shaped baffle/rib had been presented. For examples, Kumar et al. [1] experimentally investigated heat transfer and friction loss in a solar air channel with broken V-baffle. The effects of parameters, relative baffle gap, baffle height, baffle pitch, and flow attack angle, were studied. They reported that the broken V-baffle has better performance when compared with other shapes. They also concluded that the optimums of the relative baffle gap distance, relative baffle gap width, baffle height ratio, baffle pitch ratio, and flow attack angle are around 0.67, 1, 0.5, 1.5, and 60°, respectively. Jedsadaratanachai et al. [2] numerically studied laminar flow and heat transfer in a circular tube fitted with 45° V-baffle. They indicated that the 45° V-baffle can enhance heat transfer rate higher than the smooth tube. They also summarized that the maximum thermal performance is around 3.2 at the baffle height ratio of 0.2 and 0.25 for V-Upstream and V-Downstream, respectively. Jedsadaratanachai and Boonloi [3] presented the numerical investigation on flow and heat transfer behaviors in a square channel placed with 30° double V-baffle. The influences of baffle height and baffle pitch were investigated for Re = 100–1200. They found that the installation of the double V-baffle in the heating section performs upper heat transfer rate and thermal performance compared with the smooth square channel. Caliskan and Baskaya [4] studied the enhancement of heat transfer for the surfaces placed with V-rib. Jin et al. [5] numerically investigated flow pattern and heat transfer characteristic in a solar air heater added with staggered multiple V-shaped rib. The parameters, rib height, pitch, and attack angle, were varied. They pointed out that the maximum thermal performance is around 2.43. Deo et al. [6] reported performance analysis of a solar air heater duct roughened with multigap V-down ribs combined with staggered ribs. They showed that the maximum augmentations of the Nusselt number and thermal performance are around 3.34 and 2.45 times, respectively. Promthaisong et al. [7] selected the discrete broken V-rib to enhance heat transfer rate and thermal performance in a square channel heat exchanger. They concluded that the square channel with the discrete broken V-rib not only provides higher heat transfer rate, but also enhances pressure loss. Fang et al. [8] studied turbulent flow structure in a square channel with V-shaped rib on one wall. The flow attack angles for the V-shaped rib were compared. Maithani and Saini [9] reported the heat transfer and friction loss in a solar air heater duct with V-rib. The influences of the Reynolds number, number of gaps, relative gap width, relative roughness width, and flow attack angle were investigated experimentally. They detected that the enhancements of heat transfer rate and friction loss are around 3.6 and 3.67 times above the smooth duct. Jin et al. [10] numerically studied heat transfer and flow topology in a solar air heater duct with multi V-shaped ribs on absorber plate. They indicated that the rib in the duct provided better fluid mixing which is the cause for heat transfer enhancement.

Another interesting vortex generator is wavy surface. The wavy surface always selects improving heat transfer rate and thermal performance in fin-and-tube heat exchanger. The wavy surface can increase the strength of the flow in the heat exchanger. The combinations between wavy fin and the other types of the turbulators in the fin-and-tube heat exchanger had been reported by many researchers [11–16].

Boonloi and Jedsadaratanachai [17] numerically investigated laminar flow and heat transfer in a square channel fitted with V-wavy plate. The V-wavy plate is a combination of the vortex generators between V-shaped baffle and wavy surface. The researchers found that the V-wavy plate gives the heat transfer rate and efficiency close to the V-baffle. The production and maintenance of the V-wavy plate are more convenient than the V-shaped baffle. The optimum flow attack angle of the V-wavy plate in the heat exchanger channel for the laminar regime was concluded.

As [17], similar configuration of the heat exchanger channel inserted with V-wavy surface is tested for the turbulent regimes in the present work. The influences of the flow attack angle on heat transfer and flow structure are performed. The numerical results are reported in terms of flow and heat transfer mechanisms in the test channel. The thermal performance in the heat exchanger channel is also analyzed.

## 2. Physical Model and Boundary Condition

The wavy plate with square profile ($0.2H \times 0.2H$) is inserted in the middle of the square channel heat exchanger as in Figure 1. The periodic module of the numerical model is equal to the channel height ($H$). The flow attack angles of the wavy plate ($\alpha$ = 15°, 20°, 25°, 30°, 35°, 40°, 45°, 50°, 55°, and 60°) are varied for the Reynolds number in the range of 3000–10,000. The effects of arrangements for the wavy plate are studied: V-tip pointing downstream called "V-Downstream" and V-tip pointing upstream called "V-Upstream." The created domain for the square channel inserted with wavy plate is also depicted in Figure 1.

The square channel walls are set with uniform heat flux around 600 W/m$^2$, while the wavy plate is assumed as adiabatic wall (insulator). The inlet and outlet of the computational domain for the square channel inserted with wavy plate are set with periodic boundary. No slip wall condition is imposed for all surfaces of the numerical model.

## 3. Mathematical Foundation and Assumption

The tested fluid is air that keeps the thermal properties at the average bulk mean temperature. The flow and heat transfer are steady in three dimensions. The flow is turbulent and incompressible. The natural convection, radiation, viscous dissipation, and body force are not considered.

The turbulent model is realizable $k-\varepsilon$ with enhanced wall treatment as follows:

$$\begin{aligned}
&\frac{\partial}{\partial t}(\rho k) + \frac{\partial}{\partial x_j}(\rho k u_j) \\
&\quad = \frac{\partial}{\partial x_j}\left[\left(\mu + \frac{\mu_t}{\sigma_k}\right)\frac{\partial k}{\partial x_j}\right] + G_k + G_b - \rho\varepsilon - Y_M + S_k, \\
&\frac{\partial}{\partial t}(\rho\varepsilon) + \frac{\partial}{\partial x_j}(\rho\varepsilon u_j) \\
&\quad = \frac{\partial}{\partial x_j}\left[\left(\mu + \frac{\mu_t}{\sigma_\varepsilon}\right)\frac{\partial \varepsilon}{\partial x_j}\right] + \rho C_1 S\varepsilon - \rho C_2\frac{\varepsilon^2}{k + \sqrt{\nu\varepsilon}} \\
&\quad + C_{1\varepsilon}\frac{\varepsilon}{k}C_{3\varepsilon}G_b + S_\varepsilon.
\end{aligned} \tag{1}$$

where

$$\begin{aligned}
C_1 &= \max\left[0.43, \frac{\eta}{\eta + 5}\right], \\
\eta &= S\frac{k}{\varepsilon}, \\
S &= \sqrt{2S_{ij}S_{ij}};
\end{aligned} \tag{2}$$

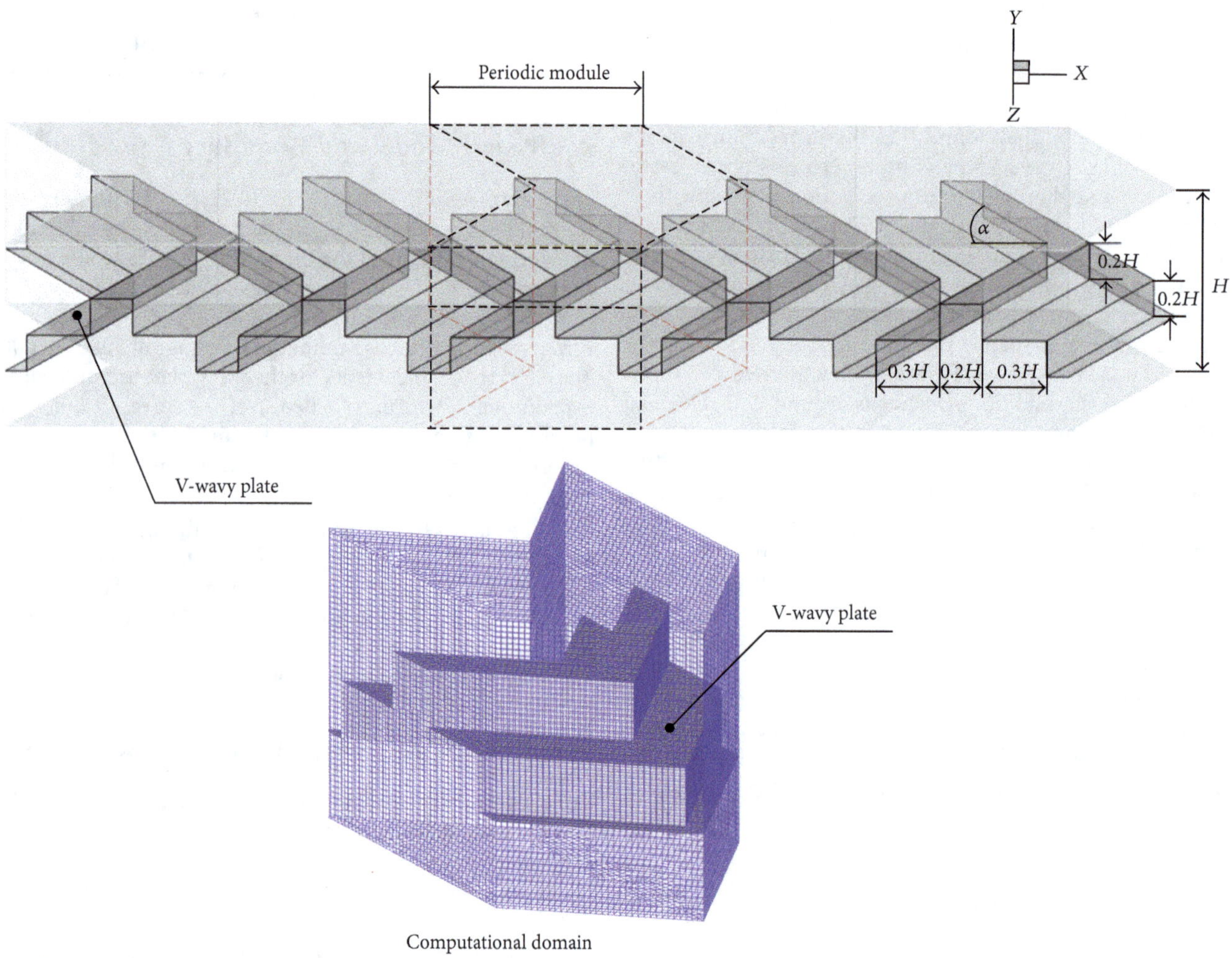


FIGURE 1: Square channel heat exchanger inserted with V-wavy plate and computational domain.

the constants in the model are given as follows:

$$C_{1\varepsilon} = 1.44,$$
$$C_2 = 1.9,$$
$$\sigma_k = 1.0,$$
$$\sigma_\varepsilon = 1.2. \quad (3)$$

The SOU numerical scheme is selected for all governing equations, decoupling with the SIMPLE algorithm using a finite volume method (FLUENT commercial code). The solutions are considered to be converged when the normalized residual is less than $10^{-9}$ and $10^{-5}$ for the energy equation and the other variables, respectively.

The important parameters are Reynolds number (Re), friction factor ($f$), Nusselt number (Nu), and thermal enhancement factor (TEF).

The Reynolds number is computed from

$$\text{Re} = \frac{\rho u_0 H}{\mu}. \quad (4)$$

$\rho$, $\mu$, and $u_0$ are density, viscosity, and inlet velocity of the fluid, respectively, while $H$ is the hydraulic diameter/height of the square channel.

The pressure loss in the heating channel is presented as friction factor. The friction factor can be calculated

$$f = \frac{(\Delta P/L)\, D_h}{2\rho \bar{u}^2}. \quad (5)$$

$\Delta P$ is the pressure drop across the periodic module, $L$, and $u$ is mean flow velocity.

The heat transfer rate in the tested section is shown in terms of Nusselt number.

The local Nusselt number is computed by

$$\text{Nu}_x = \frac{h_x D_h}{k}. \quad (6)$$

$h_x$ is local heat transfer coefficient based on bulk temperature and $k$ is thermal conductivity of the air.

The average Nusselt number can be printed by

$$\text{Nu} = \frac{1}{A} \int \text{Nu}_x dA, \quad (7)$$

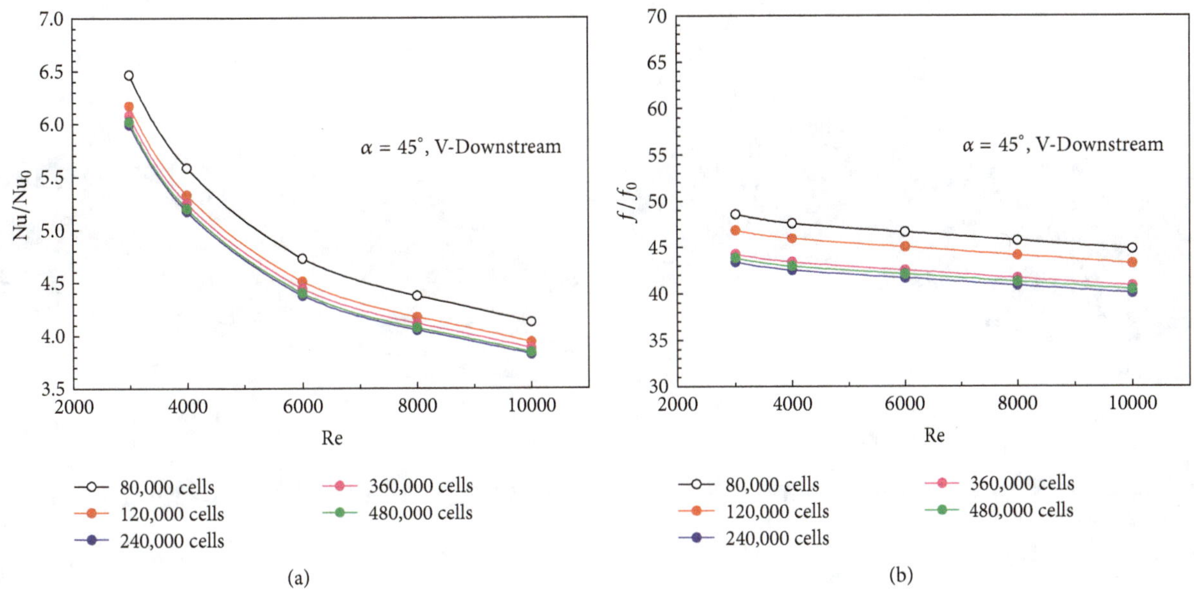


FIGURE 2: Grid independence for (a) Nusselt number and (b) friction factor of 45° V-wavy plate for V-Downstream.

where $A$ is heat transfer area of the square channel heat exchanger.

Thermal performance is presented in the form of thermal enhancement factor (TEF).

The TEF is defined as the ratio of the heat transfer coefficient of an augmented surface, $h$, to that of a smooth surface, $h_0$, under the constant pumping power condition. The TEF can be stated as follows:

$$\mathrm{TEF} = \frac{(\mathrm{Nu}/\mathrm{Nu}_0)}{(f/f_0)^{1/3}}. \quad (8)$$

## 4. Numerical Result

Numerical results for the present problem are separated into three parts: verification of the computational domain, mechanisms in the test section, and thermal performance analysis.

*4.1. Verification of the Computational Domain.* Hexahedral mesh is selected for the numerical domain. The different numbers of grids for the computational domain ($\alpha = 45°$, V-Downstream) are compared as Figures 2(a) and 2(b) for Nusselt number and friction factor, respectively. As the figures, the grid around 240000, 360000, and 480000 performs similar values on both heat transfer and friction loss. Therefore, the 240000-grid cell is applied for all cases of the present research when considering the time for investigation and accuracy of the solution.

The heat transfer rate and pressure loss in the smooth channel heat exchanger for the current problem are compared with the values from the correlations as in Figure 3. From the figure, the deviations of the heat transfer and friction loss are detected within ±3%.

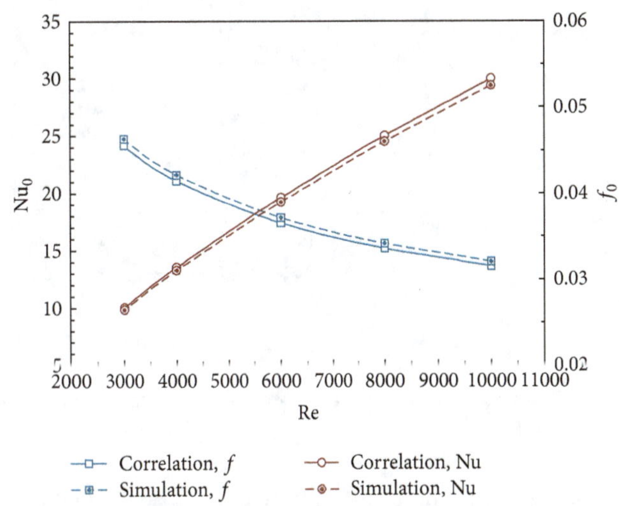


FIGURE 3: Verification of the smooth channel on Nusselt number and friction factor.

From the preliminary study, it can be concluded that the computational domain of the present study has reliability to predict flow and heat transfer in the square channel heat exchanger inserted with various flow attack angles of the V-wavy plate.

*4.2. Flow and Heat Transfer Mechanisms.* Figures 4(a) and 4(b) illustrate tangential velocity vector in transverse planes at various $x/H$ positions in the square channel heat exchanger inserted with 45° wavy plate for V-Downstream and V-Upstream, respectively, at Re = 3000. In general, the wavy plate can induce the vortex flow or swirling flow through the test section on both cases. The vortex flow in the test section disturbs the thermal boundary layer on the heat transfer

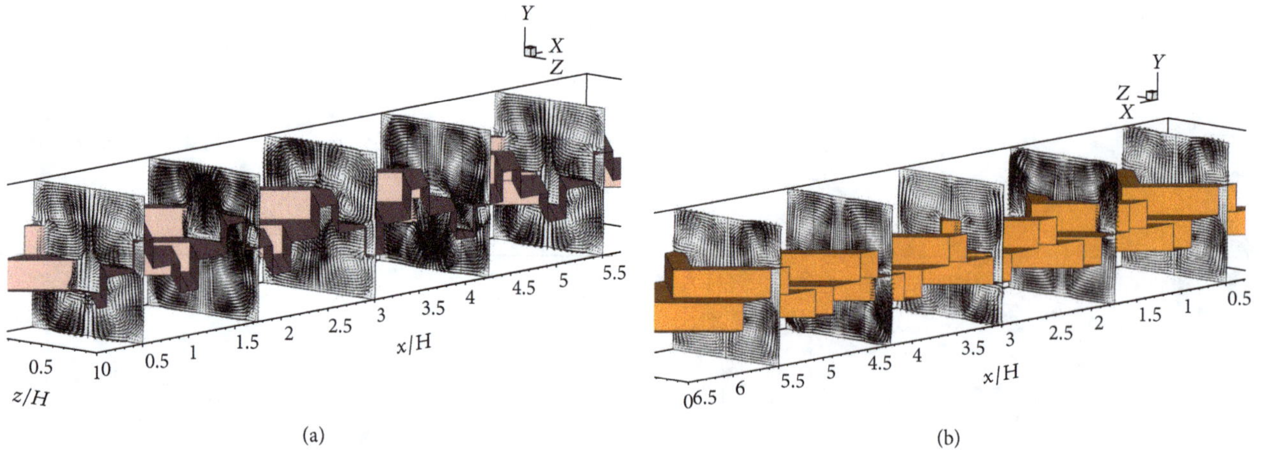


Figure 4: Tangential velocity vector in transverse planes of the square channel heat exchanger inserted with 45° V-wavy plate at Re = 3000 for (a) V-Downstream and (b) V-Upstream.

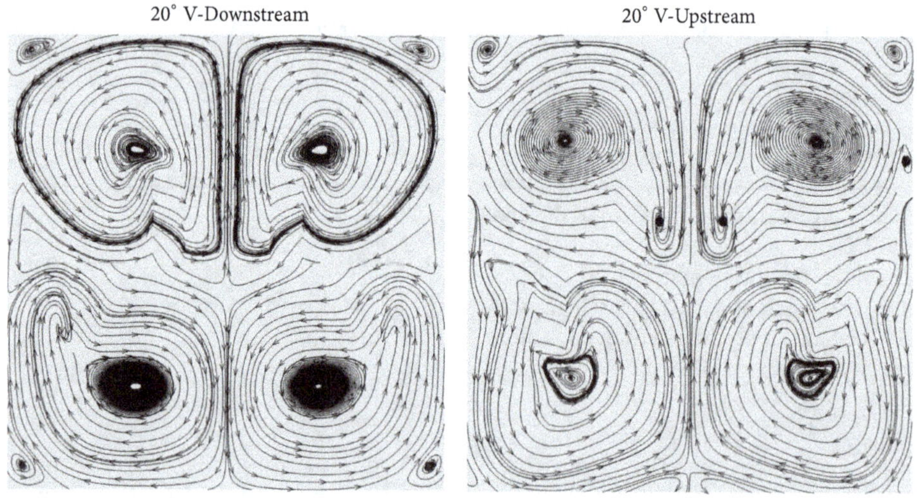


Figure 5: Streamline in transverse planes for the square channel heat exchanger inserted with 20° V-wavy plate at Re = 3000.

surface which is the cause of heat transfer augmentation. The flow includes four main vortex flows and small vortices at the four corners of the square channel. Considering the lower pair of the flow, the counterrotating flows with common-flow-down and common-flow-up are created from V-Downstream and V-Upstream, respectively (see Figure 5).

Figures 6(a) and 6(b) show longitudinal flow in the square channel inserted with 45° V-Downstream and V-Upstream, respectively, at Re = 3000. For V-Downstream, the air slides on the channel wall before it moves to the V-tip of the wavy surface. For V-Upstream, the air flows to the V-tip of the wavy plate and then slides to the side wall of the channel. The flow structure of the wavy plate is found in similar pattern for all flow attack angles, but the vortex strength is not equal.

Figures 7(a) and 7(b) present longitudinal swirling flow with local Nusselt number distribution on the channel wall for the square channel heat exchanger inserted with 45° wavy plate of V-Downstream and V-Upstream, respectively, at Re = 3000. The plots indicate that the swirling flow impinges on the channel wall. The impingement of the flow is the thermal boundary layer disturbance which is an important reason for heat transfer enhancement. The thermal boundary layer disturbance is detected in all sides of the square channel. For all flow attack angles and arrangements, the thermal boundary layer disturbance is detected, but the potency of the disturbance is not equal. The thermal boundary layer disturbance is also seen by the plot of temperature distribution in cross-sectional planes as depicted in Figure 8. The high temperature is plotted with red layer, while the low temperature is written with blue layer. The red layer near the channel wall is lightly found, while the blue layer is distributed from the center of the channel when inserting the wavy plate in the square channel. The temperature behavior for all flow attack angles and arrangements of the wavy plate in the square channel is found in similar pattern.

Figures 9(a) and 9(b) show local Nusselt number distribution on the channel walls of the square channel inserted with 45° wavy plate for V-Downstream and V-Upstream,

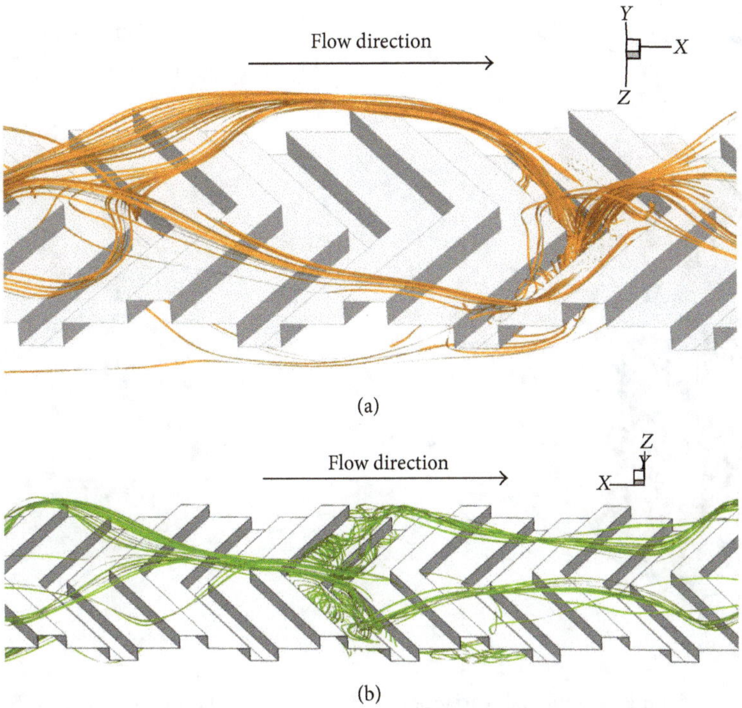


FIGURE 6: Longitudinal vortex flow of the square channel heat exchanger inserted with 45° V-wavy plate at Re = 3000 for (a) V-Downstream and (b) V-Upstream.

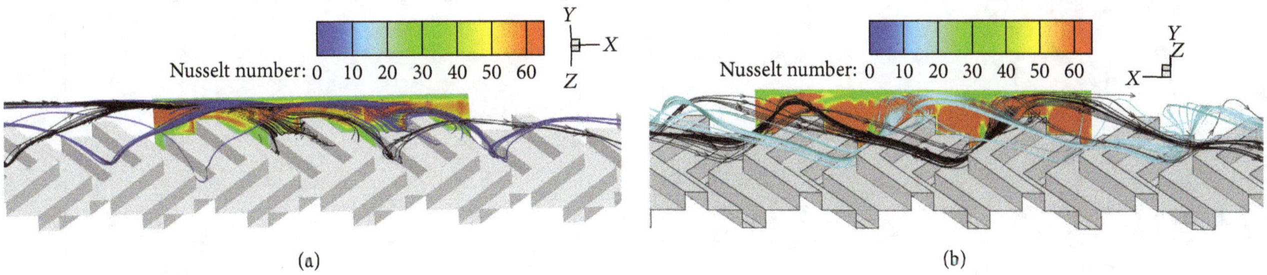


FIGURE 7: Impinging jet on the channel wall of the square channel heat exchanger inserted with 45° V-wavy plate at Re = 3000 for (a) V-Downstream and (b) V-Upstream.

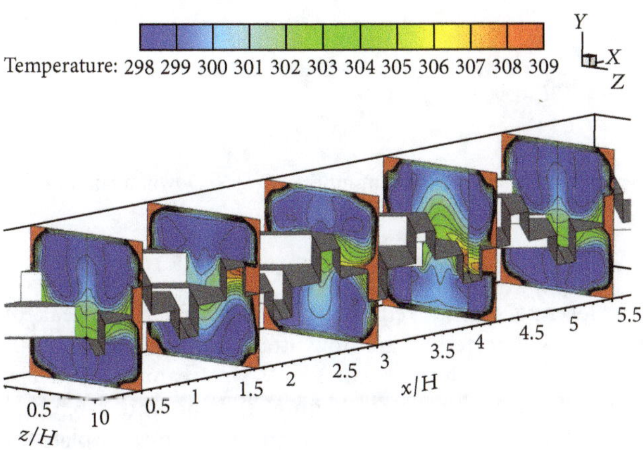


FIGURE 8: Temperature distribution in transverse planes for the square channel heat exchanger inserted with 45° wavy plate at Re = 3000 for V-Downstream.

respectively, at Re = 3000. In general, the insertion of the wavy plate improves the heat transfer rate higher than the smooth channel with no wavy plate. The figures indicate that the disturbance of the thermal boundary layer appears in all heat transfer surfaces. The heat transfer behavior of the V-Downstream and V-Upstream is not similar due to the different flow pattern. The peak of heat transfer rate is clearly found at the upper-lower walls of the square channel for V-Downstream. The disturbance of the thermal boundary layer is equally found in all sidewalls of the channel for the V-Upstream. The heat transfer characteristic is likely detected for all Re and $\alpha$.

*4.3. Thermal Performance Analysis.* Figures 10(a) and 10(b) plot the relation of the $Nu/Nu_0$ and flow attack angle of the square channel inserted with wavy plate for V-Downstream and V-Upstream, respectively. In general, the addition of the

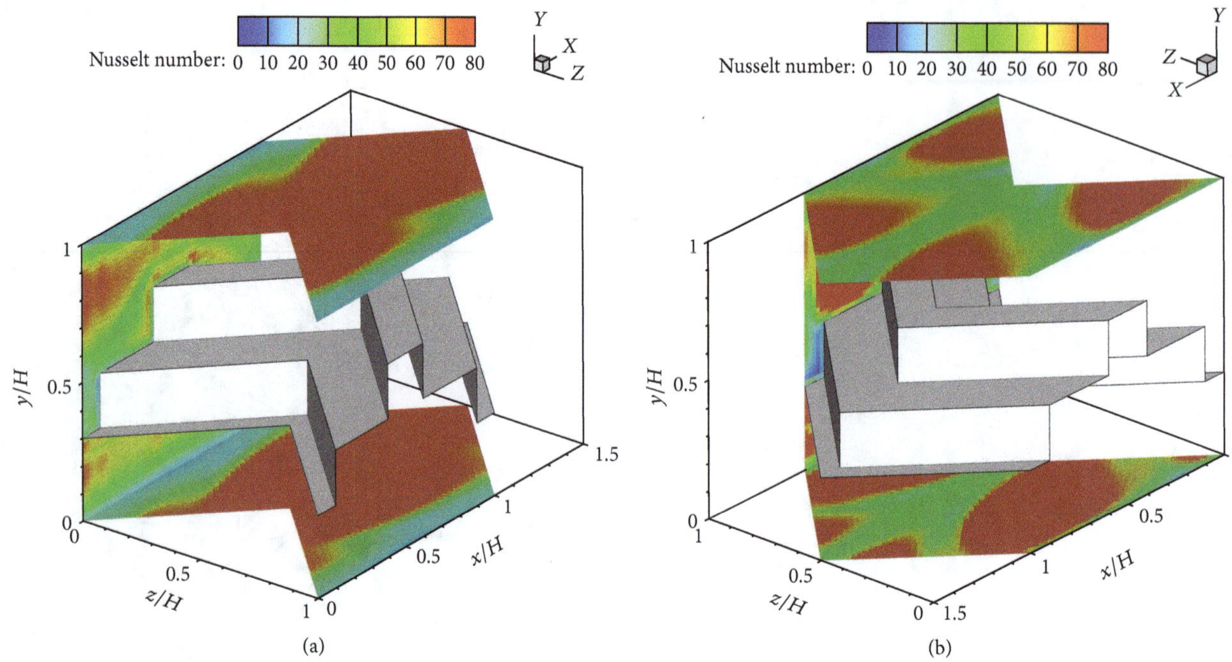


FIGURE 9: Local Nusselt number distribution on the wall surface of the square channel heat exchanger inserted with 45° V-wavy plate at Re = 3000 for (a) V-Downstream and (b) V-Upstream.

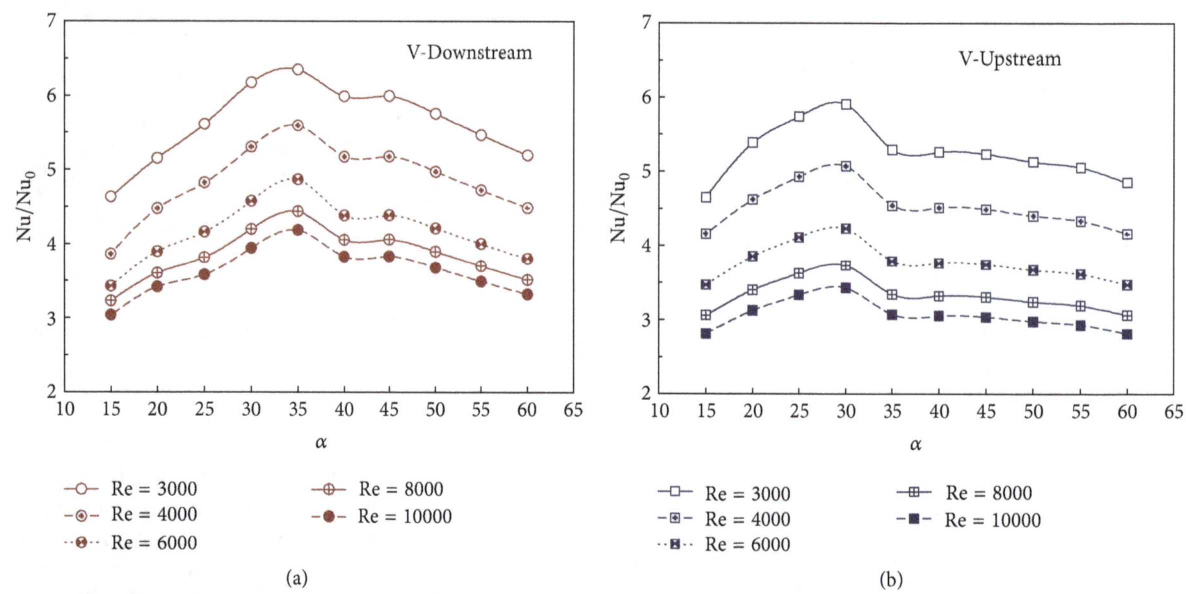


FIGURE 10: $Nu/Nu_0$ versus $\alpha$ for the square channel inserted with V-wavy plate at various Reynolds numbers of (a) V-Downstream and (b) V-Upstream.

wavy plate in the test section can improve the heat transfer rate higher than the smooth channel in all cases ($Nu/Nu_0 > 1$). $Nu/Nu_0$ increases when increasing the Reynolds number. For V-Downstream, $15° \leq \alpha \leq 35°$, the heat transfer rate tends to increase when augmenting the flow attack angle. The heat transfer rate slightly decreases when $\alpha > 35°$. For V-Upstream, $15° \leq \alpha \leq 30°$, the Nusselt number trends to increase when enhancing the flow attack angle. The heat transfer rate slightly reduces when $\alpha > 30°$. The peak of $Nu/Nu_0$ is detected at the flow attack angle of 35° and 30° wavy plate for V-Downstream and V-Upstream, respectively. In range studies, the Nusselt number is around 3–6.5 and 2.8–6 times above the smooth channel with no wavy plate for V-Downstream and V-Upstream, respectively.

The variation of the $f/f_0$ with the flow attack angle for the square channel heat exchanger inserted with wavy plate

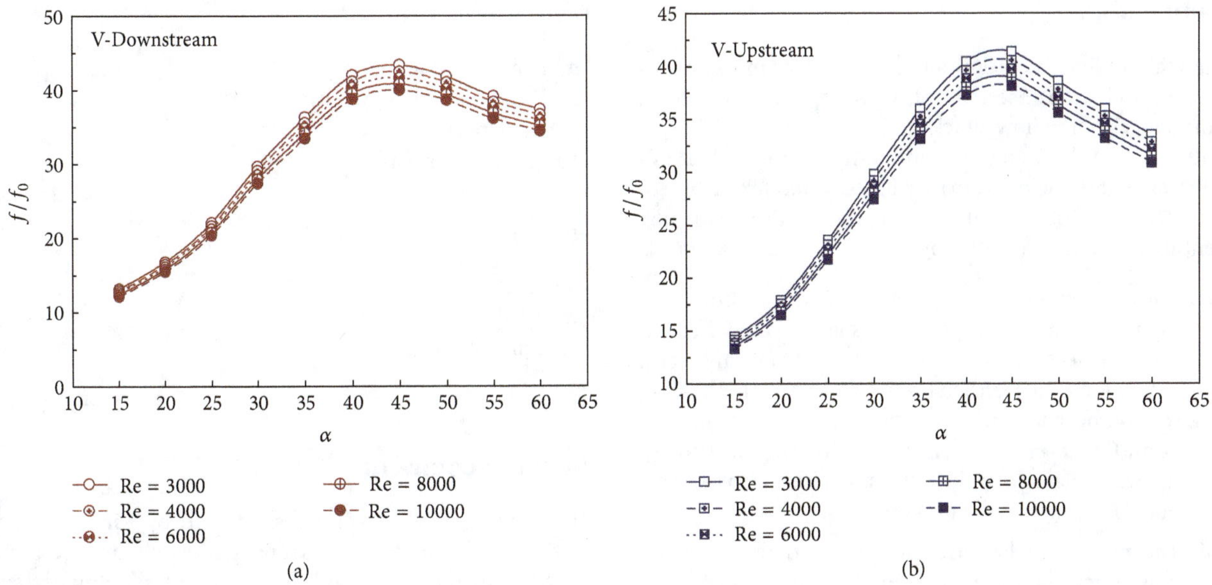


FIGURE 11: $f/f_0$ versus $\alpha$ for the square channel inserted with V-wavy plate at various Reynolds numbers of (a) V-Downstream and (b) V-Upstream.

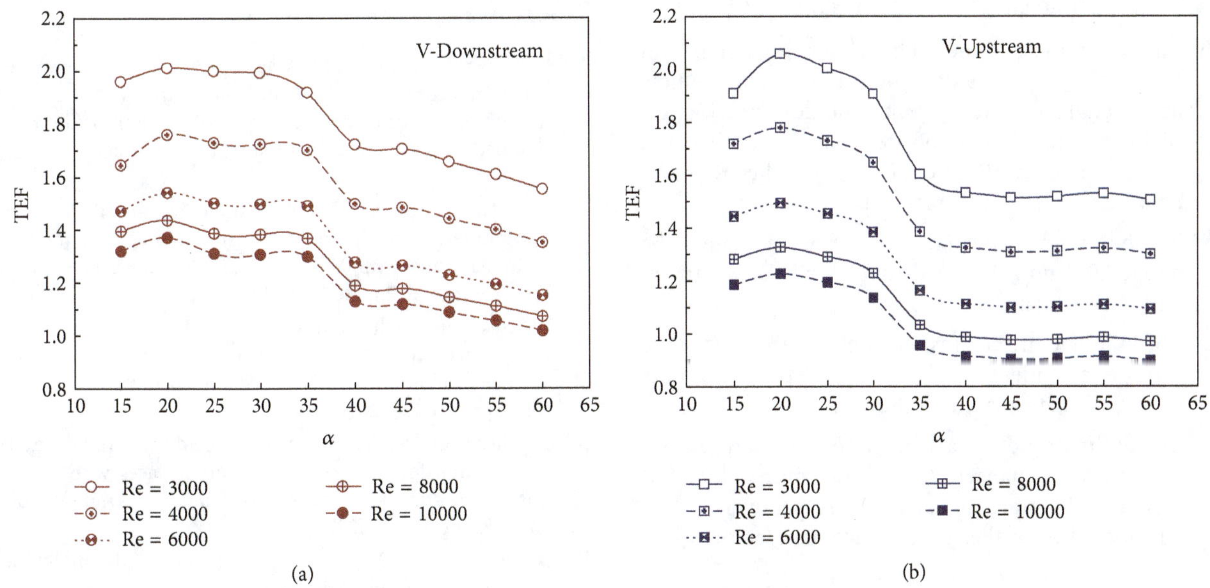


FIGURE 12: TEF versus $\alpha$ for the square channel inserted with V-wavy plate at various Reynolds numbers of (a) V-Downstream and (b) V-Upstream.

at various Reynolds numbers is depicted as Figures 11(a) and 11(b), respectively, for V-Downstream and V-Upstream. The insertion of the wavy plate in the square channel enhances the friction loss higher than the plain channel for all cases ($f/f_0 > 1$). Similar trend of the friction factor is found for both arrangements. The friction loss reduces when augmenting the Reynolds number. In range, $15^\circ \leq \alpha \leq 45^\circ$, the friction loss increases when increasing the flow attack angle. The friction factor tends to decrease when $\alpha > 45^\circ$. The wavy plate with 45° performs the highest friction loss, while the wavy plate with 15° gives the reverse result. The wavy plate with V-Downstream provides the friction loss higher than the smooth channel around 11–44 times but around 15–42 times for V-Upstream.

Figures 12(a) and 12(b) show the relation of the TEF with the flow attack angle at various Reynolds numbers for the square channel heat exchanger inserted with wavy plates for V-Downstream and V-Upstream, respectively. The TEF decreases when increasing the Reynolds number. The optimum TEF is detected at the flow attack angle around 20° for both arrangements. The TEF is around 2.02 and 2.09 for V-Downstream and V-Upstream, respectively, at Re = 3000.

## 5. Conclusion

Numerical prediction on flow and heat transfer mechanisms in the square channel inserted with V-wavy plate is presented. The influences of the flow attack angles ($\alpha = 15^\circ, 20^\circ, 25^\circ, 30^\circ, 35^\circ, 40^\circ, 45^\circ, 50^\circ, 55^\circ$, and $60^\circ$) and V-tip arrangements (V-Downstream and V-Upstream) are investigated for turbulent regime (Re = 3000–10,000). The major findings for the present investigation are as follows:

(i) The addition of the wavy plate in the test section leads to producing the vortex flow or swirling flow through the test section. The vortex flow is an important factor for heat transfer improvement in the heating section because the vortex flow disturbs the thermal boundary layer on the channel walls. The insertion of the wavy plate not only increases in heat transfer rate, but also enhances the pressure loss.

(ii) The maximum heat transfer rate is detected at the flow attack angle around $35^\circ$ and $30^\circ$ of the wavy plate for V-Downstream and V-Upstream, respectively. The wavy plate with V-Downstream and V-Upstream performs the heat transfer rate higher than the smooth channel around 3–6.5 and 2.8–6 times, respectively, for $\alpha = 15^\circ$–$60^\circ$ and Re = 3000–10,000.

(iii) The friction loss of the square channel inserted with the wavy plate is found to be maximum at the flow attack angle around $45^\circ$, while the flow attack angle of $15^\circ$ performs the opposite result for both cases. In range studies, the friction loss is higher than the smooth channel around 11–44 times.

(iv) The optimum TEF is found at $20^\circ$ wavy plate to be about 2.02 and 2.09, respectively, for V-Downstream and V-Upstream.

(v) The manufacture and installation of the V-wavy plate is more convenient than other types of the turbulators such as V-baffle placed on the channel wall.

(vi) The optimum flow attack angles of the wavy surface in the heat exchanger channel for laminar [17] and turbulent flow are $40^\circ$ and $20^\circ$, respectively, when considered at the thermal enhancement factor.

## Nomenclature

$f$: Friction factor
$H$: Channel height
$h$: Convective heat transfer coefficient ($\mathrm{W/m^2\,K}$)
$k$: Thermal conductivity (W/m K)
Nu: Nusselt number
$p$: Static pressure (Pa)
Pr: Prandtl number
Re: Reynolds number
$T$: Temperature (K)
TEF: Thermal enhancement factor
$u_i$: Velocity in $xi$-direction (m/s)
$\overline{u}$: Mean velocity in channel (m/s)
$x, y, z$: Cartesian coordinates.

*Greek Symbols*

$\alpha$: Flow attack angle, degree
$\mu$: Dynamic viscosity (Pa s)
$\rho$: Density ($\mathrm{kg/m^3}$)
$\Gamma$: Thermal diffusivity.

*Subscripts*

in: Inlet
w: Wall
0: Smooth tube.

## Acknowledgments

The funding of this work is supported by King Mongkut's Institute of Technology Ladkrabang research funds (Contract no. KREF046006). The authors would like to thank Associate Professor Dr. Pongjet Promvonge, KMITL, for suggestions.

## References

[1] R. Kumar, R. Chauhan, M. Sethi, and A. Kumar, "Experimental study and correlation development for Nusselt number and friction factor for discretized broken V-pattern baffle solar air channel," *Experimental Thermal and Fluid Science*, vol. 81, pp. 56–75, 2017.

[2] W. Jedsadaratanachai, N. Jayranaiwachira, and P. Promvonge, "3D numerical study on flow structure and heat transfer in a circular tube with V-baffles," *Chinese Journal of Chemical Engineering*, vol. 23, no. 2, pp. 342–349, 2015.

[3] W. Jedsadaratanachai and A. Boonloi, "Effects of blockage ratio and pitch ratio on thermal performance in a square channel with 30o double V-baffles," *Case Studies in Thermal Engineering*, vol. 4, pp. 118–128, 2014.

[4] S. Caliskan and S. Baskaya, "Velocity field and turbulence effects on heat transfer characteristics from surfaces with V-shaped ribs," *International Journal of Heat and Mass Transfer*, vol. 55, no. 21-22, pp. 6260–6277, 2012.

[5] D. Jin, J. Zuo, S. Quan, S. Xu, and H. Gao, "Thermohydraulic performance of solar air heater with staggered multiple V-shaped ribs on the absorber plate," *Energy*, vol. 127, pp. 68–77, 2017.

[6] N. S. Deo, S. Chander, and J. S. Saini, "Performance analysis of solar air heater duct roughened with multigap V-down ribs combined with staggered ribs," *Journal of Renewable Energy*, vol. 91, pp. 484–500, 2016.

[7] P. Promthaisong, P. Eiamsa-ard, W. Jedsadaratanachai, and S. Eiamsa-ard, "Turbulent heat transfer and pressure loss in a square channel with discrete broken V-rib turbulators," *Journal of Hydrodynamics, Series B*, vol. 28, no. 2, pp. 275–283, 2016.

[8] X. Fang, Z. Yang, B.-C. Wang, M. F. Tachie, and D. J. Bergstrom, "Highly-disturbed turbulent flow in a square channel with V-shaped ribs on one wall," *International Journal of Heat and Fluid Flow*, vol. 56, pp. 182–197, 2015.

[9] R. Maithani and J. S. Saini, "Heat transfer and friction factor correlations for a solar air heater duct roughened artificially

with V-ribs with symmetrical gaps," *Experimental Thermal and Fluid Science*, vol. 70, pp. 220–227, 2016.

[10] D. Jin, M. Zhang, P. Wang, and S. Xu, "Numerical investigation of heat transfer and fluid flow in a solar air heater duct with multi V-shaped ribs on the absorber plate," *Energy*, vol. 89, pp. 178–190, 2015.

[11] B. Lotfi, M. Zeng, B. Sundén, and Q. Wang, "3D numerical investigation of flow and heat transfer characteristics in smooth wavy fin-and-elliptical tube heat exchangers using new type vortex generators," *Energy*, vol. 73, pp. 233–257, 2014.

[12] J. Dong, L. Su, Q. Chen, and W. Xu, "Experimental study on thermal-hydraulic performance of a wavy fin-and-flat tube aluminum heat exchanger," *Applied Thermal Engineering*, vol. 51, no. 1-2, pp. 32–39, 2013.

[13] J. Dong, J. Chen, W. Zhang, and J. Hu, "Experimental and numerical investigation of thermal -hydraulic performance in wavy fin-and-flat tube heat exchangers," *Applied Thermal Engineering*, vol. 30, no. 11-12, pp. 1377–1386, 2010.

[14] J. Gong, C. Min, C. Qi, E. Wang, and L. Tian, "Numerical simulation of flow and heat transfer characteristics in wavy fin-and-tube heat exchanger with combined longitudinal vortex generators," *International Communications in Heat and Mass Transfer*, vol. 43, pp. 53–56, 2013.

[15] X. Du, L. Feng, L. Li, L. Yang, and Y. Yang, "Heat transfer enhancement of wavy finned flat tube by punched longitudinal vortex generators," *International Journal of Heat and Mass Transfer*, vol. 75, pp. 368–380, 2014.

[16] X. Du, L. Feng, Y. Yang, and L. Yang, "Experimental study on heat transfer enhancement of wavy finned flat tube with longitudinal vortex generators," *Applied Thermal Engineering*, vol. 50, no. 1, pp. 55–62, 2013.

[17] A. Boonloi and W. Jedsadaratanachai, "Laminar forced convection and heat transfer characteristics in a square channel equipped with V-wavy surface," *Journal of Mathematics and Statistics*, vol. 13, no. 3, pp. 251–260, 2017.

# 10

# A Concise Presentation of Doubly Fed Induction Generator Wind Energy Conversion Systems Challenges and Solutions

**Julius Mwaniki,[1] Hui Lin,[1] and Zhiyong Dai[2]**

[1] *Department of Electrical Engineering, School of Automation, Northwestern Polytechnical University, Xi'an 710129, China*
[2] *Automatic Control Department, School of Mechano-Electronic Engineering, Xidian University, Xi'an 710071, China*

Correspondence should be addressed to Julius Mwaniki; jukamwa@gmail.com

Academic Editor: Kamran Iqbal

There is increased worldwide wind power generation, a large percentage of which is grid connected. The doubly fed induction generator (DFIG) wind energy conversion system (WECS) has many merits and, as a result, large numbers have been installed to date. The DFIG WECS operation, under both steady state and fault conditions, is of great interest since it impacts on grid performance. This review paper presents a condensed look at the various applied solutions to the challenges of the DFIG WECS including maximum power point tracking, common mode voltages, subsynchronous resonance, losses, modulation, power quality, and faults both internal and from the grid. It also looks at approaches used to meet the increasingly stringent grid codes requirements for the DFIG WECS to not only ride through faults but also provide voltage support. These are aspects of the DFIG WECS that are critical for system operators and prospective investors and can also serve as an introduction for new entrants into this area of study.

## 1. Introduction

Wind power, projected at 500 GW at the end of 2016, is expected to supply 5% of electrical power worldwide. Growth is expected to continue, hitting 1900 GW by the end of 2020 [1]. The merits of the doubly fed induction generator (DFIG), shown in Figure 1, make it one of the most widely used generators for wind energy conversion systems (WECS). The direct drive gearless permanent magnet synchronous generator (PMSG), a newer technology, has come up to challenge its market share, but the large number of installed machines ensures the DFIG WECS cannot be overlooked in the wind industry. In addition, new projects, such as the largest wind farm in Africa at 310 MW being set up in Kenya, continue to utilize the DFIG machine.

With aerodynamic, mechanical, and electrical subsystems making up the DFIG WECS quite complex, challenges are inevitable. Wind as a prime mover only adds to these by its intermittent and uncertain nature. To complicate matters further, more stringent grid codes have been instituted due to increased wind power penetration especially as they concern fault ride through (FRT) [2, 3]. This paper seeks to present these challenges and their applied solutions. No other paper has, to the author's knowledge, tackled the issues at one go, giving an easy to understand overall picture of the DFIG WECS.

This paper is divided into five sections. Starting with an introduction in Section 1, Section 2 covers challenges faced by the DFIG WECS and Section 3 gives insight into solutions that have been applied for operation under grid faults. Section 4 presents the discussion while in Section 5 conclusions are drawn.

## 2. Challenges

Control and operation of the DFIG WECS present some unique challenges due to the large number of interacting subsystems which cut across different disciplines. These interactions determine the power supplied to the grid. The challenges to connection of wind power to the grid are threefold.

*(a) Source Induced.* The fluctuating nature of wind results in variable generation with the associated power fluctuations, voltage fluctuations, and frequency deviations. Source

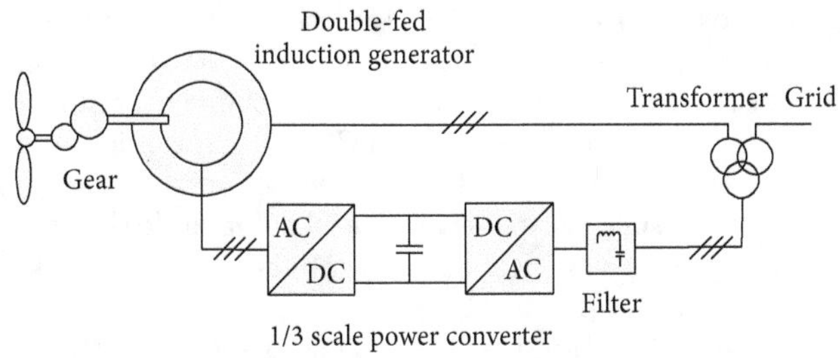


FIGURE 1: Configuration of a doubly fed induction generator wind energy conversion system [4].

induced fluctuations are critical on the medium-term basis in the timespan of seconds to minutes.

*(b) Machine Induced.* The inherent nature of the DFIG machine, as a rotating electromagnetic machine, also results in cyclical power/voltage fluctuations due to characteristics such as magnetic saturation and hysteresis. The tower also causes shadow effect as it blocks the air flows causing imbalanced forces as the rotating turbine blades come in direct line with the tower.

*(c) Grid Induced.* These originate from the grid system and include grid instability and transients.

Various challenges and proposed ways of handling them are given next.

*2.1. Maximum Power Point Tracking (MPPT).* The DFIG WECS, like any other renewable energy extraction system, seeks to maximize the extracted power by maximizing the system efficiency in a bid to shorten the installation cost return period. A shorter cost return period makes the project more appealing and competitive when viewed against the conventional energy sources [10]. This is critical considering that the maximum power extractable is limited by the strength of the renewable source which in most cases may additionally be varying. MPPT for WECS seeks to run the generator at an optimal speed relative to the wind speed as experienced by the turbine rotors. Many techniques have been proposed to achieve MPPT and include the following [10–13]:

(a) Lookup table based including power signal feedback (PSF), optimal torque (OT), and tip speed ratio (TSR) control.

(b) Nonlinear state space and state space linearization based.

(c) Neural network-fuzzy logic based.

(d) Hill climbing search (HCS) based.

(e) Hybrid hill climbing search/lookup table based including

(i) variable step size HCS,
(ii) dual step size HCS,
(iii) search-remember-reuse HCS,
(iv) modified HCS to avoid generator stall,
(v) modified PSF to avoid generator stall,
(vi) adaptive OT control,
(vii) modified OT for fast tracking,
(viii) adaptive TSR control,
(ix) limit cycle based HCS,
(x) disturbance injection based HCS,
(xi) self-tuning sensorless adaptive HCS.

However, these are not without challenges, the most common being the following:

(a) There exists requirement for additional hardware such as generator speed sensor or observer and a wind speed sensor such as an anemometer. These components introduce errors such as that due to location of the anemometer on the nacelle which is at hub height and either behind the rotor for upwind machines or before the rotor for downwind machines. The anemometer thus does not cover the whole extent of the rotors with a resultant error in measured wind speed compared to the effective wind speed experienced by the rotors.

(b) The high system inertia of the wind turbine and generator does not respond fast enough in relation to wind speed changes that could be in seconds.

(c) There is the change of system parameters due to environmental changes such as air density change with seasons, aging from thermal, mechanical, or electrical system stresses, and generator loading conditions which result in changes on power converter and generator efficiencies as the load currents determine copper losses while rotational speed determines eddy current losses. As a result, no unique curve for optimal output electrical power exists and there are differences between the peaks for the input mechanical and the output electrical powers.

(d) There exist the requirements for prior knowledge of system parameter values which suffer drift.

An optimal MPPT technique should have characteristics that include adaptive tracking, self-tuning capability, simplicity of implementation, speed of convergence, and accuracy. The self-tuning sensorless adaptive HCS is the most attractive of the techniques above as in addition to exhibiting the above characteristics, it is mechanically sensorless and has minimal memory requirement.

*2.2. Common Mode Voltage (CMV).* CMV is voltage generated between the neutral point of the machine and the ground when pulse width modulation (PWM) technique is used in a three-phase converter. It results in shaft voltage and bearing currents which cause bearing material erosion and, consequently, bearing failure [14–16].

CMV can be reduced by the following:

(a) Appropriate selection of modulation strategy: this has the unfortunate effect of possibly increasing current harmonic output depending on the strategy adopted.

(b) Shaft grounding using earthing bushes and slip rings.

(c) Use of insulated/ceramic bearings or conductive lubricants.

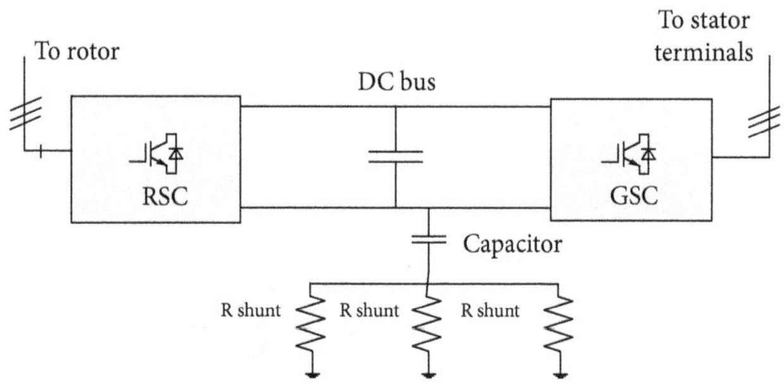


FIGURE 2: An RC bus filter [5].

(d) Increasing the converter voltage levels.

(e) Use of filters such as the RC bus filter whichconsists of a capacitor connected to the negative point of the DC bus on one end and to the earth on the other end through three resistors with a control that detects and supervises current circulation in the resistors [5]. This is shown in Figure 2.

*2.3. Subsynchronous Resonance (SSR).* This refers to resonance at subsynchronous frequencies that occurs between the mechanical parts of the DFIG and a grid system. It occurs when the DFIG is connected to a grid that is series compensated in a bid to increase system stability and power transfer capability. SSR increases as the level of series compensation in the power system increases. SSR in electrical networks appears in three ways:

(a) Torsional interactions: a situation where energy is exchanged between the grid system and the generator shaft system in an ever-increasing manner which can cause damage to the turbine-generator shaft. This is rare in wind farms as the low shaft stiffness in DFIG WECS drive train ensures low frequency torsional modes.

(b) Transient torsional amplification.

(c) Induction generator effects: when the effective stator and grid resistance at machine terminals is negative, the stator subharmonic currents interact with the rotor circuit inducing high voltages.

SSR can be damped by use of flexible AC transmission systems (FACTS) devices or intelligent control. The grid side converter (GSC) of the power electronic converter (PEC), due to its similar structure to a STATCOM, can also be controlled to damp SSR. SSR damping capability of DFIG increases with increase in wind speed [17–20].

*2.4. Power Quality.* The DFIG, being a rotating electrical machine and switching in the PEC, introduces harmonics and subharmonics that could result in voltage distortions at point of common coupling (PCC) to the grid. This effect increases as the number of DFIG WECS connected to the grid increases. The voltage distortions resulting from the rotor side converter (RSC) of the PEC cause increase in eddy current losses in the generator. Undesirable fluctuations in generated active and reactive power result from harmonics in rotor currents while harmonics in stator currents result in power quality deterioration at the PCC. The harmonics from the converter are reduced if the number of voltage levels is increased. As such, a three-level neutral point diode clamped (NPC) converter has fewer harmonics than a two-level back-to-back converter [15, 21, 22].

Compensation for reactive power, current and voltage harmonics, current and voltage imbalances, voltage flicker, voltage regulation, and neutral current is carried out using filters. An LCL filter is usually used to suppress the harmonics [4]. However, due to the large capacitance of the converter, the switching frequency is low necessitating use of a bulky filter [22]. Active filters are also used, some utilizing the extra capacity in RSC or GSC, for active filtering of harmonics by compensating the reactive power and harmonic currents. Hybrid filters, consisting of active and passive filters, are also an option [23–26].

*2.5. Modulation.* Converter switching can be modulated in different ways. Carrier based modulation is produced by triangular wave intersection with a reference modulating sinusoidal wave signal while space vector modulation (SVM) uses direct digital technique to determine the appropriate switching to be done. Sinusoidal pulse width modulation (SPWM), though simple in approach, has disadvantages as it has higher total harmonic distortion (THD) than SVM. SVM also has higher efficiency and reliability as it uses discontinuous PWM (DPWM), where switching is done only when necessary, resulting in lower converter power losses than for the continuous PWM (CPWM) applied in SPWM. DPWM, however, results in varying switching frequency which makes design of filters difficult [15, 27].

*2.6. Losses in Voltage Source Converter (VSC) Components.* Higher currents flow in the RSC than in the GSC. To accommodate the higher currents and ensure equal power device loading, the RSC cell has two parallel power devices for each device in the GSC cell. The semiconductor devices have conduction and switching power losses contributing to thermal losses with corresponding temperature rise. The losses reduce the efficiency of the VSC and subject the power electronics to thermal cyclic stresses. In subsynchronous mode, the most stressed device in the RSC is the insulated gate bipolar transistor (IGBT) while it is the freewheeling diode in the GSC. The reverse is true in supersynchronous mode.

Thermal cycling can be on

(a) long-term basis of periods in days or months from environmental disturbances such as wind speed and ambient temperature variations,

(b) medium-term basis of periods in minutes to seconds from speed control to attain maximum power production or pitch control to limit generated power to rated values,

(c) short-term basis with period ranging below one second from fast electrical disturbances such as grid faults, VSC switching, and load current alternation.

Thermal cycling can be reduced by circulating reactive power internally in the VSC without power factor distortion to the power system. This results in reduced thermal stresses on the semiconductor devices hence increasing their lifetime and reliability. However, the range of reactive power available for circulation in the RSC is limited by the generator reactive power capacity while in the GSC; it is limited by both the generator reactive power capacity and the DC-link voltage. Increase in converter voltage levels also decreases thermal losses [15, 28–30].

*2.7. Faults.* Faults can come from the DFIG WECS itself or from the grid it is connected to.

*2.7.1. Faults from the Machine.* Faults in the DFIG WECS occur in four main areas.

*(i) Turbine Faults.* These include rotor blade strikes by lightning, blade fracture due to excessive loading from wind gusts or icing as well as hydraulic system failure. Some applied solutions include lightning rods fitted on the blades as well as blades designed to stall in high winds as well as to reduce amount of ice buildup.

*(ii) Gearbox Faults.* Gearbox failures are rare, expected once or twice in the WECS lifetime, but they are the most costly as they usually involve replacement of the gearbox at around 10% of total cost of the WECS. In addition, they cause twice or thrice the downtime for failures by other components [31, 32]. These mechanical failures are due tothe following:

(i) manufacturing errors such as grind temper or material inclusions,

(ii) insufficient or absence of lubrication that causes surface related problems, such as

  (a) scuffing: the adhesive wear resulting in detachment and transfer of particles from meshing teeth,

  (b) micropitting: the creation of numerous surface cracks leading to surface fatigue,

(iii) fretting problems which are from small vibratory motions, especially when the WECS is parked [33].

Overheating of the lubrication oil can, in some cases, result in fires. Correct and sufficient lubrication, with lubricants of high quality and lifespan, coupled with efficient cooling for the lubricant is adequate to prevent most failures.

*(iii) Generator Faults.* These include

(i) open circuit in windings,

(ii) winding insulation failures leading to inter-turn winding short circuits due to overheating as a result of overload or unequal heating resulting from imbalanced voltages [34].

The integrity of the windings and the insulation is maintained by controlling operating temperatures through voltage and current regulation.

*(iv) PEC Faults.* These include the following:

(i) Semiconductor device failures: these include failure of the IGBTs, diodes, and bond wire connections. They result mainly from thermal cycling as the semiconductors switch on and off periodically and unevenly. Since they are made up of different materials, the different coefficients of expansion result in stresses due to expansion and compression forces [4]. Other faults include IGBT open and short circuits. An open circuit results in fluctuations in both active and reactive power outputs [35, 36]. IGBT fire-through, where a switch conducts before the time determined by the switching control system, forms an intermittent short circuit that negatively affects the terminal voltage [37].

(ii) DC-link capacitor failures: the bulky electrolytic DC-link capacitor is one of the weak links in the VSC and contributes 30% of failures in AC converters [38]. Imbalanced voltages result in active power oscillations which cause voltage fluctuations across the capacitor as well as second-order harmonic currents to flow through it. These raise the power loss and operating temperature, hence increasing the electrolytes rate of evaporation and shortening the capacitors' lifespan [39].

DC-link capacitor failure could be short circuit or open circuit. A short circuit fault results in a more severe effect on generator speed, voltage drop at PCC, reactive power absorbed as well as stator current spike, and oscillations than an open circuit fault. It makes the DC-link voltage to collapse, with currents flowing to supply the fault from both converters through the freewheeling diodes. The faults result in reduced active power output which drops to zero in case of unsuccessful fault clearance [21, 40].

(iii) Filter failures: filter failure results in degradation of power quality at PCC due to harmonics resulting from the VSC switching operations [4].

Whereas gearbox failures result in most downtime and cost, they are rare, accounting for only 5% of failures while PEC failures are most common, accounting for over 50% of all cases of DFIG WECS failures [4, 41].

*2.7.2. Faults from the Power System.* Faults from the power system also impact on the DFIG WECS. They include voltage sags, voltage swells, and frequency fluctuations with voltage sags being more frequent than swells. As a result, there is greater focus on low voltage ride through (LVRT) than high voltage ride through (HVRT). The major cause of voltage sags is system faults such as short circuits and faults to ground. Other causes include sudden loss of large generating units, transformers energizing, and switching in of large loads usually induction motors [11, 42–44]. However, voltage swell sometimes occurs after voltage sag due to switching

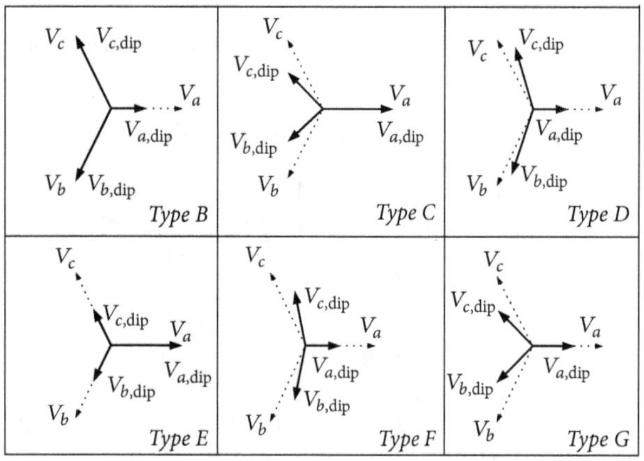


FIGURE 3: Fault types B to G [6].

off of large loads due to the voltage sag, reactive power overcompensation from capacitor banks, or single-phase short-term interruptions [45, 46].

*(i) Types of System Faults.* The five types of system faults encountered by the DFIG WECS from the grid fall into two categories [47]:

(i) Balanced/symmetrical faults: these are

(a) three-phase,
(b) three-phase to ground.

(ii) Unbalanced/asymmetrical faults: these are

(a) two-phase,
(b) two-phase to ground,
(c) one-phase to ground.

The five types of system faults have different percentages of occurrence as given next [48]:

| *Type of fault* | *% of total fault occurrence* |
|---|---|
| One-phase to ground fault | 70–85 |
| Two-phase faults | 8–15 |
| Two-phase to ground fault | 4–10 |
| Three-phase faults | 3–5 |

The five types of faults are further categorized into seven types, types A–G, depending on the changes on phase-to-neutral voltage magnitudes and phase-angle jumps if any. These are as follows [47, 49, 50]:

(i) Type A is the three-phase symmetrical fault where all three-phase-to-neutral voltages show equal drop in magnitude, with no phase jump.

(ii) Types B to G are shown in Figure 3 where dotted lines indicate voltage before faults and solid lines indicate voltage after fault.

*(ii) Effect of Transformers on Faults.* The voltage sag is sensed at the high voltage side of the connecting transformer to the PCC. This introduces transformation to the faults resulting in a different type of fault being seen at the DFIG terminals, in some cases, depending on the type of fault and the transformer connection [51, 52]. The common transformer connection types are given in Figure 4.

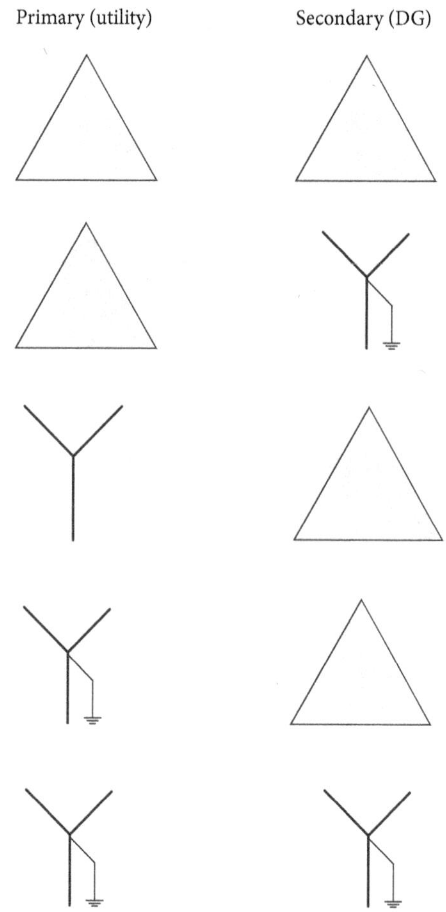


FIGURE 4: Interconnection transformer connections [7].

The effect of transformers on faults is given in Table 1 [6].

Fault types B and E do not appear at the generator terminals with $Y$-$\Delta$ connected transformer as they involve zero sequence components that are absorbed by the transformer connection. The DFIG stator isolated $Y$ or $\Delta$ winding connection has the same effect [34, 53].

*(iii) Effect of Faults on DFIG WECS.* The effect of voltage sag on the generator depends on the following [12]:

(1) Type of fault: the fault components to be dealt with depend on the type of fault. Three-phase faults only involve positive sequence fault current components with a high surge value and so are more severe. The single-phase to ground and two-phase to ground faults introduce negative and zero sequence components in addition to the positive sequence components. Two phase faults have only positive and negative sequence components.

The negative sequence voltages result in high negative sequence current components that cause double grid frequency oscillations on active power, reactive power, and

TABLE 1: Fault propagation inside a wind farm.

| Fault | Fault at PCC (HV terminals of wind farm transformer) | Fault at wind farm transformer LV terminals with $Y$-$\Delta$ connection | Fault at DFIG terminals with $Y$-$\Delta$ connected transformer | Fault at DFIG terminals with transformer of connection type other than $Y$-$\Delta$ |
|---|---|---|---|---|
| Three-phase | Type A | Type A | Type A | Type A |
| Single-phase | Type B | Type C* | Type D* | Type C* |
| Phase to phase | Type C | Type D | Type C | Type D |
| Two-phase to ground | Type E | Type F | Type G | Type F |

*In this case, the characteristic voltage is (1/3 + 2/3 $V_{sag}$) instead of being equal to $V_{sag}$.

electromagnetic torque resulting in stator winding and drive train lifetime reduction [7, 13, 47, 54]. The negative sequence stator flux rotates in opposite direction to the rotor and hence induces a high frequency component in the rotor winding at a frequency given by

$$\omega_e^- = \omega_s + \omega_r. \quad (1)$$

If the terminal voltage does not drop to zero, then there is also the component induced by the positive sequence stator flux which rotates at

$$\omega_e^+ = \omega_s - \omega_r, \quad (2)$$

where $\omega_s$ is stator frequency, $\omega_r$ is rotor rotational speed, $\omega_e$ is frequency of current induced in the rotor by stators' magnetic field, and the superscripts + and − represent the positive and negative sequence components, respectively [55].

(2) Phase-angle jump: a phase angle jump occurs when the $X/R$ ratio of the fault impedance is different from that of the source impedance. $X/R$ ratio is usually small for overhead transmission lines but big where high voltage cables are utilized. If the type of fault involves phase-angle jump, it results in higher negative sequence voltage component and consequently elevated rotor current and DC-link voltage [6].

(3) Depth of sag: this depends on distance of fault location from the generator terminals. A fault is deemed remote if it results in a voltage dip of <30%. Higher values of rotor current and DC-link voltage are experienced with increase in sag depth. The different types of faults A–G have specific depths beyond which the fault effect is most severe [47, 56].

(4) Initial point-on-wave instant of fault: the worst point-on-wave instant is when the positive and negative sequence components of stator flux are aligned in opposite directions. This results in high rotor currents because the resultant stator-forced flux is at a minimum, necessitating a large natural flux to avoid discontinuity in the stator flux trajectory since this cannot change instantaneously at the instant of fault occurrence [57].

(5) Prefault wind speed: this determines the prefault generator speed which in turn determines generator slip and resultant active power output. The d- and q-stator, RSC, and GSC currents flowing are thus obtained. The effect is critical when the machine is operating at its rated limit as the rated current is being utilized for active power generation in the stator and DC-link voltage control in the GSC. Thus, there is less reserve capacity available for generation of reactive power in both. In subsynchronous mode, power flows from the grid through the VSC to the rotor. This results in high DC-link voltage and a higher initial fault current magnitude [7].

(6) Duration of fault: longer lasting faults have greater negative effects, with voltage drops lasting longer than one second being considered as voltage interruptions. The different types of faults A–G have specific periodic durations for which the fault effect is most severe. This is the time when the current at fault recovery is farthest from prefault current [47, 53].

(7) Fault recovery process: if the voltage recovery is considered as a one-step process, large natural flux component is needed as stator-forced flux trajectory continuity must be ensured. Thus smaller natural flux components can be ensured by considering fault recovery on two or three steps. This is more realistic as the breaker phase operation is at natural zero current crossing points. Faults recovered in two steps have similar effects on the generator. The same goes for faults recovered in three steps. Smaller natural flux components increase the range of faults a generator can withstand [7, 53].

(8) Grid strength: the strength of a grid in terms of its short circuit ratio determines how severe the fault effect will be. A strong grid interconnection can withstand a higher rate of recovery, thus minimizing FRT requirement on the generator [57].

Generally rotor fault current is highest for high depth three-phase fault when the generator is delivering full power at supersynchronous speed [58].

*(iv) DFIG Response to Faults.* Both voltage sags and swells result in a large electromotive force being induced in the rotor circuit by the transient stator flux. This sets up a surge current in the rotor circuit which can destroy the PEC semiconductor devices as they cannot withstand currents of more than 200% of their nominal rating [45]. In addition, the low voltage at the PCC reduces the GSC capacity to transfer power to the grid as to square of the system fault voltage. This leads to excess power in the DC-link capacitor resulting in increased DC-link voltage. Protection systems are put in place to ensure the high currents do not destroy the RSC and the high voltage does not cause DC-link capacitor failure.

The rotor circuit is protected from these conditions by the passive crowbar.This is a hardware protection for the PEC.

It short-circuits the three phases of the generator rotor once triggered by the overvoltage of the DC bus voltage. Operation of the passive crowbar also sets into effect an emergency shutdown of the DFIG [5]. However, this is not ideal since DFIG needs to be kept connected to the grid.

## 3. Solutions to Power System Faults

Protection for DFIG has been approached in two different ways:

(i) Approaches that seek to keep the generator running and connected to the grid. This is by

  (a) avoiding tripping the machine by keeping monitored parameters, that is, rotor overspeed, stator and rotor overcurrents, and rotor and DC-link overvoltage, within limits;

  (b) maintaining a healthy voltage at the generator terminals irrespective of the fault and post-fault grid voltage value.

(ii) Approaches that seek to keep the generator connected so as to provide reactive power to the system at PCC and thus support the voltage.

The above approaches are implemented by one of the following:

(i) Addition of new devices at PCC such as FACTS devices like static var compensators (SVC), static synchronous compensator (STATCOM), static series compensator (SSC), and unified power flow compensator (UPFC).

(ii) Addition of new components to the DFIG stator circuit such as series dynamic braking resistor (SDBR), dynamic voltage restorer (DVR), and antiparallel thyristors.

(iii) Addition of new components to the DFIG rotor circuit such as crowbar circuit and SDBR.

(iv) Addition of new components to the PEC such as braking chopper circuit, battery energy storage system (BESS), and energy capacitor storage (ECS).

*3.1. Keeping the DFIG Connected to the Grid but Not Offering Support to Grid.* This is achieved by use of one or a combination of the following methods:

*(a) Active Crowbar.* An active crowbar consists of semiconductor switches, either IGBTs or diodes with resistors. When rotor overcurrent or high DC-link voltage is detected, and then a check of the grid voltage reveals a voltage sag or swell, the control system disconnects the RSC by blocking the firing pulses. The active crowbar circuit is then connected, rerouting the high rotor currents through additional crowbar resistances. The increased rotor circuit resistance causes the high rotor currents to decay. Once the rotor currents reach a safe level, the crowbar is deactivated and the RSC firing pulses are resumed [5]. During the period when the active

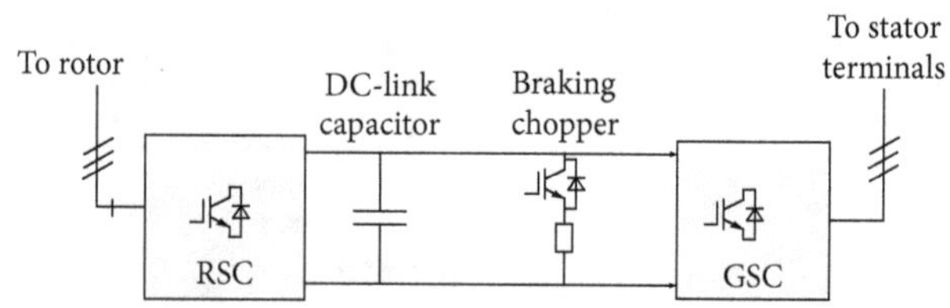


FIGURE 5: Power converter showing the chopper circuit [5].

crowbar is activated, the DFIG operates as a wound rotor induction generator (WRIG) with increased rotor resistance but no control over both active and reactive power output. As a result, reactive power cannot be availed for voltage support at a time when it is critically needed. Also, the DFIG when operating as a WRIG consumes reactive power, further delaying voltage recovery. To avoid this, the operation time of the crowbar circuit must be minimized by increasing the value of the crowbar resistance.

However, increasing the crowbar resistance beyond a certain point, while decreasing the current decay time, results in high voltages at the RSC terminals. These set up a charging current through the RSC diodes and the DC-link capacitor which charges up the capacitor, raising the DC-link voltage. If this voltage exceeds the set limits, it can trigger the active crowbar again after its first deactivation, thus delaying voltage recovery, or even the passive crowbar, leading to a machine shutdown. In addition to the rotor current and DC-link voltage, the values of the active crowbar resistances also influence the rotor speed and the active and reactive powers when the crowbar is active. The correct value of crowbar resistance is thus critical.

*(b) Braking Chopper Circuit.* It consists of an IGBT with a snubber capacitor and a chopper resistor and is connected in parallel with the DC-link capacitor. It aids in keeping the DC bus voltage under the given limits during times when the active crowbar cannot control the voltage all by itself. A braking chopper dissipates excess power in the DC bus through the chopper resistor. Figure 5 shows a braking chopper circuit.

By utilizing a crowbar circuit, with or without a DC-link chopper, the DFIG machine can be kept connected during the fault by protecting it from the rotor overcurrent, rotor overvoltage, stator overcurrent, DC-link overvoltage, and rotor overspeed [59–61].

*(c) Energy Storage System (ESS).* The use of a braking chopper circuit is enhanced by use of an energy storage system such as a BESS [62] or ECS connected across the DC bus. The ESS stores the excess power and so avoids a DC-link overvoltage. The energy so stored is later released through the GSC to the power system when normal voltage is restored thus helping improve the power quality [13].

*(d) Series Dynamic Braking Resistor (SDBR).* SDBR consists of a set of three-phase resistors connected across the three phases of either the stator circuit or the rotor circuit between the rotor and the PEC. A bypass switch keeps the resistors out

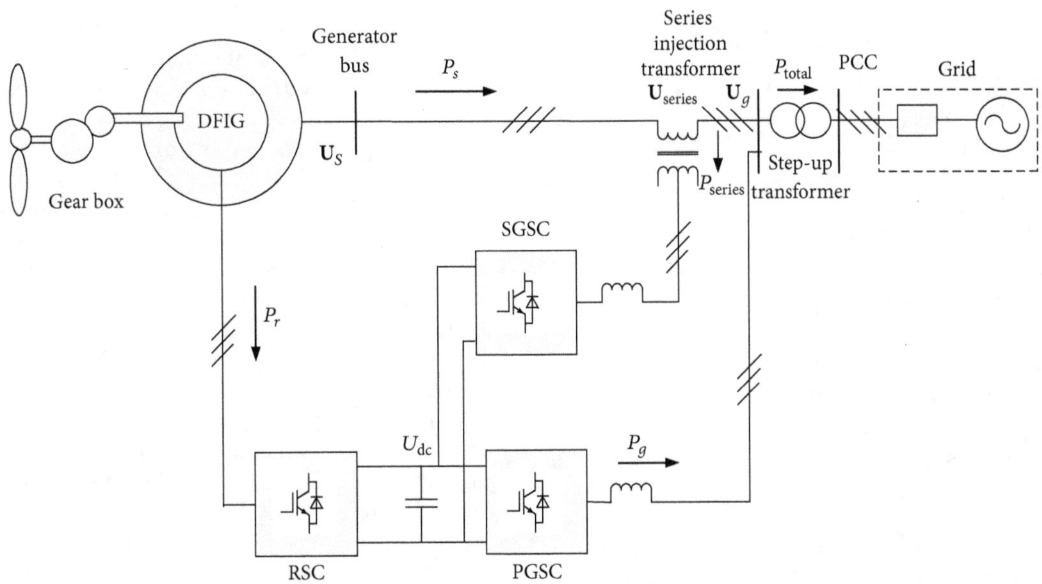


FIGURE 6: Configuration of a DFIG with SGSC [8].

of the circuit during normal operation and opens to insert them into the circuit during fault conditions. The resistors introduce a voltage divider action ensuring that the generator sees the normal voltage at its terminals. At the same time, the added resistance reduces fault currents. The resistors also dissipate part of the generated power thus avoiding rotor overspeed. A low value of the resistance, as well as its fast insertion once fault occurs and for a short duration as possible, leads to improved DFIG stability [13, 63–66].

*(e) Dynamic Voltage Restorer (DVR).* A DVR is applied in series with the stator to maintain stator terminal voltage at nominal values once fault is detected. The DFIG generates active power albeit at reduced level since MPPT is disabled. Since the export of power to the grid is curtailed, the DVR absorbs and stores or dissipates the difference between power produced by generator and power exported to grid. The power can be dissipated in its braking resistance [67, 68].

*(f) Superconducting Magnetic Energy Storage (SMES).* A SMES is connected to the PCC just like FACTS devices to store energy under normal operating or high voltage conditions. The energy so stored is later released to the PCC when grid energy goes down due to a fault hence maintaining the voltage at PCC. The DFIG thus does not experience the fault [69].

*(g) Series Grid Side Passive Impedance Network.* In these, an equivalent impedance $Z_{eq}$ is inserted into the system on occurrence of the voltage sag to ensure the stator flux $\lambda_s$, stator current $I_s$ and rotor current $I_r$, and torque $T_e$ do not change during the sag.

$$Z_{eq} = \left(\frac{V_g}{I_s}\right) + Z_g, \quad (3)$$

where $V_g$ is prefault equivalent electromotive force of grid, $Z_g$ is prefault equivalent impedance of grid, and $I_s$ is prefault equivalent stator current [70].

*(h) Virtual Resistance/Inductance.* In an approach similar to the series grid side passive impedance networks, a virtual resistance is introduced in the equivalent rotor circuit using a LVRT controller to decrease rotor fault currents and its decay time while a virtual inductance increases the leakage inductance of the rotor circuit with similar results. Torque oscillations are also reduced in both approaches [71, 72].

*(i) Flux Damping/Demagnetization.* The rotor current is controlled to counteract the dc and negative components in the stator flux that occur during a fault. To obtain the necessary current to maintain control, temporary overloading of the RSC devices to 200% is allowed [73].

*(j) Supplementary Rotor Current (SRC) Controller.* The SRC controller is used for the RSC to limit the magnitude of rotor current to within nominal values during faults by introducing a multiplication factor $k$ to the effective measured RSC currents. The value of the factor depends on the fault voltage magnitude [74].

*(k) Series Grid Side Converter (SGSC).* A SGSC is connected to the DC bus in addition to the parallel GSC as shown in Figure 6. It is utilized to eliminate the negative sequence stator voltage, ensuring balanced stator terminal voltages even under grid network voltage imbalance [8].

*3.2. Keeping the DFIG Connected to the Grid While Offering Support to Grid.* Voltage control can also be achieved by reactive power injection.

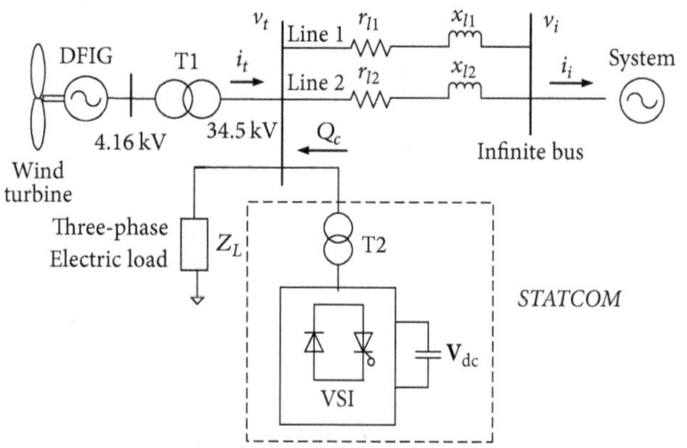


FIGURE 7: Single line diagram showing STATCOM connection for reactive power compensation [9].

*3.2.1. From External Devices.* FACTS devices are connected at the PCC to inject or absorb reactive current to keep voltage within the prescribed operational limits. They can be connected in either shunt (STATCOM, SVC), series (DVR), or hybrid manner (UPFC). This introduces the added cost of the FACTS device, maintenance costs, reduced reliability, and added control complexity [9, 66, 75–77]. Figure 7 gives an example of a STATCOM connection.

*3.2.2. From DFIG.* The added cost of the FACTS device can be avoided by utilizing the DFIG reactive power capacity. On fault occurrence, the reactive power output of the DFIG is given priority [78]. The reactive power output of the DFIG can be a total of up to 75% of its active power rating with around 20% of the active power rating being from the GSC [75]. This capability depends on rated rotor voltage and current limits and rated stator current limit [79, 80]. In addition to the aforementioned, the GSC reactive power capability must be considered [81]. Total reactive power generation is limited by rotor current while consumption is limited by stator current. For large slips, >0.31, reactive power generation is also limited by rotor voltage. There also are limitations imposed by magnetic flux saturation and the nonlinear relationships between junction temperature and rotor currents frequency [82].

Most grid codes require that the power factor at the PCC must be between 0.95 leading and 0.9 lagging. This cannot be met with DFIG alone once its active power production goes above 82% of nominal value. The most obvious approach is to increase the rating of the VSC [83] which results in increased reactive power production as well as faster active power response and less DC-link voltage oscillation after fault clearance. Also, the pitch angle requires very little adjustment. However, increasing the rating of the VSC comes at an added cost due to the increased rating [84]. Other approaches that avoid the increase in cost include the following:

(a) Since the generator rated power at supersynchronous speed is the sum of stator power and rotor power [85],

$$P_{\text{tot}} = P_s + P_r \quad (4)$$

at a certain point, to avoid exceeding rated power, the stator power has to decrease as the rotor power increases. At rated power, the stator supplies 0.7–0.8 p.u. while rotor produces 0.2–0.3 p.u [82]. This results in capacity which can be utilized for generating reactive power from the stator if the rating of the wind turbine is higher than that of the generator.

(b) Also, additional DFIG reactive power reserve capacity can be ensured by connecting at the PCC, FACTS devices such as mechanically switched capacitors, whose rating is just enough to ensure that the VSC injects a maximum of 10% of their rated reactive power capacity during steady state system operation. The reserve is then availed during transient states [86] while the cost of FACTS devices is minimized.

(c) In addition, whereas worst case scenarios consider the fully loaded DFIG at nominal active power output, it is only at 10% of the time that it supplies above 90% of this value. This gives additional capacity for reactive power production [85].

(d) With temporarily overloading of the converters' current outputs, but within manufacturers' limits, the IGBTs can switch up to 100% more current than the steady state rated current so long as it is operating within the safe operating area (SOA). This is usually given for a short duration of time typically 20 mS. As such, both RSC currents can be increased, thus increasing the active power, hence controlling rotor overspeed, while the increased GSC current ensures that active power is not sacrificed and so avoids DC-link overvoltage. The extra current capacity is used for reactive power generation, hence boosting total reactive power generation from the DFIG [55, 87].

(e) Using the chopper braking resistor to dissipate a larger percentage of the power from the rotor, as a result, more GSC capacity is availed for injecting reactive current to the power system [66].

(f) The GSC reactive power output is given priority over active power while the RSC also gives reactive current priority on fault occurrence. This results in improved terminal voltage. While prioritizing reactive power reduces the DFIG active power output, the healthy voltage positively affects other electrical components in the grid system ensuring they do not trip on over/undervoltage [78]. This is critical when these include other generators without FRT. To ensure reactive power output from the stator, RSC control must not be lost during the fault period.

## 4. Discussion

Most of the challenges resulting from the machine can be dealt with at manufacturing stage, while appropriate and timely maintenance can deal with others. The choice of an appropriate control strategy is also critical. The control system must strive to keep the thermal loading within limits and also reduce the thermal variations to ensure long lifespan especially for the electronic components. For cost effectiveness, the DFIG WECS must be used for reactive power supply during both steady and transient states especially for faults from the power system. As such, the MPPT and control strategies must incorporate this. However, coordination between the MPPT strategy and control strategy is a must if greater benefits such as reduction of component count are to be achieved. For example, the use of an MPPT strategy that requires a speed sensor minimizes the positive impact of a control strategy that does not utilize the speed sensor and vice versa.

Due to the intricate interaction between the stator and rotor circuits by magnetic flux, approaches for FRT that do not interfere with the internal setup of the DFIG WECS seem attractive. The solution must however be machine based and not at the wind farm terminals. This is as required by some grid codes such as GB 19963-2011 since FRT capability is determined based on individual machines and not on the entire wind farm. These approaches include use of SDBR and DVR on the stator circuit. However, the switching in and out of the stator side of SDBR and DVR introduces switching transients since the stator circuit carries a large percentage of the output current [88]. These voltage transients can go by up to 50% in 1–3 $\mu$S and thus introduce great stress to the DFIG windings. Future works are expected to look at reduction of switching transients where SDBR and DVR are applied for FRT.

## 5. Conclusion

The challenges faced in operation of the DFIG WECS which include faults from both the machine and the grid system are presented. Different approaches used to date to deal with these challenges have been explored with various MPPT and control strategies being presented. The paper has brought out issues to consider for an optimal solution which must coordinate the MPPT and control strategies and integrate FRT capabilities so as to satisfy the ever stringent grid connection codes.

## References

[1] http://www.wwindea.org/wwea-half-year-report-worldwind-wind-capacity-reached-456-gw.

[2] K. E. Okedu and U. Roland, "Low voltage ride-through methods for DFIG VSWT," *Journal of Emerging Trends in Engineering and Applied Sciences*, vol. 5, no. 8, pp. 221–226, 2014.

[3] D. A. Rivkin, L. D. Anderson, and L. Silk, *Wind Turbine Control Systems*, Jones Bartlett Learning, USA, 2013.

[4] F. Blaabjerg, M. Liserre, and K. Ma, "Power electronics converters for wind turbine systems," *IEEE Transactions on Industry Applications*, vol. 48, no. 2, pp. 708–719, 2012.

[5] Ingeteam, "Power converter training manual," http://www.ingeteam.com/, 2014.

[6] M. Mohseni, S. M. Islam, and M. A. S. Masoum, "Impacts of symmetrical and asymmetrical voltage sags on DFIG-based wind turbines considering phase-angle jump, voltage recovery, and sag parameters," *IEEE Transactions on Power Electronics*, vol. 26, no. 5, pp. 1587–1598, 2011.

[7] V. P. Mahadanaarachchi, *Fault analysis and protection of doubly fed induction generator-based wind farms, [MSc. thesis]*, Oklahoma State University, 2009.

[8] J. Yao, H. Li, Z. Chen et al., "Enhanced control of a DFIG-based wind-power generation system with series grid-side converter under unbalanced grid voltage conditions," *IEEE Transactions on Power Electronics*, vol. 28, no. 7, pp. 3167–3181, 2013.

[9] W. Qiao, G. K. Venayagamoorthy, and R. G. Harley, "Real-time implementation of a STATCOM on a wind farm equipped with doubly fed induction generators," *IEEE Transactions on Industry Applications*, vol. 45, no. 1, pp. 98–107, 2009.

[10] S. M. R. Kazmi, H. Goto, H.-J. Guo, and O. Ichinokura, "Review and critical analysis of the research papers published till date on maximum power point tracking in wind energy conversion system," in *Proceedings of the 2010 2nd IEEE Energy Conversion Congress and Exposition, ECCE 2010*, pp. 4075–4082, September 2010.

[11] M. A. Abdullah, A. H. M. Yatim, C. W. Tan, and R. Saidur, "A review of maximum power point tracking algorithms for wind energy systems," *Renewable and Sustainable Energy Reviews*, vol. 16, no. 5, pp. 3220–3227, 2012.

[12] B. Wu, Y. Lang, N. Zargari, and S. Kouro, *Power Conversion And Control of Wind Energy Systems*, John Wiley Sons, Inc, New Jersey, NJ, USA, 2011.

[13] O. P. Mahela and A. G. Shaik, "Comprehensive overview of grid interfaced wind energy generation systems," *Renewable and Sustainable Energy Reviews*, vol. 57, pp. 260–281, 2016.

[14] B. RUckert and W. Hofmann, "Commutation strategy for minimized common mode voltage using direct power control," in *Proceedings of the 27th Annual IEEE Applied Power Electronics Conference and Exposition (APEC '12)*, pp. 106–112, February 2012.

[15] M. Z. Sujod, "Harmonics and common mode voltage in a DFIG with two-level and three-level NPC converter using standard PWM techniques," in *Proceedings of the 39th Annual Conference of the IEEE Industrial Electronics Society (IECON '13)*, pp. 1650–1655, November 2013.

[16] A. M. Garcia, D. G. Holmes, and T. A. Lipo, "Reduction of bearing currents in doubly fed induction generators," in *Proceedings of The 41st IAS Annual Meeting Conference of the IEEE*, pp. 84–89, October 2006.

[17] S. O. Faried, I. Unal, D. Rai, and J. Mahseredjian, "Utilizing DFIG-based wind farms for damping subsynchronous resonance in nearby turbine-generators," *IEEE Transactions on Power Systems*, vol. 28, no. 1, pp. 452–459, 2012.

[18] C. Zhu, M. Hu, and Z. Wu, "Parameters impact on the performance of a double-fed induction generator-based wind turbine for subsynchronous resonance control," *IET Renewable Power Generation*, vol. 6, no. 2, pp. 92–98, 2012.

[19] L. Fan and Z. Miao, "Mitigating SSR using DFIG-based wind generation," *IEEE Transactions on Sustainable Energy*, vol. 3, no. 3, pp. 349–358, 2012.

[20] D. Muthumuni, "Investigating possible induction generator effects due to sub-synchronous resonances," https://hvdc.ca/uploads/ck/files/Application_Note_-_Investigating_Possible_Induction_Generator_Effects_Due_to_SSR_-_no_background_colour.pdf.

[21] S. Heir, *Grid Integration of Wind Energy Conversion Systems*, John Wiley Sons Ltd, 2nd edition, 2006.

[22] P. Zhan, W. Lin, J. Wen, M. Yao, and N. Li, "Design of LCL filters for the back-to-back converter in a doubly fed induction

generator," in *Proceedings of the 2012 IEEE Innovative Smart Grid Technologies - Asia, ISGT Asia*, pp. 1–6, May 2012.

[23] F. K. A. Lima, M. A. N. Nunes, E. E. C. Morais, J. L. Dantas, and C. G. C. Branco, "A new proposal for DFIG grid–side converter as double-tuned hybrid filter to improve the power quality," in *Proceedings of the 39th Annual Conference of the IEEE Industrial Electronics Society, IECON 2013*, pp. 1472–1477, November 2013.

[24] N. P. Gupta, P. Gupta, and D. Masand, "Power quality improvement using hybrid active power filter for a DFIG based wind energy conversion system," in *Proceedings of the 3rd Nirma University International Conference on Engineering (NUiCONE '12)*, pp. 1–6, December 2012.

[25] B. Singh, K. Al-Haddad, and A. Chandra, "A review of active filters for power quality improvement," *IEEE Transactions on Industrial Electronics*, vol. 46, no. 5, pp. 960–971, 1999.

[26] M. Boutoubat, L. Mokrani, and M. Machmoum, "Control of a wind energy conversion system equipped by a DFIG for active power generation and power quality improvement," *Renewable Energy*, vol. 50, pp. 378–386, 2013.

[27] A. Mudlapur, A. Raju, and U. Rao, "Evaluation of different PWM techniques for two level inverter in grid connected WECS," in *Proceedings of the 2013 2nd International Conference on Advances in Computing, Communications and Informatics (ICACCI '13)*, pp. 1753–1758, August 2013.

[28] D. Zhou, F. Blaabjerg, M. Lau, and M. Tonnes, "Thermal behavior optimization in multi-MW wind power converter by reactive power circulation," *IEEE Transactions on Industry Applications*, vol. 50, no. 1, pp. 433–440, 2014.

[29] J. Zhang, G. Chen, and X. Cai, "Thermal smooth control for Multi–MW parallel wind power converter," in *Proceedings of the 2013 IEEE International Conference of IEEE Region 10 (TENCON '13)*, pp. 1–4, October 2013.

[30] K. Ma, M. Liserre, F. Blaabjerg, and T. Kerekes, "Thermal loading and lifetime estimation for power device considering mission profiles in wind power converter," *IEEE Transactions on Power Electronics*, vol. 30, no. 2, pp. 590–602, 2015.

[31] J. Puigcorbe and A. De-Beaumont, "Wind Turbine Gearbox Reliability: the impact of rotor support, 2010," http://www.renewableenergyworld.com/articles/print/volume-13/issue-3/wind-power/wind-turbine-gearbox-reliability.html.

[32] A. M. Ragheb and M. Ragheb, "Wind Turbine Gearbox Technologies, 2011," http://www.intechopen.com/books/fundamental-and-advancedtopics-in-wind-power.

[33] Wind Power Engineering [Online], http://www.windpower-engineering.com/design/mechanical/gearboxes/gears-gearboxes-101/.

[34] W. Qiao and R. G. Harley, "Improved control of DFIG wind turbines for operation with unbalanced network voltages," in *Proceedings of the 2008 IEEE Industry Applications Society Annual Meeting (IAS '08)*, pp. 1–7, October 2008.

[35] S. Karimi, A. Gaillard, P. Poure, and S. Saadate, "FPGA–based real–time power converter failure diagnosis for wind energy conversion systems," *IEEE Transactions on Industrial Electronics*, vol. 55, no. 12, pp. 4299–4308, 2008.

[36] M. Shahbazi, M. R. Zolghadri, P. Poure, and S. Saadate, "Wind energy conversion system based on DFIG with open switch fault tolerant six–legs AC–DC–AC converter," in *Proceedings of the 2013 IEEE International Conference on Industrial Technology (ICIT '13)*, pp. 1656–1661, February 2013.

[37] A. F. Abdou, A. Abu-Siada, and H. R. Pota, "Application of STATCOM to improve the LVRT of DFIG during RSC fire–through fault," in *Proceedings of the 2012 22nd Australasian Universities Power Engineering Conference: "Green Smart Grid Systems" (AUPEC '12)*, pp. 1–6, September 2012.

[38] E. Giraldo and A. Garces, "An adaptive control strategy for a wind energy conversion system based on PWM-CSC and PMSG," *IEEE Transactions on Power Systems*, vol. 29, no. 3, pp. 1446–1453, 2014.

[39] C. Liu, D. Xu, N. Zhu, F. Blaabjerg, and M. Chen, "DC–voltage fluctuation elimination through a DC–capacitor current control for DFIG converters under unbalanced grid voltage conditions," *IEEE Transactions on Power Electronics*, vol. 28, no. 7, pp. 3206–3218, 2013.

[40] A. M. S. Yunus, A. Abu-Siada, and M. A. S. Masoum, "Impact of DC-link fault on the dynamic performance of DFIG," in *Proceedings of the 2012 IEEE Power and Energy Society General Meeting, (PES '12)*, pp. 1–6, July 2012.

[41] A. Bonanomi, "Powerful analysis of wind turbine gearboxes, 2014," http://www.powertransmissionworld.com/powerful-analysis-of-wind-turbine-gearboxes/.

[42] N. Cherfia, D. Kerdoun, and A. Boumassata, "Sliding mode control indirect strategy of the active and reactive power for the wind turbine DFIG," in *Proceedings of the Conference Internationale des Energies Renouvelables (CIER '13)*, Sousse, Tunisie, 2013.

[43] D. Zhiyong and L. Wei, "Adaptive estimation of three–phase grid voltage parameters under unbalanced faults and harmonic disturbances," *IEEE Transactions on Power Electronics*, vol. 32, no. 7, pp. 5613–5627, 2017.

[44] Z. Dai, W. Lin, and H. Lin, "Estimation of single–phase grid voltage parameters with zero steady–state error," *IEEE Transactions on Power Electronics*, vol. 31, no. 5, pp. 3867–3879, 2016.

[45] Z. Zheng, G. Yang, and H. Geng, "High voltage ride–through control strategy of grid-side converter for DFIG–based WECS," in *Proceedings of the 39th Annual Conference of the IEEE Industrial Electronics Society (IECON '13)*, pp. 5282–5287, November 2013.

[46] C. Feltes, S. Engelhardt, J. Kretschmann, J. Fortmann, F. Koch, and I. Erlich, "High voltage ride-through of DFIG-based wind turbines," in *Proceedings of the IEEE Power and Energy Society General Meeting—Conversion and Delivery of Electrical Energy in the 21st Century*, pp. 1–8, IEEE, Pittsburgh, Pa, USA, July 2008.

[47] A. Rolan, J. Pedra, and F. Córcoles, "Detailed study of DFIG-based wind turbines to overcome the most severe grid faults," *International Journal of Electrical Power and Energy Systems*, vol. 62, pp. 868–878, 2014.

[48] A. Yousef, *Wind turbines level energy storage for low voltage ride through (LVRT) support, [MSc. thesis]*, The University of Wisconsin, Milwaukee, Wis, USA, 2012.

[49] M. H. J. Bollen, *Understanding Power Quality Problems: Voltage Sags And Interruptions*, IEEE Press, New York, NY, USA, 2000.

[50] R. Prasad, *Low voltage ride–through capability for matrix converters fed adjustable–speed induction machine drives for industrial and wind appliances [Ph.D. thesis]*, University of Minnesota, Minnesota, Minn, USA, 2011.

[51] G. Abad, J. Lopez, M. Rodriguez, L. Marroyo, and G. Iwanski, *Doubly Fed Induction Machine: Modelling and Control for Wind Energy Generation*, John Wiley & Sons, 2011.

[52] R. Teodorescu, M. Liserre, and P. Rodriquez, *Grid Converters for Photovoltaic and Wind Power systems*, John Wiley & Sons Ltd, 2011.

[53] A. RolAn, F. COrcoles, and J. Pedra, "Doubly fed induction generator subject to symmetrical voltage sags," *IEEE Transactions on Energy Conversion*, vol. 26, no. 4, pp. 1219–1229, 2011.

[54] J. J. Justo, F. Mwasilu, and J.-W. Jung, "Doubly-fed induction generator based wind turbines: a comprehensive review of fault ride-through strategies," *Renewable and Sustainable Energy Reviews*, vol. 45, pp. 447–467, 2015.

[55] J. Morneau, *A comparative evaluation of low voltage ride through solutions in wind turbines using doubly fed induction generators, [MSc. thesis]*, McGill University, Montreal, Canada, 2008.

[56] G. JoOs, "Wind turbine generator low voltage ride through requirements and solutions," in *Proceedings of the IEEE Power and Energy Society 2008 General Meeting: Conversion and Delivery of Electrical Energy in the 21st Century*, pp. 1–7, July 2008.

[57] Y. M. Alsmadi, L. Xu, F. Blaabjerg, A. P. Ortega, and A. Wang, "Comprehensive analysis of the dynamic behavior of grid-connected DFIG-based wind turbines under LVRT conditions," in *Proceedings of the 7th Annual IEEE Energy Conversion Congress and Exposition, (ECCE '15)*, pp. 4178–4187, September 2015.

[58] D. Campos-Gaona, E. L. Moreno-Goytia, and O. Anaya-Lara, "Fault ride-through improvement of DFIG-WT by integrating a two-degrees-of-freedom internal model control," *IEEE Transactions on Industrial Electronics*, vol. 60, no. 3, pp. 1133–1145, 2013.

[59] K. E. Okedu, *Low Voltage Ride through Enhancement of Grid Connected Wind Farms: Augmentation of Variable Speed Wind Turbines Fault Ride through (FRT) Capability*, LAP Lambert Academic Publishing, 2012.

[60] V. Akhmatov, "Variable-speed wind turbines with doubly-fed induction generators Part IV: uninterrupted operation features at grid faults with converter control coordination," *Wind Engineering*, vol. 27, no. 6, pp. 519–529, 2003.

[61] V. Akhmatov, "Variable-speed wind turbines with doubly-fed induction generators part III: model with the back-to-back converters," *Wind Engineering*, vol. 27, no. 2, pp. 79–91, 2003.

[62] Ch. Laxmi, K. S. Latha, and Dr. Himani, "Improving the low voltage ride through capability of wind generator system using crowbar and battery energy storage system," *International Journal of Engineering Science Invention*, vol. 1, no. 7, pp. 14–19, 2013.

[63] Q. Liu, L. Yu, and G. Wu, "Comparison of control strategy for double-fed induction generator (DFIG)," in *Proceedings of the 3rd International Conference on Measuring Technology and Mechatronics Automation*, pp. 741–744, January 2011.

[64] K. E. Okedu, S. M. Muyeen, R. Takahashi, and J. Tamura, "Application of SDBR with DFIG to augment wind farm fault ride through," in *Proceedings of the 2011 International Conference on Electrical Machines and Systems, (ICEMS '11)*, August 2011.

[65] K. E. Okedu, S. M. Muyeen, R. Takahashi, and J. Tamura, "Effectiveness of current-controlled voltage source converter excited doubly fed induction generator for wind farm stabilization," *Electric Power Components and Systems*, vol. 40, no. 5, pp. 556–574, 2012.

[66] M. Nasiri, J. Milimonfared, and S. H. Fathi, "A review of low-voltage ride-through enhancement methods for permanent magnet synchronous generator based wind turbines," *Renewable and Sustainable Energy Reviews*, vol. 47, pp. 399–415, 2015.

[67] A. O. Ibrahim, T. H. Nguyen, D.-C. Lee, and S.-C. Kim, "A fault ride-through technique of DFIG wind turbine systems using dynamic voltage restorers," *IEEE Transactions on Energy Conversion*, vol. 26, no. 3, pp. 871–882, 2011.

[68] A. Moghadasi, A. Sarwat, and J. M. Guerrero, "A comprehensive review of low-voltage-ride-through methods for fixed-speed wind power generators," *Renewable and Sustainable Energy Reviews*, vol. 55, pp. 823–839, 2016.

[69] A. M. S. Yunus, M. A. S. Masoum, and A. Abu-Siada, "Application of SMES to enhance the dynamic performance of DFIG during voltage sag and swell," *IEEE Transactions on Applied Superconductivity*, vol. 22, no. 4, 2012.

[70] X. Yan, G. Venkataramanan, P. S. Flannery, Y. Wang, Q. Dong, and B. Zhang, "Voltage-sag tolerance of DFIG wind turbine with a series grid side passive-impedance network," *IEEE Transactions on Energy Conversion*, vol. 25, no. 4, pp. 1048–1056, 2010.

[71] S. Hu, X. Lin, Y. Kang, and X. Zou, "An improved low-voltage ride-through control strategy of doubly fed induction generator during grid faults," *IEEE Transactions on Power Electronics*, vol. 26, no. 12, pp. 3653–3665, 2011.

[72] D. Zhu, X. Zou, Y. Kang, L. Deng, and Q. Huang, "Inductance-simulating control for DFIG-based wind turbine to ride-through grid faults," in *Proceedings of the 31st Annual IEEE Applied Power Electronics Conference and Exposition, (APEC '16)*, pp. 3521–3525, March 2016.

[73] D. Xiang, L. Ran, P. J. Tavner, and S. Yang, "Control of a doubly fed induction generator in a wind turbine during grid fault ride-hrough," *IEEE Transactions on Energy Conversion*, vol. 21, no. 3, pp. 652–662, 2006.

[74] K. E. Okedu, S. M. Muyeen, R. Takahashi, and J. Tamura, "Wind farms fault ride through using DFIG with new protection scheme," *IEEE Transactions on Sustainable Energy*, vol. 3, no. 2, pp. 242–254, 2012.

[75] K. E. Okedu, "A Study of wind farm stabilization using DFIG or STATCOM considering grid requirements," *Journal of Engineering Science and Technology Review*, vol. 3, no. 1, pp. 200–209, 2010.

[76] A. J. Prabhakar, *Application of STATCOM for improved dynamic performance of wind farms in a power grid, [MSc. thesis]*, Missouri University of Science and Technology, 2008.

[77] K. R. Padiyar, *FACTS Controllers in Power Transmission And Distribution*, New Age International Publishers, New Delhi, India, 2007.

[78] S. Foster, L. Xu, and B. Fox, "Coordinated reactive power control for facilitating fault ride through of doubly fed induction generator-and fixed speed induction generator-based wind farms," *IET Renewable Power Generation*, vol. 4, no. 2, pp. 128–138, 2010.

[79] S. N. Singh, J. Otergaard, and B. Singh, "Reactive power capability of unified DFIG for wind power generation," in *Proceedings of the IEEE PES General Meeting, (PES '10)*, pp. 1–7, July 2010.

[80] J. Zhao, X. Li, J. Hao, and J. Lu, "Reactive power control of wind farm made up with doubly fed induction generators in distribution system," *Electric Power Systems Research*, vol. 80, no. 6, pp. 698–706, 2010.

[81] J. Tian, C. Su, and Z. Chen, "Reactive power capability of the wind turbine with Doubly Fed Induction Generator," in *Proceedings of the 39th Annual Conference of the IEEE Industrial Electronics Society, (IECON '13)*, pp. 5312–5317, November 2013.

[82] S. Engelhardt, I. Erlich, C. Feltes, J. Kretschmann, and F. Shewarega, "Reactive power capability of wind turbines based

on doubly fed induction generators," *IEEE Transactions on Energy Conversion*, vol. 26, no. 1, pp. 364–372, 2011.

[83] A. M. Kassem, K. M. Hasaneen, and A. M. Yousef, "Dynamic modeling and robust power control of DFIG driven by wind turbine at infinite grid," *International Journal of Electrical Power and Energy Systems*, vol. 44, no. 1, pp. 375–382, 2013.

[84] K. E. Okedu, "Impact of power converter size on variable speed wind turbine," *Pacific Journal of Science and Technology*, vol. 13, no. 1, pp. 176–181, 2012.

[85] B. Pokharel, *Modeling, control and analysis of a doubly fed induction generator based wind turbine system with voltage regulation, [MSc. thesis]*, Tennessee Technological University, 2011.

[86] J. M. Garcia, *Voltage control in wind power plants with doubly fed generators [Ph.D. thesis]*, Aalborg University, 2010.

[87] M. Mohseni and S. M. Islam, "Transient control of DFIG-based wind power plants in compliance with the australian grid code," *IEEE Transactions on Power Electronics*, vol. 27, no. 6, pp. 2813–2824, 2012.

[88] S. Sharma, J. P. Mishra, and S. Datta, "Sliding mode power control of a DFIG based variable speed wind energy conversion system," in *Proceedings of the 2015 Annual IEEE India Conference (INDICON)*, pp. 1–6, New Delhi, India, December 2015.

# 11

# Photoluminescence Enhancement Effect of the Layered $MoS_2$ Film Grown by CVD

**H. Li,[1,2] X. H. Zhang,[1] and Z. K. Tang[2]**

[1]*Department of Electrical and Electronic Engineering, Southern University of Science and Technology, Shenzhen 518055, China*
[2]*Institute of Applied Physics and Materials Engineering, University of Macau, Macau*

Correspondence should be addressed to X. H. Zhang; zhangxh@sustc.edu.cn and Z. K. Tang; zktang@umac.mo

Academic Editor: Qudong Wang

The layered $MoS_2$ films have been prepared by the chemical vapor deposition (CVD) method, which shows numerous $MoS_2$ convex flakes (i.e., $MoS_2$ nanosheets) on $SiO_2$/Si substrates. The convex flake islands form a uniform layered film and the single flake corresponds to one small piece of monolayer $MoS_2$ film; they show a strong photoluminescence (PL) radiation as a whole. The thickness of the layered $MoS_2$ films increases with the increases of growth time, the effect of diverse growth time on the properties of layered $MoS_2$ films has been discussed, and the PL enhancement effect is due to the PL intensity accumulation from each of the convex flakes. The strong PL radiation has been found by annealing the samples and the reason has been analyzed. The layered $MoS_2$ films could provide more photon energy than both the monolayer $MoS_2$ film and the bulk $MoS_2$. This work provides a fundamental study about the device fabrication based on layered $MoS_2$ films. The optoelectronic devices based on this study will be more energy efficient.

## 1. Introduction

Recently, transition metal dichalcogenides (TMDs), $MX_2$ (M = Mo, W; X = S, Se, and Te) have attracted extensive attention for their great potential in the fields of catalyst [1–4], nanoelectronic, and optoelectronic devices [2–15]. The transistors fabricated with the molybdenum disulfide ($MoS_2$) films demonstrate good on/off current ratio and high carrier mobility [7–10, 16–19], which make them very suitable for next generation transistors devices. Substantial works have been devoted to prepare $MoS_2$ films [10, 20], including mechanical exfoliation [21], liquid exfoliation [22, 23], physical vapor deposition (PVD), and CVD [2, 3, 11, 12]. However, synthesis of large-size $MoS_2$ single crystal films is still challenging. The thin $MoS_2$ film prepared by the mechanical exfoliation method based on scotch-tape can control the thickness, but the output is very low which makes it not suitable for electronic applications. The liquid exfoliation can generate large quantities of monolayer dispersions, but relatively small flakes limit their use in electronics or photonics applications. The approach of PVD is hard to achieve because of the high sublimation temperature (1100°C) and low oxidizing temperature of $MoS_2$. Compared with other methods, CVD is one of the most effective strategies for synthesizing large-area, high quality $MoS_2$ films [18, 19]. It was employed in a broad range of applications, such as FETs, LEDs, energy storage devices, and sensors, and $MoS_2$-based logic circuits have great potential for commercial applications in the future.

It is well known that, in bulk $MoS_2$, the energy band is 1.29 eV, while, in monolayer $MoS_2$ film, the energy band is 1.85 eV. Due to the change of band gap from bulk $MoS_2$ to monolayer $MoS_2$ film, the PL radiation appears. This phenomenon has been reported a lot [24–26], but the obvious PL radiation appeared in the layered $MoS_2$ films composed of convex flakes, and the strong PL radiation produced during the annealing process has not been reported before. In this work, the layered $MoS_2$ films is composed of many small convex flakes and exhibits relative uniformity film as a whole, so it shows a multilayer $MoS_2$ Raman characteristic and a monolayer $MoS_2$ PL characteristic simultaneously. In the layered $MoS_2$ films, the PL enhancement effects were observed and the reasons were analyzed.

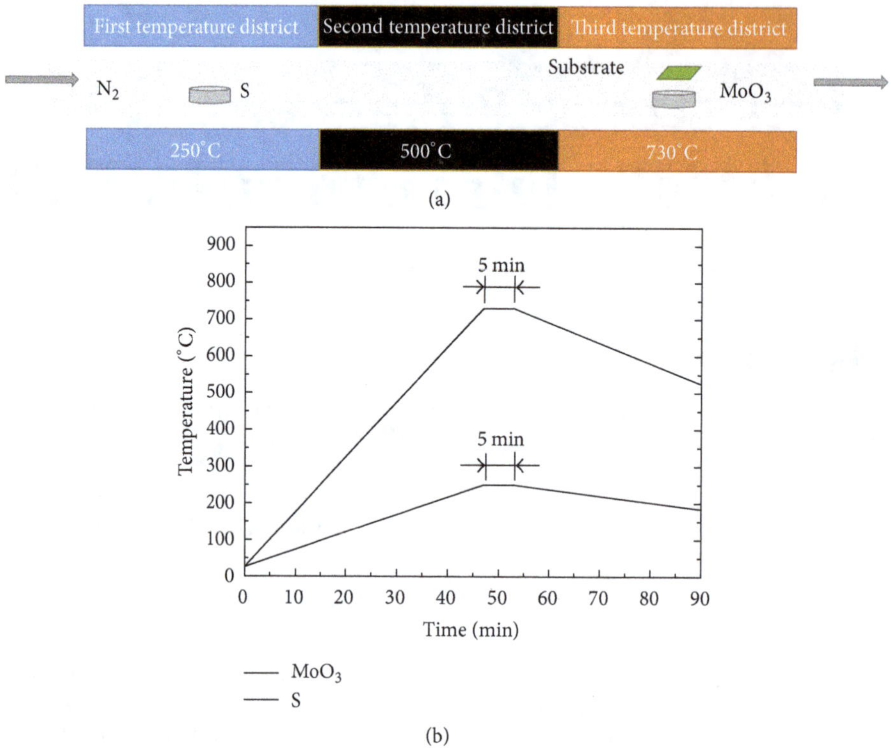


FIGURE 1: (a) A schematic of CVD setup for $MoS_2$ film growth. (b) Temperature variation curves for $MoO_3$ and S. $MoO_3$ and S with different heating rate to ensure $MoO_3$ and S sublimation at the same time. The 5 min growth time is labeled on the curves.

## 2. Experiments and Results

The $MoS_2$ films were grown with $MoO_3$ (99.95%, Sigma-Aldrich) and S (99.98%, Sigma-Aldrich) powders as precursors. $MoO_3$ and S powders were put in different temperature districts separately due to their different sublimation temperatures. The amount ratio between $MoO_3$ and S was 1:10, in which $MoO_3$ was 0.01 g and S was the 0.1 g, and the sulfur powder is completely overdosed. The size of the $SiO_2$/Si substrate was 10 mm × 10 mm with 300 nm thickness of $SiO_2$ on Si. The substrate was upside down above $MoO_3$ and $N_2$ was used as the protection gas. The $N_2$ flow velocity was 50 sccm mainly decided by the diameter of the quartz tube. Figure 1(a) shows a schematic of CVD setup for $MoS_2$ films growth. The furnace includes three temperature districts. S powder was put in the first temperature district near the $N_2$ entrance port and $MoO_3$ powder was put in the third temperature district near the $N_2$ exit port. The second temperature district was as a temperature transition zone. The S powder, the second temperature district, and the $MoO_3$ powder were raised up to 250°C, 500°C, and 730°C in 47 min, respectively, and kept at the temperature for 5 min, which aims to make sure the two precursors can be sublimation and start reaction at the same time. Then the furnace cooled down to room temperature naturally. Figure 1(b) shows the temperature variation curves for both precursors which illustrates a whole growing process of $MoS_2$ films by CVD method, and the growth time (chemical reaction time) of $MoS_2$ is 5 min. The 730°C growth temperature is the relatively low temperature for $MoO_3$ sublimation [27, 28], which aims to ensure the film does not grow to be too thick. The growth time is chemical reaction time of $MoO_3$ and S which plays a very important role for the synthesizing $MoS_2$ films. The layered $MoS_2$ films with diverse growth time (5, 10, 15, and 20 min, resp.) are synthesized, and diverse growth time leads to different growth results which will be discussed in Figure 1.

In order to observe the morphology of layered $MoS_2$ films, the Scanning Electron Microscope (SEM) images with diverse growth time have been shown in Figure 2. In Figure 2(a), the layered $MoS_2$ films were generated with 5 min growth time, a lot of convex flakes can be seen on the surface of the substrate, and they resemble standing vertically on the substrate. It should be noted that the film is actually composed of these loosed convex flakes and they form a relatively uniform film structure. Every piece convex flake corresponds to a monolayer film, and a number of convex flakes lead to strong PL radiation together. Figures 2(b) and 2(c) show the SEM images of the layered $MoS_2$ films with 10 and 15 min growth time, respectively. It is obvious to see the number of convex flakes less than the number generated in 5 min growth time; in 15 min growth time, the number of convex flakes is the least. It means most convex flakes accumulated into a thick film and deposited on the substrate led to the film becoming denser. Accompanied by thickening of the film, the PL intensity becomes weaker, which will be shown in Figure 2.

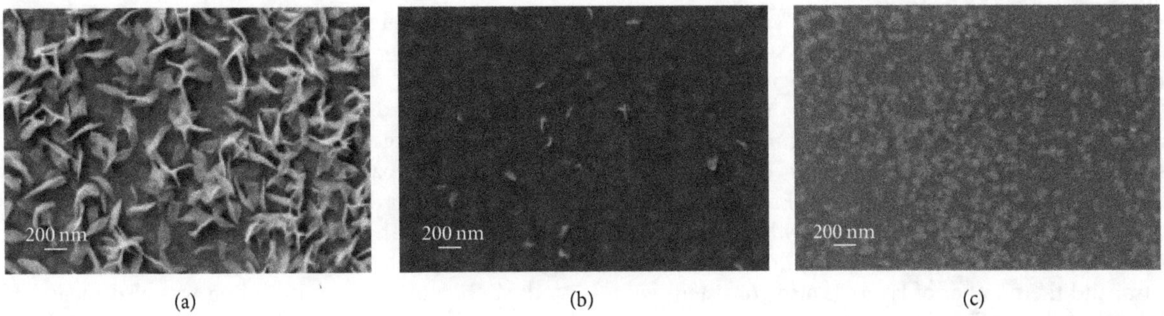


FIGURE 2: SEM images of morphology of layered $MoS_2$ films with (a) 5 min, (b) 10 min, and (c) 15 min growth time, respectively.

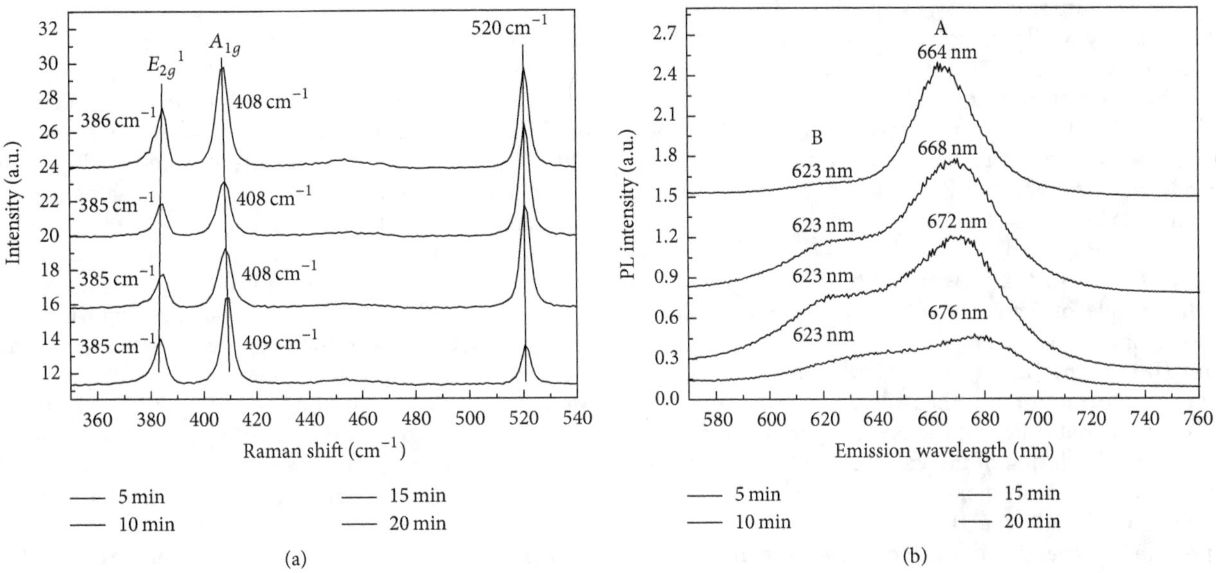


FIGURE 3: (a) Raman and (b) normalized PL spectra of layered $MoS_2$ films synthesized with diverse growth time. The samples are excited under 532 nm laser.

## 3. Analysis and Discussion

*3.1. $MoS_2$ Synthesized with Diverse Growth Time.* The Raman and PL spectra of layered $MoS_2$ films with diverse growth time have been measured and shown in Figures 3(a) and 3(b), respectively. There are two first-order Raman active modes (the in-of-plane vibration mode $E_{2g}{}^1$ and the out-of-plane vibration mode $A_{1g}$) that can be observed in Raman spectra, they are corresponding to the phonon modes of Raman scattering, and, more importantly, the two Raman active modes $E_{2g}{}^1$ and $A_{1g}$ are regularly varied with the thicknesses of thin $MoS_2$ film, so they are usually used to characterize the thickness of the layered $MoS_2$ film [2–4, 9–11]. Take the layered $MoS_2$ film with growth time of 5 minutes as an example, two Raman characteristic peaks of $MoS_2$ are located at 386 cm$^{-1}$ and 408 cm$^{-1}$ (Figure 3(a)), and the difference between the two peaks ($\Delta = A_{1g\ \text{Ram shift}} - E_{2g}{}^1{}_{\text{Ram shift}}$) is 22 cm$^{-1}$, indicating 2~3 layers' $MoS_2$ films according to the previous study [24–27, 29–31]. The peak at 520 cm$^{-1}$ is the characteristic peak of Si from $SiO_2$/Si substrate.

The layered $MoS_2$ films with 10 min growth time has two characteristic peaks at 385 cm$^{-1}$ and 408 cm$^{-1}$, the value of $\Delta$ is 23 cm$^{-1}$, according to 3~4 layers' $MoS_2$ films, and the result is the same as the layered $MoS_2$ films with 15 min growth time. When the growth time increased to 20 min, the two characteristic peaks of the layered $MoS_2$ films are at 385 cm$^{-1}$ and 408 cm$^{-1}$, $\Delta$ equal to 24 cm$^{-1}$, according 4~5 layers' $MoS_2$ films. When the growth time increased, the frequency of the $E_{2g}{}^1$ mode decreases (red shift) and that of $A_{1g}$ increases (blue shift), the value of $\Delta$ is increased, and the number of layers increased. It is due to the fact that restoring force between interlayer S-S bonds in layered $MoS_2$ is enhanced, resulting in an increase in frequency of $A_{1g}$ mode, while the in-of-plane vibration is weakened, resulting in a red shift of the $E_{2g}{}^1$ mode. When the growth time of $MoS_2$ increased, the intensity of Si peak at 520 cm$^{-1}$ reduced, also indirectly certified the thickness of the layered $MoS_2$ films is increased.

In order to further prove the thickness of the samples, the Atomic Force Microscope (AFM) images of $MoS_2$ synthesized with diverse growth time are shown in Figure 4. Due to the convex flake structure, only the thickness of the thin sheets is investigated. Take Figure 4(a) as an example, the thickness of the film shown in the right figure is measured along the blue line in the left picture. Then the thickness is read by the height difference in the right figure. It just gives a relative thickness for fixed position, but it is still able to give an intuitive expression of the sample thickness. As we can see in Figure 4, with the extension of growth time, the film thickness increases, consistent with the results of Raman and PL spectra, and the number of layers is also consistent with results derived from the Raman spectra.

Bulk $MoS_2$ is one kind of indirect bandgap material and the monolayer $MoS_2$ is a direct bandgap material [32, 33]; it is a salient feature of $MoS_2$. Bulk $MoS_2$ shows negligible PL intensity, while monolayer $MoS_2$ film exhibits the strongest PL intensity. It is due to the fact that when the indirect band gap changes to the direct band gap, monolayer $MoS_2$ film leads to increment of radiation photon energy compared with the bulk $MoS_2$, which means the quantum efficiency enhances and the PL radiation increases. So the PL intensity is inversely dependent on the thickness of $MoS_2$ nanosheets. Normalized PL spectra of layered $MoS_2$ films with diverse growth time has been shown in Figure 3(b), take the sample of 5 min growth time as an example, it has two emission wavelengths at 664 nm (A exciton) and 623 nm (B exciton) [28–31], respectively, arising from the direct excitonic transitions at the Brillouin zone K point, and the energy difference between these two peaks arises from the spin-orbital splitting of the valence band. According to previous researches, only monolayer or two layers' $MoS_2$ film can emit the PL radiation [17, 34–36]. However, through the preparation method described above, the multilayer $MoS_2$ films still show a strong PL radiation (characterized by Raman spectra showed above). It is due to the fact that convex flakes structure contributes to the PL radiation accumulation, resulting in the PL enhancement effect. Every convex flake is approximate to one monolayer $MoS_2$ film, a lot of convex flakes stacking on the substrate approximate to many monolayer $MoS_2$ films, and they radiate PL together from the same small area on the substrate excited by 532 nm laser, leading to PL radiation enhancement. So the layered $MoS_2$ films prepared by this way have a higher $MoS_2$ content and provide a large number of monolayer $MoS_2$ nanosheets simultaneously.

When the growth time is increased, the PL intensity decreased, which means the thickness of the layered $MoS_2$ films increased. What is more, the peak wavelength shifts to red with the increasing of growth time (664 nm at 5 min, 668 nm at 10 min, 672 nm at 15 min, and 676 nm at 20 min). When the thickness of layered $MoS_2$ films increased, the layered $MoS_2$ films changed from direct bandgap to indirect bandgap gradually, the energy band gap decreased, so the wavelength had a red shift. The thickness of the layered $MoS_2$ films increased with the growth time, which is consistent with Raman spectra. That is to say, when the growth time becomes longer, more $MoS_2$ nanosheets are deposited on the substrate, the space among the convex flakes becomes smaller, the film becomes denser, and the PL radiation weakened. Different positions of the $MoS_2$ domain were measured and similar results were obtained, indicating the continuity and stability of samples.

*3.2. $MoS_2$ Synthesized with Different Annealing Temperatures.* Annealing process usually played a very important role in material preparation [19]. So the layered $MoS_2$ films synthesized with different annealing temperatures have also been studied. The annealing temperature changed from 780°C to 840°C with 20°C temperature spaces, respectively, and $MoS_2$ film without annealing was also measured for a comparison. The Raman spectra of layered $MoS_2$ films at different annealing temperatures are shown in Figure 5(a). Without annealing, the difference of two Raman characteristic peaks is 21 $cm^{-1}$, corresponding to the two atom layers. When the sample is annealed, the difference between the two Raman peaks increases as the annealing temperature increases, which means that the film becomes thicker. But the PL radiation does not become weaker correspondingly. The PL spectra are shown in Figure 5(b), and the PL intensity of annealing samples is much larger than the sample without annealing. At the beginning, with the increasing of the annealing temperature, the PL intensity increases and reaches the maximum value at 800°C, it has a great improvement compared with the $MoS_2$ film without annealing. Then the PL intensity decreases with the further increasing of annealing temperature, but it is still higher than the PL intensity of $MoS_2$ film without annealing. A stronger PL enhancement effect has been shown in the annealing samples because more defeats were induced to the samples and more convex flakes were generated on the substrates. The layered $MoS_2$ film has the strongest PL radiation at 800°C, the optimal annealing temperature means the best temperature for the convex flakes generation.

In order to further understand the mechanism behind the different growth conditions, the X-ray diffraction (XRD) spectra are measured for both samples with diverse growth time and different annealing temperatures. As we can see in Figure 6(a), with the increase in growth time, more crystal orientations appear, and the intensity of (002) peak increases, and the results proved that the samples transform from layered $MoS_2$ to bulk $MoS_2$, resulting in an enlargement of the difference value between the two Raman characteristic peaks, consistent with the Raman spectra shown in Figure 5(a), which means the film becomes thicker. In Figure 6(b), after the samples have been annealed, the (100) peak disappears and the (102) peak appears, and different crystalline orientations may cause the change in the PL intensity. When the annealing temperature increases, the full width at half maximum (FWHM) of (002) peak increases, indicating that the grain size becomes smaller. The intensity of (002) peak is proportional to degree of bulk material [37]. At 800°C annealing temperature, the intensity of (002) peak is weakest, and the PL intensity is strongest, which shows the highest degree for film formation, which is consistent with the PL spectra in Figure 5(b). More importantly, PL intensity of layered $MoS_2$ is proportional to defect density and impurity concentration [38]. The introduction of more defects in the annealing process is also one reason for PL radiation enhancement.

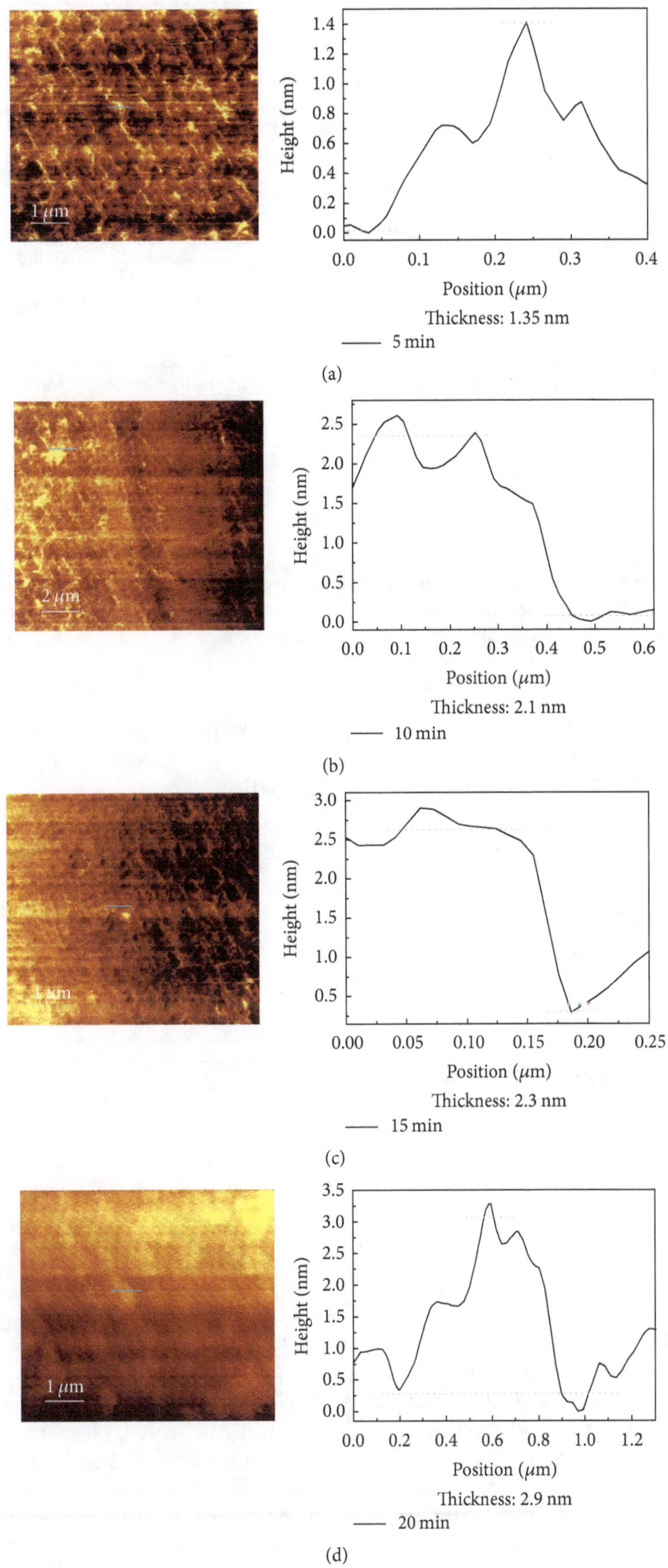


FIGURE 4: AFM images of $MoS_2$ synthesized with (a) 5 min, (b) 10 min, (c) 15 min, and (d) 20 min growth time.

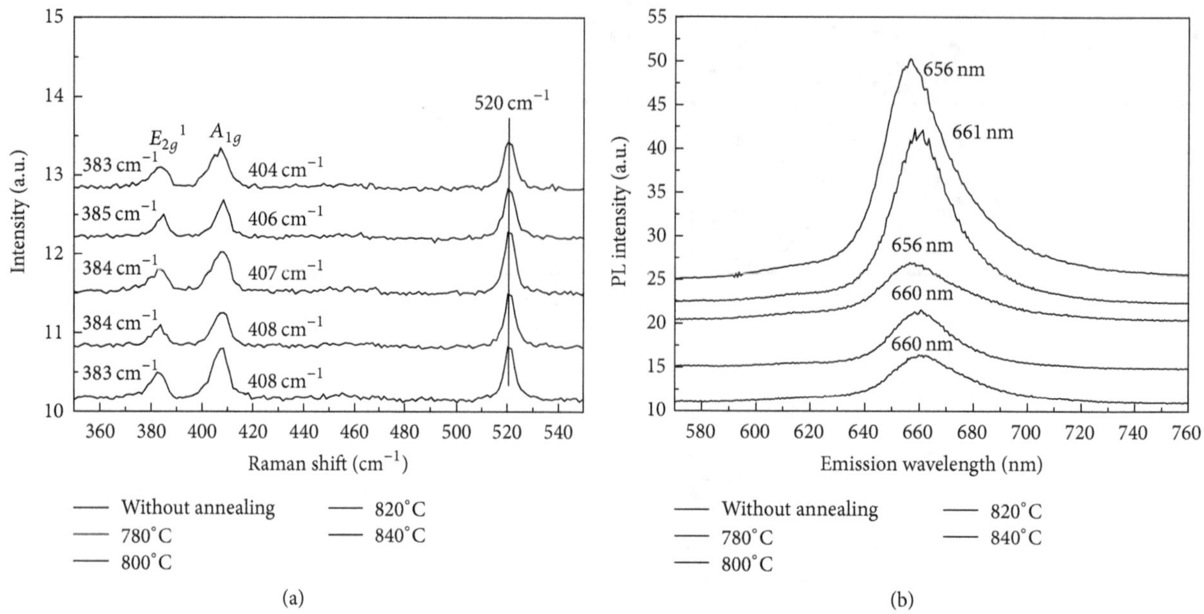


FIGURE 5: (a) Raman and (b) PL spectra of layered $MoS_2$ film synthesized at different annealing temperatures. The annealing temperature changed from 780°C to 840°C with 20°C temperature spaces, respectively. The $MoS_2$ film synthesized without annealing is also shown as a comparison. All measurements were carried out at the same experiment conditions.

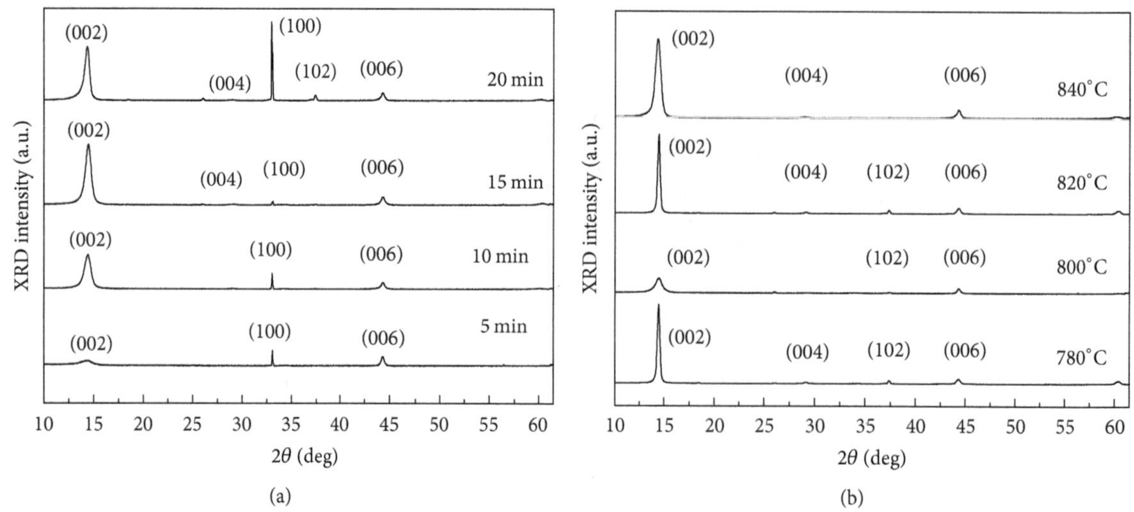


FIGURE 6: XRD spectrum of $MoS_2$ synthesized with (a) diverse growth time and (b) different annealing temperatures.

## 4. Conclusions

In this paper, the layered $MoS_2$ films on $SiO_2/Si$ substrates have been prepared by CVD method. It is composed of lots of $MoS_2$ convex flakes, each convex flake corresponding to a monolayer $MoS_2$ film which has a PL radiation, and lots of convex flakes showed strong PL radiation as a whole. The effect of diverse growth time on the quality of the layered $MoS_2$ films has been discussed, and the strongest PL radiation has been observed on the layered $MoS_2$ films with 5 min growth time. A strong PL enhancement effect was found after the samples were annealed, and the strongest PL intensity has been obtained under the growth condition of 730°C and annealing at 800°C. It provides a foundation study for preparation of layered $MoS_2$ film by CVD. In the previous study, it usually could not observe so strong PL radiation when the difference between the two characteristic Raman peaks exceeded 20 cm$^{-1}$, but, due to the $MoS_2$ convex flakes structure, the monolayer and multilayer $MoS_2$ films coexist on the substrate, and the samples have a higher $MoS_2$ content and provided a monolayer component simultaneously. So it could provide more charge carriers and 2D component when it is used in optoelectronics devices. This work provides a new choice about the device fabrication based on the layered $MoS_2$ films generated by CVD method. It could provide more photon energy due to more charge carriers that emerged through the direct band transition. The optoelectronic devices based on this study will be more energy efficient and environmentally friendly and are of interest to the development of sustainable energy.

## Acknowledgments

The authors would like to acknowledge the support of Natural Science Foundation of Guangdong Province under Grant no. 2014A030313728.

## References

[1] Y. Li, H. Wang, L. Xie, Y. Liang, G. Hong, and H. Dai, "$MoS_2$ nanoparticles grown on graphene: an advanced catalyst for the hydrogen evolution reaction," *Journal of the American Chemical Society*, vol. 133, no. 19, pp. 7296–7299, 2011.

[2] S. Balendhran, S. Walia, H. Nili et al., "Two-dimensional molybdenum trioxide and dichalcogenides," *Advanced Functional Materials*, vol. 23, no. 32, pp. 3952–3970, 2013.

[3] R. Ganatra and Q. Zhang, "Few-layer $MoS_2$: a promising layered semiconductor," *ACS Nano*, vol. 8, no. 5, pp. 4074–4099, 2014.

[4] S. Wang, Y. Rong, Y. Fan et al., "Shape evolution of monolayer $MoS_2$ crystals grown by chemical vapor deposition," *Chemistry of Materials*, vol. 26, no. 22, pp. 6371–6379, 2014.

[5] Y.-H. Chang, W. Zhang, Y. Zhu et al., "Monolayer $MoSe_2$ grown by chemical vapor deposition for fast photodetection," *ACS Nano*, vol. 8, no. 8, pp. 8582–8590, 2014.

[6] S. Wu, C. Huang, G. Aivazian, J. S. Ross, D. H. Cobden, and X. Xu, "Vapor-solid growth of high optical quality $MoS_2$ monolayers with near-unity valley polarization," *ACS Nano*, vol. 7, no. 3, pp. 2768–2772, 2013.

[7] M.-Y. Lin, C.-E. Chang, C.-H. Wang et al., "Toward epitaxially grown two-dimensional crystal hetero-structures: single and double $MoS_2$/graphene hetero-structures by chemical vapor depositions," *Applied Physics Letters*, vol. 105, no. 7, Article ID 073501, 2014.

[8] D. J. Late, B. Liu, H. S. S. R. Matte, V. P. Dravid, and C. N. R. Rao, "Hysteresis in single-layer $MoS_2$ field effect transistors," *ACS Nano*, vol. 6, no. 6, pp. 5635–5641, 2012.

[9] Y.-H. Lee, X.-Q. Zhang, W. Zhang et al., "Synthesis of large-area $MoS_2$ atomic layers with chemical vapor deposition," *Advanced Materials*, vol. 24, no. 17, pp. 2320–2325, 2012.

[10] Q. H. Wang, K. Kalantar-Zadeh, A. Kis, J. N. Coleman, and M. S. Strano, "Electronics and optoelectronics of two-dimensional transition metal dichalcogenides," *Nature Nanotechnology*, vol. 7, no. 11, pp. 699–712, 2012.

[11] Y. Cao, X. Luo, S. Han et al., "Influences of carrier gas flow rate on the morphologies of $MoS_2$ flakes," *Chemical Physics Letters*, vol. 631-632, pp. 30–33, 2015.

[12] X. Lu, M. I. B. Utama, J. Lin et al., "Large-area synthesis of monolayer and few-layer $MoSe_2$ films on $SiO_2$ substrates," *Nano Letters*, vol. 14, no. 5, pp. 2419–2425, 2014.

[13] A. Gurarslan, Y. Yu, L. Su et al., "Surface-energy-assisted perfect transfer of centimeter-scale monolayer and few-layer $MoS_2$ films onto arbitrary substrates," *ACS Nano*, vol. 8, no. 11, pp. 11522–11528, 2014.

[14] S. H. Baek, Y. Choi, and W. Choi, "Large-area growth of uniform single-layer $MoS_2$ thin films by chemical vapor deposition," *Nanoscale Research Letters*, vol. 10, article 388, 2015.

[15] Y.-H. Lee, L. Yu, H. Wang et al., "Synthesis and transfer of single-layer transition metal disulfides on diverse surfaces," *Nano Letters*, vol. 13, no. 4, pp. 1852–1857, 2013.

[16] J. H. Kim, J. Lee, J. H. Kim, C. C. Hwang, C. Lee, and J. Y. Park, "Work function variation of $MoS_2$ atomic layers grown with chemical vapor deposition: the effects of thickness and the adsorption of water/oxygen molecules," *Applied Physics Letters*, vol. 106, Article ID 251606, 2015.

[17] K.-K. Liu, W. Zhang, Y.-H. Lee et al., "Growth of large-area and highly crystalline $MoS_2$ thin layers on insulating substrates," *Nano Letters*, vol. 12, no. 3, pp. 1538–1544, 2012.

[18] Y. Yu, C. Li, Y. Liu, L. Su, Y. Zhang, and L. Cao, "Controlled scalable synthesis of uniform, high-quality monolayer and few-layer $MoS_2$ films," *Scientific Reports*, vol. 3, article 1866, 2013.

[19] X. Wang, H. Feng, Y. Wu, and L. Jiao, "Controlled synthesis of highly crystalline $MoS_2$ flakes by chemical vapor deposition," *Journal of the American Chemical Society*, vol. 135, no. 14, pp. 5304–5307, 2013.

[20] M. Chhowalla, H. S. Shin, G. Eda, L.-J. Li, K. P. Loh, and H. Zhang, "The chemistry of two-dimensional layered transition metal dichalcogenide nanosheets," *Nature Chemistry*, vol. 5, no. 4, pp. 263–275, 2013.

[21] K. S. Novoselov, D. Jiang, F. Schedin et al., "Two-dimensional atomic crystals," *Proceedings of the National Academy of Sciences of the United States of America*, vol. 102, no. 30, pp. 10451–10453, 2005.

[22] J. N. Coleman, M. Lotya, A. O'Neill et al., "Two-dimensional nanosheets produced by liquid exfoliation of layered materials," *Science*, vol. 331, pp. 568–571, 2011.

[23] V. Nicolosi, M. Chhowalla, M. G. Kanatzidis, M. S. Strano, and J. N. Coleman, "Liquid exfoliation of layered materials," *Science*, vol. 340, no. 6139, Article ID 1226419, 2013.

[24] M. Xu, T. Liang, M. Shi, and H. Chen, "Graphene-like two-dimensional materials," *Chemical Reviews*, vol. 113, no. 5, pp. 3766–3798, 2013.

[25] R. Ionescu, W. Wang, Y. Chai et al., "Synthesis of atomically thin $MoS_2$ triangles and hexagrams and their electrical transport properties," *IEEE Transactions on Nanotechnology*, vol. 13, no. 4, pp. 749–754, 2014.

[26] A. M. Van Der Zande, P. Y. Huang, D. A. Chenet et al., "Grains and grain boundaries in highly crystalline monolayer molybdenum disulphide," *Nature Materials*, vol. 12, no. 6, pp. 554–561, 2013.

[27] S. Najmaei, Z. Liu, W. Zhou et al., "Vapour phase growth and grain boundary structure of molybdenum disulphide atomic layers," *Nature Materials*, vol. 12, no. 8, pp. 754–759, 2013.

[28] Y. Lee, S. Park, H. Kim, G. H. Han, Y. H. Lee, and J. Kim, "Characterization of the structural defects in CVD-grown monolayered $MoS_2$ using near-field photoluminescence imaging," *Nanoscale*, vol. 7, no. 28, pp. 11909–11914, 2015.

[29] Y. Huang, J. Guo, Y. Kang, Y. Ai, and C. M. Li, "Two dimensional atomically thin $MoS_2$ nanosheets and their sensing applications," *Nanoscale*, vol. 7, no. 46, pp. 19358–19376, 2015.

[30] D. Qiu, D. U. Lee, S. W. Pak, and E. K. Kim, "Structural and optical properties of $MoS_2$ layers grown by successive two-step chemical vapor deposition method," *Thin Solid Films*, vol. 587, pp. 47–51, 2015.

[31] D. W. Kim, J. M. Ok, W.-B. Jung et al., "Direct observation of molybdenum disulfide, $MoS_2$, domains by using a liquid crystalline texture method," *Nano Letters*, vol. 15, no. 1, pp. 229–234, 2015.

[32] X. Huang, Z. Zeng, and H. Zhang, "Metal dichalcogenide nanosheets: preparation, properties and applications," *Chemical Society Reviews*, vol. 42, pp. 1934–1946, 2013.

[33] M. Ye, D. Winslow, D. Zhang, R. Pandey, and Y. K. Yap, "Recent advancement on the optical properties of two-dimensional molybdenum disulfide ($MoS_2$) thin films," *Photonics*, vol. 2, no. 1, pp. 288–307, 2015.

[34] A. Splendiani, L. Sun, Y. Zhang et al., "Emerging photoluminescence in monolayer $MoS_2$," *Nano Letters*, vol. 10, no. 4, pp. 1271–1275, 2010.

[35] H. Liu and D. Chi, "Dispersive growth and laser-induced rippling of large-area singlelayer $MoS_2$ nanosheets by CVD on c-plane sapphire substrate," *Scientific Reports*, vol. 5, article 11756, 2015.

[36] X. Ling, Y.-H. Lee, Y. Lin et al., "Role of the seeding promoter in $MoS_2$ growth by chemical vapor deposition," *Nano Letters*, vol. 14, no. 2, pp. 464–472, 2014.

[37] D. Gao, M. Si, J. Li et al., "Ferromagnetism in freestanding $MoS_2$ nanosheets," *Nanoscale Research Letters*, vol. 8, article 129, 2013.

[38] H. Nan, Z. Wang, W. Wang et al., "Strong photoluminescence enhancement of $MoS_2$ through defect engineering and oxygen bonding," *ACS Nano*, vol. 8, no. 6, pp. 5738–5745, 2014.

# 12

# Optimal Power Flow Techniques under Characterization of Conventional and Renewable Energy Sources

**Baseem Khan**[1] **and Pawan Singh**[2]

[1] *School of Electrical and Computer Engineering, Hawassa University Institute of Technology, Hawassa, Ethiopia*
[2] *School of Informatics, Hawassa University Institute of Technology, Hawassa, Ethiopia*

Correspondence should be addressed to Baseem Khan; baseem.khan04@gmail.com

Academic Editor: Daniela Proto

The exhaustive knowledge of optimal power flow (OPF) methods is critical for proper system operation and planning, since OPF methods are utilized for finding the optimal state of any system under system constraint conditions, such as loss minimization, reactive power limits, thermal limits of transmission lines, and reactive power optimization. Incorporating renewable energy sources optimized the power flow of system under different constraints. This work presents a comprehensive study of optimal power flows methods with conventional and renewable energy constraints. Additionally, this work presents a progress of optimal power flow solution from its beginning to its present form. Authors classify the optimal power flow methods under different constraints condition of conventional and renewable energy sources. The current and future applications of optimal power flow programs in smart system planning, operations, sensitivity calculation, and control are presented. This study will help the engineers and researchers to optimize power flow with conventional and renewable energy sources.

## 1. Introduction

The OPF in power system is an optimization problem under various constraints. It is practically significant and well-explored subfield of constrained optimization. The important feature of OPF is the presence of the load flow equations in the set of equality constraints. Carpentier [1] introduced the OPF problem in 1979. Carpentier introduced OPF as an extension to the problem of optimal economic dispatch (ED) of generation in traditional power systems. Carpentier's key contribution was the inclusion of the electric power flow equations in the ED formulation. OPF techniques are traditional and metaheuristic based. Figure 1 shows the classification of traditional and metaheuristic techniques.

Further, due to large integration of renewable energy sources in conventional power system, incorporation of uncertainties in OPF calculation is essential.

OPF largely depends on static optimization method for minimizing a scalar optimization function. Dommel and Tinney [2] introduced OPF in 1968, for minimization purpose, in which the first-order gradient algorithm is subject to equality and inequality constraints. Momoh et al. [3] utilized OPF to resolve the problems of deregulated electrical industry. Further, authors have used OPF to solve problems of vertical electricity market.

In this paper, authors present a comprehensive study of optimal power flow methods. Section 2 presents the review of OPF methods with conventional energy sources. In Section 3, authors reviewed OPF methods with renewable energy sources. Section 4 presents a comparative analysis between different OPF methods followed by the conclusion.

## 2. OPF for Power System with Conventional Energy Sources

There are various methods to solve the problem of optimal power flow with conventional energy sources. Some of them are as follows.

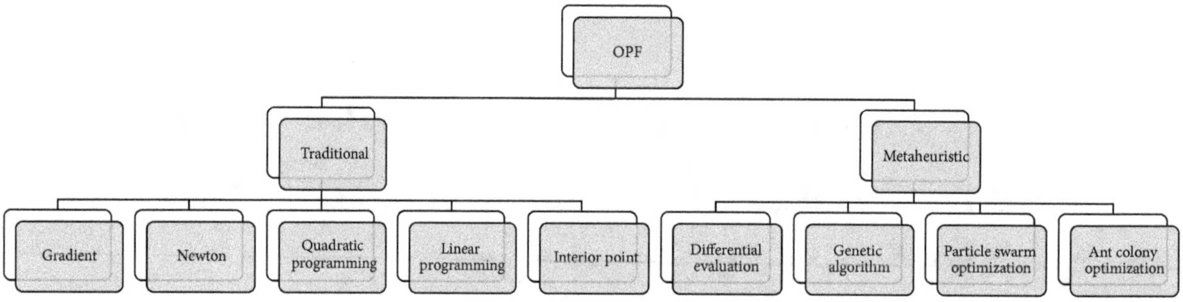


FIGURE 1: Classification of OPF methods.

*2.1. Mathematical Methods.* Classical mathematical methods (gradient, Newton's, linear, and integer programming, etc.) allow the finding of optimal solutions of real-world problems.

*2.1.1. Gradient Method.* In Jan 2004, Sun et al. [4] proposed a penalty-based scheme for OPF with transient stability, in which adjoint equation technique calculates the gradient of the penalty term related to the stability constraints. This adjoint equation approach greatly reduces the computational cost.

*2.1.2. Newton Method.* In 1984, David et al. [5] presented a method to solve classical OPF problem with a nonseparable objective function. The classical OPF is a problem with controllable variables, which is regulated to optimize an objective function, by satisfying physical and operating limits. In 2007, Pizano-Martinez et al. [6] presented a Newton's algorithm based model of a voltage source converter-high voltage direct current (VSC-HVDC) system, which is fit for OPF solutions. The VSC-HVDC's capability to deliver the independent control of converter's AC voltage magnitudes and phase angles is well represented by the model, which permits active and reactive power control separately for system regulation. In 2012, Zhao et al. [7] proposed a variant of primal-dual interior point method (PDIPM) algorithm based on the continuous Newton's method to improve the robustness of PDIPM. Newton's algorithm in PDIPM is exchanged with continuous Newton's method. The developed method indicates good stability and exceptional ability to converge. In 2000, Ambriz-Perez et al. [8] presented advanced load flow models for the static VAR compensator (SVC). The models are merged into current load flow (LF) and OPF Newton's algorithm. A complete SVC susceptance and firing angle models are suitable for conventional and optimal power flow analysis. In 2009, Milano [9] presented a continuous version of Newton's method for solving the power flow problem. It has two main contributions: the first, a general framework for applying efficient numerical integration techniques to solve ill conditioned or badly initialized power flow cases and the second, a formal classification of the existing numerical schemes for solving the power flow problem.

*2.1.3. Linear, Nonlinear, and Semidefinite Programming.* In 1998, Wei et al. [10] presented a new interior point nonlinear programming algorithm for OPF, which is based on the uneasy Kurush-Kuhn-Tucker (KKT) conditions of the original problem but not on a logarithmic barrier function method. Through the concept of centeringdirection, the interior point method is stretched to classical power flow (C-PF) and approximate OPF (A-OPF) problems. In 2008, Bai et al. [11] presented a new solution using the semidefinite programming (SDP) technique to solve the OPF. The developed method includes reformulating the OPF problems into SDP model and developing an algorithm of interior point method (IPM) for SDP. Based on SDP, the OPF problem is solved by primal-dual interior point algorithms, which possess super linear convergence. In 1994, Granville [12] presented an application of an interior point method based on the primal-dual algorithm to optimal reactive power dispatch (ORPD), which is a large-scale nonconvex nonlinear programming problem with nonlinear constraints. In 2005, Zhang et al. [13] presented a nonlinear interior point optimal power flow (OPF) method based on a current mismatch formulation in rectangular coordinates.

*2.1.4. Uncategorised Mathematical Techniques.* In 2015, Abdelouadoud et al. [14] presented a second-order cone (SOC) relaxation algorithm to solve OPF based on a branch flow model of a radial and balanced distribution system. In 2016, Baran and Fernandes [15] presented a three-phase OPF, which includes the mutual impedances in order to minimize the losses of system. In 2016, Garces [16] presented a quadratic approximation for OPF in power distributions systems. The proposed method is based on a linearized load flow, which is effective for power distribution systems as well as three-phase unbalanced operation. In 2015, Emily Manoranjitham and Shunmugalatha [17] presented a process to solve OPF in the power system, which utilized UPFC. This UPFC recovers power transfer capability and transient stability as well as diminishing the transmission loss and fuel cost of generation. In 2016, Xu et al. [18] presented the preventive–corrective security-constrained optimal power flow (PCSCOPF) to accomplish the best coordination between the preventive control (PC) and corrective control (CC) by considering the probabilistic nature of the contingencies and cost of CC as well as other binding constraints. In 2015, Zhang et al. [19] presented the optimal model of carbon energy combined flow (OCECF) which is solved by a new estimated multiobjective solution (AIMS) $Q(\lambda)$ learning. The carbon emissions, fuel cost, active power loss, voltage deviation, and carbon emission loss are selected as the optimization objectives. In 2015, Zhao et al. [20] proposed a voltage stability constrained dynamic optimal reactive power flow (VSC-DORPF) model. It decreases daily network losses, improves voltage quality, and enhances voltage stability of power system. In 2016, Wang et al. [21] presented a mathematical technique to solve the corrective risk-based security-constrained optimal power flow (CRB-SCOPF) model, which considers the system's corrective capabilities after contingency occurred. The proposed method applies Lagrangian relaxation to the system risk constraints and then applies benders decomposition to the remaining Lagrangian subproblem.

*2.2. Metaheuristic Algorithms.* Metaheuristic techniques are powerful and flexible search methodologies that have successfully tackled practically difficult problems. Heuristic and metaheuristic algorithms seek to produce good quality solutions in reasonable computation times that are good enough for practical purposes.

*2.2.1. Genetic Algorithm.* In 2002, Bakirtzis et al. [22] presented the OPF, which is a nonlinear, nonconvex, large-scale, static optimization problem based on an enhanced genetic algorithm (EGA). For the solution of OPF, both continuous and discrete control variables are used. In 1995, Lee et al. [23] presented an enhanced simple genetic algorithm, used for reactive power system planning. Further, a new population selection and generation method is proposed, which uses Bender's cut method. In 1998, Lee [24] presented a comparative study for three evolutionary algorithms (EAs) to solve the optimal reactive power planning (ORPP) problem. The ORPP problem is decomposed into $P$ and $Q$ optimization modules, and the evolutionary algorithms optimize each module in an iterative manner to obtain the global solution. In 2006, Todorovski and Rajicic [25] presented a new method to solve the OPF problem by using genetic algorithm. It depends on the application of new initialization procedure, which utilizes voltage angles at generator-buses as control variables to achieve voltages at load-buses with less computation.

*2.2.2. Particle Swarm Optimization.* In 2002, Abido [26] presented particle swarm optimization (PSO) algorithm based an efficient and reliable approach to solve the OPF problem. The proposed approach employs the global and local investigation capabilities of PSO to examine the optimal setting of control variables. Different objective functions have been considered to minimize the fuel cost, improve the voltage profile, and enhance voltage stability. In 2006, Vlachogiannis and Lee [27] presented three types of PSO algorithms: the enhanced general passive congregation (GPAC), local passive congregation (LPAC) with limitation factor approach based on the passive congregation operator, and the CA based on the coordinated aggregation operator. The proposed PSO algorithms and the conservative interior point OPF based algorithm competed in the optimization problems of reactive power and voltage control. In 2005, Esmin et al. [28] presented an approach to optimize the power loss by using PSO algorithm and improved the PSO to the hybrid PSO (HPSO) algorithm. The proposed approach employs local and global capabilities to search the solution of optimal loss reduction by installing the shunt reactive power compensator. In 2008, Zhang and Liu [29] presented a formulation of multiobjective reactive power and voltage control problem. The multiobjective problem has been solved by means of fuzzy optimization strategy and fuzzy adaptive particle swarm optimization (FAPSO). The factors are active power loss, voltage deviation, and the voltage stability index of the system. In 2009, Al-Rashidi and El-Hawary [30] presented a complete formulation of dissimilar particle swarm optimization (PSO), which is applied to solve optimization problems in power systems. In 2016, Su et al. [31] presented an optimization based sequential strategy and multiobjective optimization based real-time strategy for the optimal placement and control of delta-connected switched capacitors. Considering load variations, a complete optimization for capacitor placement is formulated to increase the net annual returns from network loss reduction and capacity release. Further, a sequential strategy based on loss sensitivity analysis is presented for efficient capacitor placement on large-scale unbalanced distribution networks. In 2008, Valle et al. [32] presented a comprehensive overview of the basic concepts of PSO and its variants. In addition, a comprehensive study on different power system problems, which is solved by PSO, is presented. In 2016, Singh et al. [33] employed an aging leader and challengers PSO (ALC-PSO) to solve different types of OPF problem of power system with several types of complexities and capacities.

*2.2.3. Uncategorised Metaheuristic and Search Algorithms.* In 2011, Niknam et al. [34] presented a multiobjective improved shuffle frog-leaping algorithm (SLFA) to solve the OPF emission problem in power systems considering both economic and environmental issues. The proposed technique produces optimal values, taking into account different objectives, including the best cost and the best emission. In 2014, Ghasemi et al. [35] presented chaotic invasive weed optimization (CIWO) algorithms based on chaos and examine its performance for optimal settings of OPF and its control variables. In 2016, Abaci and Yamacli [36] presented a differential search based optimization method to solve various types of problems including complex, single, and multiobjective functions within the constraints concerning optimal power flow (OPF). In 2016, Acharjee [37] presented the self-adaptive differential evolutionary (SADE) algorithm for increasing and controlling the power flow using unified power flow controller (UPFC) under practical security constraints (SCs). In 2016, Ardeshiri Lajimi and Amraee [38] presented an artificial neural network (ANN) based transient stability constrained optimal power flow (TSCOPF) formulation by using imperialist competitive algorithm (ICA). In order to increase the rotor angle transient stability of power system against large disturbances, critical clearing time is used as the transient stability index. In 2016, Ayan and Kiliç [39] presented the solution of OPF problem for biterminal high-voltage direct current (HVDC) power systems by using back-tracking search algorithm (BSA). To show the applicability and efficiency of BSA, three different test systems are utilized. In 2016, Daryani et al. [40] presented an adaptive group search optimization (AGSO) algorithm for solving optimal power flow (OPF) problem. In 2015, Mahdad and Srairi [41] presented a new approach of power system planning which is based on hybrid firefly algorithm (FFA) and pattern search (PS) algorithm and supported with brainstorming rules to minimize total fuel cost, power losses, and voltage deviation. In 2015, Mukherjee [42] presented chaotic krill herd algorithm (CKHA) to solve the OPF problem of power system with different objective functions and to increase the performance of basic KHA method. In 2016, Mukherjee and Mukherjee [43] presented a novel opposition based Krill herd algorithm (OKHA), a metaheuristic algorithm, to resolve the OPF problem of power system incorporated with flexible AC transmission systems (FACTS) devices. In 2016, Pandiarajan and Babulal [44] presented a fuzzy logic based harmony search algorithm (FHSA) to solve optimal power flow problem of security enhancement in power system. The minimization of fuel cost and severity index objectives are considered. The fuel cost is minimized by changing generator active power, generator bus voltage magnitude, transformer taps, and VAR of shunts. In 2016, Prasad and Mukherjee [45] presented a symbiotic organisms search (SOS) metaheuristic algorithm to solve the OPF problem of power system equipped with FACTS devices. In 2015, Ramesh Kumar and Premalatha [46] presented an adaptive real coded biogeography-based optimization (ARCBBO) method to solve different objective functions of OPF problems with several physical and operating constraints. The different objectives are fuel cost minimization, active power loss minimization, emission minimization, voltage profile, and voltage stability enhancement. In 2015, Venkateswara Rao and Nagesh Kumar [47] utilized BAT algorithm to minimize real power losses in a power system. Further optimal power flow problem is solved with unified power flow controller. In 2015, Yuan et al. [48] presented an upgraded artificial bee colony algorithm with quantum theory and chaotic local search operator (QCABC) to solve the OPF problem.

## 3. OPF for Power System with Renewable Energy Sources

*3.1. Distributed Generation (DG).* In 2011, Amanifar and Hamedani Golshan [49] presented the PSO algorithm to find the optimal locations and sizes of DGs, with an objective to minimize the total cost of the system, real power loss, and the number of DGs to be installed. In 2005, Harriso and Wallace [107] presented government led targets and incentives for increasing the capacity of distributed generation connecting to distribution networks. In 2012, Pazheri et al. [50] presented economic/environmental dispatching (EED) problem formulation for a hybrid system, which includes thermal generating units, solar, wind, and renewable storage. Analysis is carried out using MATLAB simulation for a high irradiation solar region. In 2014, Nick et al. [51] presented the optimal allocation of dispersed storage systems (DSSs) in active distribution networks (ADNs) by describing a multiobjective optimization problem to find the optimal trade-off between technical and economic goals. In 2010, Atwa et al. [52] proposed a probabilistic planning technique for optimally allocating different types of DG (i.e., wind DG, solar DG, and biomass DG) in the distribution system to minimize annual energy losses. In 2013, Georgilakis and Hatziargyriou [108] presented a comprehensive description of different models and optimization methods to solve the optimal DG placement (ODGPP) problem, for examining as well as categorising current and future research approaches in this field. In 2014, Gill et al. [53] presented a broad dynamic optimal power flow (DOPF) framework for the active network management (ANM) schemes. ANM technologies are focused on intertemporal effects. The DOPF is modelled for renewable energy curtailment, energy storage, and flexible demand. In 2015, Jabr et al. [54] presented a sparse formulation for the affinely adjustable robust counterpart (AARC) of the multiperiod OPF problem with RES and storage. In 2012, Moradi and Abedini [55] presented a combined method to solve location and capacity problems for DG. This method utilized GA and PSO to determine the location and capacity of DG, respectively. In 2011, Sortomme and El-Sharkawi [109] presented the potential profits and effects of unidirectional vehicle to gird (V2G). In 2007, Yang et al. [56] presented the hybrid solar wind system optimization sizing model (HSWSO) to optimize the capacity sizes of hybrid solar wind power generation systems employing a battery bank. In 2011, Atwa and El-Saadany [57] presented a probabilistic planning technique for optimally fulfilling wind-based distributed generation (DG) in distribution systems to minimize annual energy loss. In 2007, Ashok [110] discussed various system apparatus for hybrid energy system to improve a universal model and found an optimal combination of energy components for supplying rural community. In 2016, Bouhouras et al. [111] presented the optimization concerning the placement and sizing of DG units in distribution networks (DNs) for loss minimization. In 2015, Azizipanah-Abarghooee et al. [112] presented a probabilistic OPF problem for the system consisting of the thermal units (TUs), wind power plants (WPPs), photovoltaic cell (PV), and combined heat and power (CHP). In 2016, Wei et al. [58] proposed an integration technique of DG based on stochastic optimal power flow (S-OPF). A low-cost optimal model of DG integration for Distribution System Operators (DSOs) is modelled. In 2013, Lin and Chen [59] presented a distributed and parallel OPF (DPOPF) algorithm. Additionally, a Petri net (PN) based computational synchronization mechanism is proposed to solve OPF smart grid transmission system (OPFSG) problem. In 2015, Summers et al. [60] presented a stochastic-multiperiod OPF problem, for which a family of convex rough calculations is used to trade-off cost against security in different ways. In 2016, Tazvinga et al. [113] presented an optimal model of a photovoltaic-diesel-battery (PDB) based hybrid energy management system to reduce both fuel costs and battery wear costs. It also discovers the optimal power flow, photovoltaic power availability, battery bank state of charge, and load power demand. In 2016, Chen et al. [61] presented a PSO based optimal power management method to consider uncertain driving conditions for plug-in hybrid electric vehicles (HEVs). In 2015, Ma et al. [114] presented a technoeconomic analysis of the standalone hybrid solar-wind-pumped storage system for an isolated microgrid. In 2017, Sichilalu et al. [62] presented an optimal control (OC) model of a heat pump water heater (HPWH), which is supplied by a wind generator-photovoltaic-grid system. The objective function is energy cost minimization, which considers the time-of-use electricity tariff.

*3.2. Microgrid.* In 2011, Sanseverino et al. [63] presented an execution monitoring and replanning approach to solve the optimal generation dispatch problem in a smart grid through the minimization of carbon emissions and production costs as well as improve the quality. In 2015, Shi et al. [64] presented an online energy management strategy (EMS) for real-time operation of microgrids. In 2013, Dall'Anese et al. [65] presented the numerical tests. The main role of the proposed approach is to achieve the globally optimal solution of the original nonconvex optimal power flow. In 2012, Battistelli et al. [115] presented an optimization tool for energy management within small energy systems incorporated with V2G systems. In 2011, Chen et al. [66] presented a smart energy management system (SEMS) to organize optimally the power production of DG sources and energy storage system (ESS), and, additionally, minimize the operational costs of microgrids. In 2012, Liang et al. [67] presented a wide-area measurement based dynamic stochastic optimal power flow (DSOPF) control algorithm using the adaptive critic design (ACD) technique. In 2013, Levron et al. [68] presented the optimal energy management system of storage devices in grid-connected microgrids. Stored energy is controlled to balance the loads and renewable sources as well as minimize the total cost of energy at the point-of-common-coupling (PCC). In 2011, Moghaddam et al. [69] presented a multiobjective adaptive modified PSO (AMPSO) optimization algorithm to solve the multioperation management problem in a typical MG with renewable energy sources (RESs). In 2010, Morais et al. [70] presented the operational optimization of an isolated system by a virtual power producer (VPP). The main objective is to decide the best VPP management strategy to minimize the generation costs and optimize storage charging and discharging time subjected to all operational constraints. In 2011, Tanaka et al. [116] presented an approach for optimal operation of a smart grid to minimize the interconnection-point power flow fluctuation. To accomplish the planned optimal operation, distributed controllable loads such as battery and heat pump are used. In 2015, Alavi et al. [71] proposed an optimal approach for operating a MG in islanded mode. The uncertainties associated with output of renewable resources and demands are demonstrated by point estimate method (PEM) and robust optimization (RO), respectively. In 2014, Bahmani-Firouzi and Azizipanah-Abarghooee [72] presented an effective outline for MG operation management studies with respect to operation, maintenance, and financial points. The fixed and maintenance cost of battery energy storage (BES) was considered in the optimization of MG studies. A robust and effective metaheuristic improved bat algorithm (IBA) approach is used to develop corrective strategies and perform least cost dispatches. In 2015, Bracale et al. [117] presented an optimal control approach for a DC microgrid (DCMG) that included dispatchable (such as microturbine) and nondispatchable (such as photovoltaic generator) generation units, a storage system, and controllable/noncontrollable loads. This approach is designed to minimize the daily total energy costs. In 2015, Gholami et al. [73] employed GA with new settings and operators, which is capable of capacitor placement in both islanded and grid-connected modes at different load levels. In 2012, Khorramdel and Raoofat [74] presented a stochastic programming approach for reactive power scheduling of a MG under the uncertainty of wind farms. A multiobjective function with goals of loss minimization, reactive power reserve, and voltage security margin maximization are optimized using a four-stage multiobjective nonlinear programming. In 2016, Lv et al. [75] proposed a multiobjective bilevel optimization algorithm. The upper-level model defines the optimal dispatch of distribution network (DN) to accomplish minimum power loss and voltage profile. The lower-level model considers microgrids (MGs) operating cost as objective to decide the optimal operation scheme of distributed generator DGs. In 2015, Mclarty et al. [118] presented two optimization approaches for the dispatch of a multichiller cooling plant with cold-water thermal storage. The optimization objective is to decrease both cost and emission while considering real operational constraints of a plant. In 2015, Riva Sanseverino et al. [119] presented a novel OPF algorithm for islanded MG. The algorithm yields minimum losses and stable operating point with relevant droop parameters, which are used for primary voltage and frequency regulation. In 2015, Sharma et al. [76] presented a quasi-oppositional swine influenza model based optimization with quarantine (SIMBO-Q) to reduce the total operation cost of MG considering optimal size of battery energy storage.

*3.2.1. Microgrid with Renewable Energy Sources and/or Battery Energy Storage System.* Presently, batter storage system is a vital component in microgrid technology. For optimizing the investment in data centre's storage capacity, Thompson et al. [120] presented a method. In [77], Sharma et al. utilized gray wolf optimization method for finding the optimum capacity of energy storage. In [121], Krishnamurthy and Kwasinski discussed microgrid's power supply resiliency under severe conditions. Xu et al. [122] presented an engineering experience with energy storage system. For optimizing storage capacity in microgrid, Liu et al. [78] proposed an optimal coordinated planning scheme. In [123], Khodabakhsh and Sirouspour developed two different methodologies for online rolling horizon optimal control of storage system. Shen et al. [79] presented an energy management scheme for microgrid, which contains battery storage, diesel generators, PV, and wind. For optimal working of an electric vehicle parking deck, Guo et al. [80] presented a bilevel structure. For frequency control of microgrid, Hassanzadehfard et al. [124] employed battery banks as long-term storages and ultracapacitors as short-term storages. To calculate the optimal power rating and capacity of energy storage, Alharbi and Bhattacharya [125] developed an optimization model.

*3.2.2. Electric Vehicle Technology.* The latest development in the field of electric vehicle technology has a great impact on microgrid operation. For renewable energy sources and vehicle to grid system integration, Melhem et al. [81] proposed a residential energy management system. Yu et al. [82] investigated a model to find the impact of electric vehicle technology on demand response mobility. For integrating electric vehicle in smart gird, Laureri et al. [126] developed an optimization technique. Paterakis et al. [83] presented an optimization model for minimizing energy procurement cost of smart household. Li et al. [84] presented an online methodology for performing cost aware scheduling. To charge electric vehicle, Yao et al. [127] presented a real-time charging scheme.

*3.3. Solar.* In 2012, Lin et al. [85] utilized an active power limitation strategy to diminish PV power injection during peak solar irradiation, to avoid voltage violation. In 2015, Martín [86] presented the process of a concentrated solar plant, which is based on regenerative Rankine cycle. Additionally, dry cooling technologies are optimized by using mathematical programming techniques. In 2014, Bianchi et al. [128] presented a PV/battery hybrid system, to endorse the optimal design of hybrid system in terms of PV module number, PV module tilt, and number and capacity of batteries. In 2015, Burlafinger et al. [87] utilized selective absorbers for concentrated solar power (CSP) plants to minimize thermal losses and increase the efficiency. Additionally, solar thermal absorber's potential is optimized by using selective absorption properties and temperature. In 2016, Guo et al. [88] presented the optimal ratio of turbine pressure drop to the available total pressure difference and examined it broadly using theoretical analysis and 3D numerical simulations, for solar chimney power plant (SCPP). In 2015, Ha and Vakiloroaya [129] addressed the modelling and control problem of a fully developed hybrid solar-assisted, split-system air-conditioner to improve the performance of an existing solar air-conditioning system. In 2013, Lizarraga-Garcia et al. [130] presented an optimization case study for the time-variable operation of the concentrated solar power on demand (CSPonD) solar energy receiver and thermal energy storage. In the study, the factors considered are time-variable electricity prices and electricity buy-back from the grid. In 2017, Mokheimer et al. [131] presented an integrated solar gas turbine cogeneration plant (ISGCP) which generates electricity and produces process steam at a constant rate of 81.44 kg/s at 45.88 bar and 394°C throughout the year. In addition to this, the generation of electricity is investigated for different gas turbine generating capacities. In 2014, Nazir et al. [89] presented the power flow optimization of the PV system connected to the grid. Optimization has been performed by calculating the root of the active power equation $P$ (ma, $\alpha$) and reactive power equation $Q$ (ma, $\alpha$) using Newton Raphson method. In 2016, Ntsaluba et al. [90] presented a flow-rate optimization of solar water heating system (SWHS) with specific focus placed on pump flow-rate optimization. In 2015, Wang et al. [91] presented an approach to analyse the optimal capacity and economic feasibility of a hybrid energy storage system (HESS) supporting the dispatch of a 30 MW photovoltaic (PV) power plant.

*3.4. Wind.* In 2014, Rabiee et al. [92] presented a complete OPF formulation for a power system with uncertain wind power injection through line-commutated converter high-voltage DC (LCC-HVDC) links, voltage source converter (VSC-HVDC), and doubly fed induction generators' (DFIGs). The objective is to maximize the toughness of total costs against the intermittent wind power generation using info-gap decision theory (IGDT). In 2016, Ke et al. [93] presented a new probabilistic optimal power flow (POPF) model with chance constraints that reflects the uncertainties of wind power generation (WPG) and loads. In 2014, Bienstock et al. [94] presented a technique that depends on chance constraints to limit the probability that any line (or generator) is overloaded, together with an optimal online control that accounts for controllable generator response to renewable fluctuation. In 2011, Brekken et al. [132] presented the sizing and control procedures for a zinc–bromine flow battery-based energy storage system. The results show that through an effective control and coordination of energy storage systems, the expectedness of wind plant outputs is increased and the cost of combination associated with reserve necessities is decreased. In 2013, Ghofrani et al. [95] presented an optimal placement of energy storage in a deregulated power system to minimize the hourly social cost. By the means of historical data and curve fitting, both wind and load are modelled stochastically. A GA-enhanced market-based probabilistic optimal power flow (POPF) with energy storage integration and wind generation maximizes wind power utilization over the scheduling period. In 2009, Jabr and Pal [96] presented a stochastic model of wind generation in an OPF dispatching program. The model faces the coordination of wind and thermal power while accounting for the expected penalty cost for not using all available wind power and the expected cost of utilizing power reserves because of wind power shortage. In 2016, Sedghi et al. [133] presented the planning technique for defining optimal location, capacity, and power rating of the batteries while minimizing the cost function under the technical constraints. The objective function involves the monetary factors, such as investment, operation, and reliability costs, as well as the technical penalty factor. In 2014, Sebastián et al. [97] presented a probabilistic AC optimal power flow (POPF) that takes into account load variation, wind's stochastic behaviour, and variable line's thermal rating which is usually used as a deterministic value in several studies. In 2015, Aien et al. [98] reviewed the probabilistic techniques used for probabilistic optimal power flow (POPF) and proposed a method by using the unscented transformation (UT) technique. In 2013, Arabali et al. [134] presented a technical outline to evaluate the energy storage application, optimal placement, and economic advantage, for the social cost and transmission congestion. In 2014, Chang et al. [99] presented an evolutionary particle swarm optimization (EPSO) approach for solving the OPF problem of a wind-thermal power system. In 2015, Jadhav and Roy [135] presented an optimization system to calculate the optimal operation of system containing electric vehicles and offshore wind farm connected to the grid through HVDC link. In 2015, Li et al. [100] presented the probabilistic optimal power flow for wind

Table 1: Comparative analysis of different OPF techniques.

| OPF problem classification | Constraints | | | | | | Assumptions |
|---|---|---|---|---|---|---|---|
| | Voltage and angle | Transmission line power limit | Reactive power limits | Losses | Generation costs | Contingency | |
| AC | Yes | Yes | Yes | Yes | No | No | Slack bus selection |
| DC | No | Yes | No | Not definite | Yes | No | Voltage magnitudes are fixed |
| Decoupled | Yes | Yes | Yes | Yes | Yes | No | Interaction between active and reactive power is not considered |
| Security-constrained economic dispatch | Yes | Yes | No | Yes | Yes | Yes | Bus voltages are fixed |
| Economic dispatch | No | No | No | Conditional | Yes | No | No transmission constraints |
| Security-constrained | Yes | Yes | Yes | Yes | Yes | Yes | Assumptions regarding postfault flows are there |
| Optimal reactive power dispatch | Yes | Yes | Yes | Yes | Yes | No | Convexity assumption of generators' cost function |
| Metaheuristic based OPF | Yes | Yes | Yes | Yes | Yes | No | Inherent inadequacy that needs further attention in the future including the lack of transparency, knowledge extraction, and model uncertainty |

Table 2: Comparative analysis of traditional and metaheuristic based OPF methods.

| S. number | Properties | Traditional technique | Metaheuristic techniques |
|---|---|---|---|
| (1) | Model formulation | Model is used to simulate a system when the system is not extremely complicated. | Metaheuristic techniques try to imitate, in a very simplistic way, the human cognition capability. |
| (2) | Prerequisites | Use the principles (e.g., physical laws) to derive the relationships of the system. | Dependent on data to determine the structure and parameters. |
| (3) | Nonlinearity integration | Possible if prior knowledge of the nature of the nonlinearity exists. | The prior information of the nature of the nonlinearity is not required. |
| (4) | Hypothesis | Simplifying the problem or incorporating assumptions to make model more understandable. | No need to simplify the model and/or add assumptions. |
| (5) | Precision & accuracy | Provide highest level of accuracy. | Provide greater precision and accuracy due to less chances of error. |
| (6) | Efficiency | Needs more time to solve the complex problem. | Utilized to carry out repetitive and time-consuming tasks, efficiently. |
| (7) | Renewable energy integration | Requires complex mathematical modelling. | With metaheuristic model renewable energy, source incorporation is easy. |

turbine-integrated power system with wind uncertainty and correlated loads. In 2014, Lin and Lin [101] presented a risk-limiting optimal power flow (RLOPF) problem for systems with high penetration of wind power, to address the problem of possibly violating the security constraints in power systems due to the instability of wind power generations. In 2014, Luo et al. [136] presented an energy storage sizing method for reliability requirement. Additionally, a bilevel control strategy for the isolated grids is proposed. In 2016, Mohammadi et al. [102] designed a state feedback controller (SFC) for damping oscillations and tracking the set point input as well as discarding the disturbances correctly for a variable wind system. The suggested method reflects damping and performance together by including a performance index into linear quadratic (LQ) criterion. In 2016, Mohseni-Bonab et al. [103] presented the stochastic multiobjective optimal reactive power dispatch (SMO-ORPD) problem for wind integrated power system, taking into account the uncertainties of system load and wind power generations. In 2014, Panda and Tripathy [104] presented an OPF solution for modified power system in which three conventional generators are replaced by wind-energy conversion systems (WECS). To justify the limitation of reactive power generation capability of WECS, genetic algorithm (GA) and a modified bacteria

Table 3: Summarization of OPF methods for power system with conventional energy sources.

| | OPF for power system with conventional energy sources | | | | | | |
|---|---|---|---|---|---|---|---|
| | Mathematical methods | | | Metaheuristic algorithms | | | |
| OPF methodologies | Gradient method | Newton's method | Linear, nonlinear, and semidefinite programming | Genetic algorithm | Particle swarm optimization | BAT algorithm | Weed optimization |
| Approach | Penalty-based scheme | Newton's algorithm | (i) Nonlinear programming<br>(ii) Semidefinite programming | (i) Enhanced genetic algorithm (EGA)<br>(ii) Evolutionary algorithms | (i) Hybrid PSO (HPSO)<br>(ii) Fuzzy adaptive particle swarm optimization (FAPSO)<br>(iii) Dissimilar particle swarm optimization (PSO) | BAT algorithm | Chaotic invasive weed optimization (CIWO) |
| Techniques | (i) Adjoint equation technique | (i) Voltage source converter-high voltage direct current (VSC-HVDC) system<br>(ii) Primal-dual interior point method<br>(iii) SVC susceptance and firing angle models<br>(iv) Numerical integration techniques | (i) Interior point method<br>(ii) Primal-dual interior point algorithms | (i) Continuous and discrete control variables<br>(ii) Benders' cut method<br>(iii) $P$ and $Q$ optimization modules<br>(iv) Voltage angles at generator-buses as control variables | (i) Enhanced general passive congregation (GPAC), local passive congregation (LPAC)<br>(ii) The CA based coordinated aggregation operator<br>(iii) Fuzzy optimization strategy, aging leader, and challengers PSO (ALC-PSO) | (i) Unified power flow controller | Based on chaos |
| Applications | Transient stability analysis | (i) System regulation<br>(ii) PDIPM robustness improvement<br>(iii) Conventional and optimal power flow analysis<br>(iv) Ill conditioned or badly initialized power flow cases | (i) Classical power flow (C-PF) and approximate OPF (A-OPF) problems<br>(ii) Reformulation of the OPF problems<br>(iii) Optimal reactive power dispatch<br>(iv) Current mismatch formulation | (i) Nonlinear, nonconvex, large-scale, static optimization problem<br>(ii) Reactive power system planning<br>(iii) Compute voltages at load-buses with less computation | (i) Minimization of fuel cost, improving the voltage profile, and enhancing voltage stability<br>(ii) Reactive power and voltage control<br>(iii) Optimal loss reduction<br>(iv) Multiobjective reactive power and voltage control problem<br>(v) Power system optimization problems | (i) Minimize real power losses | (i) Optimal settings of OPF and its control variables |
| References | [4] | [5–9] | [10–13] | [22–25] | [26–30, 33] | [47] | [35] |

TABLE 4: Summarization of OPF methods for power system with renewable energy sources.

| Approach | Techniques | Application | References |
|---|---|---|---|
| | OPF for power system with renewable energy sources | | |
| Distributed generation (DG) | (i) PSO<br>(ii) Economic/environmental dispatch (EED)<br>(iii) Multiobjective optimization<br>(iv) Probabilistic technique<br>(v) Dynamic optimal power flow<br>(vi) Affinely adjustable robust counterpart<br>(vii) GA and PSO<br>(viii) Hybrid solar wind system optimization sizing model<br>(ix) Probabilistic planning technique<br>(x) Stochastic optimal power flow (S-OPF)<br>(xi) Distributed and parallel OPF<br>(xii) Stochastic-multiperiod OPF<br>(xiii) PSO based optimal power management method<br>(xiv) Optimal control (OC) model | (i) Optimal locations and sizes<br>(ii) Optimal dispatch<br>(iii) Optimal trade-off<br>(iv) Optimal allocation of different types of DGs<br>(v) Active network management (ANM) schemes<br>(vi) Multiperiod OPF<br>(vii) Optimal location and capacity of DG<br>(viii) Optimizing the capacity sizes of hybrid solar wind power generation systems<br>(ix) Minimizing annual energy loss<br>(x) Optimal model of DG integration<br>(xi) OPF smart grid transmission system<br>(xii) Find trade-off cost against security<br>(xiii) Plug-in hybrid electric vehicles<br>(xiv) Energy cost minimization | [49–62] |
| Microgrid | (i) Execution monitoring and replanning<br>(ii) Online energy management strategy (EMS)<br>(iii) Nonconvex optimal power flow<br>(iv) Smart energy management system (SEMS)<br>(v) Dynamic stochastic optimal power flow (DSOPF)<br>(vi) Optimal energy management system<br>(vii) Multiobjective adaptive modified PSO (AMPSO)<br>(viii) Virtual power producer (VPP)<br>(ix) Point estimate method (PEM) and robust optimization (RO)<br>(x) Improved bat algorithm (IBA)<br>(xi) GA<br>(xii) Stochastic programming<br>(xiii) Multiobjective bilevel optimization<br>(xiv) Quasi-oppositional swine influenza model based optimization with quarantine (SIMBO-Q) | (i) Optimal generation dispatch problem<br>(ii) Real-time operation of microgrids<br>(iii) Global optimal solution<br>(iv) Optimal power production of DG sources<br>(v) Wide-area measurement<br>(vi) Minimizing the total cost of energy<br>(vii) Multioperation management problem<br>(viii) Minimizing the generation costs and optimize storage charging and discharging time<br>(ix) Optimal operation of MG in islanded mode<br>(x) Corrective strategies<br>(xi) Capacitor placement<br>(xii) Reactive power scheduling of a MG<br>(xiii) Minimizing power loss, optimal operation scheme<br>(xiv) Reducing total operation cost of MG | [63–76] |
| Microgrid with renewable energy sources and/or battery energy storage system | (i) Gray wolf optimization<br>(ii) Optimal coordinated planning strategy<br>(iii) Energy management scheme for MG<br>(iv) Bilevel structure for the optimal working of a MG | (i) Finding optimal capacity of BES<br>(ii) Optimization of energy sources capacity<br>(iii) EV parking deck | [77–80] |
| Electric vehicle technology | (i) Residential energy management<br>(ii) V2G mobile energy system<br>(iii) Bidirectional power flow<br>(iv) Stochastic optimization | (i) Renewable power sources and V2G integration<br>(ii) Demand response management<br>(iii) Minimizing the total energy procurement cost<br>(iv) Cost aware scheduling of EV | [81–84] |
| Solar | (i) Active power limitation strategy<br>(ii) Mathematical programming techniques<br>(iii) Selective absorption<br>(iv) 3D numerical simulations<br>(v) Newton Raphson method<br>(vi) pump flow-rate optimization<br>(vii) Optimal dispatch approach | (i) Diminishing PV power injection during peak solar irradiation<br>(ii) Optimization of dry cooling technologies<br>(iii) Minimizing thermal losses<br>(iv) Optimal ratio of turbine pressure drop<br>(v) Power flow optimization<br>(vi) Flow-rate optimization of solar water heating system (SWHS)<br>(vii) Optimal capacity and economic feasibility | [85–91] |

Table 4: Continued.

| OPF for power system with renewable energy sources | | | |
|---|---|---|---|
| Approach | Techniques | Application | References |
| Wind | (i) Info-gap decision theory (IGDT)<br>(ii) Probabilistic optimal power flow (POPF)<br>(iii) Optimal online control<br>(iv) Historical data and curve fitting stochastically, GA-enhanced market-based probabilistic optimal power flow (POPF)<br>(v) Stochastic wind generation model<br>(vi) Probabilistic AC optimal power flow (POPF)<br>(vii) Unscented transformation (UT) technique<br>(viii) Evolutionary particle swarm optimization (EPSO)<br>(ix) Probabilistic optimal power flow<br>(x) Risk-limiting optimal power flow (RLOPF)<br>(xi) Linear quadratic (LQ) criterion<br>(xii) Stochastic multiobjective optimal reactive power dispatch (SMO-ORPD)<br>(xiii) Genetic algorithm (GA) and a modified bacteria foraging algorithm<br>(xiv) Modified bacteria foraging algorithm<br>(xv) Artificial bee colony optimization algorithm (GABC) | (i) OPF with uncertain wind power injection<br>(ii) OPF with chance constraints of wind power generation (WPG)<br>(iii) Controllable generator response to renewable fluctuation<br>(iv) Minimizing the hourly social cost<br>(v) Optimal dispatch<br>(vi) Optimal power flow with wind's stochastic behaviour<br>(vii) Probabilistic optimal power flow (POPF)<br>(viii) OPF problem of a wind-thermal power system<br>(ix) OPF with wind uncertainty and correlated loads<br>(x) Security constraints violation problem<br>(xi) Damping and performance<br>(xii) Optimal reactive power dispatch<br>(xiii) Determining the optimal schedule<br>(xiv) Multiobjective optimization problem<br>(xv) OPF with probabilistic nature of wind power | [92–106] |

foraging algorithm are employed independently, for determining the optimal schedule. In 2015, Panda and Tripathy [105] presented a modified bacteria foraging algorithm, which is capable of handling multiobjective optimization problems. In 2015, Roald et al. [137] presented a technique of modelling risk in power system operation, accounting for the system availability of redispatch. The resulting risk measure formulates risk-based constraints for the postcontingency line flows. In 2015, Roy and Jadhav [106] presented an OPF study in view of probabilistic nature of wind power. The wind power intermittency is modelled by the parameters of Weibull probability function. The optimization problems are solved by artificial bee colony optimization algorithm (GABC).

## 4. Comparative Analysis of OPF Methods

This section presents a comparative analysis of different OPF methods, which are reviewed in the previous sections. Table 1 presents a comparative analysis of different OPF techniques.

Table 2 presents a comparison between traditional and metaheuristic technique based OPF methods used for the optimization of renewable energy sources.

A detailed summarization of the OPF methods for power system with conventional energy sources is presented in Table 3.

Table 4 presents a summarization of OPF methods for power system with renewable energy sources.

## 5. Conclusions

Optimal power flow is an optimizing tool for power system operation analysis, scheduling, and energy management. Use of the optimal power flow is becoming more important because of its capabilities to deal with various situations. This problem involves the optimization of an objective function that takes various forms while satisfying a set of operational and physical constraints. Hence, in this work authors present a comprehensive review of solution techniques and methods used for optimization of power flows. Further, techniques used for optimization of systems incorporating renewable energy sources such as microgrid, storage system, electrical vehicle, wind, and solar are also reviewed in this work. Different metaheuristic techniques used for OPF are further discussed. Different OPF problems are discussed with respect to the constraints applied and assumptions made. The traditional and metaheuristic based OPF techniques are compared with respect to different properties of OPF techniques. Different mathematical and metaheuristic algorithms used for OPF power system with conventional and renewable energy sources are discussed. A summarization of different techniques used for OPF with traditional and renewable energy sources is presented based on their adopted approach, techniques, and applications.

There are still many constraints and there is nonlinearity, which should be incorporated in future OPF problem. Problems related to mathematical validation, deregulated market constraints, contingencies incorporation, and renewable sources integration are latest challenges for future OPF problems. This study will assist researchers in comparing and selecting an appropriate OPF technique, to find the optimal state of any system under system constraint. This work may also be supportive for the commercial utilization of OPF. Additionally, other better techniques may also be considered for further study. The futuristic enrichment of the current study may be to develop a OPF technique, which can provide better results.

## References


[1] J. Carpentier, "Optimal power flows," *International Journal of Electrical Power & Energy Systems*, vol. 1, no. 1, pp. 3–15, 1979.

[2] H. W. Dommel and W. F. Tinney, "Optimal Power Flow Solutions," *IEEE Transactions on Power Apparatus and Systems*, vol. 87, no. 10, pp. 1866–1876, 1968.

[3] J. A. Momoh, R. J. Koessler, M. S. Bond et al., "Challenges to optimal power flow," *IEEE Transactions on Power Systems*, vol. 12, no. 1, pp. 444–447, 1997.

[4] Y. Sun, Y. Xinlin, and H. F. Wang, "Approach for optimal power flow with transient stability constraints," *IEE Proceedings Generation, Transmission and Distribution*, vol. 151, no. 1, pp. 8–18, 2004.

[5] D. I. Sun, B. Ashley, B. Brewer, A. Hughes, and W. F. Tinney, "Optimal power flow by Newton approach," *IEEE Transactions on Power Apparatus and Systems*, vol. 103, no. 10, pp. 2864–2880, 1984.

[6] A. Pizano-Martinez, C. R. Fuerte-Esquivel, H. Ambriz-Pérez, and E. Acha, "Modeling of VSC-based HVDC systems for a Newton-Raphson OPF algorithm," *IEEE Transactions on Power Systems*, vol. 22, no. 4, pp. 1794–1803, 2007.

[7] X. Zhao, Y. Yang, and H. Wei, "An interior point method based on continuous Newton's method for optimal power flow," in *Proceedings of the IEEE Innovative Smart Grid Technologies*, pp. 1–6, May 2012.

[8] H. Ambriz-Perez, E. Acha, and C. R. Fuerte-Esquivel, "Advanced SVC models for Newton-Raphson load flow and Newton optimal power flow studies," *IEEE Transactions on Power Systems*, vol. 15, no. 1, pp. 129–136, 2000.

[9] F. Milano, "Continuous Newton's method for power flow analysis," *IEEE Transactions on Power Systems*, vol. 24, no. 1, pp. 50–57, 2009.

[10] H. H. Wei, S. J. Kubokawa, and R. Yokoyama, "An interior point nonlinear programming for optimal power flow problems with a novel data Structure," *IEEE Transactions on Power Systems*, vol. 13, no. 3, pp. 870–877, 1998.

[11] X. Bai, H. Wei, K. Fujisawa, and Y. Wang, "Semidefinite programming for optimal power flow problems," *International Journal of Electrical Power & Energy Systems*, vol. 30, no. 6-7, pp. 383–392, 2008.

[12] S. Granville, "Optimal reactive dispatch through interior point methods," *IEEE Transactions on Power Systems*, vol. 9, no. 1, pp. 136–146, 1994.

[13] X. P. Zhang, S. G. Petoussis, and K. R. Godfrey, "Nonlinear interior-point optimal power flow method based on a current mismatch formulation," pp. 795–805.

[14] S. Y. Abdelouadoud, R. Girard, F. P. Neirac, and T. Guiot, "Optimal power flow of a distribution system based on increasingly tight cutting planes added to a second order cone relaxation," *International Journal of Electrical Power & Energy Systems*, vol. 69, pp. 9–17, 2015.

[15] A. R. Baran and T. S. P. Fernandes, "A three-phase optimal power flow applied to the planning of unbalanced distribution networks," *International Journal of Electrical Power & Energy Systems*, vol. 74, pp. 301–309, 2016.

[16] A. Garces, "A quadratic approximation for the optimal power flow in power distribution systems," *Electric Power Systems Research*, vol. 130, pp. 222–229, 2016.

[17] G. Emily Manoranjitham and A. Shunmugalatha, "Application of firefly algorithm on optimal power flow control incorporating simplified impedance UPFC model," *International Journal of Electrical Power & Energy Systems*, vol. 71, pp. 358–363, 2015.

[18] Y. Xu, H. Yang, R. Zhang, Z. Y. Dong, M. Lai, and K. P. Wong, "A contingency partitioning approach for preventive-corrective security-constrained optimal power flow computation," *Electric Power Systems Research*, vol. 132, pp. 132–140, 2016.

[19] X. Zhang, T. Yu, B. Yang, L. Zheng, and L. Huang, "Approximate ideal multi-objective solution Q($\lambda$) learning for optimal carbon-energy combined-flow in multi-energy power systems," *Energy Conversion and Management*, vol. 106, pp. 543–556, 2015.

[20] J. Zhao, L. Ju, Z. Dai, and G. Chen, "Voltage stability constrained dynamic optimal reactive power flow based on branch-bound and primal-dual interior point method," *International Journal of Electrical Power & Energy Systems*, vol. 73, pp. 601–607, 2015.

[21] Q. Wang, J. D. McCalley, T. Zheng, and E. Litvinov, "Solving corrective risk-based security-constrained optimal power flow with Lagrangian relaxation and Benders decomposition," *International Journal of Electrical Power & Energy Systems*, vol. 75, pp. 255–264, 2016.

[22] A. G. Bakirtzis, P. N. Biskas, C. E. Zoumas, and V. Petridis, "Optimal power flow by enhanced genetic algorithm," *IEEE Transactions on Power Systems*, vol. 17, no. 2, pp. 229–236, 2002.

[23] K. Y. Lee, X. Bai, and Y.-M. Park, "Optimization Method for Reactive Power Planning by Using a Modified Simple Genetic Algorithm," *IEEE Transactions on Power Systems*, vol. 10, no. 4, pp. 1843–1850, 1995.

[24] K. Y. Lee, "Optimal reactive power planning using evolutionary algorithms: a comparative study for evolutionary programming, evolutionary strategy, genetic algorithm, and linear programming," *IEEE Transactions on Power Systems*, vol. 13, no. 1, pp. 101–108, 1998.

[25] M. Todorovski and D. Rajicic, "An initialization procedure in solving optimal power flow by genetic algorithm," *IEEE Transactions on Power Systems*, vol. 21, no. 2, pp. 480–487, 2006.

[26] M. A. Abido, "Optimal power flow using particle swarm optimization," *International Journal of Electrical Power & Energy Systems*, vol. 24, no. 7, pp. 563–571, 2002.

[27] J. G. Vlachogiannis and K. Y. Lee, "A comparative study on particle swarm optimization for optimal steady-state performance of power systems," *IEEE Transactions on Power Systems*, vol. 21, no. 4, pp. 1718–1728, 2006.

[28] A. A. A. Esmin, G. Lambert-Torres, and A. C. Zambroni de Souza, "A hybrid particle swarm optimization applied to loss power minimization," *IEEE Transactions on Power Systems*, vol. 20, no. 2, pp. 859–866, 2005.

[29] W. Zhang and Y. T. Liu, "Multi-objective reactive power and voltage control based on fuzzy optimization strategy and fuzzy adaptive particle swarm," *International Journal of Electrical Power & Energy Systems*, vol. 30, no. 9, pp. 525–532, 2008.

[30] M. R. Al-Rashidi and M. E. El-Hawary, "A survey of particle swarm optimization applications in electric power systems," *IEEE Transactions on Evolutionary Computation*, vol. 13, no. 4, pp. 913–918, 2009.

[31] X. Su, M. A. S. Masoum, and P. J. Wolfs, "PSO and improved BSFS based sequential comprehensive placement and real-time multi-objective control of delta-connected switched capacitors in unbalanced radial MV distribution networks," *IEEE Transactions on Power Systems*, vol. 31, no. 1, pp. 612–622, 2016.

[32] Y. del Valle, G. K. Venayagamoorthy, S. Mohagheghi, J.-C. Hernandez, and R. G. Harley, "Particle swarm optimization: basic concepts, variants and applications in power systems," *IEEE Transactions on Evolutionary Computation*, vol. 12, no. 2, pp. 171–195, 2008.

[33] R. P. Singh, V. Mukherjee, and S. P. Ghoshal, "Particle swarm optimization with an aging leader and challengers algorithm for the solution of optimal power flow problem," *Applied Soft Computing*, vol. 40, pp. 161–177, 2016.

[34] T. Niknam, M. R. Narimani, M. Jabbari, and A. R. Malekpour, "A modified shuffle frog leaping algorithm for multi-objective optimal power flow," *Energy*, vol. 36, no. 11, pp. 6420–6432, 2011.

[35] M. Ghasemi, S. Ghavidel, E. Akbari, and A. A. Vahed, "Solving non-linear, non-smooth and non-convex optimal power flow problems using chaotic invasive weed optimization algorithms based on chaos," *Energy*, vol. 73, pp. 340–353, 2014.

[36] K. Abaci and V. Yamacli, "Differential search algorithm for solving multi-objective optimal power flow problem," *International Journal of Electrical Power & Energy Systems*, vol. 79, pp. 1–10, 2016.

[37] P. Acharjee, "Optimal power flow with UPFC using security constrained self-adaptive differential evolutionary algorithm for restructured power system," *International Journal of Electrical Power & Energy Systems*, vol. 76, pp. 69–81, 2016.

[38] R. Ardeshiri Lajimi and T. Amraee, "A two stage model for rotor angle transient stability constrained optimal power flow," *International Journal of Electrical Power & Energy Systems*, vol. 76, pp. 82–89, 2016.

[39] K. Ayan and U. Kiliç, "Optimal power flow of two-terminal HVDC systems using backtracking search algorithm," *International Journal of Electrical Power & Energy Systems*, vol. 78, pp. 326–335, 2016.

[40] N. Daryani, M. T. Hagh, and S. Teimourzadeh, "Adaptive group search optimization algorithm for multi-objective optimal power flow problem," *Applied Soft Computing*, vol. 38, pp. 1012–1024, 2016.

[41] B. Mahdad and K. Srairi, "Security optimal power flow considering loading margin stability using hybrid FFA-PS assisted with brainstorming rules," *Applied Soft Computing*, vol. 35, article no. 3031, pp. 291–309, 2015.

[42] A. Mukherjee, "Solution of optimal power flow using chaotic krill herd algorithm," *Chaos, Solitons & Fractals*, vol. 78, pp. 10–21, 2015.

[43] A. Mukherjee and V. Mukherjee, "Solution of optimal power flow with FACTS devices using a novel oppositional krill herd

algorithm," *International Journal of Electrical Power & Energy Systems*, vol. 78, pp. 700–714, 2016.

[44] K. Pandiarajan and C. K. Babulal, "Fuzzy harmony search algorithm based optimal power flow for power system security enhancement," *International Journal of Electrical Power & Energy Systems*, vol. 78, pp. 72–79, 2016.

[45] D. Prasad and V. Mukherjee, "A novel symbiotic organisms search algorithm for optimal power flow of power system with FACTS devices," *Engineering Science & Technology*, vol. 19, no. 1, pp. 79–89, 2016.

[46] A. Ramesh Kumar and L. Premalatha, "Optimal power flow for a deregulated power system using adaptive real coded biogeography-based optimization," *International Journal of Electrical Power & Energy Systems*, vol. 73, article no. 3472, pp. 393–399, 2015.

[47] B. Venkateswara Rao and G. V. Nagesh Kumar, "Optimal power flow by BAT search algorithm for generation reallocation with unified power flow controller," *International Journal of Electrical Power & Energy Systems*, vol. 68, pp. 81–88, 2015.

[48] X. Yuan, P. Wang, Y. Yuan, Y. Huang, and X. Zhang, "A new quantum inspired chaotic artificial bee colony algorithm for optimal power flow problem," *Energy Conversion and Management*, vol. 100, pp. 1–9, 2015.

[49] O. Amanifar and M. E. Hamedani Golshan, "Optimal Distributed Generation Placement and Sizing for Loss and THD Reduction and Voltage Profile Improvement," *Technical and Physical Problems of Engineering (IJTPE)*, vol. 3, no. 2, 2011.

[50] F. R. Pazheri, M. F. Othman, N. H. Malik, and S. O. K, "Economic and Environmental Dispatch at Highly Potential Renewable Area with Renewable Storage," *International Journal of Environmental Science and Development*, pp. 177–182, 2012.

[51] M. Nick, R. Cherkaoui, and M. Paolone, "Optimal allocation of dispersed energy storage systems in active distribution networks for energy balance and grid support," *IEEE Transactions on Power Systems*, vol. 29, no. 5, pp. 2300–2310, 2014.

[52] Y. M. Atwa, E. F. El-Saadany, M. M. A. Salama, and R. Seethapathy, "Optimal renewable resources mix for distribution system energy loss minimization," *IEEE Transactions on Power Systems*, vol. 25, no. 1, pp. 360–370, 2010.

[53] S. Gill, I. Kockar, and G. W. Ault, "Dynamic optimal power flow for active distribution networks," *IEEE Transactions on Power Systems*, vol. 29, no. 1, pp. 121–131, 2014.

[54] R. A. Jabr, S. Karaki, and J. A. Korbane, "Robust Multi-Period OPF with Storage and Renewables," *IEEE Transactions on Power Systems*, vol. 30, no. 5, pp. 2790–2799, 2015.

[55] M. H. Moradi and M. Abedini, "A combination of genetic algorithm and particle swarm optimization for optimal DG location and sizing in distribution systems," *International Journal of Electrical Power & Energy Systems*, vol. 34, no. 1, pp. 66–74, 2012.

[56] H. Yang, L. Lu, and W. Zhou, "A novel optimization sizing model for hybrid solar-wind power generation system," *Solar Energy*, vol. 81, no. 1, pp. 76–84, 2007.

[57] Y. M. Atwa and E. F. El-Saadany, "Probabilistic approach for optimal allocation of wind-based distributed generation in distribution systems," *IET Renewable Power Generation*, vol. 5, no. 1, pp. 79–88, 2011.

[58] C. Wei, Y. Fu, Z. Li, and Y. Jiang, "Optimal DG penetration rate planning based on S-OPF in active distribution network," *Neurocomputing*, vol. 174, pp. 514–521, 2016.

[59] S.-Y. Lin and J.-F. Chen, "Distributed optimal power flow for smart grid transmission system with renewable energy sources," *Energy*, vol. 56, pp. 184–192, 2013.

[60] T. Summers, J. Warrington, M. Morari, and J. Lygeros, "Stochastic optimal power flow based on conditional value at risk and distributional robustness," *International Journal of Electrical Power & Energy Systems*, vol. 72, pp. 116–125, 2015.

[61] Z. Chen, R. Xiong, and J. Cao, "Particle swarm optimization-based optimal power management of plug-in hybrid electric vehicles considering uncertain driving conditions," *Energy*, vol. 96, pp. 197–208, 2016.

[62] S. Sichilalu, T. Mathaba, and X. Xia, "Optimal control of a wind–PV-hybrid powered heat pump water heater," *Applied Energy*, vol. 185, pp. 1173–1184, 2017.

[63] E. R. Sanseverino, M. L. Di Silvestre, M. G. Ippolito, A. De Paola, and G. Lo Re, "An execution, monitoring and replanning approach for optimal energy management in microgrids," *Energy*, vol. 36, no. 5, pp. 3429–3436, 2011.

[64] W. Shi, N. Li, C. Chu, and R. Gadh, "Real-Time Energy Management in Micro grids," *IEEE Transactions on Smart Grid*, no. 99, pp. 1-1, 2015.

[65] E. Dall'Anese, H. Zhu, and G. B. Giannakis, "Distributed optimal power flow for smart microgrids," *IEEE Transactions on Smart Grid*, vol. 4, no. 3, pp. 1464–1475, 2013.

[66] C. Chen, S. Duan, T. Cai, B. Liu, and G. Hu, "Smart energy management system for optimal microgrid economic operation," *IET Renewable Power Generation*, vol. 5, no. 3, pp. 258–267, 2011.

[67] J. Liang, G. K. Venayagamoorthy, and R. G. Harley, "Wide-area measurement based dynamic stochastic optimal power flow control for smart grids with high variability and uncertainty," *IEEE Transactions on Smart Grid*, vol. 3, no. 1, pp. 59–69, 2012.

[68] Y. Levron, J. M. Guerrero, and Y. Beck, "Optimal power flow in microgrids with energy storage," *IEEE Transactions on Power Systems*, vol. 28, no. 3, pp. 3226–3234, 2013.

[69] A. A. Moghaddam, A. Seifi, T. Niknam, and M. R. Alizadeh Pahlavani, "Multi-objective operation management of a renewable MG (micro-grid) with back-up micro-turbine/fuel cell/battery hybrid power source," *Energy*, vol. 36, no. 11, pp. 6490–6507, 2011.

[70] H. Morais, P. Kádár, P. Faria, Z. A. Vale, and H. M. Khodr, "Optimal scheduling of a renewable micro-grid in an isolated load area using mixed-integer linear programming," *Journal of Renewable Energy*, vol. 35, no. 1, pp. 151–156, 2010.

[71] S. A. Alavi, A. Ahmadian, and M. Aliakbar-Golkar, "Optimal probabilistic energy management in a typical micro-grid based-on robust optimization and point estimate method," *Energy Conversion and Management*, vol. 95, pp. 314–325, 2015.

[72] B. Bahmani-Firouzi and R. Azizipanah-Abarghooee, "Optimal sizing of battery energy storage for micro-grid operation management using a new improved bat algorithm," *International Journal of Electrical Power & Energy Systems*, vol. 56, pp. 42–54, 2014.

[73] R. Gholami, M. Shahabi, and M. R. Haghifam, "An efficient optimal capacitor allocation in DG embedded distribution networks with islanding operation capability of micro-grid using a new genetic based algorithm," *International Journal of Electrical Power & Energy Systems*, vol. 71, pp. 335–343, 2015.

[74] B. Khorramdel and M. Raoofat, "Optimal stochastic reactive power scheduling in a microgrid considering voltage droop scheme of DGs and uncertainty of wind farms," *Energy*, vol. 45, no. 1, pp. 994–1006, 2012.

[75] T. Lv, Q. Ai, and Y. Zhao, "A bi-level multi-objective optimal operation of grid-connected microgrids," *Electric Power Systems Research*, vol. 131, pp. 60–70, 2016.

[76] S. Sharma, S. Bhattacharjee, and A. Bhattacharya, "Operation cost minimization of a Micro-Grid using Quasi-Oppositional Swine Influenza Model Based Optimization with Quarantine," *Ain Shams Engineering Journal*, 2015.

[77] S. Sharma, S. Bhattacharjee, and A. Bhattacharya, "Grey wolf optimisation for optimal sizing of battery energy storage device to minimise operation cost of microgrid," *IET Generation, Transmission & Distribution*, vol. 10, no. 3, pp. 625–637, 2016.

[78] Z. Liu, Y. Chen, Y. Luo, G. Zhao, and X. Jin, "Optimized planning of power source capacity in microgrid, considering combinations of energy storage devices," *Applied Sciences (Switzerland)*, vol. 6, no. 12, article no. 416, 2016.

[79] J. Shen, C. Jiang, Y. Liu, and X. Wang, "A Microgrid Energy Management System and Risk Management under an Electricity Market Environment," *IEEE Access*, vol. 4, pp. 2349–2356, 2016.

[80] Y. Guo, J. Xiong, S. Xu, and W. Su, "Two-Stage Economic Operation of Microgrid-Like Electric Vehicle Parking Deck," *IEEE Transactions on Smart Grid*, vol. 7, no. 3, pp. 1703–1712, 2016.

[81] F. Y. Melhem, N. Moubayed, and O. Grunder, "Residential energy management in smart grid considering renewable energy sources and vehicle-to-grid integration," in *Proceedings of the 2016 IEEE Electrical Power and Energy Conference, EPEC 2016*, can, October 2016.

[82] R. Yu, W. Zhong, S. Xie, C. Yuen, S. Gjessing, and Y. Zhang, "Balancing Power Demand Through EV Mobility in Vehicle-to-Grid Mobile Energy Networks," *IEEE Transactions on Industrial Informatics*, vol. 12, no. 1, pp. 79–90, 2016.

[83] N. G. Paterakis, O. Erdinc, I. N. Pappi, A. G. Bakirtzis, and J. P. S. Catalao, "Coordinated Operation of a Neighborhood of Smart Households Comprising Electric Vehicles, Energy Storage and Distributed Generation," *IEEE Transactions on Smart Grid*, vol. 7, no. 6, pp. 2736–2747, 2016.

[84] G. Li, D. Wu, J. Hu, Y. Li, M. S. Hossain, and A. Ghoneim, "HELOS: Heterogeneous Load Scheduling for Electric Vehicle-Integrated Microgrids," *IEEE Transactions on Vehicular Technology*, vol. 66, no. 7, pp. 5785–5796, 2017.

[85] C.-H. Lin, W.-L. Hsieh, C.-S. Chen, C.-T. Hsu, and T.-T. Ku, "Optimization of photovoltaic penetration in distribution systems considering annual duration curve of solar irradiation," *IEEE Transactions on Power Systems*, vol. 27, no. 2, pp. 1090–1097, 2012.

[86] M. Martín, "Optimal annual operation of the dry cooling system of a concentrated solar energy plant in the south of Spain," *Energy*, vol. 84, pp. 774–782, 2015.

[87] K. Burlafinger, A. Vetter, and C. J. Brabec, "Maximizing concentrated solar power (CSP) plant overall efficiencies by using spectral selective absorbers at optimal operation temperatures," *Solar Energy*, vol. 120, pp. 428–438, 2015.

[88] P. Guo, J. Li, Y. Wang, and Y. Wang, "Evaluation of the optimal turbine pressure drop ratio for a solar chimney power plant," *Energy Conversion and Management*, vol. 108, pp. 14–22, 2016.

[89] R. Nazir, K. Kanada, Syafii, and P. Coveria, "Optimization active and reactive power flow for PV connected to grid system using Newton Raphson method," in *Proceedings of the 2nd International Conference on Sustainable Energy Engineering and Application, ICSEEA 2014*, pp. 77–86, idn, October 2014.

[90] S. Ntsaluba, B. Zhu, and X. Xia, "Optimal flow control of a forced circulation solar water heating system with energy storage units and connecting pipes," *Journal of Renewable Energy*, vol. 89, pp. 108–124, 2016.

[91] G. Wang, M. Ciobotaru, and V. G. Agelidis, "Optimal capacity design for hybrid energy storage system supporting dispatch of large-scale photovoltaic power plant," *Journal of Energy Storage*, vol. 3, pp. 25–35, 2015.

[92] A. Rabiee, A. Soroudi, and A. Keane, "Information gap decision theory based OPF with HVDC connected wind farms," *IEEE Transactions on Power Systems*, vol. 30, no. 6, pp. 3396–3406, 2014.

[93] D. Ke, C. Y. Chung, and Y. Sun, "A novel probabilistic optimal power flow model with uncertain wind power generation described by customized Gaussian mixture model," *IEEE Transactions on Sustainable Energy*, vol. 7, no. 1, pp. 200–212, 2016.

[94] D. Bienstock, M. Chertkov, and S. Harnett, "Chance-constrained optimal power flow: risk-aware network control under uncertainty," *SIAM Review*, vol. 56, no. 3, pp. 461–495, 2014.

[95] M. Ghofrani, A. Arabali, M. Etezadi-Amoli, and M. S. Fadali, "A framework for optimal placement of energy storage units within a power system with high wind penetration," *IEEE Transactions on Sustainable Energy*, vol. 4, no. 2, pp. 434–442, 2013.

[96] R. A. Jabr and B. C. Pal, "Intermittent wind generation in optimal power flow dispatching," *IET Generation, Transmission & Distribution*, vol. 3, no. 1, pp. 66–74, 2009.

[97] G. J. Sebastián, C. J. Alexander, and G. Mauricio, "Stochastic AC Optimal Power Flow Considering the Probabilistic Behavior of the Wind, Loads and Line Parameters," *Ingeniería, Investigación y Tecnología*, vol. 15, no. 4, pp. 529–538, 2014.

[98] M. Aien, M. Rashidinejad, and M. F. Firuz-Abad, "Probabilistic optimal power flow in correlated hybrid wind-PV power systems: A review and a new approach," *Renewable & Sustainable Energy Reviews*, vol. 41, pp. 1437–1446, 2015.

[99] Y.-C. Chang, T.-Y. Lee, C.-L. Chen, and R.-M. Jan, "Optimal power flow of a wind-thermal generation system," *International Journal of Electrical Power & Energy Systems*, vol. 55, pp. 312–320, 2014.

[100] X. Li, J. Cao, and D. Du, "Probabilistic optimal power flow for power systems considering wind uncertainty and load correlation," *Neurocomputing*, vol. 148, pp. 240–247, 2015.

[101] S.-Y. Lin and A.-C. Lin, "RLOPF (risk-limiting optimal power flow) for systems with high penetration of wind power," *Energy*, vol. 71, pp. 49–61, 2014.

[102] A. Mohammadi, S. Tavakoli, and S. M. Barakati, "On power tracking and alleviation by a new controller for fulfilment of the damping and performance requisites for a variable speed wind system: An optimal approach," *International Journal of Electrical Power & Energy Systems*, vol. 75, pp. 187–193, 2016.

[103] S. M. Mohseni-Bonab, A. Rabiee, and B. Mohammadi-Ivatloo, "Voltage stability constrained multi-objective optimal reactive power dispatch under load and wind power uncertainties: A stochastic approach," *Journal of Renewable Energy*, vol. 85, pp. 598–609, 2016.

[104] A. Panda and M. Tripathy, "Optimal power flow solution of wind integrated power system using modified bacteria foraging algorithm," *International Journal of Electrical Power & Energy Systems*, vol. 54, pp. 306–314, 2014.

[105] A. Panda and M. Tripathy, "Security constrained optimal power flow solution of wind-thermal generation system using modified bacteria foraging algorithm," *Energy*, vol. 93, pp. 816–827, 2015.

[106] R. Roy and H. T. Jadhav, "Optimal power flow solution of power system incorporating stochastic wind power using Gbest guided artificial bee colony algorithm," *International Journal of Electrical Power & Energy Systems*, vol. 64, pp. 562–578, 2015.

[107] G. P. Harrison and A. R. Wallace, "Optimal power flow evaluation of distribution network capacity for the connection of distributed generation," in *Proceedings of the IEE Proc.-Gener. Transm. Distrib.*, pp. 115–122, 2005.

[108] P. S. Georgilakis and N. D. Hatziargyriou, "Optimal distributed generation placement in power distribution networks: models, methods, and future research," *IEEE Transactions on Power Systems*, vol. 28, no. 3, pp. 3420–3428, 2013.

[109] E. Sortomme and M. A. El-Sharkawi, "Optimal charging strategies for unidirectional vehicle-to-grid," *IEEE Transactions on Smart Grid*, vol. 2, no. 1, pp. 131–138, 2011.

[110] S. Ashok, "Optimised model for community-based hybrid energy system," *Journal of Renewable Energy*, vol. 32, no. 7, pp. 1155–1164, 2007.

[111] A. S. Bouhouras, K. I. Sgouras, P. A. Gkaidatzis, and D. P. Labridis, "Optimal active and reactive nodal power requirements towards loss minimization under reverse power flow constraint defining DG type," *International Journal of Electrical Power & Energy Systems*, vol. 78, pp. 445–454, 2016.

[112] R. Azizipanah-Abarghooee, T. Niknam, M. Malekpour, F. Bavafa, and M. Kaji, "Optimal power flow based TU/CHP/PV/WPP coordination in view of wind speed, solar irradiance and load correlations," *Energy Conversion and Management*, vol. 96, pp. 131–145, 2015.

[113] H. Tazvinga, B. Zhu, and X. Xia, "Optimal power flow management for distributed energy resources with batteries," *Energy Conversion and Management*, vol. 102, 2016.

[114] T. Ma, H. Yang, L. Lu, and J. Peng, "Optimal design of an autonomous solar-wind-pumped storage power supply system," *Applied Energy*, vol. 160, pp. 728–736, 2015.

[115] C. Battistelli, L. Baringo, and A. J. Conejo, "Optimal energy management of small electric energy systems including V2G facilities and renewable energy sources," *Electric Power Systems Research*, vol. 92, pp. 50–59, 2012.

[116] K. Tanaka, K. Uchida, K. Ogimi et al., "Optimal operation by controllable loads based on smart grid topology considering insolation forecasted error," *IEEE Transactions on Smart Grid*, vol. 2, no. 3, pp. 438–444, 2011.

[117] A. Bracale, P. Caramia, G. Carpinelli, E. Mancini, and F. Mottola, "Optimal control strategy of a DC micro grid," *International Journal of Electrical Power & Energy Systems*, vol. 67, pp. 25–38, 2015.

[118] D. McLarty, C. Civit Sabate, J. Brouwer, and F. Jabbari, "Microgrid energy dispatch optimization and predictive control algorithms; A UC Irvine case study," *International Journal of Electrical Power & Energy Systems*, vol. 65, pp. 179–190, 2015.

[119] E. Riva Sanseverino, N. Nguyen Quang, M. L. Di Silvestre, J. M. Guerrero, and C. Li, "Optimal power flow in three-phase islanded microgrids with inverter interfaced units," *Electric Power Systems Research*, vol. 123, pp. 48–56, 2015.

[120] C. C. Thompson, P. E. Konstantinos Oikonomou, A. H. Etemadi, and V. J. Sorger, "Optimization of Data Center Battery Storage Investments for Microgrid Cost Savings, Emissions Reduction, and Reliability Enhancement," *IEEE Transactions on Industry Applications*, vol. 52, no. 3, pp. 2053–2060, 2016.

[121] V. Krishnamurthy and A. Kwasinski, "Effects of Power Electronics, Energy Storage, Power Distribution Architecture, and Lifeline Dependencies on Microgrid Resiliency during Extreme Events," *IEEE Journal of Emerging and Selected Topics in Power Electronics*, vol. 4, no. 4, pp. 1310–1323, 2016.

[122] X. Xu, M. Bishop, O. Donna G, and H. Chen, "Application and modeling of battery energy storage in power systems," *CSEE Journal of Power and Energy Systems*, vol. 2, no. 3, pp. 82–90, 2016.

[123] R. Khodabakhsh and S. Sirouspour, "Optimal Control of Energy Storage in a Microgrid by Minimizing Conditional Value-at-Risk," *IEEE Transactions on Sustainable Energy*, vol. 7, no. 3, pp. 1264–1273, 2016.

[124] H. Hassanzadehfard, S. M. Moghaddas-Tafreshi, and S. M. Hakimi, "Optimization of grid-connected microgrid consisting of PV/FC/UC with considered frequency control," *Turkish Journal of Electrical Engineering & Computer Sciences*, vol. 23, no. 1, pp. 1–16, 2015.

[125] H. Alharbi and K. Bhattacharya, "Optimal sizing of battery energy storage systems for microgrids," in *Proceedings of the 2014 Electrical Power and Energy Conference, EPEC 2014*, pp. 275–280, November 2014.

[126] F. Laureri, L. Puliga, M. Robba, F. Delfino, and G. Odena Bultò, "An optimization model for the integration of electric vehicles and smart grids Problem definition and experimental validation," in *Proceedings of the 2nd IEEE International Smart Cities Conference, ISC2 2016*, September 2016.

[127] L. Yao, W. H. Lim, and T. S. Tsai, "A Real-Time Charging Scheme for Demand Response in Electric Vehicle Parking Station," *IEEE Transactions on Smart Grid*, vol. 8, no. 1, pp. 52–62, 2017.

[128] M. Bianchi, L. Branchini, C. Ferrari, and F. Melino, "Optimal sizing of grid-independent hybrid photovoltaic–battery power systems for household sector," *Applied Energy*, vol. 136, pp. 805–816, 2014.

[129] Q. P. Ha and V. Vakiloroaya, "Modeling and optimal control of an energy-efficient hybrid solar air conditioning system," *Automation in Construction*, vol. 49, pp. 262–270, 2015.

[130] E. Lizarraga-Garcia, A. Ghobeity, M. Totten, and A. Mitsos, "Optimal operation of a solar-thermal power plant with energy storage and electricity buy-back from grid," *Energy*, vol. 51, pp. 61–70, 2013.

[131] E. M. A. Mokheimer, Y. N. Dabwan, and M. A. Habib, "Optimal integration of solar energy with fossil fuel gas turbine cogeneration plants using three different CSP technologies in Saudi Arabia," *Applied Energy*, vol. 185, pp. 1268–1280, 2017.

[132] T. K. A. Brekken, A. Yokochi, A. von Jouanne, Z. Z. Yen, H. M. Hapke, and D. A. Halamay, "Optimal energy storage sizing and control for wind power applications," *IEEE Transactions on Sustainable Energy*, vol. 2, no. 1, pp. 69–77, 2011.

[133] M. Sedghi, A. Ahmadian, and M. Aliakbar-Golkar, "Optimal storage planning in active distribution network considering uncertainty of wind power distributed generation," *IEEE Transactions on Power Systems*, vol. 31, no. 1, pp. 304–316, 2016.

[134] A. Arabali, M. Ghofrani, and M. Etezadi-Amoli, "Cost analysis of a power system using probabilistic optimal power flow with energy storage integration and wind generation," *International Journal of Electrical Power & Energy Systems*, vol. 53, pp. 832–841, 2013.

[135] H. T. Jadhav and R. Roy, "Stochastic optimal power flow incorporating offshore wind farm and electric vehicles," *International Journal of Electrical Power & Energy Systems*, vol. 69, pp. 173–187, 2015.

[136] Y. Luo, L. Shi, and G. Tu, "Optimal sizing and control strategy of isolated grid with wind power and energy storage system," *Energy Conversion and Management*, vol. 80, pp. 407–415, 2014.

[137] L. Roald, M. Vrakopoulou, F. Oldewurtel, and G. Andersson, "Risk-based optimal power flow with probabilistic guarantees," *International Journal of Electrical Power & Energy Systems*, vol. 72, pp. 66–74, 2015.

# 13

# Design Consideration for High Step-Up Nonisolated Multicellular dc-dc Converter for PV Micro Converters

**Yusuke Hayashi, Yoshikatsu Matsugaki, and Tamotsu Ninomiya**

*Green Electronics Research Institute, Kitakyushu (GRIK), 1-8-409 Hibikino, Wakamatsu-ku, Kitakyushu, Fukuoka 808-0135, Japan*

Correspondence should be addressed to Yusuke Hayashi; hayashi@grik.jp

Academic Editor: Hongjie Jia

High step-up nonisolated multicellular dc-dc converter has been newly proposed for PV microconverters. The multicellular converter consists of the nonisolated step-up cell converters using bidirectional semiconductor switches, and these cell converters are connected in Input Parallel Output Series (IPOS). The voltage transformation ratio of the step-up converter is $N/(1 - D)$ in case all the transistors in $N$ cell converters are operated at the duty ratio of $D$. The proposed multicellular dc-dc converter also accomplishes high efficiency because of no magnetic coupling such as the high frequency transformer and the coupled inductor. Laboratory prototype has been fabricated to show the feasibility of the proposed converter. Design consideration for the 20 V–40 V to 384 V, 240 W nonisolated multicellular dc-dc converter has been also conducted, and the potential to achieve the efficiency of 98% has been shown. The proposed multicellular converter contributes to realizing the environmentally aware data centers for future low carbon society.

## 1. Introduction

Environmentally aware data centers have been proposed by NTT (Nippon Telegraph and Telephone Corporation) to realize highly electrified low carbon society [1, 2]. The environmentally aware data center reduces the conversion loss by introducing 380 V dc distribution system and minimizes the environmental impact by installing the solar PV power systems into the 380 V dc distribution system. Highly efficient and ultracompact (high-power-density) power electronics converters are necessary to utilize the renewable energy sources effectively.

Microconverter concept is one of attractive solutions for the effective use of the solar PV generations. The low input voltage and low output power dc-dc converter (microconverter) is installed into each solar module and all of microconverters are controlled to achieve MPPT (Maximum Power Point Tracking). Influence of the partially shaded or soiled solar cells on the total output power of the solar power systems is minimized because the output power of solar modules is controlled individually.

One of issues for the microconverter concept is the high step-up voltage transformation ratio (voltage gain: output voltage/input voltage) of 10–20. It is difficult to develop the high step-up microconverter by the general nonisolated boost chopper circuit because of the parasitic resistance [3]. High frequency transformers and coupled inductors based on the magnetic coupling are employed to accomplish the high step-up voltage transformation ratio [4–8]. The magnetic coupling achieves the high transformation ratio; however, this causes the nonnegligible power dissipation.

In this paper, the high step-up nonisolated dc-dc converter is newly proposed for PV microconverters. The proposed dc-dc converter is based on the multicellular converter topology, and the boost chopper circuits with the bidirectional switches for cell converters are connected in IPOS. The voltage gain is $N/(1 - D)$ in case transistors in $N$ cell converters are operated at the duty ratio of $D$. The number of cell converter contributes to achieving high transformation ratio.

In Section 2, the next generation dc distribution system for the data center is introduced. The circuit configuration, the control scheme, the fundamental behavior, and the feasibility of the nonisolated multicellular dc-dc converter are described in Section 3. In Section 4, design consideration for the proposed multicellular converter is

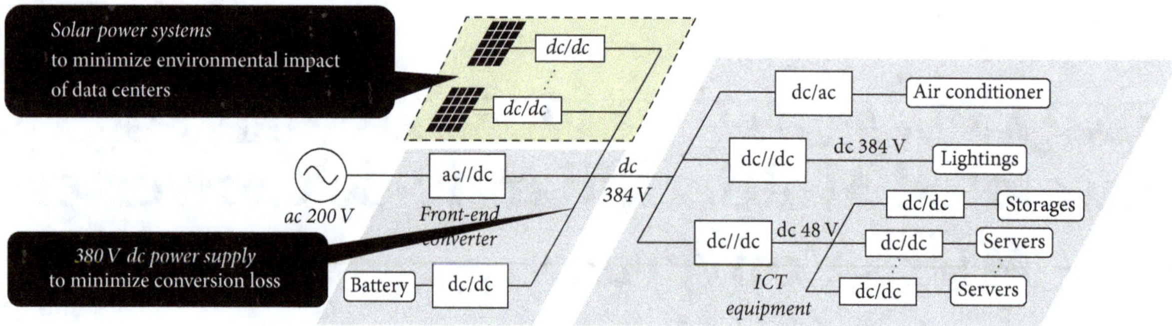


FIGURE 1: Configuration of next generation dc distribution system for environmentally aware data centers [9].

conducted and the potential to achieve high efficiency is presented.

## 2. Environmentally Aware Data Center for Low Carbon Society

*2.1. Dc Distribution System in Environmentally Aware Data Center.* The simplified schematic diagram of the dc distribution system for the environmentally friendly data centers is shown in Figure 1 [9, 13]. This system is based on the 380 V dc distribution system which goes beyond the conventional 48 V dc distribution system, and this achieves higher efficiency providing smaller supply current [14]. In the 380 V dc distribution system, the power factor correction (PFC) converters, the dc-dc transformers (DCX), and point-of-load (POL) converters are introduced to supply the electric power from ac 200 V to ICT (information and communication technology) equipment. Intensive studies have been conducted to realize highly efficient PFCs, DCXs, and POLs [15–18], and the potential to improve the total conversion efficiency in the dc distribution system from 80% to 94% has been discussed [9, 16].

In the dc distribution system for the environmentally aware data center, the environmental impact will be also minimized by installing the renewable energy sources such as solar PV power generations. The 380 V dc distribution system contributes smooth introduction of the solar power systems because of the dc interface. To utilize the generated electric power effectively, highly efficient and ultracompact dc-dc converters for PV generation are also indispensable.

*2.2. Power Electronics Converters for PV Generation.* There are mainly two solutions to connect PV cells to the dc distribution system via dc-dc converters. Figure 2 shows the schematic for two solutions. One is the centralized converter solution where low voltage PV modules are connected in series and the bundled higher voltage is applied to the single central converter. The other is the microconverter solution where the low-power dc-dc converters are placed beside PV modules and the output powers of PV modules are controlled independently. Characteristics of two solutions are summarized as follows.

(i) Highly efficient dc-dc converter can be developed in the centralized converter solution because of the low voltage transformation ratio. However, the generated electric power from PV cells is not provided to the distribution system effectively in case some solar modules are shaded or soiled.

(ii) Influence of the partially shaded or soiled solar modules on total generated electric power from PV cells can be minimized in the microconverter solution. The efficiency of the dc-dc converter in the microconverter solution is lower than the converter in the centralized solution because of the high voltage transformation ratio.

Microconverter solution is attractive for environmentally aware data centers in terms of the effective utilization of renewable energy sources. Developing highly efficient high step-up dc-dc converter is one of the key issues.

The voltage transformation ratio of, for example, 9.6 to 19.2, is required for the dc-dc converter in case the PV modules whose output voltages are, for example, 20 V to 40 V are connected to the 384 V dc distribution line. It is difficult to realize the general boost chopper circuit with the voltage transformation ratio over 10, because the parasitic resistance in the circuit causes the voltage drop [3]. Therefore, the circuit topologies based on the magnetic coupling such as the high frequency transformer and the coupled inductor are employed to realize the high voltage gain converters [4–8]. The magnetic coupling causes the significant power dissipation and prevents achieving high efficiency.

## 3. High Step-Up Nonisolated Multicellular dc-dc Converter

*3.1. Circuit Configuration and Control System of Nonisolated Multicellular Converter.* The nonisolated multicellular dc-dc converter is proposed to realize highly efficient high step-up dc-dc converter. Figure 3 shows the circuit configuration of the proposed converter. The multicellular converter consists of the nonisolated cell converters and these cell converters are connected in IPOS. Configuration of each cell converter is based on the general boost chopper circuit and the bidirectional switches substitute for the free-wheeling diode to achieve the isolation barrier. The bidirectional switches affect the converter performance significantly, and the ultralow loss GaN bidirectional transistors under the research

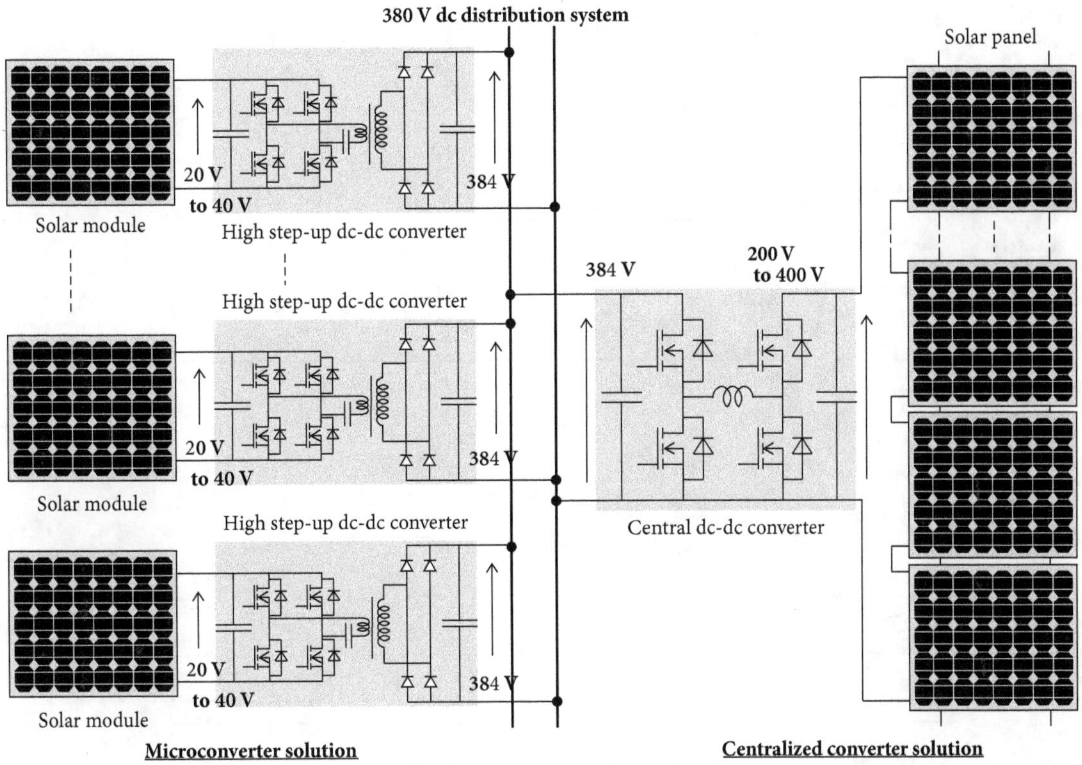


FIGURE 2: Centralized converter solution and microconverter solution for connecting PV modules into dc distribution system.

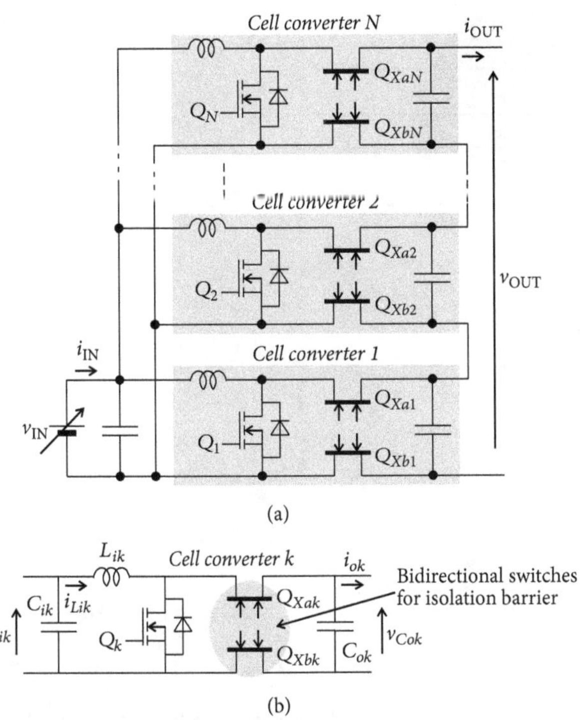


FIGURE 3: Configuration of high step-up non-isolated multicellular dc-dc converter. (a) IPOS connected cell converters and (b) single cell converter.

and development are attractive for the proposed converter [19].

Figure 4 shows the control system of the proposed multicellular dc-dc converter for the microconverter solution. The control system consists of the controller for the MPPT (Maximum Power Point Tracking) to adjust the duty ratio $D$ and the controller for PWM (Pulse Width Modulation) to generate the gate signals for transistors $Q_k$ $(k = 1, 2, \ldots, N)$ in the cell converters. The flowchart for the well-known P & O (Perturbation and Observation) algorithm is shown in Figure 4(a) and this control algorithm is generally applied to the boost converter for PV applications [10, 11, 20].

The isolated dc-dc converter with the high frequency transformers is often used to develop the IPOS multicellular converter. On the other hand, the proposed converter consists of the nonisolated dc-dc cell converters to simplify the circuit configuration. The isolation barrier has to be achieved by the intrinsic capacitances of the bidirectional switches $Q_{Xak}$, $Q_{Xbk}$ $(k = 1, 2, \ldots, N)$, taking the control strategy into account. The gate signals for transistors $Q_k$ during one cycle are shown in Figure 4(b). The duty ratio for all the transistors $Q_k$ is $D$, and the bidirectional switches $Q_{Xak}$, $Q_{Xbk}$ are operated complementally. In the proposed circuit, one of cell converters can provide the electric power to the load at an arbitrary time, and the remaining $N-1$ cell converters have to be isolated by turning off the bidirectional switches to prevent the short circuit among cell converters. The interleaved

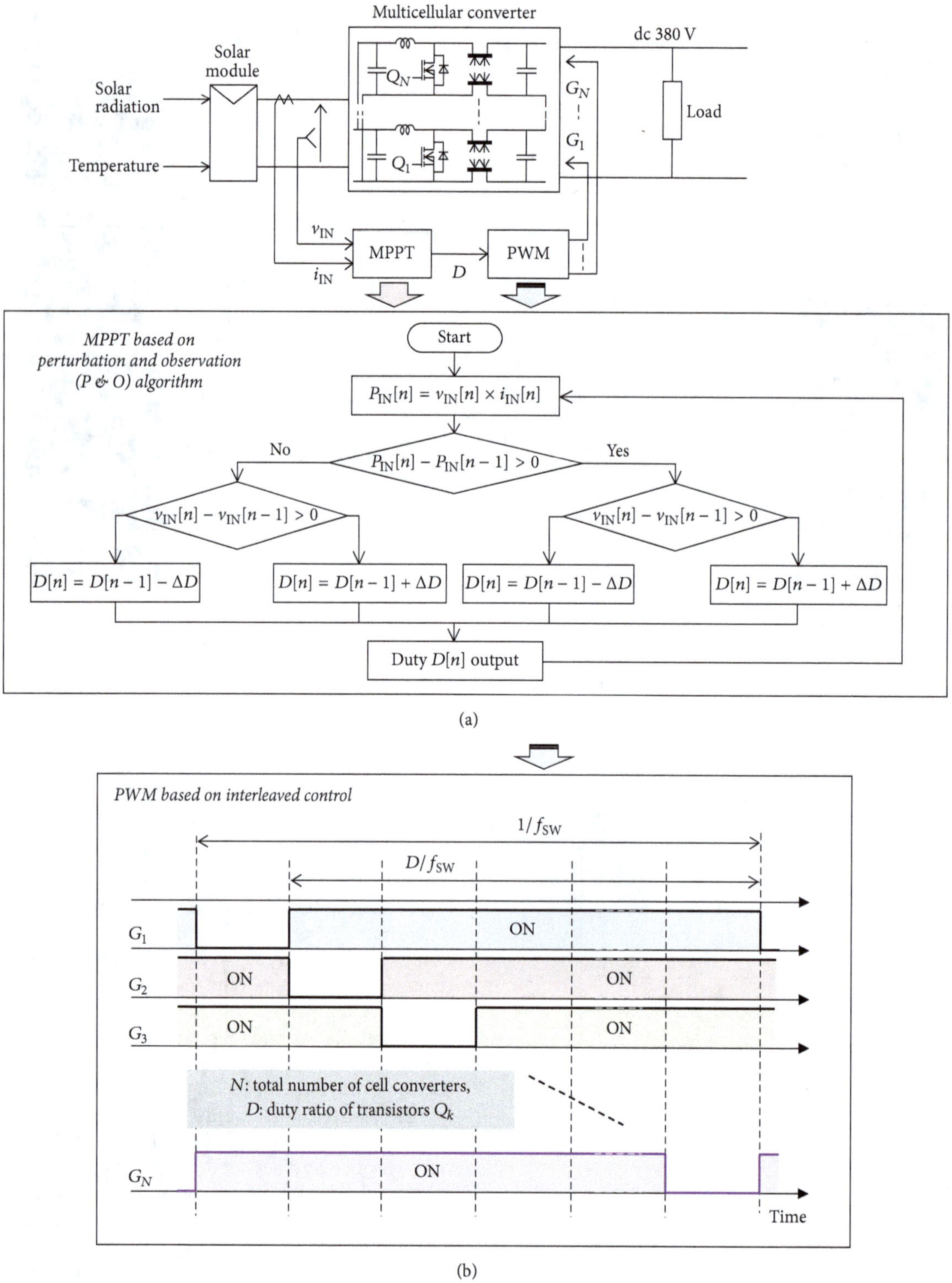


FIGURE 4: Control system of nonisolated multicellular converter for microconverter solution. (a) Control algorithm for MPPT and (b) PWM control.

control is applied to control the cell converters, and the duty ratio is limited as follows.

$$\frac{N-1}{N} \leq D \leq 1. \tag{1}$$

All the cell converters operating at the duty ratio in (1) behave as the general boost choppers independently, and the whole multicellular converter achieves high voltage gain realizing the isolation barrier among cell converters. Details of the converter characteristics are described in the next section.

*3.2. Analysis for Nonisolated Multicellular Converter Based on Space-State Averaging Method.* The circuit configuration of the cell converter is based on the typical boost chopper topology, and the fundamental characteristics of the proposed converter are explained by the space-state averaging method [3, 12, 21]. The transfer characteristics of each cell converter shown in Figure 3(b) are formulated as follows.

$$\frac{d}{dt}\begin{bmatrix} i_{Lik} \\ v_{Cok} \end{bmatrix} = \begin{bmatrix} 0 & \dfrac{-D'}{L_{ik}} \\ \dfrac{D'}{C_{ok}} & \dfrac{-1}{RC_{ok}} \end{bmatrix} \begin{bmatrix} i_{Lik} \\ v_{Cok} \end{bmatrix} + \begin{bmatrix} \dfrac{1}{L_{ik}} \\ 0 \end{bmatrix} v_{ik}. \quad (2)$$

In the above equation, the spate variables of $i_{Lik}$, $v_{Cok}$ mean the current through the input inductor $L_{ik}$ and the applied voltage to the output capacitor $C_{ok}$, respectively. The input voltage of the cell converter $k$ is $v_{ik}$ and its output current to the resistive load $R$ is $i_{ok}$. The transistor $Q_k$ is operated at the duty ratio of $D$ ($D' = 1-D$). Here the integer $k$ ($k = 1, 2, \ldots, N$) means the number of IPOS connected $N$ cell converters in Figure 3(a). Parasitic resistances in the switches $Q_k$, $Q_{Xak}$, $Q_{Xbk}$, the inductor $L_{ik}$, and the capacitor $C_{ok}$ are not considered here to simplify the analysis.

In Figure 3, the input voltage $v_{IN}$ is applied to all cell converters and the output current $i_{OUT}$ flows through all cell converters. In the IPOS connection topology, the following relationships are obtained.

$$\begin{aligned} v_{IN} &= v_{i1} = v_{i2} = \cdots = v_{iN} \\ i_{IN} &= i_{Li1} + i_{Li2} + \cdots + i_{LiN} \\ v_{OUT} &= v_{Co1} + v_{Co2} + \cdots + v_{CoN} \\ i_{OUT} &= i_{o1} = i_{o2} = \cdots = i_{oN}. \end{aligned} \quad (3)$$

In case that the common design methodology is applied to all cell converters, the equalized inductance and capacitance values are employed for all cell converters as follows, without taking the influence of the variation in the real products into account.

$$\begin{aligned} L_i &= L_{i1} = L_{i2} = \cdots = L_{iN} \\ C_o &= C_{o1} = C_{o2} = \cdots = C_{oN}. \end{aligned} \quad (4)$$

Based on the above relations, the transfer characteristics in (2) are summarized as follows.

$$\sum_{k=1}^{N} \frac{d}{dt}\begin{bmatrix} i_{Lik} \\ v_{Cok} \end{bmatrix} = \sum_{k=1}^{N} \begin{bmatrix} 0 & \dfrac{-D'}{L_i} \\ \dfrac{D'}{C_o} & \dfrac{-1}{RC_o} \end{bmatrix} \begin{bmatrix} i_{Lik} \\ v_{Cok} \end{bmatrix} + \begin{bmatrix} \dfrac{1}{L_i} \\ 0 \end{bmatrix} N v_{IN} \quad (5)$$

$$\therefore \frac{d}{dt}\begin{bmatrix} i_{IN} \\ v_{OUT} \end{bmatrix} = \begin{bmatrix} 0 & \dfrac{-D'}{L_i} \\ \dfrac{D'}{C_o} & \dfrac{-1}{RC_o} \end{bmatrix} \begin{bmatrix} i_{IN} \\ v_{OUT} \end{bmatrix} + \begin{bmatrix} \dfrac{1}{L_i} \\ 0 \end{bmatrix} N v_{IN}. \quad (6)$$

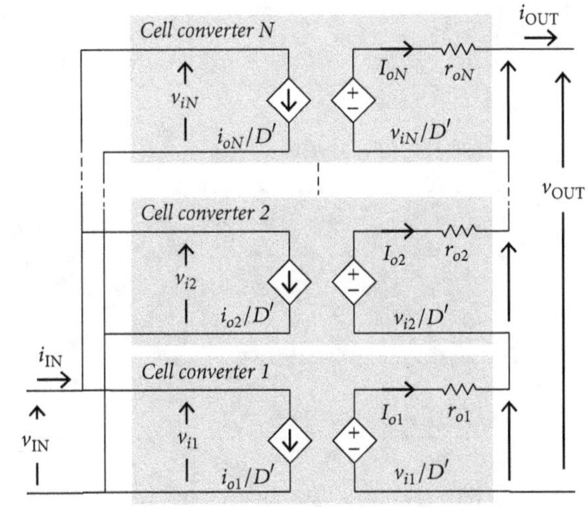


FIGURE 5: Simplified equivalent circuit of multicellular dc-dc converter based on IPOS connection topology.

Equation (6) means that the multicellular converter in Figure 3(a) behaves as the single step-up converter whose input voltage is $N \cdot v_{IN}$. The relationship between the input voltage $V_{IN}$ and the output voltage $V_{OUT}$ of the multicellular converter is described as follows through the Laplace transformation.

$$\frac{V_{OUT}}{V_{IN}} = \frac{N}{D'} \frac{1}{1 + (2\delta/\omega_0)\, s + (1/\omega_0^2)\, s^2} \quad (7)$$

$$\omega_0 = \frac{D'}{\sqrt{L_i C_o}}, \qquad \delta = \frac{1}{2D'R}\sqrt{\frac{L}{C}} \quad (8)$$

$$\lim_{s\to 0} s \cdot V_{OUT} = \frac{N}{D'} \cdot V_{IN}. \quad (9)$$

Equation (9) shows the voltage transformation ratio of the proposed multicellular converter under the steady state operation condition. The cycle of the output perturbation for the photovoltaics is generally at a level of several minutes. The high frequency converters achieve high-speed response and the converters accomplish the ratio in (9) during the perturbation. The obtained transformation ratio is $N$ times higher than the ratio of the conventional step-up converter under the corresponding duty ratio $D$.

*3.3. Output Voltage Unbalance among IPOS Connected Cell Converters.* In the real circuit operation condition, the parasitic resistances in switches $Q_k$, $Q_{Xak}$, $Q_{Xbk}$, the inductors $L_{ik}$, and the capacitors $C_{ok}$ cause the output voltage unbalance among cell converters. The influence of the parasitic resistances on the unbalanced voltage is discussed here.

The equivalent circuit of the multicellular step-up converter is shown in Figure 5. The cell converters can be regarded as the dc voltage transformer from (9), and the cell converter $k$ ($k = 1, 2, \ldots, N$) is simply composed of the dependent voltage source, the dependent current source, and

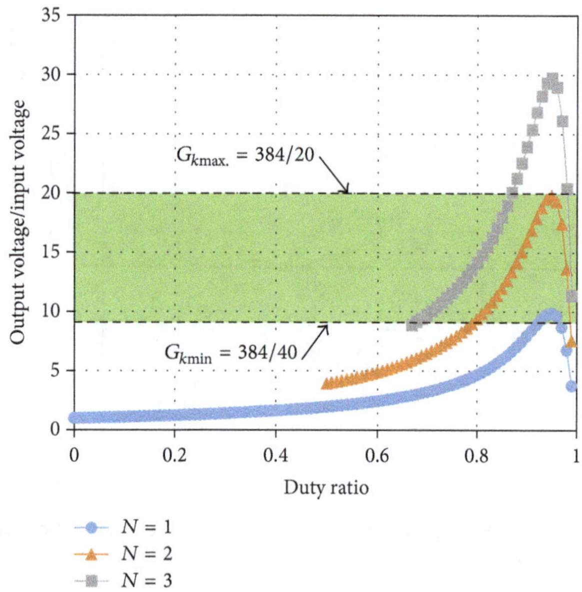


FIGURE 6: Voltage transformation ratio of nonisolated step-up multicellular dc-dc converter taking parasitic resistances into account.

the equivalent output resistance $r_{ok}$ [18, 22]. The parasitic resistances caused by the switches and the inductors are bundled to the resistance $r_{ok}$. From Figure 5, the voltage and the current in each cell converter have the following relationship.

$$\frac{1}{D'}v_{ik} - r_{ok} \cdot i_{ok} = v_{ok}. \tag{10}$$

Based on (3), the output voltage $v_{ok}$ of the cell converter $k$ is obtained as follows.

$$\frac{N}{D'}v_{\mathrm{IN}} - \sum_{k=1}^{N} r_{ok} \cdot i_{\mathrm{OUT}} = v_{\mathrm{OUT}} = N \cdot R \cdot i_{\mathrm{OUT}} \tag{11}$$

$$i_{\mathrm{OUT}} = \frac{1}{D'}\frac{1}{R + \left(\sum_{k=1}^{N} r_{ok}\right)/N}v_{\mathrm{IN}} \tag{12}$$

$$\therefore v_{ok} = \frac{1}{D'}v_{\mathrm{IN}} - r_{ok} \cdot i_{\mathrm{OUT}} = \frac{1}{D'}\left(1 - \frac{r_{ok}}{R + \sum_{k=1}^{N} r_{ok}/N}\right)v_{\mathrm{IN}}. \tag{13}$$

From (13), the output voltages of the cell converters have the unbalance caused by the variation in the equivalent output resistances $r_{ok}$. However, the influence of the resistance $r_{ok}$ is negligible in the case of highly efficient cell converter because the load resistance $R$ is sufficiently larger than the resistance $r_{ok}$. This means that the output cell voltages $v_{ok}$ converge, and the unbalanced voltage among cell converters is negligible in case all the cell converters are operated at the duty ratio $D$.

*3.4. Voltage Transformation Ratio of Nonisolated Multicellular Converter under Real Circuit Operation Condition.* Figure 6 shows the relationship between the duty ratio $D$ of the transistor $Q_k$ ($k = 1, 2, \ldots, N$) and the actual voltage transformation ratio $G$ of the proposed multicellular converter, taking the parasitic resistances in the cell converters into account. As the duty ratio approaches 1.0, the transformation ratio diminishes because of the voltage drop caused by the parasitic resistances.

The calculation results of the voltage transformation ratio for three kinds of multicellular converter ($N = 1, 2, 3$ in Figure 3(a)) are shown in Figure 6. The transformation ratio $G$ is calculated as follows based on (11) and (12).

$$v_{\mathrm{OUT}} = N \cdot R \cdot i_{\mathrm{OUT}} = \frac{N}{D'}\frac{R}{\left(R + \sum_{k=1}^{N} r_{ok}/N\right)}v_{\mathrm{IN}} \tag{14}$$

$$\therefore G = \frac{v_{\mathrm{OUT}}}{v_{\mathrm{IN}}} = \frac{N}{D'}\frac{1}{\left(1 + \left(\sum_{k=1}^{N} r_{ok}/N\right)/R\right)} = \frac{N}{D'}\frac{1}{(1 + Z_o/R)}. \tag{15}$$

In (15), the symbol of $R$ means the output resistance based on the rated output power of the cell converter. The average parasitic resistance of all the cell converters is $Z_o$. The voltage gain $G$ of the IPOS multicellular converter is simply calculated by stacking the voltage gains of the cell converters without taking the variation in the real product into account. In the steady state averaging method, the average resistance $Z_o$ is formulated by using the resistance $r_s$ from the main switch $Q_k$, the resistance $r_D$ from the bidirectional switches $Q_{xak}$, $Q_{xbk}$, and the resistance $r_L$ from the input inductor $L_{ik}$ as follows [3].

$$Z_o = \frac{Dr_s + (1 - D)r_D + r_L}{(1 - D)^2} \cong \frac{r_L}{(1 - D)^2}. \tag{16}$$

In (16), the resistances in the cell converter are bundled to $r_L$ to simplify the calculation. Here, the parasitic resistance $r_L$ of 0.5 Ω and the output resistance $R$ of 205 Ω were applied to calculate the voltage transformation ratio, assuming the cell converter with the output voltage of 128 V (=384 V/3) and the output power of 80 W (240 W/3).

Figure 6 means that the nonisolated multicellular converter using 3 cell converters achieves the voltage transformation ratio of 9.6 to 19.2 for the PV microconverter.

## 4. Simulation Analysis and Experimental Verification of Nonisolated Multicellular Converter

*4.1. Simulation Analysis for 20 V–40 V to 384 V Multicellular dc-dc Converter Using Three Nonisolated Cell Converters.* Figure 7 shows the circuit configuration of the nonisolated multicellular dc-dc converter for circuit simulation analysis. Two antiseries connected power transistors $Q_{Xak}$, $Q_{Xbk}$ in Figure 7 substituted for the single bidirectional switch shown in Figure 3 because of the research and development phase and no circuit simulation models for the single bidirectional switch.

Table 1 shows the parameters for the simulation. The input voltage of the multicellular converter $V_{\mathrm{in}}$ is from 20 V

TABLE 1: Parameters for circuit simulation of nonisolated multicellular converter.

| Multicellular converter | |
|---|---|
| Input voltage $V_{IN}$ | 20 V to 40 V |
| Output voltage $V_{OUT}$ | 384 V |
| Output power $P_{OUT}$ | 240 W |
| Connection topology | Input Parallel Output Series |
| Number of cells | 3 ($k = 1, 2, 3$) |
| **Single cell converter** | |
| Input voltage $V_{IN}$ | 20 V to 40 V |
| Output voltage $V_{ok}$ | 128 V |
| Output power $P_{ok}$ | 80 W |
| Transistor $Q_k$ | GaN-FET (50 mΩ, 200 V) $C_{oss}$ = 110 pF @ 100 V |
| Bidirectional switches $Q_{Xak}$, $Q_{Xbk}$ | Two GaN-HEMTs (150 mΩ, 600 V) $C_{oss}$ = 145 pF @ 100 V connected in antiseries |
| Input capacitor $C_{ik}$ | MLCC (52.3 μF, 63 V) |
| Output capacitor $C_{ok}$ | MLCC (1.37 μF, 200 V) |
| Input inductor $L_{ik}$ | Mn-Zn ferrite (68.8 μH) |
| Switching frequency $f_{SW}$ | 100 kHz |
| Parasitic inductance | 22.5 nH |

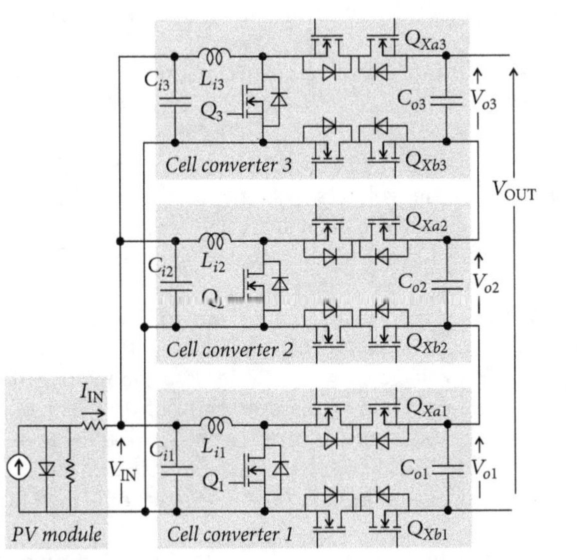


FIGURE 7: Circuit configuration of non-isolated multicellular dc-dc converter using three cell converters for circuit simulation.

to 40 V, taking the output perturbation of the PV module into account [4–7]. The output voltage of the multicellular converter $V_{out}$ is nominally 384 V to connect to the 380 V dc distribution system. The rated output power of the converter $P_{out}$ is 240 W here. The output voltage of the cell converter $V_{ok}$ is nominally 128 V and the output power of the cell converter $P_{ok}$ is 80 W because of three cell converters. The GaN-FET (50 mΩ, 200 V) was utilized for the main transistor $Q_k$ and the GaN-HEMTs (150 m Ω, 650 V) were employed for the bidirectional switches $Q_{Xak}$, $Q_{Xbk}$ in each cell converter. The multilayer ceramic capacitors (MLCCs) were utilized for both the input and the output capacitors $C_{ik}$, $C_{ok}$.

The capacitances $C_{ik}$, $C_{ok}$ were designed to suppress the voltage ripple of the input and the output voltages within 3% of their rated voltages, respectively. The input inductance $L_{ik}$ was designed to achieve the zero current switching (ZCS) under the critical current mode (CCM) of the boost chopper cell converter at the input voltage of 40 V and the output power of 80 W. Here, the subscript $k$ for the symbols means the number of the cell converters and $k$ has the integer of 1, 2, 3.

The output characteristics of the PV module are given by the published datasheet. Figure 8 shows the normalized $I$-$V$ and $P$-$V$ characteristics of the commercially available PV module, taking the influence of the solar radiation into account. The operation conditions to achieve the MPPT are extracted from Figure 8. In case the normalized voltage of 1.0 corresponds to 40 V and the normalized current of 1.0 means 8.0 A, the maximum output power of the PV module for the irradiation of 100% becomes 244 W under the condition that the voltage $V_{IN}$ and the current $I_{IN}$ are 33.2 V and 7.36 A, respectively. The maximum output power for the solar radiation of 60% becomes 140 W under the voltage $V_{IN}$ of 32.0 V. Here, the effect of the temperature of the PV module on its $I$-$V$ characteristics is not considered.

Figures 9(a) and 9(b) show the simulation results of 32.0 V–384 V, 140 W operation and 33.4 V–384 V, 240 W operation, respectively. PSIM software was utilized for the simulation. The operation conditions in Figures 9(a) and 9(b) corresponded to the conditions which achieved the maximum power points MPP2, MPP1 under the irradiations of 60% and 100%, respectively, in Figure 8. In Figure 9(a), the transistors $Q_k$ ($k = 1, 2, 3$) were operated at the duty ratio of 0.75 to boost the input voltage $V_{IN}$ of 32.0 V to the output voltage of the cell converter $V_{ok}$ of 128 V. The output voltage $V_{out}$ of 384 V was obtained because three cell converters were connected in IPOS. In Figure 9(b), the duty ratio for transistors was 0.74.

The interleaved control was applied to the proposed multicellular converter to prevent the short circuit among cell converters. The off-states of the drain to source voltages $V_{Q1}$, $V_{Q2}$, $V_{Q3}$ for transistors $Q_1$, $Q_2$, $Q_3$ appeared alternately, and this meant that the electric power was provided from one cell converter to the load at an arbitrary time.

The interleaved control enables shrinking the input inductor. The input inductors $L_{ik}$ were designed to achieve CCM at the input voltage of 40 V and the inductor currents $I_{Lik}$ have large ripple for the average current as shown in Figure 9. These ripples caused by the inductor currents $I_{Lik}$ were bundled and canceled at the input side and the reduced ripple was confirmed in the input current $I_{IN}$.

The applied voltages to the bidirectional switches $Q_{Xak}$, $Q_{Xbk}$ ($k = 1, 2, 3$) were also shown in Figure 9. The bidirectional switches were utilized for the isolation barrier. Maximum applied voltage to the switches $Q_{Xak}$, $Q_{Xbk}$ corresponded to the output voltage $V_{out}$, and the applied voltages varied for the switching conditions of transistors $Q_k$.

The applied voltages to the transistors and the bidirectional switches have overshoots. The voltage overshoots

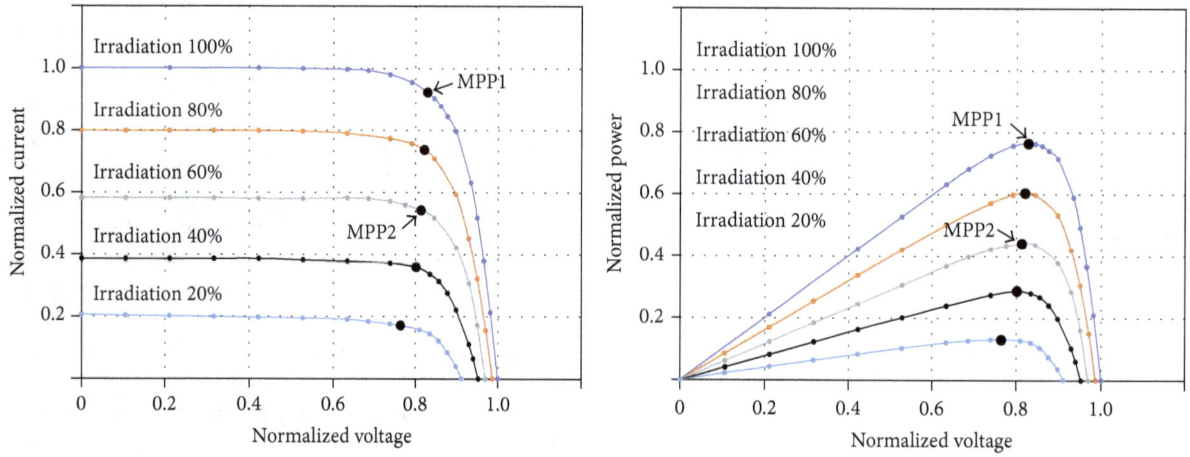


FIGURE 8: $I$-$V$ and $P$-$V$ characteristics of PV module.

TABLE 2: Parameters for experiment.

| Multicellular converter | |
|---|---|
| Nominal input voltage $V_{IN}$ | 12 V |
| Nominal output voltage $V_{OUT}$ | 53 V |
| Resistive load $R_{OUT}$ | 20 Ω |
| Connection topology | Input Parallel Output Series |
| Number of cells $N$ | 2 ($k = 1, 2$) |
| Voltage gain $G$ | 4.44 |
| **Single cell converter** | |
| Input voltage $V_{ik}$ | 12 V |
| Output voltage $V_{ok}$ | 26.6 V |
| Duty ratio $D$ | 0.55 |
| Voltage gain $G_k$ | 2.22 |
| Switching frequency $f_{SW}$ | 100 kHz |
| Main transistor $Q_k$ | 100 V, 2.8 mΩ (IRF7769 from IR) |
| Input inductor $L_{ik}$ | 10 $\mu$H |
| Output capacitor $C_{ok}$ | 44 $\mu$F |

depend on the intrinsic and parasitic parameters caused by the circuit dimensions and the device structures. In this simulation, the 110 pF for GaN-FETs, 145 pF for GaN-HEMTs and 22.5 nH for the circuit parasitic inductances were considered in each cell converter.

*4.2. First Laboratory Prototype of Nonisolated Multicellular Converter Using Two Cell Converters.* The feasibility of the nonisolated multicellular dc-dc converter is verified in this section. Figure 10 shows the circuit configuration of the nonisolated multicellular converter for the experiment. This multicellular converter consists of two cell converters and these cell converters are connected in IPOS. The detailed circuit parameters are shown in Table 2. The nominal input voltage $V_{IN}$ was 12 V and the input voltages of cell converters $V_{i1}$, $V_{i2}$ corresponded to the input voltage $V_{IN}$. The voltage transformation ratio of cell converters $G_k$ ($k = 1, 2$) was 2.22 because the duty ratio of the main switches $Q_1$, $Q_2$ was 0.55. The voltage gain $G$ of the multicellular boost converter was 4.44 because the output terminals of two cell converters were connected in series.

In the experiment, Si-MOSFETs (100 V, 2.8 mΩ from IR) were employed for the main switches $Q_1$, $Q_2$. The body diodes of the Si-MOSFETs and the Si-SBD (Shottoky Barrier Diode) substituted for the bidirectional switches $Q_{Xak}$, $Q_{Xbk}$ ($k = 1, 2$) shown in Figure 7 because the applied voltages to the switches $Q_{Xak}$, $Q_{Xbk}$ were unidirectional in the case of the number of the cell converters $N$ was 2.

Figures 11 and 12 show the experimental apparatus and the experimental result, respectively. The legends for the voltage waveforms $V_{IN}$, $V_{o1}$, $V_{o2}$, $V_{OUT}$ and the current waveforms $I_{IN}$, $I_{Li1}$, $I_{Li2}$, $I_{OUT}$ shown in Figure 12 corresponded to the legend in Figure 10. The voltages $V_{Q1}$, $V_{Q2}$ in Figure 12 means the applied voltages to the main switches $Q_1$, $Q_2$.

In Figure 12, the input voltage $V_{IN}$ was 11.1 V and the output voltages of cell converters $V_{o1}$, $V_{o2}$ were approximately 24.8 V. The voltage transformation ratios of the cell converters $G_1$, $G_2$ were 2.23 (=24.8/11.1) and these ratios were nearly equal to the voltage gain of 2.22 in case the main switches $Q_1$, $Q_2$ were operated at the duty ratio of 0.55. The output voltage of the multicellular converter $V_{OUT}$ resulted in 49.6 V because of the IPOS connection of two cell converters. The voltage transformation ratio $G$ of 4.46 was confirmed under operating at the duty ratio of 0.55.

The inductor currents $I_{Li1}$, $I_{Li2}$ had the current ripples of ±50% for the average inductor currents of 6.3 A. In the proposed converter, these current ripples were cancelled because the main switches in the cell converters were controlled alternately. The input and the output currents $I_{IN}$, $I_{OUT}$ were 12.7 A and 2.6 A, respectively, and the current ripples for $I_{IN}$, $I_{OUT}$ were suppressed. The off-state voltages of $V_{Q1}$, $V_{Q2}$ were 24 V and corresponded to the input voltage $V_{IN}$. This means that the dc voltage between the input side and the output side was successfully blocked by the switches $Q_{Xak}$, $Q_{Xbk}$ ($k = 1, 2$).

The fundamental circuit behavior of the proposed multicellular converter was confirmed here. The multicellular converter consisted of two cell converters because the unidirectional devices could be utilized in the case of $N =$

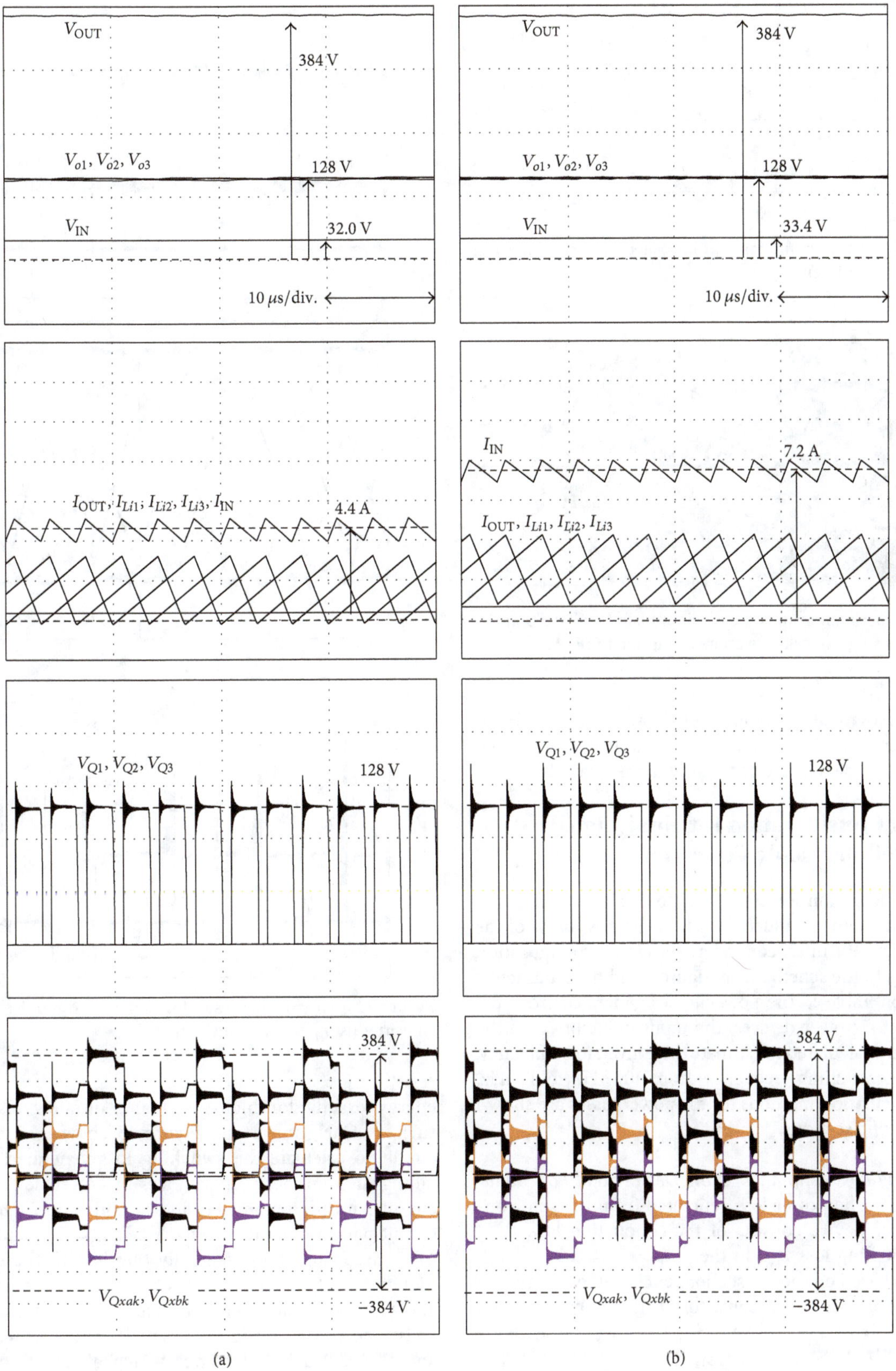


FIGURE 9: Simulation result for nonisolated multicellular dc-dc converter. (a) 32.0 V–384 V 140 W operation and (b) 33.4 V–384 V 240 W operation.

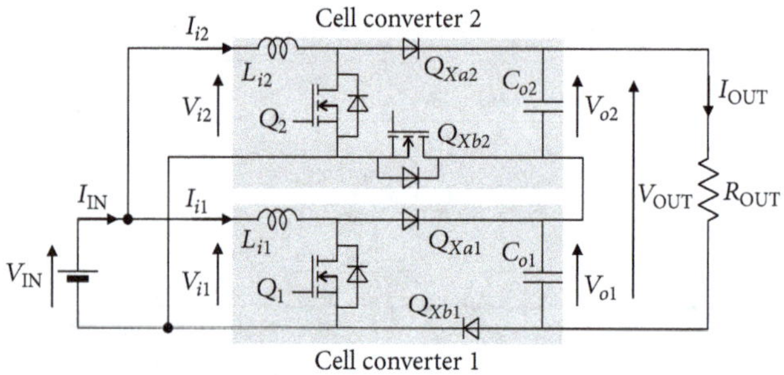


Figure 10: Circuit configuration of nonisolated multicellular boost converter for experiment.

Figure 11: Experimental apparatus of nonisolated multicellular dc-dc converter using two cell converters connected in IPOS.

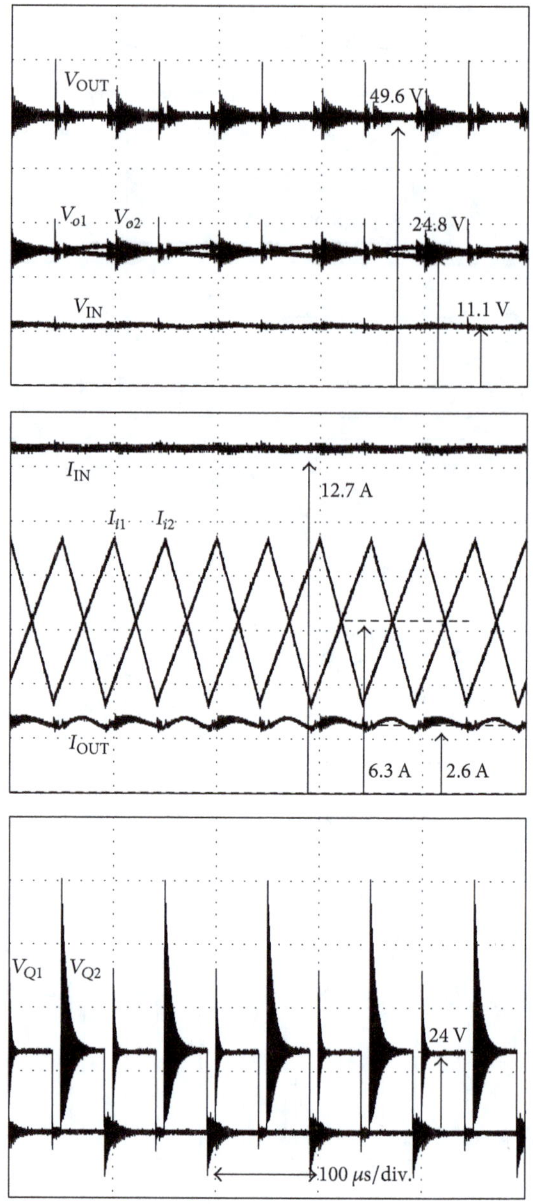


Figure 12: Experimental result for nonisolated multicellular boost converter using two cell converters.

2. The employment of bidirectional switches enables to operate the multicellular converter with large number of cell converters as shown in the above simulation analysis.

## 5. Design Consideration for Nonisolated Multicellular dc-dc Converter

Design consideration for 20 V–40 V to 384 V, 240 W multicellular converter is conducted to show the possibility of the highly efficient PV microconverter. The circuit configuration corresponds to the schematic in Figure 7 and the parameters are based on Table 1. The power losses generated from the semiconductor power devices, the input inductors, and the capacitors are calculated to estimate the conversion efficiency, varying the switching frequency from 100 kHz to 1 MHz. Detailed calculation methods for the power losses are shown in the following sections.

*5.1. Switching Loss Estimation Using Device Total Loss Simulator (DTLS).* The power loss generated from the semiconductor power device consists of the conduction loss $P_{\text{cond}}$ and the switching loss $P_{\text{sw}}$. In the proposed circuit, the cell converter is based on the boost chopper circuit topology, and the conduction loss of the transistor $Q_k$ is simply calculated by the formula of $P_{\text{cond}} = D \cdot I_{Lik}^2$. The conduction losses generated from the bidirectional switches $Q_{Xak}$, $Q_{Xbk}$ are also calculated in the same manner as the transistor $Q_k$.

The switching loss $P_{\text{sw}}$ generated from $Q_k$ is calculated by using the device total loss simulator (DTLS) [23, 24]. The DTLS was developed to estimate the switching loss energy exactly for high-speed and ultralow loss devices such as GaN transistors. Figure 13 shows the equivalent circuit to calculate the switching loss generated from $Q_k$, taking the nonlinear capacitances in power device and the circuit parasitic parameters into account. In Figure 13, the circuit parasitic inductances are $L_{s1}$, $L_{s2}$, $L_{s3}$, and $L_{s4}$, and these inductances are generally bundled to the equivalent inductor $L_{\text{st}}$ $(=L_{s1} + L_{s2} + L_{s3} + L_{s4})$. The symbols of $C_{\text{sH}}$ and $C_{\text{sL}}$ mean the circuit parasitic capacitances. The gate resistance and the inductance are $R_g$ and $L_{\text{sg}}$, respectively, and the common inductance which affects both the gate circuit and the main circuit is $L_{\text{sc}}$. Not only the device parameters but also the circuit parasitic parameters have to be designed in advance to estimate the exact switching loss.

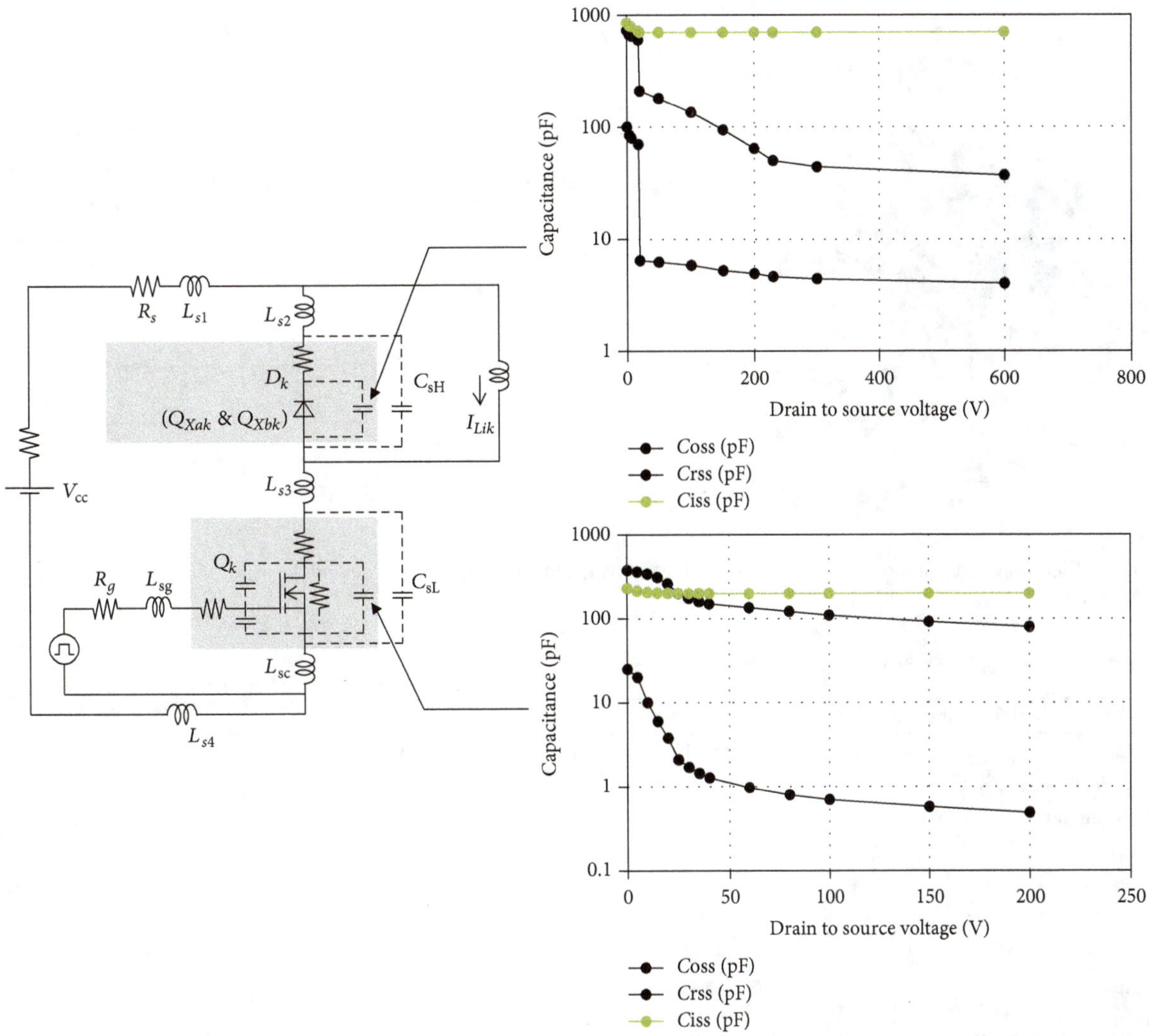


FIGURE 13: Equivalent circuit for switching loss energy calculation in device total loss simulator [10].

In DTLS, the turn-on energy $E_{on}$ and the turn-off energy $E_{off}$ generated from $Q_k$ are calculated as follows.

$$E_{on} = \int i_d(t) \times v_{ds}(t)\,dt + \frac{1}{2}C_{Dk}V_{cc}^2 + \frac{1}{2}L_{st}I_{Lik}^2 \quad (17)$$

$$E_{off} = \int i_d(t) \times v_{ds}(t)\,dt + \frac{1}{2}C_{oss}V_{cc}^2. \quad (18)$$

The drain to source voltage $v_{ds}$ and the drain current $i_d$ of the transistor $Q_k$ are basically calculated based on the numerical analysis, taking the nonlinearity of device capacitances into account. As shown in (17) and (18), a part of the switching loss energy is described by the stored energies in the circuit parasitic inductance and the device capacitances. The nonlinear analysis which causes the accumulated error is reduced to estimate the switching loss energies in DTLS.

To confirm the availability of DTLS, the switching loss energy of the GaN-HEMT (600 V, 150 mΩ) was measured here. Figure 14 shows the measurement setup and the measured turn-off waveforms for the drain current of 2 A to 12 A. The circuit parasitic inductance $L_{st}$ of the setup was 22.5 nH and the gate resistance $R_g$ of 3.0 Ω was connected externally in the experimental setup.

Both the measured switching loss energies of GaN-HEMT and the calculated energies by DTLS were compared in Figure 15. Detailed circuit parameters were shown in Table 3, and these parameters corresponded in the experimental condition in Figure 14. The circles filled in blue were the turn-on energies by the experiment and the circles filled in white were the measurement results of the turn-off energies in Figure 14. The blue solid line means the turn-on energy and the blue dashed line means the turn-off energy calculated by DTLS. The calculation results had good agreement with the experimental results. The validity for the commercially available 200 V class GaN-HFET has been already confirmed [23]; therefore, DTLS is available to estimate the switching loss of datasheet-level high-speed and ultralow loss power transistors exactly.

The switching loss energy from the GaN-FET (200 V, 50 mΩ) suitable for the cell converter was calculated and the calculation results were also shown in Figure 15. The solid and the dashed lines in green means the loss energies for

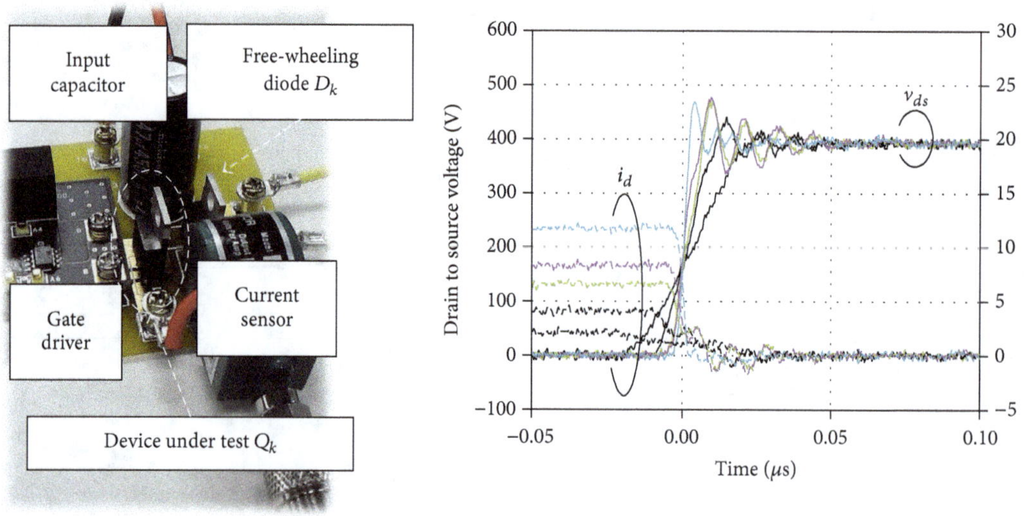


FIGURE 14: Measurement result of turn-off waveforms for 600 V 150 mΩ GaN-HEMT under 384 V, 2 A to 12 A operation conditions.

TABLE 3: Parameters for switching energy calculation.

| Circuit parasitic parameters | |
|---|---|
| Total parasitic inductance $L_{st}$ | 22.5 nH |
| Parasitic capacitances $C_{sH}$, $C_{sL}$ | 10 pF, 30 pF |
| Common inductance $L_{sc}$ | 1 nH |
| Gate inductance $L_{sg}$ | 3 nH |
| Gate resistance $R_g$ | 3 Ω |
| GaN-HEMT $Q_{Xak}$, $Q_{Xbk}$ for free-wheeling diode $D_k$ | |
| On-resistance | 150 mΩ |
| Output capacitance | 110 pF @ 100 V |
| Breakdown voltage | 600 V |
| GaN-FET $Q_k$ | |
| On-resistance | 50 mΩ |
| Output capacitance | 110 pF @ 100 V |
| Breakdown voltage | 200 V |

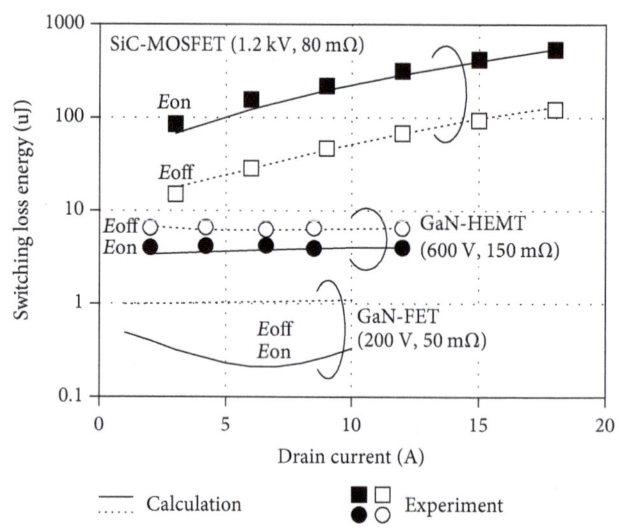


FIGURE 15: Turn-on and turn-off switching energies of GaN-FET in nonisolated multicellular converter calculated by DTLS.

the GaN-FET. The circuit parasitic parameters were based on Table 3 and the turn-on and the turn-off switching energies below 1 $\mu$J are expected. The switching loss of GaN-FET for cell converter is calculated based on Figure 15, taking the turn-on and the turn-off currents obtained from the circuit simulation.

*5.2. Input Inductor Design for Cell Converter.* The power loss generated from the inductor consists of the copper loss $P_{Cu}$ and the core loss $P_{core}$. The copper loss is generally calculated by the formula of $P_{Cu} = r_w(f_{sw}) \cdot I_{Lik}^2$. Here, $r_w$ is the resistance of the inductor winding and the resistance $r_w$ depends on the switching frequency $f_{sw}$ of the converter because of the skin effect. Details to estimate the core loss is described below.

The input inductance of the cell converter $L_{ik}$ is designed to achieve the CCM under the 40 V–128 V, 80 W operation to shrink the inductor here. The inductance value of 68.8 $\mu$H at the switching frequency of 100 kHz is obtained by the following equation.

$$V_{Lik} = L_{ik} \frac{\Delta I_{Lik}}{\Delta t}$$
$$L_{ik} = \frac{V_{in} \cdot D}{2 \cdot I_{Lik} \cdot f_{sw}} = \frac{40\,\mathrm{V} \cdot 0.688}{2 \cdot 2\,\mathrm{A} \cdot 100\,\mathrm{kHz}} = 68.8\,\mu\mathrm{H}. \quad (19)$$

The relationship between the inductance and the magnetic core shape is approximately formulated as follows.

$$L_{ik}\,(\mathrm{H}) = \mu_0 \mu_r N^2 \frac{S_e\,(\mathrm{m}^2)}{l_e\,(\mathrm{m})} \quad (20)$$

$$\mathrm{Vol}_{core}\,(\mathrm{m}^3) \cong l_e \cdot S_e. \quad (21)$$

TABLE 4: Parameters for inductor design installed into cell converter.

| Magnetic material | |
|---|---|
| Material | Mn-Zn ferrite |
| Permeability $\mu_r$ | 2,300 |
| Max. flux density $B_m$ | 0.5 T |
| Residual magnetization $B_r$ | 0.12 T |
| Coercivity $H_c$ | 14.3 A/m |
| Circuit specifications | |
| Switching frequency $f_{sw}$ | 100 kHz to 1 MHz |
| Inductance $L_{ik}$ | 68.8 $\mu$H to 6.88 $\mu$H |
| Magnetic core dimensions | |
| Core shape | PQ core |
| Width $X$ | 10 mm to 35 mm |
| Depth $Y$ | 6.7 mm to 23.4 mm |
| Height $Z$ | 8.6 mm to 30.2 mm |
| Gap length $l_g$ | 0 mm to 25 mm |

Here, the permeability of vacuum is $\mu_0$, and the relative permeability of the magnetic material is $\mu_r$. The turn number of the inductor winding is $N$. The magnetic length and the cross section area of the magnetic core are $l_e$ and $S_e$, respectively. The magnetic core volume $\text{Vol}_{\text{core}}$ is approximately calculated as (21). The core loss $P_{\text{core}}$ is generally estimated as follows.

$$P_{\text{core}}\,(\text{W}) = f_{\text{sw}}\,(\text{Hz}) \cdot E_{\text{core}}\left(\text{J/m}^3\right) \cdot \text{Vol}_{\text{core}}\left(\text{m}^3\right). \quad (22)$$

As shown in the above equation, the core loss depends on the dimensions of the magnetic core $\text{Vol}_{\text{core}}$. There are various solutions to realize the designated inductor $L_{ik}$. To extract the suitable core dimension to realize the compact and low loss inductor, design consideration for the inductor is necessary.

Table 4 shows the parameters for the input inductor design. The Mn-Zn ferrite with the permeability of 2,300 was assumed for the magnetic core material of the inductor [25]. This is based on the commercially available magnetic material. The maximum magnetic flux density $B_m$, residual magnetization $B_r$, and the coercivity $H_c$ in Table 4 are extracted from the published datasheet. Figure 16 shows the core loss energy $E_{\text{core}}$ of the magnetic core material.

The PQ core was also assumed for the dimensions of the magnetic core. The core width was varied from 10 mm to 35 mm to design the inductors which were operated at 100 kHz to 1 MHz. The core depth $Y$ and the core height $Z$ are proportion to the core width $X$, keeping the dimensions of existing PQ core shape. The magnetic length $l_e$ and the cross section area of the core $S_e$ are found by the dimensions $X$, $Y$, $Z$.

Figure 17 shows the calculation result of the efficiency and the power density of the inductor for 80 W cell converter. The power loss from the inductor and the inductor volume were translated to the efficiency and the power density to normalize the performance of the inductor. The copper loss $P_{\text{Cu}}$ and the core loss $P_{\text{core}}$ were considered to calculate the inductor loss, and the envelope volume determined by the core dimensions $X$, $Y$, $Z$ was also taken into account for

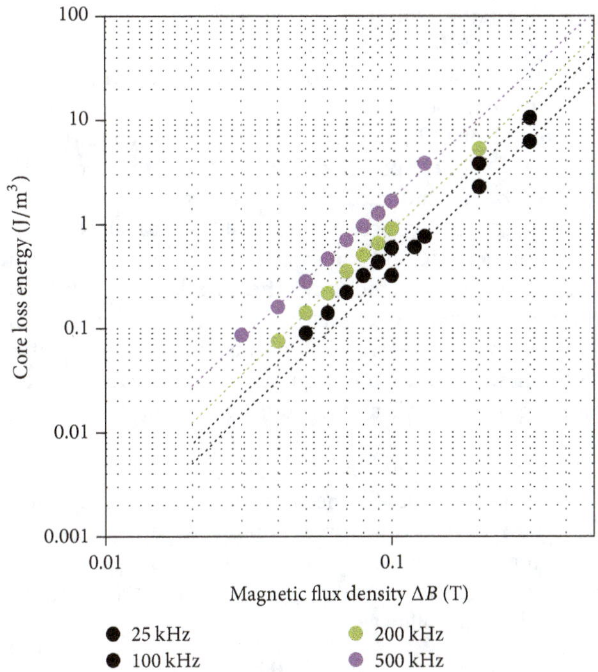


FIGURE 16: Core loss energy of magnetic core material [11].

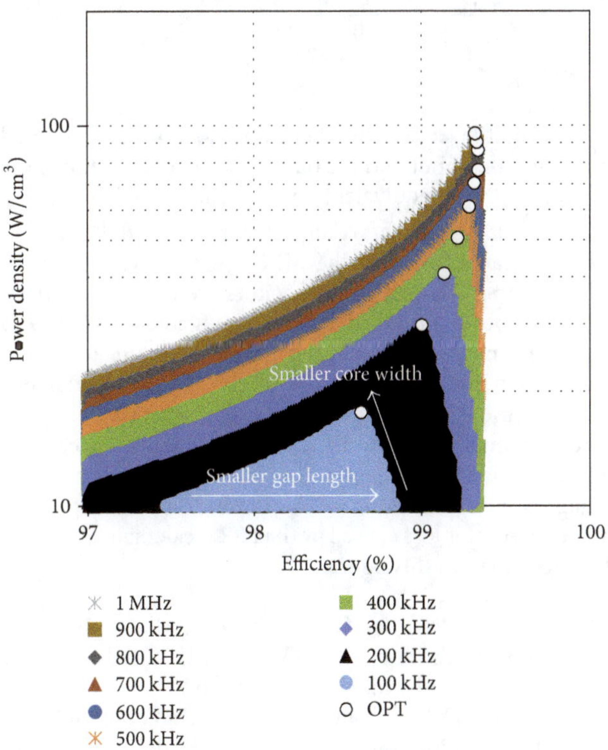


FIGURE 17: Calculated efficiency and power density of inductor.

the inductor volume. The heat sink volume is not considered here.

As the gap length $l_g$ becomes shorter, the inductor loss decreases. The material characteristics for the saturation determine the minimum gap length. As the core width $X$ becomes shorter, the core volume shrinks. The turn number of the inductor winding $N$ increases, and the efficiency decreases because of the copper loss. Based on the aforementioned trend, the inductor has the operation condition to maximize its power density for each switching frequency. The symbol of OPT in Figure 17 means the conditions to achieve the highest power density for each frequency. The efficiency of the inductor at OPT is improved in case the switching frequency increases from 100 kHz to 700 kHz. The smaller inductance and the smaller core volume contribute to achieving higher efficiency. The core loss over 700 kHz operation affects the inductor loss significantly and the efficiency at OPT decreases. In this study, the operating condition to achieve the highest power density for each switching frequency is applied to the converter design.

*5.3. Input and Output Capacitor Design for Cell Converter.* The capacitance values for the input and the output capacitors $C_{ik}$, $C_{ok}$ are designed to suppress the voltage ripples within 3% of the rated input and output voltages $V_{\text{in}}$, $V_{ok}$. The capacitances of 56.3 $\mu$F and 1.37 $\mu$F under 100 kHz operation are calculated as follows.

$$C_{ik} = \frac{I_{Lik} \cdot D}{\Delta V_{ik} \cdot f_{\text{sw}}} = \frac{4\,\text{A} \cdot 0.844}{0.03 \cdot 20\,\text{V} \cdot 100\,\text{kHz}} = 56.3\,\mu\text{F}$$

$$C_{ok} = \frac{I_{\text{out}} \cdot D}{\Delta V_{ok} \cdot f_{\text{sw}}} = \frac{0.625\,\text{A} \cdot 0.844}{0.03 \cdot 128\,\text{V} \cdot 100\,\text{kHz}} = 1.37\,\mu\text{F}. \quad (23)$$

The multilayer ceramic capacitors (MLCC) are assumed to develop the input and the output capacitors to reduce the loss caused by the equivalent series resistance (ESR).

Figure 18 shows the relationship between the ESR and the capacitance value of the MLCC for the various rated voltages. The ESRs of the capacitors were obtained from the design tool provided by the manufacturer [26]. The ESRs were plotted for the capacitances of 0.1 $\mu$F to 100 $\mu$F whose rated voltages were 6.3 V to 1 kV. As the capacitance increases, the ESR decreases because the capacitor volume is generally proportional to the capacitance and the electrode area depends on the capacitance. The influence of the rated voltage of the capacitor on its ESR is not clear from Figure 18, and the capacitor loss caused by the ESR is calculated by using the fitted curve in this figure.

*5.4. Efficiency Calculation for Cell Converter.* Figure 19 shows the conversion efficiency of 32.0 V–128 V, 47 W cell converter in the case of the irradiation of 60% and the efficiency of 33.2 V–128 V, 80 W cell converter for the irradiation of 100%, taking the operation condition to achieve MPPT in Section 4. The conduction loss $P_{\text{cond}}$ and the switching loss $P_{\text{sw}}$ were considered for the transistor $Q_k$. The symbol of $P_{sr}$ means the conduction loss generated from the bidirectional switch $Q_{Xak}$, $Q_{Xbk}$. The copper loss $P_{\text{Cu}}$ and the core loss $P_{\text{core}}$ were also considered for the inductor $L_{ik}$. The losses generated by ESR in the input and the output capacitors were $P_{\text{Ci}}$, $P_{\text{Co}}$.

From Figure 19, the maximum efficiency over 98.0% was expected for the 32.0 V–128 V converter under the 60% irradiation in case the switching frequency was smaller than

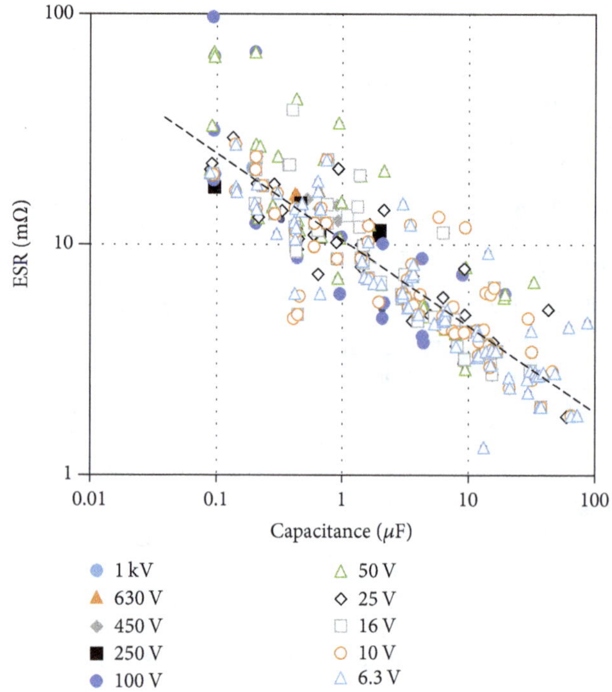


FIGURE 18: Equivalent series resistance of multilayer ceramic capacitor [12].

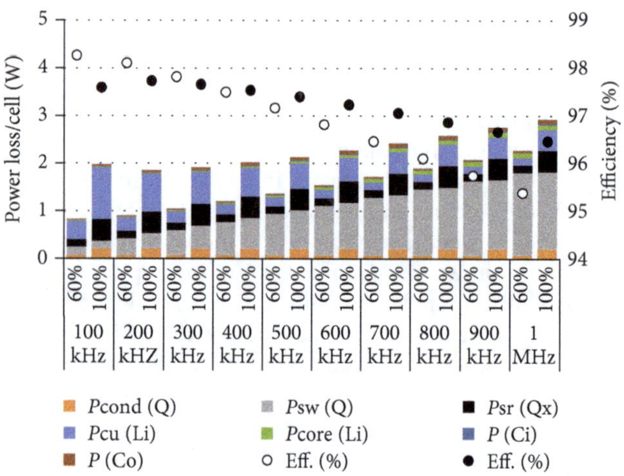


FIGURE 19: Calculated power loss and efficiency for cell converter. (a) 32.0 V–128 V, 47 W operation with 60% irradiation and (b) 33.2 V–128 V, 80 W operation with 100% radiation.

200 kHz. The switching loss from the power devices was dominant over 300 kHz. The maximum efficiency of 97.8% was also observed around 200 kHz for the 33.2 V–128 V cell converter with 100% irradiation.

The performance of the cell converter ideally corresponds to the performance of the multicellular converter [22]. Based on the multicellular converter topology, the maximum efficiency of 98% is expected to the nonisolated multicellular dc-dc converter. Now, the efficiencies of 93% to 97.5% have been reported for the microconverter [4–8]. This means that the proposed converter topology has the potential to achieve

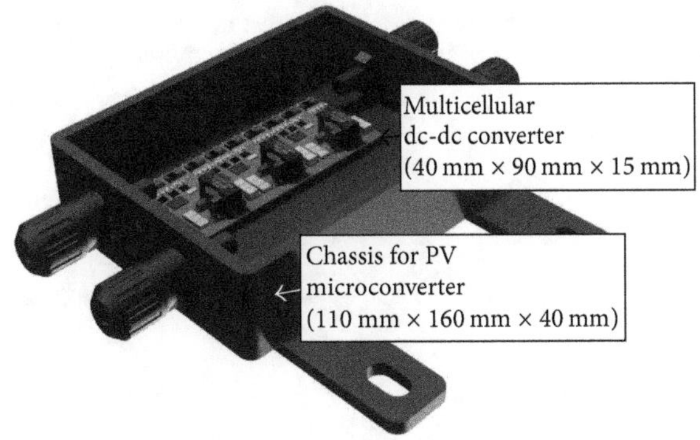


Figure 20: Illustration for PV microconverter based on multicellular converter topology.

higher efficiency and has the possibility to contribute to realizing low carbon society.

Figure 20 shows the illustration of the PV microconverter based on the proposed multicellular dc-dc converter. The electric components designed in the previous sections were arranged in the chassis for the microconverter. In Figure 20, the converter volume was small enough to house it into the chassis, taking the noise filter and microcontroller into account. The details of the thermal design, the noise analysis, and the experimental verification will be operated in near future.

## 6. Conclusions

High step-up nonisolated multicellular dc-dc converter was proposed to realize the highly efficient PV microconverter. The circuit configuration and the control scheme were introduced and the fundamental behavior of the proposed converter was also confirmed by the circuit simulation. Laboratory prototype using two cell converters was fabricated and the feasibility of the proposed multicellular converter was also confirmed. Design consideration for the 20 V–40 V to 384 V, 240 W multicellular converter was carried out and the potential to achieve the efficiency of 98% was observed.

Feasibility of the proposed converter topology using three cell converters with the bidirectional switches will be verified in near future. The detailed thermal design and the noise analysis will be also operated.

The proposed topology contributes to realizing the low carbon society through the smooth introduction of PV modules to dc distribution network.

## References

[1] Y. Sugiyama, ""Green ICT toward Low Carbon Society", Design for Innovative Value Towards a Sustainable Society," in *Proceedings of EcoDesign 2011, 7th International Symposium on Environmentally Conscious Design and Inverse Manufacturing*, pp. 739–742, 2011.

[2] T. Babasaki, T. Tanaka, Y. Nozaki, T. Tanaka, T. Aoki, and F. Kurokawa, "Developing of higher voltage direct-current power-feeding prototype system," in *Proceedings of the 31st International Telecommunications Energy Conference, INTELEC 2009*, Incheon, Republic of Korea, October 2009.

[3] K. Harada, T. Ninomiya, and W. J. Gu, *The Fundamentals of Switched-Mode Converters*, Corona Publishing Co., LTD., 1992.

[4] S. Allimbhasha and G. Nagaraju, "Advanced hyrid transformer high boost DC-DC converter for photovoltaic module applications," *International Journal of Professional Engineering Studies*, vol. VIII, no. 2, 2017.

[5] B. Gu, J. Dominic, J. S. Lai, Z. Zhao, and C. Liu, "High boost ratio hybrid transformer DC-DC converter for photovoltaic module applications," in *Proceedings of the 27th Annual IEEE Applied Power Electronics Conference and Exposition, APEC 2012*, pp. 598–606, Orlando, FL, USA, February 2012.

[6] H.-D. Gui, Z. Zhang, X.-F. He, and Y.-F. Liu, "A high voltage-gain LLC micro-converter with high efficiency in wide input range for PV applications," in *Proceedings of the 29th Annual IEEE Applied Power Electronics Conference and Exposition, APEC 2014*, pp. 637–642, Fort Worth, TX, USA, March 2014.

[7] G. Sundar, N. Karthick, and S. Rama Reddy, "High Step-Up DC—DC Converter for AC Photovoltaic Module with MPPT Control," *Journal of Electrical Engineering*, vol. 65, no. 4, pp. 248–253, 2014.

[8] S. Jain, B. S. Satish, and J.-S. J. Lai, "High gain resonant boost converter for PV micro-converter system," in *Proceedings of the 2016 IEEE Energy Conversion Congress and Exposition, ECCE 2016*, pp. 1–6, Milwaukee, WI, USA, September 2016.

[9] T. Ninomiya, Y. Ishizuka, R. Shibahara, and S. Abe, "EnergySaving Technology Using NextGeneration Power Electronics," in *Proceedings of the IEE-Japan (IEEJ) Industry Applications Society Conference*, pp. I15–120, Chiba, Japan, 2012 (Japanese).

[10] S. E. Babaa, M. Armstrong, and V. Pickert, "Overview of Maximum Power Point Tracking Control Methods for PV Systems," *Journal of Power and Energy Engineering*, vol. 02, no. 08, pp. 59–72, 2014.

[11] A. Attou, A. Massoum, and M. Saidi, "Photovoltaic Power Control Using MPPT and Boost Converter," *Balkan Journal of Electrical & Computer Engineering*, vol. 2, no. 1, pp. 23–27, 2014.

[12] R. Middlebrook and S. Cuk, "A general unified approach to modelling switching-converter power stages," in *Proceedings of the 1970 IEEE Power Electronics Specialists Conference*, pp. 18–34, April 1970.

[13] M. Yamasaki, T. Kanai, M. Mino, and H. Ohashi, "Potentialities and Problems of the SiC Devices to Apply Practical Use," in *Proceedings of the Japan Industry Applications Society Conference*, pp. 147–150, Okinawa, Japan, 2011 (Japanese).

[14] A. Fukui, T. Takeda, K. Hirose, and M. Yamasaki, "HVDC power distribution systems for telecom sites and data centers," in *Proceedings of the 2010 International Power Electronics Conference - ECCE Asia -, IPEC 2010*, pp. 874–880, Sapporo, Japan, June 2010.

[15] L. Zhou, Y. Wu, J. Honea, and Z. Wang, "High-efficiency true bridgeless totem pole PFC based on GaN HEMT: Design challenges and cost-effective solution," in *Proceedings of the 2015 International Exhibition and Conference for Power Electronics, Intelligent Motion, Renewable Energy and Energy Management, PCIM Europe 2015*, Nuremberg, Germany, May 2015.

[16] Y. Hayashi, D. Takai, A. Matsumoto, and T. Ise, "Highly Efficient dc–dc Transformer Based on Multicell Converter Topology for

Next Generation dc Distribution System," *Electrical Engineering in Japan (English translation of Denki Gakkai Ronbunshi)*, vol. 198, no. 4, pp. 98–110, 2017.

[17] M. Ahmed, C. Fei, F. C. Lee, and Q. Li, "High-efficiency high-power-density 48/1V sigma converter voltage regulator module," in *Proceedings of the 32nd Annual IEEE Applied Power Electronics Conference and Exposition, APEC 2017*, pp. 2207–2212, Tampa, FL, USA, March 2017.

[18] http://www.vicorpower.com/.

[19] T. Morita, M. Yanagihara, H. Ishida et al., "650V 3.1mΩcm2 GaN-based monolithic bidirectional switch using normally-off gate injection transistor," in *Proceedings of the 2007 IEEE International Electron Devices Meeting, IEDM*, pp. 865–868, Washington, DC, USA, December 2007.

[20] C.-T. Tsai and S.-H. Chen, "PV power-generation system with a phase-shift PWM technique for high step-up voltage applications," *International Journal of Photoenergy*, vol. 2012, Article ID 838231, 11 pages, 2012.

[21] R. W. Erickson and D. Maksimović, *Fundamentals of Power Electronics*, Springer, 2001.

[22] Y. Hayashi and T. Ninomiya, "Analysis and prototyping of multicellular DC-DC transformer for environmentally friendly data centers," *Journal of Energy and Power Engineering*, vol. 10, no. 5, pp. 313–323, 2016.

[23] A. Nakajima, K. Takao, and H. Ohashi, "GaN power transistor modeling for high-speed converter circuit design," *IEEE Transactions on Electron Devices*, vol. 60, no. 2, pp. 646–652, 2013.

[24] Y. Hayashi, K. Takao, T. Shimizu, and H. Ohashi, "High power density design methodology," in *Proceedings of the Power Conversion Conference - Nagoya (PCC '07)*, pp. 569–574, IEEE, Nagoya, Japan, April 2007.

[25] https://www.mag-inc.com/.

[26] http://ds.murata.co.jp/software/simsurfing/.

# 14

# Energy Efficiency Audit Based on Wireless Sensor and Actor Networks: Air-Conditioning Investigation

**Abdel-Rahman Al-Qawasmi**[1] **and I. Tlili**[2]

[1]*Electrical Engineering Department, College of Engineering, Majmaah University, Majmaah 11952, Saudi Arabia*
[2]*Energy and Thermal Systems Laboratory, National Engineering School of Monastir, Street Ibn El Jazzar, 5019 Monastir, Tunisia*

Correspondence should be addressed to Abdel-Rahman Al-Qawasmi; a.alqawasmi@mu.edu.sa

Academic Editor: Yuezhong Meng

This paper investigates the energy performance of new design air-conditioning system combined with energy audit in KSA governmental building since air conditioning is responsible for more than 60 % of electricity consumption. Alternative solutions are provided. Based on the results, the old conditioners will be replaced for 113 rooms in three floors of the MUAB. The applied methodology uses Wireless Sensor and Actuator Networks (WSAN) for sensing and measuring technique characteristics of the conditioners and compares them with the ones available on the market, which were selected according to ASHRAE Standard 62.1-2007 specified by the manufacturers. Energy balance examination proves that operational conditioners have greater electric consumption and power factor comparing to the known standards. Energy audit applied in the MUAB provided a saving of 10% in the average consumption of electric energy. More than that an exploration is performed to insert Stirling engine in air conditioning cycle proving that it can be a promoted solution and can reduce significantly energy consumption; an alpha Stirling engine has been tested numerically for the same temperature and pressure reigned in the refrigeration cycle proving that it can work with energy efficiency around 42%.

## 1. Introduction

Energy necessities have amplified weirdly according to the prompt expansion of the global economy. The awareness that fossil fuel possessions required for the generation of energy are becoming scarce and the climate change is toughly connected to carbon emissions leads engineers and researchers to give more consideration and concern to energy saving and environmental protection.

Table 1 includes the temperature, the humidity, and the wind speed data collected in different 20 stations in Saudi Arabia in 2006 which demonstrate that wind energy can be a promising source of renewable energy in KSA as shown in this table which mentions the maximum wind velocity extents at 16.5 m/s in Guriat station, and it is well known that power is a function of the cube of the wind velocity. The measured wind speeds at all the stations are averaged; the maximum wind speed reaches up to 16.5 m/s in Guriat station, while the minimum wind speed is 7.7 m/s in Gizan. The power available from the wind is a function of the cube of the wind speed. The rising electric energy request and current crisis of the energetic sector lead to meaningful increase of the electric energy. The most consuming part of electricity comes from air conditioning which reaches around 60 % of the total building consumption. In order to lessen electric energy costs, several researchers and engineers have developed and investigated diversities of theoretical and experimental model of air conditioning, and many prototypes have been designed and built [1–7]. On the other hand, the community behavior should be changed and required two foundations: (i) habit change when using electric equipment and (ii) technology innovation replacing old equipment by efficient equipment [8].

Replacement of the ineffective apparatus as traditional air conditioning by other energetic efficient technologies is highly requested. Air conditioners are the primary energy consumer in the KSA due to the hot weather (70% of the year). Figure 1 shows the temperature changes over month

TABLE 1: Weather data from meteorological station in KSA.

| *S. N.* | *Station* | *Pressure (mb)* | *Rain (mm)* | *temperature* | | | *Relative humidity (%)* | *Wind speed* | |
|---|---|---|---|---|---|---|---|---|---|
| | | | | *Mean* | *min* | *max* | *Mean* | *mean* | *max* |
| 1 | Riyadh | 942.4 | 70.0 | 26.7 | 0.0 | 47.8 | 26.2 | 3.09 | 8.8 |
| 2 | Jeddah | 1007.3 | 55.0 | 28.2 | 9.8 | 49.0 | 61.4 | 3.71 | 11.3 |
| 3 | Dhahran | 1006.7 | 125.0 | 26.4 | 25 | 49.0 | 52.5 | 4.38 | 11.8 |
| 4 | Guriat | 954.8 | 36.5 | 19.5 | -8.0 | 47.6 | 43.5 | 4.22 | 16.5 |
| 5 | Al jouf | 936.1 | 34.0 | 22.0 | -6.0 | 46.7 | 32.1 | 4.02 | 15.9 |
| 6 | Gizan | 1007.7 | 90.0 | 30.2 | 11.8 | 45.3 | 68.4 | 3.24 | 7.7 |
| 7 | Abha | 794.0 | 119.9 | 18.6 | 0.0 | 34.1 | 54.6 | 2.94 | 14.9 |
| 8 | Tabouk | 926.0 | 36.0 | 22.1 | -3.5 | 46.4 | 34.0 | 2.73 | 15.5 |
| 9 | Yanbu | 1007.8 | 73.2 | 27.7 | 4.7 | 49.0 | 53.8 | 3.76 | 10.3 |
| 10 | Hail | 901.3 | 47.5 | 22.4 | -9.4 | 44.5 | 33.2 | 3.24 | 10.8 |

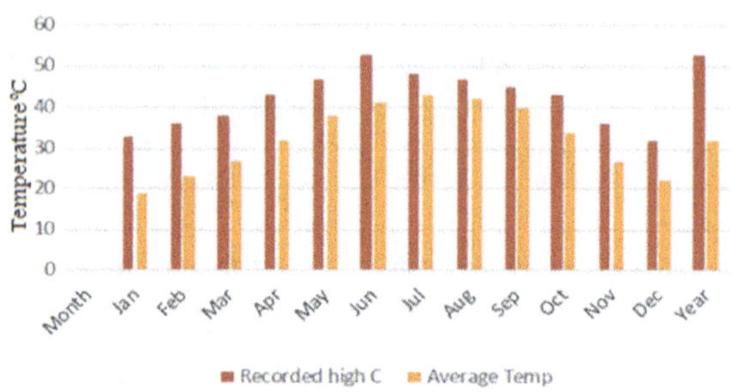


FIGURE 1: Temperature data for Riyadh, Saudi Arabia [8].

during one year. The average of high recorded temperature is 42.6 Celsius and the average of the temperature averages is 32.2 Celsius.

Due to the complex execution, the conditioners retrofit is a significant challenge due to the high cost and the implementation of alternative solutions such as saving plans and power consumption scheduling solutions. Conditioning technologies have improved in the last years. More efficient conditioners are manufactured with a high-energy efficiency. Moreover, environmental pollution of air conditioning should be concerned and the noise level the air conditioner emits is increasingly important; for this reason introducing Stirling engine in the refrigerator cycle will reduce vibration and noise [9]. Split conditioners consume a very huge amount of power and are also characterized by low quality production [10]. In order to revise, modify, and enforce the Saudi standard SASO 2663/2012, the technical team of the Saudi Energy Efficiency Program adopted an approach based on Consensus and Unanimity.

On the other hand, a pure mechanical engineering solution should be applied to air conditioning to reducing energy consumption as replacing compressor by Stirling engine. Stirling engines are significant for low pollution and small noise, since the vibration is very low, with higher durability, simple structure, and capability to self-start; for these reasons, it can be used for air conditioning. Being a regenerative external heated engine with higher performance, it is characterized by an extensive variety of energy resources and it can be a motivated solution for energy consumption in air conditioning [11–16].

D.G. Thombare et al. [17] deliver and investigate the important efforts taken for the development of the Stirling cycle engine and techniques used for engine analysis. They conclude that for fruitful operation of engine system with higher efficiency a cautious design of heat exchangers, appropriate assortment of drive mechanism, and engine configuration are essential. Recently Stirling cycle engine working with relatively low temperature with air of helium as working fluid is potentially attractive engines of the future air conditioning particularly in case of low-temperature differential Stirling engines with alpha configuration. Variations of engine torque, shaft power, and brake thermal efficiency at various heat inputs with engine speed and engine performance are presented by B. Kongtragool et al. [18] who designed and made a gamma-configuration for low-temperature differential Stirling engine; they verified the engine with numerous heat suppliers. They conclude that the engine performance will increase with increasing heat input. The engine torque, shaft power, brake thermal efficiency, speed, and heater temperature also increase with increasing heat input. P. Puech et al. [19] provide a theoretical exploration on the thermodynamic analysis of a Stirling engine with linear and sinusoidal variations of the component volume; they show that the dead volume strongly amplifies the imperfect regeneration effect, and the engine performance with perfect regeneration does not depend on the regenerator dead volume. The mean pressure power formula is very adequate and appropriate for the calculation of Stirling engine especially a gamma-configuration, low-temperature differential Stirling engine power output as shown by S. Wongwises et al. [20]; in their study thy provide a study on power output determination of a gamma-configuration Stirling engine with low-temperature differential. B. Cullen et al. [21] developed a model for theoretical decoupled Stirling cycle engine that is to be used as a tool for analysis of the ideal Stirling cycle engine and the limits on its real-world realization.

This paper will discuss two ways for reducing air-conditioning consumption; in the first part authors will

TABLE 2: Energy classification and concert requirements for air conditioners.

| Air Conditioner appliance type | Cooling capacity limit (CC) (Btu/h) At testing conditions T1 (35°C) | Mandatory EER (Btu/h)/watt Phase 1: 7 September 2013 T1 (35°C) | T3 (46°C) | Mandatory EER (Btu/h)/watt Phase 2: 1 January 2015 T1 (35°C) | T3 (46°C) |
|---|---|---|---|---|---|
| Window type | CC <18,000 | 8.5 | 6.12 | 9.8 | 7.06 |
| | 18,000 ≤CC< 24,000 | 8.5 | 6.12 | 9.7 | 6.98 |
| | CC ≥ 24,000 | 8.5 | 6.12 | 8.5 | 6.12 |
| Split type and other types | All Capacities | 9.5 | 6.84 | 11.5 | 8.28 |

TABLE 3: Star rating of air-conditioner quality.

| EER limits | Star Rating |
|---|---|
| EER > 10 | 6 |
| 9.5 < EER ≤ 10 | 5 |
| 9 < EER ≤ 9.5 | 4 |
| 8.5 < EER ≤ 9 | 3 |

perform an energy audit for air-conditioning system used in the most governmental building with economic influence. In the second part authors will expose an original mechanical solution to lessening energy depletion by introducing a Stirling engine in the cycle in parallel with compressor and then perform an energetic investigation.

## 2. Regulation and Endorsement

The choice of the appropriate conditioner depends on different parameters; consequently a new regulation established by the ministry of electricity requires clear information from manufactures to the consumers. The Saudi standard SASO 2663/2015 provided an energy labeling to show the minimum energy performance requirements for air conditioners. Table 2 shows the main air conditioners parameters for Windows and split types [22].

Starting from 7/9/2013, SASO will not issue licenses to air conditioners that do not comply with phase 1 of the modified standard and starting 1/1/2015 for phase 2. SASO provides quality methods to follow to recognize the efficiency of electrical devices. Table 3 provides the star rating to be applied from 3 stars and above only. This starting system is a measure used by SASO to control the quality of the air conditioners.

Conditioners characterized by some difficulties in maneuver created by environmental conditions such as dust and high temperature difference between day and night. The energy consuming is based on various complex parameters that are difficult to follow or to control. Therefore, in this paper, the energy consumption is the primary parameter in the study and other parameters will remain constants.

Figure 2 presents the new label and the main parameters that should be considered in the replacement process.

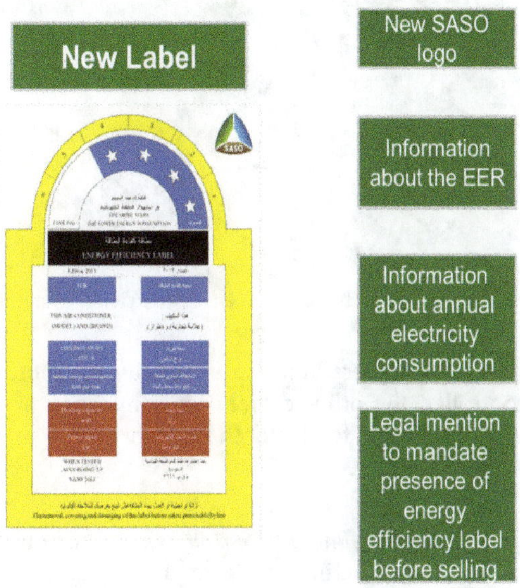


FIGURE 2: New label of new star energy efficiency standard [8].

As seen, the adopted parameter is the annual electricity consumption. In this research, the electricity consumption is measured and collected during nine months. Every air conditioner had its own data file gathered by sensors and its data analyzed as individual.

## 3. Energy Efficiency Systems Using WSAN

The designed WSAN monitoring system architecture contains three phases as shown in Figure 3. The three phases are as follows.

*3.1. Sensing and Actuating Phase (Pervasive Layer).* In this phase, sensors provide the Graphical User Interface (GUI) data about the energy consumption in different rooms in different floors within the building. The sensing process is a continuous process that is provided every half an hour. The sensed device was mainly a conditioner, printer, light of the offices, and classrooms.

All sensors are linked to each other using wireless transmission, and it has a WSN specification that has a

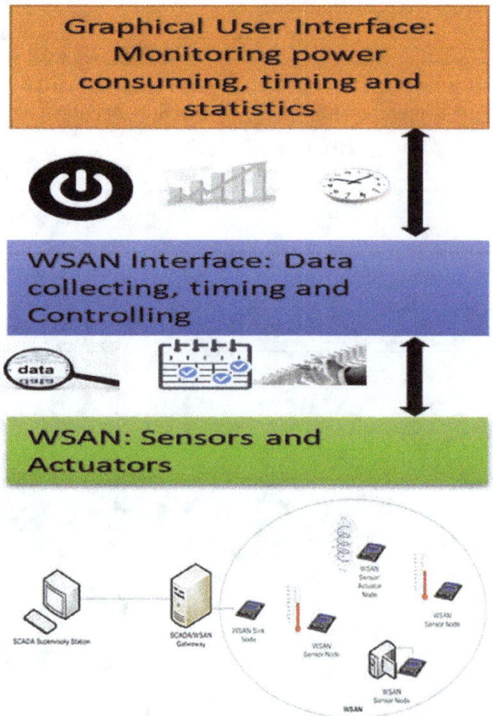


FIGURE 3: The architecture of energy efficiency system.

range up to 10 meters among devices. It is a coordinator that controlled the other devices for connection and data exchange. It is similar to Bluetooth Protocol using Master and slaves network architecture [23].

*3.2. WSAN Interface Phase (Controlling).* In this phase, the WSAN interface is acting as a gateway for supporting the controlling mechanism provided by actors and gathering data by sensors. In this phase, an integration to wireless networking protocol is performed.

*3.3. Implementation Layer (Applications).* The software used is suitable for a web browsing and smartphones. It is connected to the system to support secure access.

The adaptability of any wireless system is the critical issue to achieve a high dynamic environment. Some factors can be applied to attain high-energy efficiency, e.g., [24].

## 4. Design Specification for Air Conditioning Based on Stirling Engine

Several researches concerning Stirling engine were carried out because of their advantages; however, in the best authors knowledge no investigation dealt with air conditioning including Stirling engine. We apply here this model with the best choice configuration of Stirling engine and an optimization of pertinent specification for higher performance. The most advantages of the Yoke Ross mechanism are being distinguished by the nonsinusoidal volume variations. Different studies prove that alpha Stirling engine Figure 4 has different advantage compared to beta or gamma Stirling

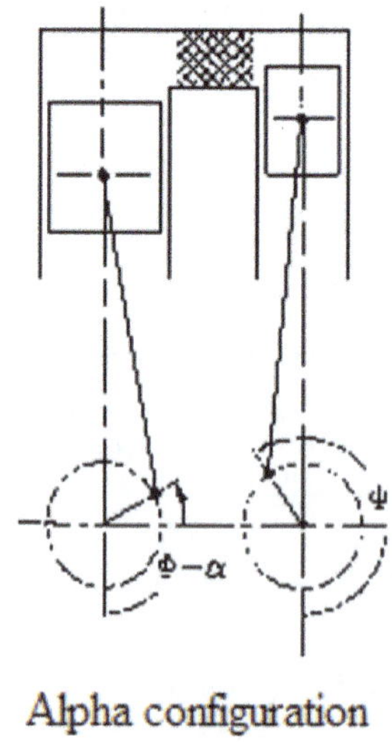


FIGURE 4: Alpha Stirling engine configuration.

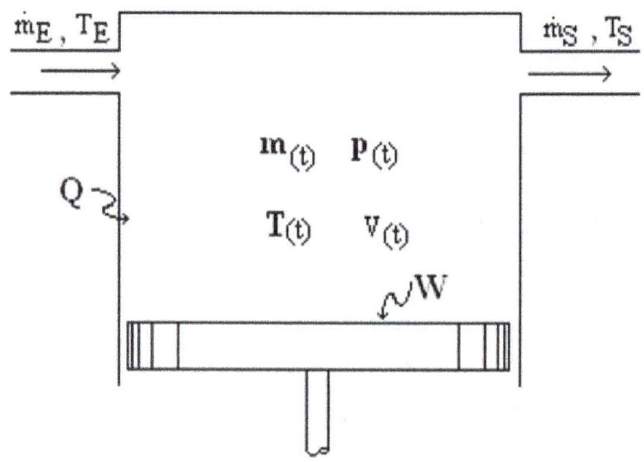


FIGURE 5: Generalised cell.

engine. A numerical simulation which takes account of thermal and mechanical losses has been investigated recently [1–7]. The results obtained are more realistic and are very suitable with the working environment. The model is used to determine different thermodynamics parameters of the Yoke Ross Stirling engine with specification as displayed in Table 6 and to predict the engine performance inside air conditioning.

## 5. Equations of Thermodynamic Model

The temperatures in different section are given as below using law of gas perfect as shown by the generalised cell in Figure 5

$$T_c = \frac{P_c \cdot V_c}{R \cdot m_c} \quad (1)$$

$$T_f = \frac{P_f \cdot V_f}{R \cdot m_f} \quad (2)$$

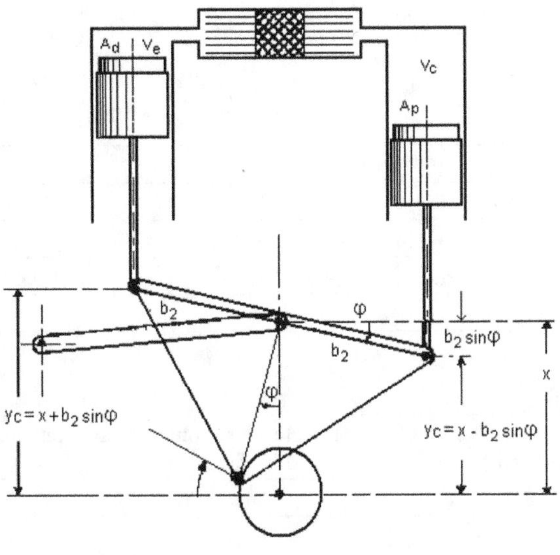


FIGURE 6: Yoke Ross Stirling engine configuration.

$$T_h = \frac{P_h \cdot V_h}{R \cdot m_h} \tag{3}$$

$$T_d = \frac{P_d \cdot V_d}{R \cdot m_d} \tag{4}$$

The temperatures for the two regenerator cell are, respectively,

$$T_{r1} = \frac{P_{r1} \cdot V_{r1}}{R \cdot m_{r1}} \tag{5}$$

$$T_{r2} = \frac{P_{r2} \cdot V_{r2}}{R \cdot m_{r2}} \tag{6}$$

The boundary temperatures between the tree major sections are expressed as follows [2]:

$$T_{r-f} = \frac{3T_{r1} - T_{r2}}{2} \tag{7}$$

$$T_{r-r} = \frac{T_{r1} + T_{r2}}{2} \tag{8}$$

$$T_{r-h} = \frac{3T_{r2} - T_{r1}}{2} \tag{9}$$

Therefore, the engine thermal efficiency is

$$\eta = \frac{W}{Q_h} \tag{10}$$

And the total engine volume $V_T$ represents the sum of all section volume and the model involves Yoke Ross Stirling engine configuration as shown in Figure 6.

The pertinent other parameter for the thermodynamics model introducing energy conservation is as follows [3]:

(i) Energy conservation equation [4, 5] is

$$\delta\dot{Q} + C_p T_E \dot{m}_E - C_p T_S \dot{m}_S = P\frac{dV}{dt} + C_v \frac{d(mT)}{dt} \tag{11}$$

## 6. Research Methodology

The Energy Efficiency Program developed by the KSA in the vision 2030 concentrated on the investment in the energy efficiency measuring and power saving technologies. It is one of the primary objectives "Business Environment, Restructure Our Economic Cities, Create Special Zones and Deregulate the Energy Market to Make It More Competitive" [25].

The WSAN technique that presented in this paper connects WSANs with Ethernet/Internet to perform an energy saving algorithm that can be performed online to implement some steps that measured the energy consumption. This technique is managed and controlled by the distribution of WSANs in the lecture halls, laboratories, and faculty members' offices in the assigned building using the communication protocols such as ZigBee and 6LoWPAN. Such protocols are integrated into a large number of home automation devices that are usually integrated with computation and sensing; WSAN devices are available commercially with low cost and radio communication capabilities [26] systems, most with small size and low power features.

In the proposed energy saving technique, these devices contain nodes with joined sensors computing energy and mutual power adjustment and instruments to measure pertinent other parameter [25–27].

The energy consumption of the air conditioner in room 802 shows different values based on the time of day and the month.

The average cost of the air conditioner that satisfies the Saudi standard SASO 2663/2015 (24,000 BTU) unit is around 2500 SAR (667$).

Recently refrigerants 407C and 410A replaced the old one R-22 refrigerant and even they have less efficiency compared to R-22 refrigerant. The decreasing is about 10% at 46°C.

Air conditioner power consumption trumps that of most appliances. The power consumption of air conditioners averages 228 watts (for a 24,000 BTU unit) in most KSA buildings per month. That would cost the following: the average energy consumption for survey dwellings based on the type of air-conditioning systems: $kWh/m^2/year$ for Central 221.5, Mini-Split 144.3, Window-type and 183.3 Window-type and Mini-Split 156.8 [28].

Table 4 shows applying New Consumption for all Categories of Service According to Council of Ministers' Decree dated 12/12/2017; this consumption tariff will be applied from 1/1/2018. For private educational facilities and private medical facilities, there is one category.

The total price of the energy consumed by the air conditioner (Type 24,000 BTU Mini-Split unit) with SASO Compliance Standards is *144.3 ($kWh/m^2/year$) X8$m^2$X (0.048\$/kWh)=55.8\$*. The total price of the energy consumed by the air conditioner as shown in Table 5 (Type 24,000 BTU Mini-Split unit) monitored by WSAN is *1602 ($kWh/16m^2/year$) (0.048\$/kWh)=76\$*.

The saving difference between the two air conditioners is 20$ and for 113 rooms, and the total saving is 2326. Assuming 5-year plan cost compensation, the total saving is 11632.67$.

TABLE 4: Energy consumption over 9 months for room 802.

| Time of Day<br>Month | 10:00 kWh | 11:00 kWh | 11:30 kWh | 12:00 kWh | 12:30 kWh | The average |
|---|---|---|---|---|---|---|
| June | 1738 | 1660 | 1829 | 1891 | 1785 | 1780.6 |
| July | 1733.1 | 1591 | 1596.7 | 1585.4 | 1624 | 1626.04 |
| August | 1643 | 1567.5 | 1644.9 | 1666.4 | 1729.2 | 1650.2 |
| September | 1353 | 1423.7 | 1428 | 1267 | 1290.5 | 1352.44 |
| The average = 1602.32 kWh | | | | | | |

TABLE 5: Consumption tariff.

| Consumption categories (Kwh) | Governmental (Halalah / kwh) | Industrial (Halalah / kwh) | Private educational facilities, private medical facilities (Halalah / kwh) |
|---|---|---|---|
| 1-6000 | 32 (0.088 $) | 18 (0.048$) | 18 (0.048$) |

## 7. Results of the Analysis

In this section, some preliminary results are shown by monitoring the energy consumption of air conditioner in 802 in the University during one academic year tracing the change in the consumption during the different seasons. Figure 7 shows the energy consumption for 3 months (July, August, and September). Other months are not considered due to the low average energy consumption during the winter and spring days.

The retrofiring process is very expensive and the saving difference based on the electricity costs will be compensated only after 75145/2326=32.3 years.

The alternative solution is the merged power saving plan using optimization energy saving plans that can save up to 10% and the total saving will be 113X16$=1808$ per year. Using the difference in energy consumption cots, three air conditioners can be replaced and this will give additional 60 $ per year.

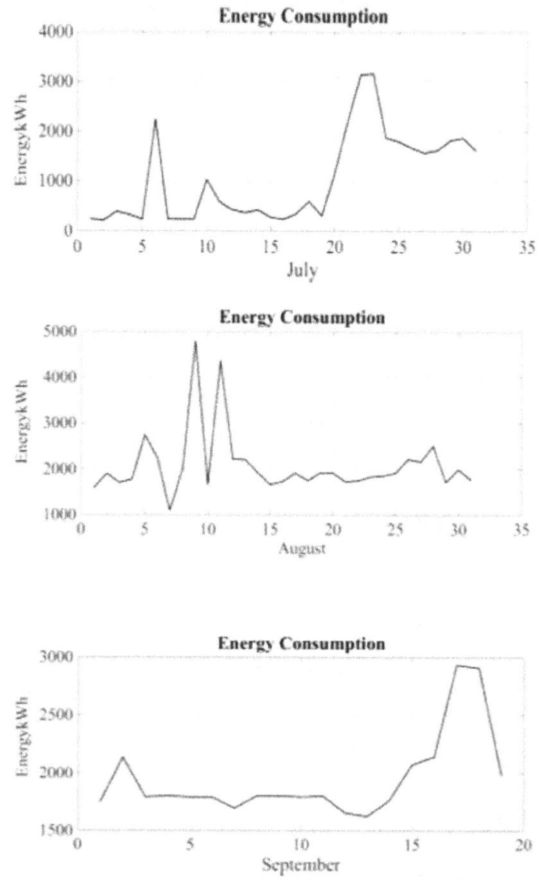


FIGURE 7: The energy consumption of the room 802 in 3 months.

*7.1. Temperature Variations in Different Compartment.* The temperature variations and the temperature profile of the model are shown in Figure 8. In the regenerator, we notice that the temperature variation of the matrix ($T_{par1}$, $T_{par2}$) is about 10°C what shows that the quantity of heat stored is low. The difference between the matrix and the associated gas temperature (Trl, $T_{r2}$) is about 20°C which proves that the regenerator has good effectiveness. Figures 9 and 10 display, respectively, the pressure variation along the cycle and the P-V diagram. The gas in the heater and the cooler, however, shows a very much higher temperature variation, respectively, of more than 100°C and 50°C over the cycle.

Different researches prove that regenerator represents the most important section in Stirling engine; therefore, a regenerator with higher efficiency leads to significant increase in the overall energy efficiency of the engine.

Tlili et al. [10–14] study the effect of the heat capacity of regenerator matrices and the conductivity on the engine performance; they demonstrate that an optimal value for both heat capacity of regenerator matrices and the conductivity leads to the highest energy efficiency of the Stirling engine. In our case, Figure 11 depicts that rise of thermal conductivity on matrices c1 and c2 lessens significantly the engine; this can be explained by the fact that the irreversibilities related to the conduction have been increased [12]. Figure 12 represents that the heat capacity of the regenerator matrix leads to enhancing the engine performances. We can conclude that the engine

TABLE 6: Specifications of the Ross Yoke Stirling engine.

| | |
|---|---|
| Crank length | 7.6 mm |
| displacement | Ymin = 17.75 mm |
| Dead volume | Vmc = 1.14 cm3 |
| Piston Diameter | 38 mm |
| Swept volumes | 24.42 cm3 |
| Volumes of the exchangers | Vcol = 5 cm3<br>Vr = 6.75 cm3<br>Vhot = 4.78 cm3 |
| Angular phase shift | 90° |
| Average pressure | 1 bar |
| Mass of working fluid | 0.0276 g |
| Hot temperature | 323 K |
| Cold temperature | 283 K |

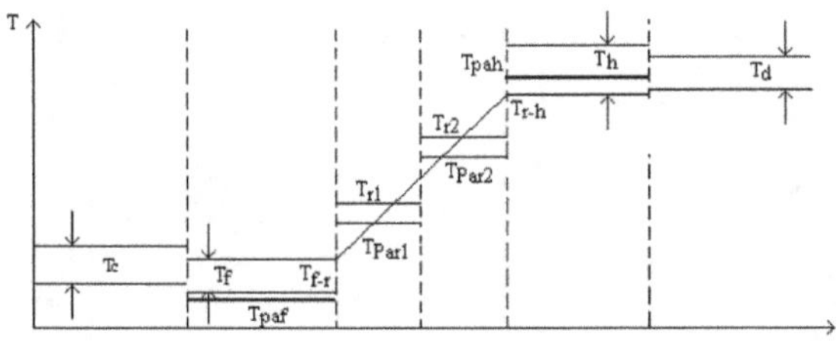


FIGURE 8: Temperature distribution.

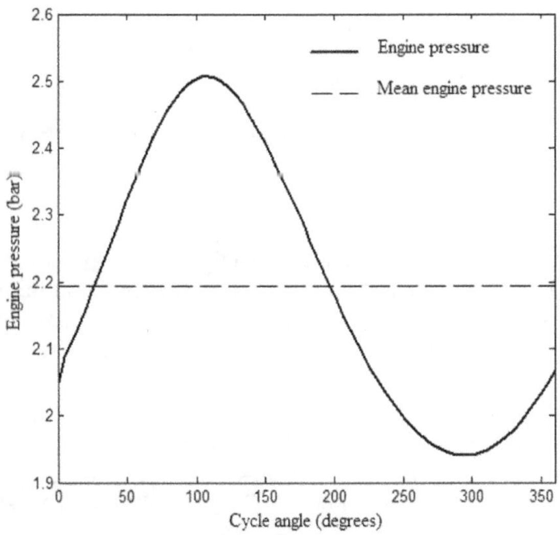


FIGURE 9: Engine pressure variation.

performance depends on the material nature of matrices c1 and c2.

### 7.2. Performance Optimization of Alpha Stirling Engine

*7.2.1. Influence of Working Fluid Mass.* Tlili et al. [15, 16] study the effect of working fluid mass on engine performance; it is well known that an augmenting on mass affects meaningfully gas velocity, load, and pressure, since all the sections are closed volume. Consequently, irreversibilities related to pressure drop and energy losses augment [17]; nevertheless, the engine power grows and it is found that an optimum value of working fluid mass gives an extreme engine efficiency around 42% for m = 1 g, as revealed in Figure 13. This can be explained by the increasing in pressure loss and the capacity restriction of matrices c1 and c2 in terms of conductivity. The results found show that for m = 4 g we obtain a perfect performance for the engine.

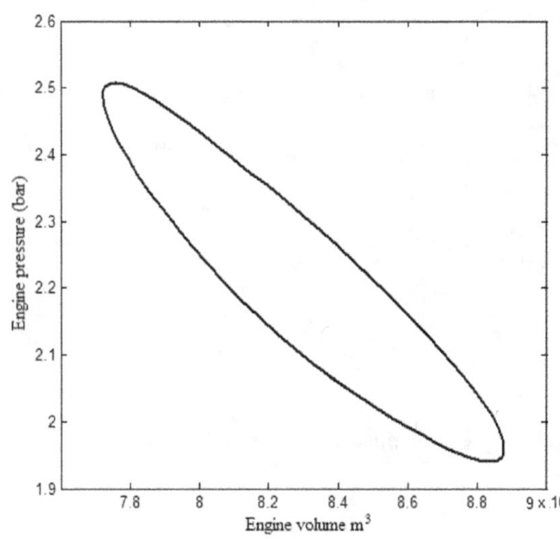


FIGURE 10: Pressure volume diagram.

Also, we notice that the increase of the total mass of gas induces an increase continuously in heat exchanged. It is also noticed that the temperature also supports the heat exchanged.

*7.2.2. Effect of Heater Temperature.* Figure 14 depicts the variation of engine performance with heater temperature. It is perceived that an increasing in heater temperature leads to an augmentation on engine performances, whereas the engine

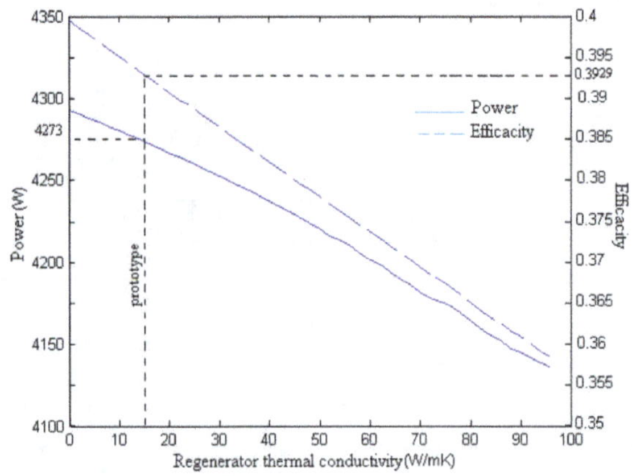


FIGURE 11: Power and efficiency variation with regenerator section conductivity.

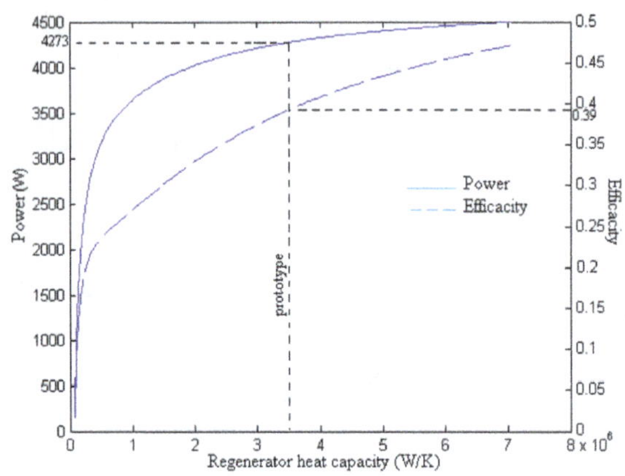


FIGURE 12: Power and efficiency variation with regenerator section heat capacity.

power enhances with the operating frequency. Furthermore, it is well known that an increasing in heater temperature leads to engine thermal losses [3]; while in our case the engine performance increases, this can be interpreted by the fact the heat transferred in the regenerator between working fluid and matrices has been increasing meaningfully. It can be seen that growth of the heater temperature makes an unceasingly growth in heat exchanged. It should be pointed out that the development technology related to material will enhance the performance of Stirling engines and as a consequence in air conditioning systems.

## 8. Conclusion

The production and consumption of energy are responsible for a large portion of the environmental problems worldwide. In addition, the need to boost the efforts for further development and promotion of renewable energy sources and energy efficiency has been felt world over in light of rising exhaustible energy resources prices. Since energy security, economic growth, and environment protection are the national energy policy drivers of Saudi Arabia, it continues its commitment toward the utilization of more efficient energy systems based

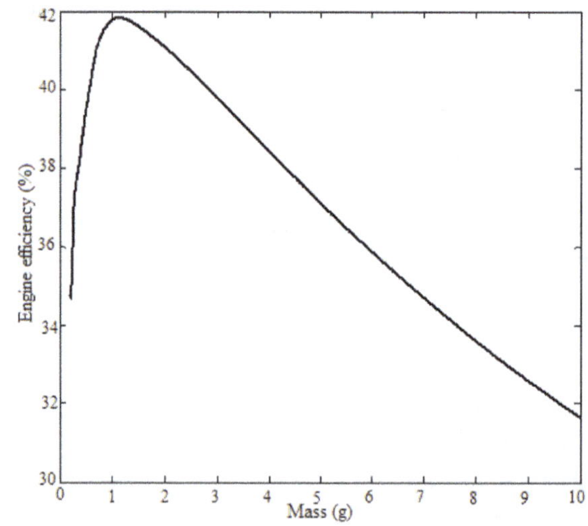


FIGURE 13: Engine efficiency variation with working fluid mass.

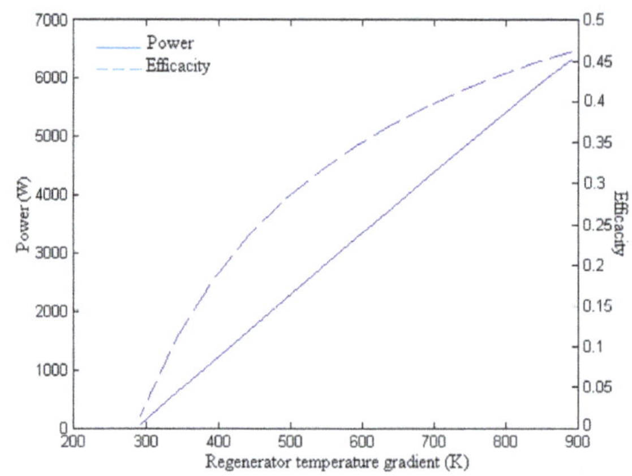


FIGURE 14: Power and efficiency variation with regenerator the temperature gradient.

on the long-term importance of sustainable energy sources as in playing a supportive role to reducing energy consumption. That is why the Kingdom is at the forefront of a number of research initiatives into exploring more potential and deployment of energy efficiency technologies, in particular showing itself to be a leader in this issue. However, barriers to the deployment of more efficient energy on a wide scale still remain. This paper investigates the energy efficiency and the economic impact provided by air conditioning in KSA governmental building (Majmaah University Administrative Building MUAB), since air conditioning is responsible for more than 60 % of electricity consumption. Alternative solutions are provided. More than that an investigation performed to insert Stirling engine in air conditioning cycle proves that it can be a promoted solution and can reduce significantly energy consumption; an alpha Stirling engine has been tested for the same temperature and pressure reigned in the refrigeration cycle proving that it can work with energy efficiency around 42%. To carry out this study, we presented initially the constraints of the Yoke Ross Stirling engine into a thermodynamic model; the attained results were very close to other results [4–7]. Then, we examine the effect of pertinent parameter of the proposed engine; the main results show that the lessening of the matrix porosity and conductivity of the regenerator raise the performance. The increasing of the working fluid mass leads to augmenting the engine power and pressure; nevertheless, the effectiveness reaches the maximum. Also, it is demonstrated that when the exchanger piston section increased and its stroke diminished, the engine output increases and the efficiency reaches the maximum. Finally, the model optimizing these parameters show that Yoke Ross Stirling engine can work between $T_h$= 330 K and $T_c$ = 280 K. Applying this study to prototypes of air conditioning will lead to the determination of their design optimal parameters and consequently to enhance performance.

## Nomenclature

$A$: Area, $m^2$
$Cp$: Specific heat at constant pressure, Jkg-1$K^{-1}$
$Cv$: Specific heat at constant volume, $Jkg^{-1}$ $K^{-1}$
$D$: Diameter, m
$\varepsilon$: Regenerator efficiency
$k$: Thermal conductivity, $Wm^{-1}$ $K^{-1}$
$L$: Length, m
$M$: Mass of working gas in the engine, kg
$\dot{m}$: Mass flow rate, $kgs^{-1}$
$m$: Mass of gas in different component, kg
$P$: Pressure, Pa
$Q$: Heat, J
$\dot{Q}$: Power, W
$R$: Gas constant, $JkgK^{-1}$
$T$: Temperature, K
$V$: Volume, $m^3$
$W$: Work, J
$c$: Compression space
$ch$: Load
$cd$: Conduction
$d$: Expansion space
$E$: Entered
ext: Outside
$f$: Cooler
$h$: Heater
$pis$: Piston
$r$: Regenerator
$\theta$: Crank angle, rd
$\rho$: Density, $kgm^{-3}$
$\omega$: Angular frequency, $rads^{-1}$.

## Acknowledgments

The first author Abdel-Rahman Al-Qawasmi acknowledges with thanks the Deanship of Scientific Research (DSR) at Majmaah University, Majmaah, Saudi Arabia, for technical and financial support through Vote no. 37/66 for this research project.

## References

[1] I. Tlili, Y. Timoumi, and S. Ben Nasrallah, "Numerical simulation and losses analysis in a stirling engine," *International Journal of Heat and Technology*, vol. 24, no. 1, pp. 97–105, 2006.

[2] I. Tlili, Y. Timoumi, and S. Ben Nasrallah, "Thermodynamic analysis of the Stirling heat engine with regenerative losses and internal irreversibilities," *International Journal of Engine Research*, vol. 9, no. 1, pp. 45–56, 2007.

[3] Y. Timoumi, I. Tlili, and S. Ben Nasrallah, "Performance optimization of Stirling engines," *Journal of Renewable Energy*, vol. 33, no. 9, pp. 2134–2144, 2008.

[4] M. H. Ahmadi, M. A. Ahmadi, F. Pourfayaz et al., "Designing a powered combined Otto and Stirling cycle power plant through multi-objective optimization approach," *Renewable & Sustainable Energy Reviews*, 2016.

[5] L. Angui et al., "The effect of air-conditioning parameters and deposition dust on microbial growth in supply air ducts," *Energy and Buildings*, vol. 42, no. 4, pp. 449–454, April 2010.

[6] I. Tlili, "Renewable energy in Saudi Arabia: Current status and future potentials," *Environment, Development and Sustainability*, vol. 17, no. 4, pp. 859–886, 2015.

[7] Reggae, Country Energy Profile: Saudi Arabia - Clean Energy Information Portal, http://www.reegle.info/countries/saudi-arabia-energy-profile/SA.

[8] https://en.wikipedia.org/wiki/Climate_of_Saudi_Arabia.

[9] A. Sa'ed and I. Tlili, "Numerical investigation of working fluid effect on stirling engine performance," *International Journal of Thermal and Environmental Engineering*, vol. 10, no. 1, pp. 31–36, 2015.

[10] I. Tlili, "Finite time thermodynamic evaluation of endoreversible Stirling heat engine at maximum power conditions," *Renewable & Sustainable Energy Reviews*, vol. 16, no. 4, pp. 2234–2241, 2012.

[11] I. Tlili, "Thermodynamic study on optimal solar stirling engine cycle taking into account the irreversibilities effects," *Energy Procedia*, vol. 14, pp. 584–591, 2012.

[12] I. Tlili, "A numerical investigation of an alpha Stirling engine," *International Journal of Heat & Technology*, 2012.

[13] I. Tlili and S. A. Musmar, "Thermodynamic evaluation of a second order simulation for Yoke Ross Stirling engine," *Energy Conversion and Management*, vol. 68, pp. 149–160, 2013.

[14] I. Tlili, Y. Timoumi, and S. B. Nasrallah, "Analysis and design consideration of mean temperature differential Stirling engine for solar application," *Journal of Renewable Energy*, vol. 33, no. 8, pp. 1911–1921, 2008.

[15] Y. Timoumi, I. Tlili, and S. B. Nasrallah, "Design and performance optimization of GPU-3 Stirling engines," *Energy*, vol. 33, no. 7, pp. 1100–1114, 2008.

[16] Y. Timoumi, I. Tlili, and S. B. Nasrallah, "Reduction of energy losses in a stirling engine," *International Journal of Heat and Technology*, vol. 25, no. 1, pp. 84–93, 2007.

[17] D. G. Thombare and S. K. Verma, "Technological development in the Stirling cycle engines," *Renewable & Sustainable Energy Reviews*, vol. 12, no. 1, pp. 1–38, 2008.

[18] B. Kongtragool and S. Wongwises, "Performance of low-temperature differential Stirling engines," *Journal of Renewable Energy*, vol. 32, no. 4, pp. 547–566, 2007.

[19] P. Puech and V. Tishkova, "Thermodynamic analysis of a Stirling engine including regenerator dead volume," *Journal of Renewable Energy*, vol. 36, no. 2, pp. 872–878, 2011.

[20] B. Kongtragool and S. Wongwises, "Investigation on power output of the gamma-configuration low temperature differential Stirling engines," *Journal of Renewable Energy*, vol. 30, no. 3, pp. 465–476, 2005.

[21] B. Cullen and J. McGovern, "Development of a theoretical decoupled Stirling cycle engine," *Simulation Modelling Practice and Theory*, vol. 19, no. 4, pp. 1227–1234, 2011.

[22] https://www.airconco.com/infographic/decibel-scale/.

[23] Saudi Energy Efficiency program, working session on the implementation of the modified Saudi standard SASO 2663/2012 Energy labeling and minimum energy performance requirements for air-conditioners May 2013.

[24] Digi International Inc, "ZigBee Wireless Standard," 2014. http://www.digi.com/technology/rf-articles/wireless-zigbee.

[25] http://vision2030.gov.sa/en.

[26] H. Grindvoll et al., "A wireless sensor network for intelligent building energy management based on multi-communication standards–a case study," *Journal of Information Technology in Construction*, vol. 17, p. 43, 2012.

[27] P. J. M. Havinga and G. J. M. Smit, "Energy-efficient wireless networking for multimedia applications," *Wireless Communications and Mobile Computing*, vol. 1, no. 2, pp. 165–184, 2001.

[28] F. Alrashed and M. Asif, "Trends in residential energy consumption in Saudi Arabia with particular reference to the Eastern province," *Journal of Sustainable Development of Energy, Water and Environment Systems*, vol. 2, no. 4, pp. 376–387, 2014.

# 15

# Analysis and Design of an Energy Efficient Differential Coherent Chaos based System in Additive White Gaussian Noise Environment

**Nizar Al Bassam[1] and Oday D. Jerew[2]**

[1]*Middle East College, Muscat, Oman*
[2]*Asia Pacific International College, Sydney, NSW 2008, Australia*

Correspondence should be addressed to Nizar Al Bassam; nazarhooby@yahoo.co.uk

Academic Editor: Daesik Hong

A new differential coherent chaos based scheme is proposed and named as Differential Chaos ON-OFF Keying (DCOOK). The proposed scheme provides reduction in bit energy and better bit error performance at large spreading factor. This is achieved by presenting each transmitted bit by either identical segment or no transmission within same bit duration. The receiver performs simple correlation between the received signal and its delayed version to determine the transmitted information utilizing the low correlation between the noise signals. The bit error rate (BER) performance of the proposed scheme is evaluated analytically using Gaussian Approximation (GA) method and compared with the simulation results. The results show excellent agreement between the derived expression and simulation. Moreover, the BER of DCOOK scheme is compared with the standard chaos systems: Chaos ON-OFF Keying (COOK), Differential Chaos Shift Keying (DCSK), and Correlation Delay Shift Keying (CDSK). The comparison results show that DCOOK scheme can always achieve superior performance compared to COOK and CDSK schemes and even exceeds the performance of DCSK scheme at typical spreading factor values.

## 1. Introduction

Unlike pseudo random signals which are limited in number and are periodic, chaotic signals can theoretically produce infinite number of chaotic signals which are nonperiodic. Moreover, chaotic signal has broadband spectrum and has the ability to resist fading. All these properties make them of particular interest in communication systems, specifically in spread spectrum communications.

Chaotic communications systems can be classified into coherent, noncoherent, and differentially coherent systems [1]. To recover the information in coherent systems, chaotic receiver should generate replica copy of the chaotic carrier signals which is used by the transmitter. However, this is difficult to achieve in noisy channels due to nonperiodicity property of the chaos signals. In noncoherent systems, synchronization is not required and the information is decoded by estimating one of the unique signal parameters like a power as in noncoherent Chaos Shift Keying (CSK) scheme [2] or Chaotic ON-OFF Keying (COOK) scheme [3]. However, optimum threshold determination is the major drawback of noncoherent systems which causes a poor performance with the noise existence [3, 4].

In differential coherent systems, each information bit is presented by two identical time slots. First time slot is called a reference signal and is used to demodulate the information carrying signal in the second time slot as in Differential Chaos Shift Keying (DCSK) and Correlation Delay Shift Keying (CDSK) [5]. This system design requires a double bit energy and rate. To overcome these drawbacks, many differential coherent systems and their analytical BER performances are proposed and tested in AWGN environment [6–8].

Combining reference signal and information bearing signal in one-time slots using Walsh code is proposed in [9, 10] and initial condition modulation in [11]. Both schemes improve the spectral efficiency; however, systems require complex receiver design. To enhance bit rate of DCSK systems, information carrying signal is modulated with two information bits as in [12]. In addition, modulation of the reference signal is suggested to double the bit rate in [13]. Furthermore, bit error rate of DCSK scheme is enhanced by using multiple repeated sequences in one duration instead of continuous sample function. On the other hand, the receiver requires moving average calculation each time over one bit duration and the filtered segment is correlated with its delayed version [14]. Efficient utilization of transmitted signal energy is proposed in multicarrier-DCSK (MC-DCSK) [15]. The scheme requires narrow band modulators with high degree of accuracy which is complex to design.

In this paper, we introduce a simple and energy efficient differentially coherent scheme: Differential Chaos ON-OFF Keying (DCOOK). In DCOOK scheme, two identical chaotic segments are transmitted for bit "1." On the other hand no transmission is required for the same bit duration for bit "0." Thus, only half bit energy is utilized to compete the performance of standard chaos based system. Our contributions in this work are summarized as follows:

(1) The proposed DCOOK scheme offers the following:

   (i) Notable improvement in BER (up to 1.5 dB) compared to standard DCSK and CDSK schemes using only half of their bit energy.
   (ii) Superior BER performance compared to COOK scheme.
   (iii) Double bit rate compared to DCSK scheme.

(2) Theoretical value of BER is estimated using baseband model of the proposed scheme. The expression is verified by simulation in AWGN channel. Clear matching is obtained between the estimated values and simulation results.

The remainder of this paper is organized as follows. Standard chaos based systems including DCSK, CDSK, and COOK are described in Section 2. Section 3 gives an explicit description of the proposed system. Section 4 derives the analytical BER performance. The comparison between the theoretical BER results and the simulation results is presented in Section 5. It also compares the BER performance of the proposed system with the DCSK, CDSK, and COOK schemes. Finally, Section 6 draws the conclusions.

## 2. Standard Chaos Based Systems Description

*2.1. DCSK.* For each bit of information, the DCSK transmitter outputs a chaotic sequence $x_i$ at time instant $i$. Each transmitted signal consists of two sample functions with equal time duration and with length of $M$, where $M$ represents the spreading factor. The first sample function is called reference signal, while the second is called information

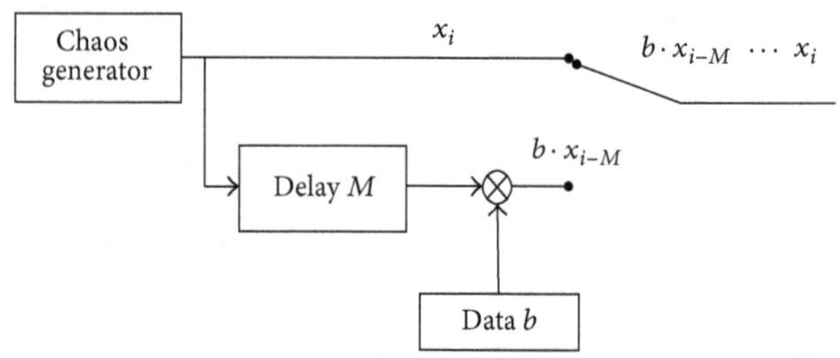


FIGURE 1: DCSK transmitter.

bearing signal. The information bearing is only reference signal delayed by $M$ and multiplied by information bit $b$ where $b \in \{-1, 1\}$. The transmitter of DCSK is shown in Figure 1. The transmitted sequence $s_i$ for a single bit is given by

$$s_i = \begin{cases} x_i & 0 < i \le M, \\ b \cdot x_{i-M} & M < i \le 2M. \end{cases} \tag{1}$$

The average bit energy of DCSK signal can be calculated as

$$E_b = 2MV(x), \tag{2}$$

where $V(x)$ represents the variance.

The DCSK receiver is shown in Figure 2, and the received signal $r_i$ is multiplied by its delayed version. The delay is set to be half bit duration. Thus, the correlator output at the end of bit duration is given by the sum

$$\begin{aligned} Z_{\text{DCSK}} &= \sum_{i=1}^{M} r_i r_{i-M} = \sum_{i=1}^{M} (s_i + \zeta_i)(s_{i-M} + \zeta_{i-M}) \\ &= b\sum_{i=1}^{M} x_i^2 + \sum_{i=1}^{M} x_i (\zeta_{i-M} + b\zeta_i) + \sum_{i=1}^{M} \zeta_i \zeta_{i-M}, \end{aligned} \tag{3}$$

where $\zeta_i$ is the Gaussian noise sample such that $E(\zeta) = 0$ and variance $V(\zeta) = E(\zeta^2)$. The receiver extracts the information $\tilde{b}$ according to the following rule:

$$\tilde{b} = \begin{cases} 1 & Z_{\text{DCSK}} \ge 0, \\ -1 & Z_{\text{DCSK}} < 0. \end{cases} \tag{4}$$

*2.2. CDSK.* The CDSK modulated signal is generated by adding the chaotic sequence $x_i$ and of the delayed chaotic sequence multiplied by the information signal $b_\ell$ as shown in Figure 3. The transmitted signal $s_i$ for single bit duration can be written as $s_i = x_i + b_\ell x_{i-M}$ where $\ell$ represents the bit counter.

The CDSK receiver utilizes the low correlation value between chaotic sequence $x_i$ and its shifted version $x_{i-M}$ to differentiate between the transmitted information. Hence, average bit energy can be given also by (2). By replacing $M$ by

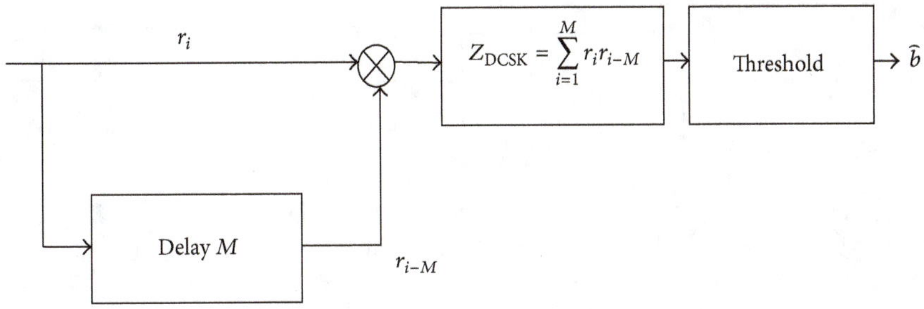


FIGURE 2: DCSK receiver.

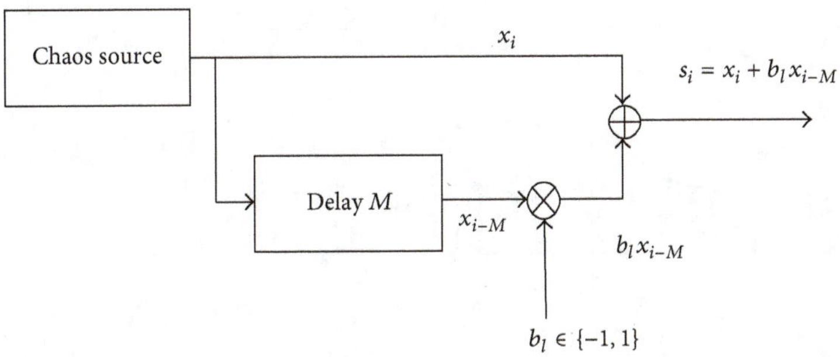


FIGURE 3: CDSK transmitter.

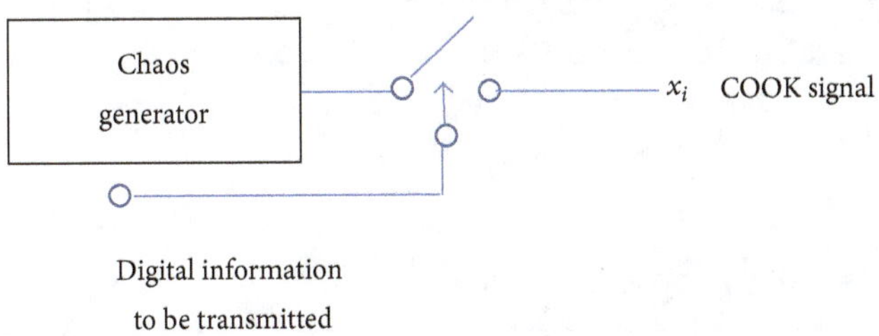


FIGURE 4: COOK transmitter.

$L$, the CDSK receiver becomes similar to the DCSK [5] and the correlator output $Z_{CDSK}$ can be formulated as

$$\begin{aligned} Z_{CDSK} &= \sum_{i=1}^{M} r_i r_{i-M} \\ &= \sum_{i=1}^{M} \left(x_i + b_\ell x_{i-M} + \zeta_i\right)\left(x_{i-M} + b_{\ell-M} x_{i-2M} + \zeta_{i-M}\right) \\ &= b_\ell \sum_{i=1}^{M} x_{i-M}^2 + \sum_{i=1}^{M} x_i x_{i-M} + b_{\ell-M} \sum_{i=1}^{M} x_i x_{i-2M} \\ &\quad + \sum_{i=1}^{M} x_i \zeta_{i-M} + b_\ell b_{\ell-M} \sum_{i=1}^{M} x_{i-M} x_{i-2M} \\ &\quad + b_\ell \sum_{i=1}^{M} x_{i-M} \zeta_{i_M} + \sum_{i=1}^{M} x_{i-M} \zeta_i + b_{\ell-M} \sum_{i=1}^{M} x_{i-2M} \zeta_i \\ &\quad + \sum_{i=1}^{M} \zeta_i \zeta_{i-M}. \end{aligned} \tag{5}$$

Although the systems does not require separate time slot to send the information bearing signal which increases the bit rate, this advantage is defeated by the low BER performance of the system due to multiple intrasignal interference terms in the correlator output.

*2.3. COOK.* In noncoherent COOK scheme, only one basis function is used as shown in Figure 4. Bit "1" is sent by continuous transmission of the chaotic signal $x_i$ within bit duration. Bit "0" is presented by null transmission within same duration. The modulation process is simple as controlling turning off and on the chaos generator. Clearly, the bit energy of transmitted signal takes either a positive value or zero depending on the symbol sent.

The receiver structure of COOK scheme is illustrated in Figure 5. Each incoming signal $r_i$ is multiplied by itself and averaged over the spreading sequence $M$ which represents simple energy detection process. However, setting threshold at the output is the major disadvantage of the COOK scheme; namely, that threshold value of the decision circuit depends on the noise level.

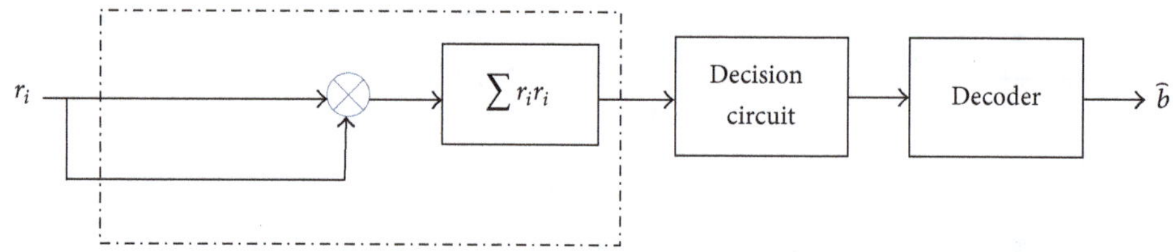


FIGURE 5: COOK receiver.

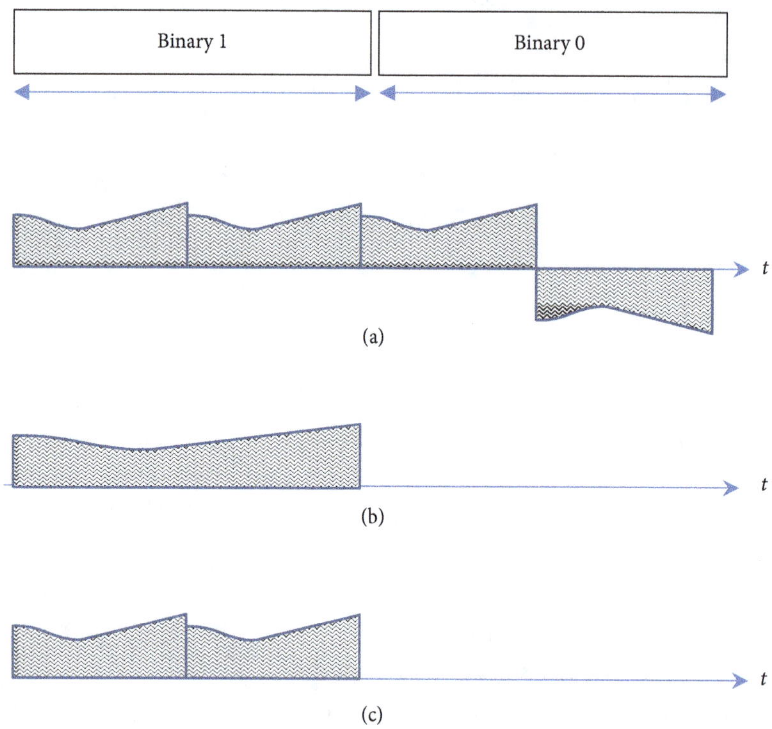


FIGURE 6: Signal format for chaos based systems. (a) Differentially coherent schemes (CDSK and DCSK). (b) COOK scheme. (c) Proposed scheme (DCOOK).

## 3. Proposed System Description

In this section, signal format of different chaos based systems including the proposed system is given before details of the DCOOK scheme are presented.

*3.1. Signal Format.* Different signal formats for different chaos based systems are illustrated in Figure 6. In CDSK and DCSK schemes, each bit is transmitted by sending a reference signal followed by information bearing signal. The information bearing signal is selected according to the transmitted bit and is either identical or inverted version from the reference signal as shown in Figure 6(a). The signal format of COOK scheme is shown in Figure 6(b); the transmitter sends one chaotic segment for bit "1" and no signal to be sent within the same bit duration for bit "0."

In our system, the source generates two identical, successive chaotic segments for bit "1" as shown in Figure 6(c). For bit "0," the source will stop emitting the chaotic signals for the same bit duration. Hence, average bit energy of DCOOK scheme is reduced to half compared to DCSK and CDSK schemes.

*3.2. Transmitter and Receiver Description.* In order to generate the signal in Figure 6(c), a baseband modulator is designed as shown in Figure 7, which appears to be a slightly modified version of DCSK transmitter [5].

For bit "1," chaos source is used to generate sample $x_i$ at each $i$th instant within the first half of bit duration. In the next half of bit duration that is after $M$ samples, the switch is connected to the delay element to send the same copy in the next half of the bit duration. For bit "0" transmissions, the chaotic source will be disabled and no transmission is required. Hence, the transmitted signal $s_i$ at any instant for a single information bit can be formulated as

$$s_i = \begin{cases} bx_i & 0 < i \le M, \\ bx_{i-M} & M < i \le 2M, \end{cases} \quad (6)$$

where $b \in \{1, 0\}$.

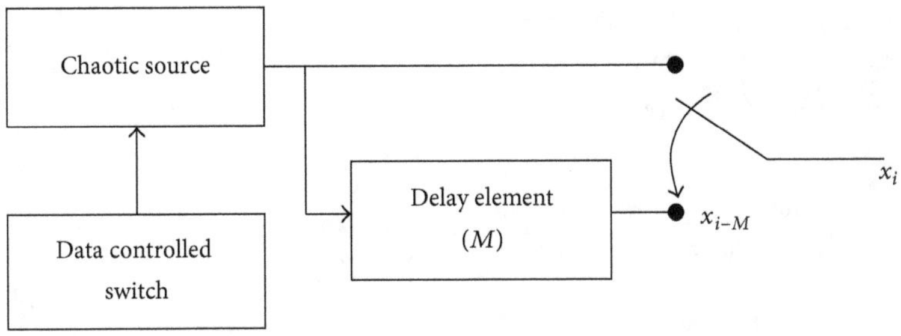


FIGURE 7: DCOOK transmitter.

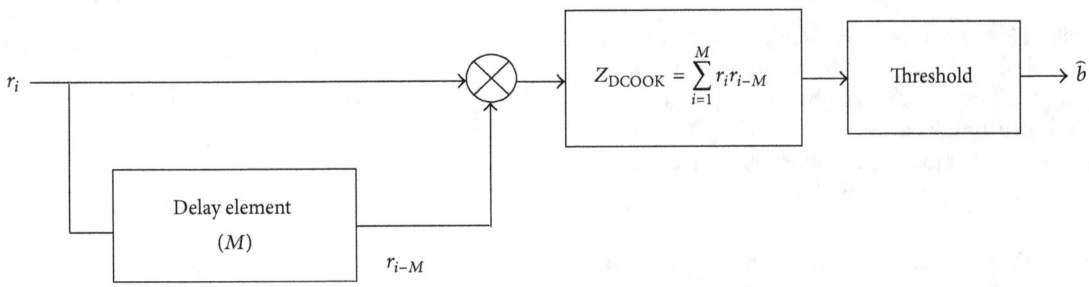


FIGURE 8: DCOOK receiver.

Figure 8 shows the DCOOK receiver; since coherent receiver requires the synchronized replicas of the chaotic carriers, which is actually quite difficult to achieve [1], differential coherent method is used to recover the information. Under the assumption that transmitter and receiver are synchronized in time, the received signal $r_i$ is multiplied by the received signal delayed by $M$, $r_{i-M}$, and correlator output is averaged over the last half of bit duration. Thus, the correlator output $Z_{\text{DCOOK}}$ can be written as

$$Z_{\text{DCOOK}} = \sum_{i=1}^{M} r_i r_{i-M} \tag{7}$$

In this paper, the channel under investigation is AWGN, and then the correlator output can be given as

$$\begin{aligned} Z_{\text{DCOOK}} &= \sum_{i=1}^{M} (s_i + \zeta_i)(s_{i-M} + \zeta_{i-M}) \\ &= \sum_{i=1}^{M} (bx_{i-M} + \zeta_i)(bx_{i-M} + \zeta_{i-M}) \\ &= b\sum_{i=1}^{M} x_{i-M}^2 + b\sum_{i=1}^{M} x_{i-M}\zeta_{i-M} + b\sum_{i=1}^{M} x_{i-M}\zeta_i \\ &\quad + \sum_{i=1}^{M} \zeta_i \zeta_{i-M}. \end{aligned} \tag{8}$$

Analyzing correlator output in (8), it can be easily noticed that there are two different possible outputs according to the value of $b$. The first term in (8) contains the useful signal energy which varies from one bit to another due to chaotic nature. The remaining are the zero-mean disturbance components, which are formed by the noise-signal and noise-noise products. It is important to keep in mind that these components are effective only if $b = 1$; otherwise, the correlator input will be only noise-noise term.

Correlator output is decoded by passing the correlator output $Z_{\text{DCOOK}}$ to a threshold detector which operates according to the following rule:

$$\tilde{b} = \begin{cases} 1 & Z_{\text{DCOOK}} \geq \alpha, \\ 0 & Z_{\text{DCOOK}} < \alpha. \end{cases} \tag{9}$$

The delay element might be challenging for the transceiver integration in our proposed scheme. Fortunately, lots of programmable, picosecond-to-nanosecond-delayed digital delay lines can offer high-data-rate applications. For instance, sampling shift strategies equipped with ultrafast and low-power-consumption sampling techniques and various digital CMOS delay elements can be suggested [12].

## 4. BER Performance Evaluation

The performance of DCOOK scheme is evaluated analytically using Gaussian Approximation (4), which provides an accurate estimation of BER at large value of $M$. For small value of $M$, an integration method is preferred. This is due to the fact that bit energy is not constant for each transmitted bit [16]. The rest of the analyses will be based on the following [4, 5, 12]:

(1) Symmetric tent map, defined by the equation $x_{n+1} = 1 - 2|x_n|$, is applied to generate the discrete chaotic

signal. The chaotic sequence is stationary and is uniformly distributed from $(-1, 1)$ with zero-mean and computed variance $V(x) = 1/3$ and $V(x^2) = 4/45$ [4].

(2) Correlation between each chaotic segment $x_i$ and its time shifted version $x_{i+k}$ is for large value of $k$ such that $E(x_i x_{i+k}) = 0$.

(3) $\zeta_i$ is not only statistically independent from $x_i$ for any $(i, j)$, but also statistically independent from $\zeta_j$ for any $i \neq j$.

(4) At bit "1," the transmitter sends $2M$ consecutive $x_i$ samples, while for bit "0," null transmission is required. Therefore, average bit energy for DCOOK scheme can be given as $E_b = (2MV(x) + 0)/2 = MV(x)$. This is exactly half bit energy for the differentially coherent systems such as DCSK and CDSK.

By applying the previous assumptions and properties, the conditional probability distribution of $\Pr(Z_{\text{DCOOK}} \mid b = 1)$ and $\Pr(Z_{\text{DCOOK}} \mid b = 0)$ can be regarded as Gaussian distributed [5, 17]. However, average value and variance are different. $Z_{\text{DCOOK}}$ histogram is shown in Figure 9.

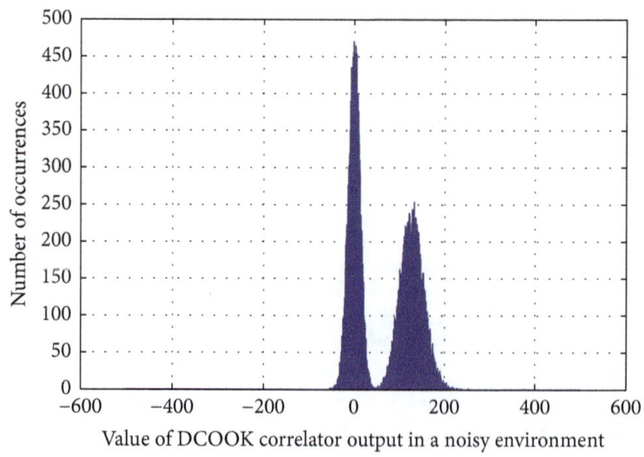


FIGURE 9: Histogram of noisy received signal at $E_b/N_o = 16$ dB and $M = 100$.

Therefore, the performance can be fully characterized by calculating the means and variances as bit "1" and bit "0" transmission, respectively, as follows:

$$\begin{aligned} E(Z_{\text{DCOOK}} \mid b = 1) &= M_s V(x) = E_b \\ V(Z_{\text{DCOOK}} \mid b = 1) &= \sum_{i=1}^{M} V(x_{i-M}^2) \\ &\quad + \sum_{i=1}^{M} V(x_{i-M}) V(\zeta_{i-M}) \\ &\quad + \sum_{i=1}^{M} V(x_{i-M}) V(\zeta_i) \\ &\quad + \sum_{i=1}^{M} V(\zeta_{i-M}) V(\zeta_i) \\ &= \frac{4}{5} MV(x) V(x) \\ &\quad + 2MV(x) V(\zeta) \\ &\quad + MV(\zeta) V(\zeta). \end{aligned} \tag{10}$$

Select $V(\zeta) = N_o/2$ where $N_o$ is the noise power spectral density, and then (10) can be rewritten as

$$V(Z_{\text{DCOOK}} \mid b = 1) = \frac{4}{5M}\frac{E_b^2}{N_o} + E_b N_o + M\frac{N_o^2}{4}. \tag{11}$$

Similarly, we can find

$$\begin{aligned} E(Z_{\text{DCOOK}} \mid b = 0) &= 0 \\ V(Z_{\text{DCOOK}} \mid b = 0) &= M\frac{N_o^2}{4}. \end{aligned} \tag{12}$$

Assuming $\Pr(0) = \Pr(1)$ and for simplicity, we will chose threshold of the decision circuit in the middle between the average values of the correlator output $\alpha = MV(x)/2 = E_b/2$. Then the average BER can be calculated as

$$\begin{aligned} &\text{BER} \\ &= \frac{1}{2}\Pr(Z_{\text{DCOOK}} \le \alpha \mid b = 1) \\ &\quad + \frac{1}{2}\Pr(Z_{\text{DCOOK}} > \alpha \mid b = 0) \\ &\Pr(Z_{\text{DCOOK}} \le \alpha \mid b = 1) \\ &= \frac{1}{\sqrt{2\pi V(Z_{\text{DCOOK}} \mid b = 1)}} \\ &\quad \cdot \int_{-\infty}^{\alpha} e^{-(Z_{\text{DCOOK}} - E_b)^2 / 2V(Z_{\text{DCOOK}} \mid b=1)} \\ &= Q\left(\frac{E_b}{N_o}\left(\sqrt{\frac{1}{(16/5M)(E_b/N_o) + 4 + M(N_o/E_b)}}\right)\right) \\ &\Pr(Z_{\text{DCOOK}} > \alpha \mid b = 0) \\ &= \frac{1}{\sqrt{2\pi V(Z_{\text{DCOOK}} \mid b = 0)}} \int_{\alpha}^{\infty} e^{-(Z_{\text{DCOOK}})^2 / 2V(Z_{\text{DCOOK}} \mid b=0)} \\ &= Q\left(\frac{E_b}{N_o}\sqrt{\frac{1}{M}}\right), \end{aligned} \tag{13}$$

where $Q(\cdot)$ is the Q-Function and is given by

$$Q(y) = \frac{1}{\sqrt{2\pi}} \int_{y}^{\infty} e^{-x^2/2} dx. \tag{14}$$

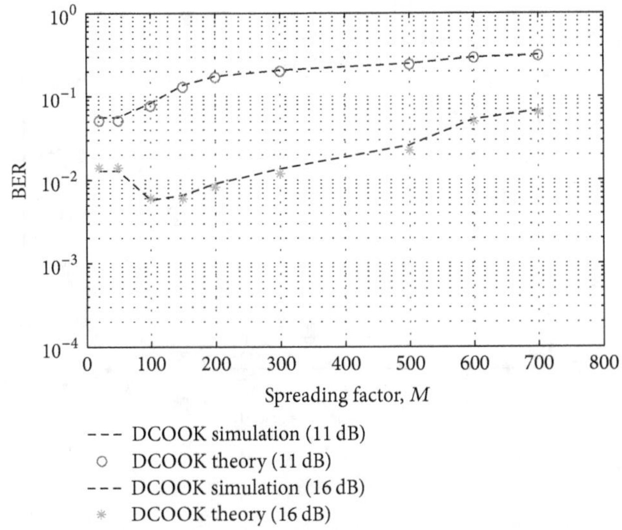


FIGURE 10: BER performance versus spreading factor $M$.

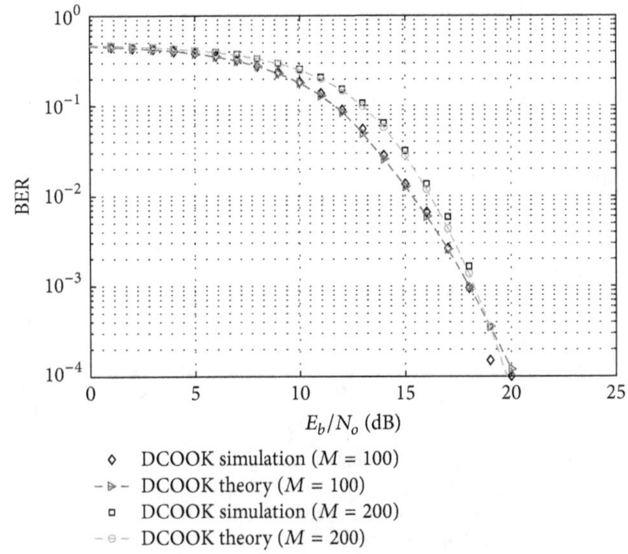


FIGURE 11: Simulated and theoretical BER performance of DCOOK at $M = 100$ and $M = 200$.

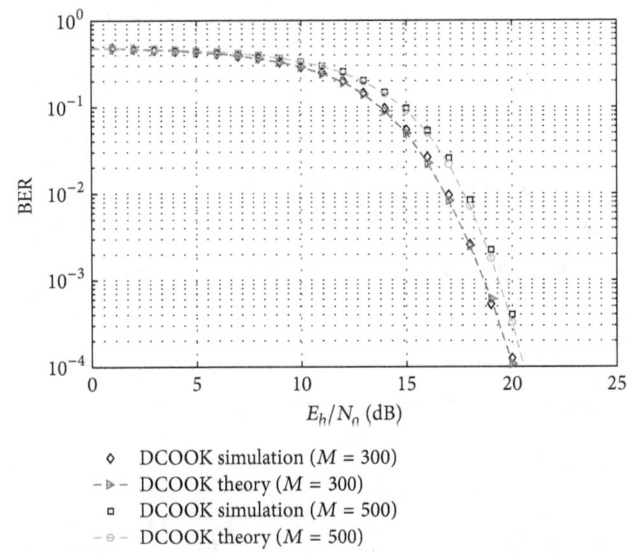


FIGURE 12: Simulated and theoretical BER performance of DCOOK at $M = 300$ and $M = 500$.

The final expression for BER can be formulated by

$$\text{BER} = \frac{1}{2}\left[Q\left(\frac{E_b}{N_o}\left(\sqrt{\frac{1}{16E_b/5MN_o + 4 + MN_o/E_b}}\right)\right) + Q\left(\frac{E_b}{N_o}\sqrt{\frac{1}{M}}\right)\right]. \tag{15}$$

## 5. Simulation Results and Discussion

In this section, the theoretical expression to estimate the BER performance of DCOOK scheme is validated by simulation and with different scenarios. Additionally, the BER performance of DCOOK scheme, after threshold optimization, is compared with different chaos based system in AWGN channel environment.

According to (15), the BER of DCOOK scheme is calculated and validated by simulation. Keep the signal level at $E_b/N_o = 11$ dB and $E_b/N_o = 16$ dB and vary the spreading factor $M$. An excellent matching between the prediction and simulation result as shown in Figure 10 can be noticed. It can also be observed that the proposed scheme has the optimal performance as both curves reach their bottom points at fixed $E_b/N_o$ with $M$ varying from 10 to 100. This behavior is caused by the fluctuations in $E_b$ at small $M$ and the noise-noise cross correlation at large $M$. On the other hand, the derived equation is used to predict the BER at various $E_b/N_o$ keeping $M$ constant at $M = 100$, 200, 300, 500, 600, and 700. An agreement between the predicted value of BER and simulation result can be easily noticed in Figures 11, 12, and 13.

The BER performance of the proposed scheme is tested against the standard DCSK, CDSK, and COOK schemes in AWGN channel environment. Since the correlator output has different energy profiles for 0 and 1 as described earlier, then threshold optimization is required to enhance the performance of DCOOK scheme. One way to choose the optimum threshold is by testing different values of threshold within the range $(0, E_b/2)$.

At the optimized threshold, DCOOK scheme shows superior BER performance in comparison with the standard COOK scheme as shown in Figures 14, 15, and 16 at $M = 500$, 600, and 700, respectively. The reason behind this performance is the threshold optimization capability of DCOOK scheme over the COOK scheme. Obviously, in Figure 14, CDSK scheme seems always to be 1.5 dB worse than DCOOK scheme at BER of $1 \times 10^{-4}$ and this continue for all $M$, the reason of which is that there are much more intrasignal interference terms in the CDSK correlator output.

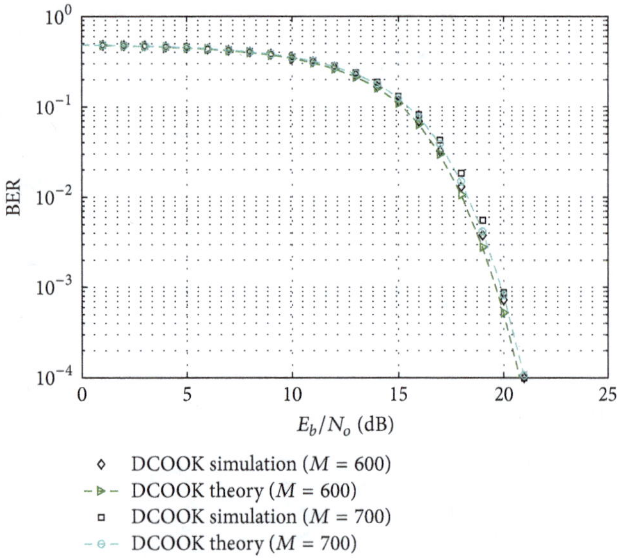


FIGURE 13: Simulated and theoretical BER performance of DCOOK at $M = 600$ and $M = 700$.

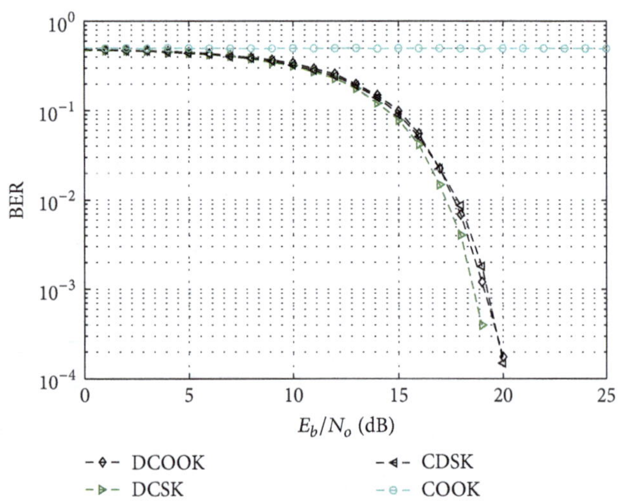


FIGURE 14: BER performance of DCOOK scheme versus COOK, DCSK, and CDSK schemes at $M = 500$.

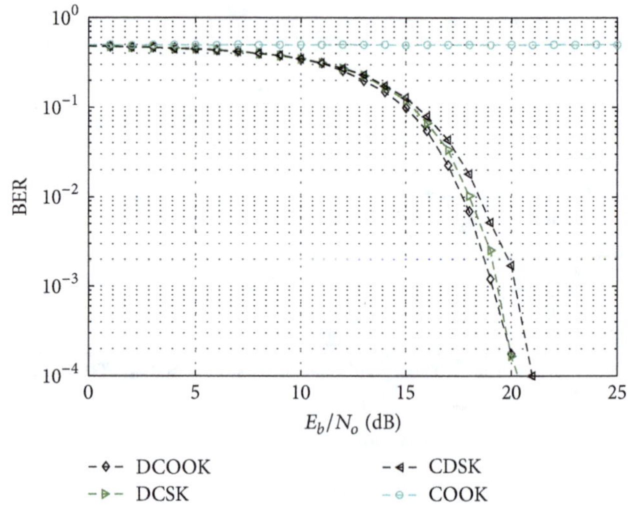


FIGURE 15: BER performance of DCOOK scheme versus COOK, DCSK, and CDSK scheme at $M = 600$.

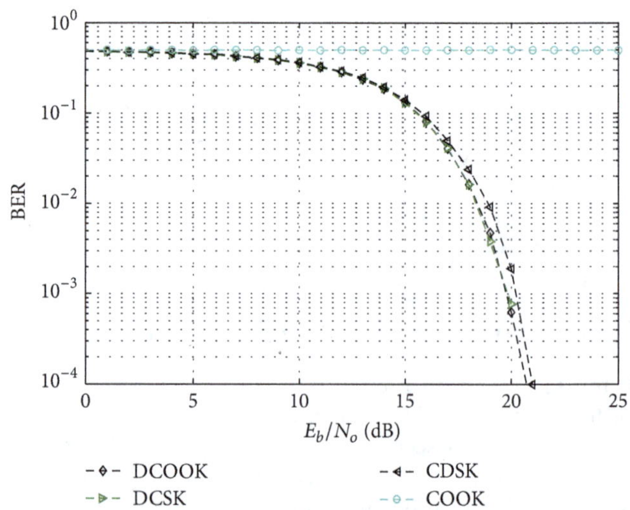


FIGURE 16: BER performance of DCOOK scheme versus COOK, DCSK, and CDSK schemes at $M = 700$.

It is interesting to note that DCOOK scheme outperforms the BER performance of DCSK scheme by 1 dB at $M = 600$, while keeping the same performance at $M = 500$ and $M = 700$ using only half of bit energy used by DCSK scheme. As $M$ increases, BERs for all schemes become worse, which can be attributed to the increasing negative contribution from noise-noise cross terms in their corresponding correlator outputs.

## 6. Conclusion

In this paper, we present a new differential coherent system, DCOOK, which saves bit energy by half and improves BER. Each information bit is presented by sending either two identical chaos signals or no transmission. The performance of the proposed scheme is evaluated, and the bit error rate expressions are derived at Gaussian Channel condition. The theoretical BER expression is verified by simulation over a considerable range of spreading factor. Excellent agreement between the derived expression and simulation result is clearly observed. The BER performance of the proposed system is compared to DCSK, CDSK, and COOK schemes.

Our analysis revealed that, at half bit energy, the proposed scheme outperforms DCSK and CDSK at average of 1.25 db while DCOOK shows superior performance compared to COOK at wide range of spreading factor. Additionally, DCOOK offers double data rate compared to DCSK. Our future work will focus on evaluation of the proposed scheme over the Rayleigh fading channel.

## References


[1] G. Kolumban, M. P. Kennedy, and L. O. Chua, "The role of synchronization in digital communications using chaos. II. chaotic modulation and chaotic synchronization," *IEEE Transactions on Circuits and Systems I: Fundamental Theory and Applications*, vol. 45, no. 11, pp. 1129–1140, November 1998.

[2] M. Long, Y. Chen, and F. Peng, "Bit error rate improvement for chaos shift keying chaotic communication systems," *IET Communications*, vol. 6, no. 16, pp. 2639–2644, 2012.

[3] G. Kis, Z. Jako, M. Kennedy, and G. Kolumban, "Chaotic communications without synchronization," in *Proceedings of the International Conference on Telecommunications, IEE*, pp. 49–53, Edinburgh, UK, 1998.

[4] F. C. Lau and C. K. Tse, "Performance Analysis Methods for Non-Coherent Differential Chaos-Shift-Keying Systems," in *Chaos-Based Digital Communication Systems*, Signals and Communication Technology, pp. 63–96, Springer Berlin Heidelberg, Berlin, Heidelberg, 2003.

[5] M. Sushchik, L. S. Tsimring, and A. R. Volkovskii, "Performance analysis of correlation-based communication schemes utilizing chaos," *IEEE Transactions on Circuits and Systems I: Fundamental Theory and Applications*, vol. 47, no. 12, pp. 1684–1691, 2000.

[6] N. Al Bassam and O. Jerew, "Design and implementation of a chaotic scheme in additive white Gaussian noise channel," *Journal of Computer Networks and Communications*, vol. 2016, Article ID 5976282, 2016.

[7] F. J. Escribano, G. Kaddoum, A. Wagemakers, and P. Giard, "Design of a new differential chaos-shift-keying system for continuous mobility," *IEEE Transactions on Communications*, vol. 64, no. 5, pp. 2066–2078, 2016.

[8] H. Yang, G.-P. Jiang, and J. Duan, "Phase-separated DCSK: a simple delay-component-free solution for chaotic communications," *IEEE Transactions on Circuits and Systems II: Express Briefs*, vol. 61, no. 12, pp. 967–971, 2014.

[9] G. Kaddoum and F. Gagnon, "Design of a high-data-rate differential chaos-shift keying system," *IEEE Transactions on Circuits and Systems II: Express Briefs*, vol. 59, no. 7, pp. 448–452, 2012.

[10] W. Xu, L. Wang, and G. Chen, "Performance analysis of the CS-DCSK/BPSK communication system," *IEEE Transactions on Circuits and Systems I*, vol. 61, no. 9, pp. 2624–2633, 2014.

[11] M. Xu and H. Leung, "A novel high data rate modulation scheme based on chaotic signal separation," *IEEE Transactions on Communications*, vol. 58, no. 10, pp. 2855–2860, 2010.

[12] H. Yang and G.-P. Jiang, "High-efficiency differential-chaos-shift-keying scheme for chaos-based noncoherent communication," *IEEE Transactions on Circuits and Systems II: Express Briefs*, vol. 59, no. 5, pp. 312–316, 2012.

[13] H. Yang and G.-P. Jiang, "Reference-modulated DCSK: a novel chaotic communication scheme," *IEEE Transactions on Circuits and Systems II: Express Briefs*, vol. 60, no. 4, pp. 232–236, 2013.

[14] G. Kaddoum and E. Soujeri, "NR-DCSK: a noise reduction differential chaos shift keying system," *IEEE Transactions on Circuits and Systems II: Express Briefs*, vol. 63, no. 7, pp. 648–652, 2016.

[15] G. Kaddoum, F.-D. Richardson, and F. Gagnon, "Design and analysis of a multi-carrier differential chaos shift keying communication system," *IEEE Transactions on Communications*, vol. 61, no. 8, pp. 3281–3291, 2013.

[16] G. Kaddoum, P. Chargé, D. Roviras, and D. Fournier-Prunaret, "A methodology for bit error rate prediction in chaos-based communication systems," *Circuits, Systems, and Signal Processing*, vol. 28, no. 6, pp. 925–944, 2009.

[17] N. I. Chernov, "Limit theorems and Markov approximations for chaotic dynamical systems," *Probability Theory and Related Fields*, vol. 101, no. 3, pp. 321–362, 1995.

16

# Electrical Power Analysis Using the Scattering Bond Graph

**Roberto Tapia-Sánchez**

*División de Estudios de Posgrado, Facultad de Ingeniería Eléctrica, Universidad Michoacana de San Nicolás de Hidalgo, 58000 Morelia, Mich, Mexico*

Correspondence should be addressed to Roberto Tapia-Sánchez; rtsanchez@dep.fie.umich.mx

Academic Editor: Lucian Dascalescu

The perspective of the well known concept of active and reactive power in electrical systems is addressed in this paper. This concept is presented from the point of view of the bond graph methodology. The scattering matrix concept is used in order to deal with these formulations. The concept of active and reactive power is presented with some examples derived from single-phase AC circuits. The formulations and simulations of these circuits are presented.

## 1. Introduction

Power concept is of central importance in electrical power systems. Reactive and active power concepts have been used since the beginning of the alternate current electrical generation. These power concepts can be encountered in the electric circuit theory, and they have been used naturally in the electrical power systems.

*Reactive power* (Q) is defined to represent the quantity of electric power due to the load current that is not in phase with the source voltage. The average of this reactive power during one period of line frequency is zero. It means that this power does not contribute to energy transfer from the source to the load. Apparent power ($A$) gives the idea of how much power can be delivered or consumed if the voltage and current are sinusoidal and perfect in phase. The power factor (PF) gives a relation between the average power actually delivered or consumed in a circuit and the apparent power at the same point. The higher the power factor, the better the circuit utilization. Consequently, the power factor is more efficient electrically as well as economically.

In this paper, the flow of power in an alternate current (AC) circuit is investigated. By using the concept of scattering and bond graph, the *instantaneous power* is resolved into two components. A plot of these components is obtained using the 20Sim software to observe that single-phase AC networks not only consume energy at an average rate but also borrow and return energy to its source. This leads to the basic definitions of average power and reactive power. The volt-ampere, which is a mathematical formulation based on the phasor forms of voltage and current, is introduced using the concepts of incident and reflected waves.

It is known that, through the bond graph methodology, any physical system can be modeled (i.e., mechanical, electrical, and thermodynamic) in the same reference frame. This methodology presents some properties that can be directly applied to the model [1]. A bond graph consists of subsystems linked together by half arrows, representing power bonds [2]. These bonds exchange instantaneous power at places called ports. The variables are forced to be identical when two ports are connected; the power variables are assumed to be functions of time. The different power variables are classified in a universal scheme and are called either effort $e(t)$ or flow $f(t)$. Their product $P(t) = e(t) \cdot f(t)$ is the instantaneous power flowing between the ports. As the instantaneous power is only presented in a bond, can it be possible to get the reactive power and apparent power, which are presented in an electrical power system? This is the question that this paper seeks to answer.

In order to answer this question, the relation between scattering matrix and bond graph is explored. Some publications have dealt with this formulation [3–6]. In the first publication [3], the concept of scattering matrix has been introduced by using the forewave and backwave concepts.

The authors emphasize that wave scattering is an alternative approach to model physical systems. The property of the scattering operators and matrices, which deals with causal relations and conservation laws, has been indicated. In [4], the scattering matrix concept and the bond graph have been combined, and the constitutive relations of every single element have been addressed from the point of view of the scattering formalism. A bond graph transformation is proposed in order to formulate a reduced bond graph. This reduced bond graph manages two reduced power variables instead of the traditional effort and flow. From the reduced bond graph (transformed) both the transfer function and the scattering relation can be extracted. In [5], the name "bond graph scattering" appears as a result of the proposal, and the authors use the quadrupole concept in order to get the transfer function between two ports. Finally in [6], a more general view of the bond graph scattering is presented by adding the energy notion and their use in the study of linear and nonlinear, multienergy representation of physical systems.

In this paper, the bond graph scattering concept will be used in order to split up the instantaneous power into the two types of power ($P$ and $Q$) commonly used when an electrical system is analyzed. Then, the power balance is demonstrated by means of examples focusing on the single-phase systems. Next, the transmission power between two voltage sources is considered, and the dependency of real power on voltage phase angle and of reactive power on voltage magnitude is established.

The outline of the paper is as follows: in the first part, the traditional electrical power is reviewed. The scattering bond graph is presented in Section 3. Section 4 describes some examples focusing on the single- phase circuits. Section 5 deals with a power flow study case. Finally, conclusions of the conducted investigation are drawn in Section 6.

## 2. Review of Electrical Power

In the AC circuits, energy storage elements (inductors and capacitances) may result in periodic reversals of the direction of energy flow. The averaged portion of power flow over a complete cycle resulting in a net transfer of energy in one direction is known as real or *active power* ($P$). The portion of power flow due to stored energy, which returns to the source in each cycle, is known as reactive power ($Q$).

The power theory based on active, reactive, and apparent power definition has been sufficient for the analysis and design of power systems. Nevertheless, some papers have been published showing that the conventional concept of reactive and apparent power loses its usefulness in nonsinusoidal cases [7]. Two approaches to power definitions under nonsinusoidal conditions were introduced in [8–10]. In [8, 9], the power was defined in the frequency domain, and in [10], it was defined in the time domain.

The introduction of power electronics technology introduced new conditions of power theory. This is because electronic power converters may bring out reactive power as well as harmonic current from power networks. Thus, the conventional power theory based on average and RMS (root mean square) values [9] of voltages and sources cannot be applicable to the analysis and design of power converters and power networks.

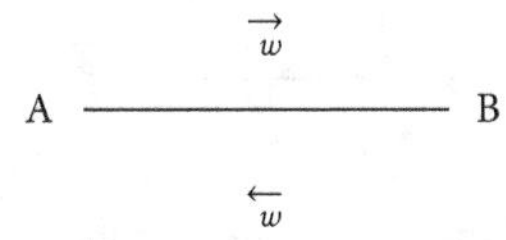


FIGURE 1: Bond between two elements.

The researches [11–13] presented the basic concepts to control the reactive power in order to compensate it.

In [14], the *p-q* theory defines a set of instantaneous power in the time domain. This theory is focusing on a three-phase circuit, and it always considers three-phase systems together, not as a superposition or sum of three single-phase circuits.

## 3. Scattering Bond Graph (StBG)

This section presents a recapitulation of the scattering bond graph and also gives a more general standpoint of its use.

As mentioned in [3], system description requires knowing its elements and interconnections. In order to model any physical system, it is necessary to define the system boundary. In case of a boundary breaking up the system into two subsystems, the interaction between subsystems can be described by a duplex pair of oppositely directed signals. Whatever bond exists between two parts of a system may be equivalently represented as a pair of directed signals.

The forewave ($\vec{w}$) and backwave ($\overleftarrow{w}$) have been addressed in [3]. Two systems A and B are joined by a single line, which represents the interaction of the two waves. There is a power exchange between the two systems; part of the complete power could be moving from A to B and part from B to A. Figure 1 shows this interaction between the two systems.

If $P$ is defined as the net power flowing from A to B, then $\vec{P}$ is defined as the fore-power and $\overleftarrow{P}$ as the back-power. Each of these bilateral power flows represents the integration at the port of a scattering flux, which is defined as the product of local scattering density and local velocity of energy propagation. As the forewave and backwave have units of root power, they can be represented using a single power bond.

The net power flow is considered positive in the fore-direction. Then, the net power can be expressed as a function of the waves as follows:

$$\begin{aligned} P &= \vec{P} - \overleftarrow{P} = \frac{\vec{w}^2}{2} - \frac{\overleftarrow{w}^2}{2} \\ &= \frac{1}{\sqrt{2}} (\vec{w} + \overleftarrow{w}) \cdot \frac{1}{\sqrt{2}} (\vec{w} - \overleftarrow{w}) = e \cdot f. \end{aligned} \quad (1)$$

Equation (1) shows the relationship between the scattering variables and the normalized effort and flow variables.

In this paper is proposed the graphical representation of the scattering bond as shown in Figure 2.

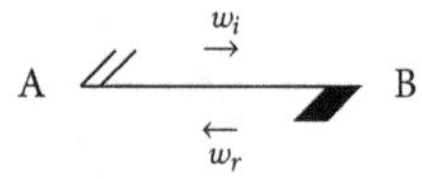


FIGURE 2: Scattering bond.

TABLE 1: Scattering source equations.

| | Effort source | Flow source |
|---|---|---|
| Se or Sf, $\overrightarrow{w_i}$, $\overleftarrow{w_r}$ | $w_i = \sqrt{2}\cdot V - w_r$ | $w_i = \sqrt{2}\cdot I + w_r$ |
| $\overrightarrow{w_i}$, $\overleftarrow{w_r}$, Se or Sf | $w_r = \sqrt{2}\cdot V - w_i$ | $w_r = -\sqrt{2}\cdot I + w_i$ |

The forewave and backwave are called incident ($w_i$) and reflected ($w_r$) waves, respectively (as it was made in [4]). The bold half-arrow indicates the sense of the incident wave, while the double half-arrow the sense of the reflected wave. The junction, ports, and elements have the same representation as in a traditional bond graph; only their mathematical relations change. Figure 3 shows the basic elements used in order to formulate a StBG.

The source elements (Se and Sf) give their variable ($V$ for voltage and $I$ for the current) via the incident wave. The original values are normalized by the operator $\sqrt{2}$ as is common in the scattering representation [15, 16]. These sources can be visualized from an energetic conservation point of view [6], by using the relationship given in (1). Then, the sources of effort and flow can be interpreted as four scattering sources. Table 1 summarizes these sources.

The relations shown in Table 1 make references to an electrical circuit because voltage ($V$) and current ($I$) variables have been considered as the effort and flow, respectively. It is important to note that the combination of these four sources arises because the incident wave defines the sign convention.

The scattering variables ($w_i$, $w_r$) are defined at each port of the system and always adopt the positive power convection in a system.

As shown in Figure 3, the constitutive relation for $C$, $I$, and $R$ elements is normalized by a resistance $R_0$. The normalization by $R_0$ corresponds to an internal resistance or impedance considered in a system (i.e., source, resistance, and capacitance). This resistance makes it possible to quantify the power transferred between two systems.

By considering a linear capacitance $C$, the relation between $w_i$ and $w_r$ is given by $w_i/w_r = Z/R_0$, where $R_0$ is the normalized resistance and $Z$ is the impedance of the element: $Z = 1/Cs$ ($s$ being the Laplace operator). From this impedance and the scattering relation $[w_r] = S[w_i]$ given between two ports, it is simple to find the relation shown in Figure 3 by replacing the $s$ operator by $\rho$. The constitutive relation of this element contains a zero in the right half-plane and is nonminimum phase element, although, clearly, it is normally considered to be physically real. The compliance element is obtained in the same manner.

The scattering relation for a one-port resistor has the extreme values of −1 and +1 when $R > R_0$, $S_R < 0$ and when $R < R_0$, $S_R > 0$, respectively. Then, when $R = R_0$, $S_R$ is zero in both cases and the resistor is said to be perfectly matched [3, 17–19].

This paper is focusing on single-phase AC circuits; nevertheless, the StBG can be used in all physical systems including those that do not have alternate input sources.

## 4. Single-Phase AC Circuits

Electrical power systems commonly use an impedance load connected to a source so that the analysis can be simplified by using the phasor concept. This concept focuses on complex values having a magnitude and a phase in order to calculate the power using those vector quantities.

In the bond graph methodology, the impedance element has already been formulated [19]. In this publication, another element (besides the nine basic elements) is proposed as the impedance element, namely, $Z$-element. With the use of this $Z$-element, the notion of the interaction between different elements is lost because two or more basic elements can be grouped into one.

Next is proposed the use of the scattering bond graph in order to keep close to the structure of the analyzed circuit. Later, two basic examples of a single-phase electrical system are presented.

*4.1. Circuit VR.* The basic AC electrical circuit is composed only of a resistive load ($R$) connected in parallel to the alternative voltage source ($V$). Figure 4 shows the StBG of the $VR$ circuit.

As shown in Figure 4, the two variables in each scattering bond are the incident and the reflected waves ($w_i$, $w_r$). The mathematical formulation is made by considering the relations given earlier and is as follows.

The source MSe and the $R$ element have the mathematical relations given by (2) and (3), respectively.

Consider

$$w_{i1} = \sqrt{2}v - w_{r1}, \tag{2}$$

$$w_{r2} = Sw_{i2} = \frac{r-1}{r+1}w_{i2}, \tag{3}$$

where $r = R/R_0$. As the two components are joined by a *1-junction*, the scattering relation corresponds to

$$S = \begin{bmatrix} 0 & -1 \\ -1 & 0 \end{bmatrix}. \tag{4}$$

Then, considering these relations, the incident and reflected waves in each scattering bond are given by

$$\begin{aligned} w_{r2} &= -\left(\frac{r-1}{2r}\right)\sqrt{2}v, \\ w_{i2} &= -\left(\frac{r+1}{2r}\right)\sqrt{2}v, \end{aligned} \tag{5}$$

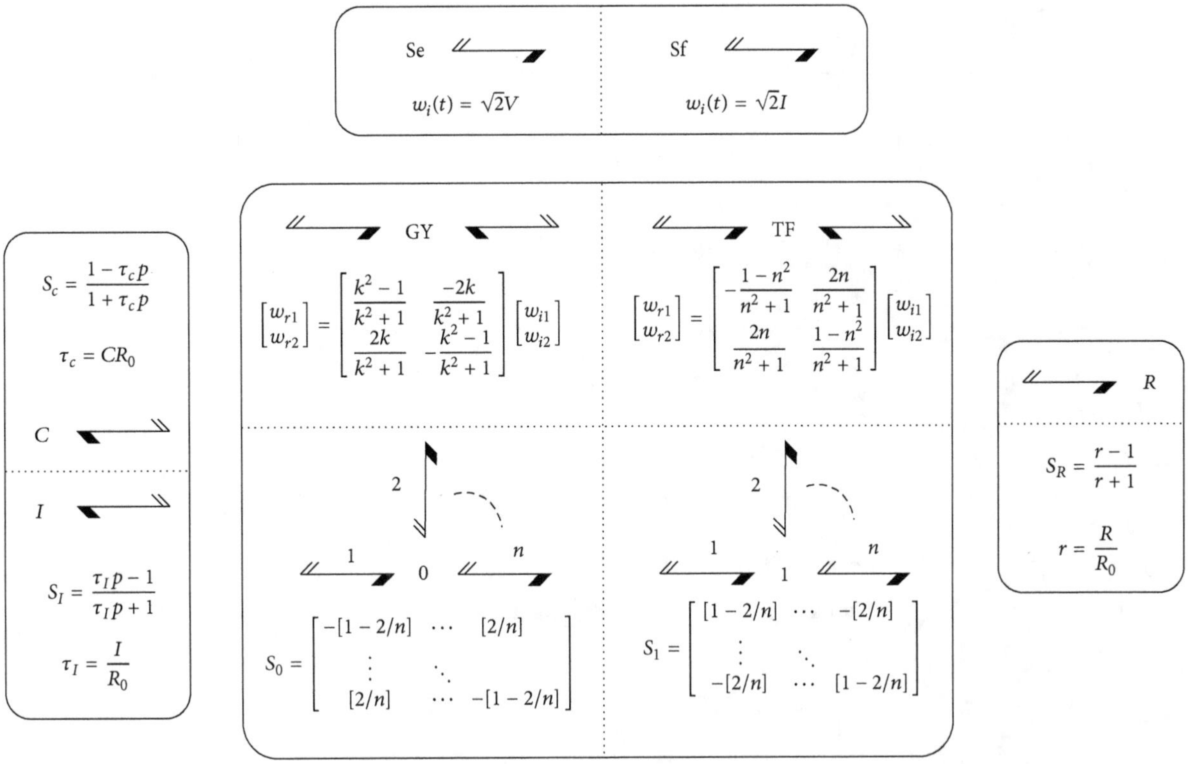

FIGURE 3: Scattering bond graph relations.

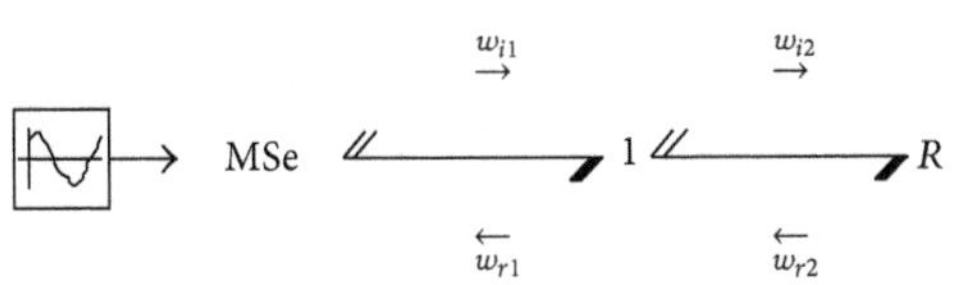

FIGURE 4: StBG of the VR circuit.

$$w_{r1} = \left(\frac{r+1}{2r}\right)\sqrt{2}v,$$
$$w_{i1} = \left(\frac{r-1}{2r}\right)\sqrt{2}v. \quad (6)$$

Equations (5) and (6) represent the scattering waves of the StBG of Figure 4; they do not match the traditional effort and flow present in a bond graph. In any case, the original variables of effort and flow can be recovered [6] from the new variables using (1).

The objective to use the scattering variables is to obtain the active and reactive power directly by regarding the incident and reflected waves. Then, the power delivered by each element is given by

$$P_t = P_i - P_r = \frac{(w_i)^2}{2} - \frac{(w_r)^2}{2}, \quad (7)$$

where $P_t$ is the total power and $P_i$ and $P_r$ are the incident and reflected power, respectively. By substituting (5) in (7), the total power dissipated by the $R$ element is calculated as follows:

$$P_t = \frac{\left(-\left((r+1)/2r\right)\sqrt{2}V\right)^2}{2} - \frac{\left(-\left((r-1)/2r\right)\sqrt{2}V\right)^2}{2} = \frac{V^2}{r}. \quad (8)$$

Equation (8) agrees with the usual power expressed in a resistive element.

In order to simulate the StBG of Figure 4, a library for the 20Sim software has been developed. The parameters used for the simulation are source $v = 100 * \sin(2\pi 50)$ and $R = 1.25\,\Omega$. Figure 5 shows the incident and reflected waves in each scattering bond.

The responses $w_{i1}$ and $w_{r1}$ correspond to the source MSe and $w_{i2}$ and $w_{r2}$ to the resistive $R$-element. It is shown that responses are the same for the two elements from the point of view of magnitude peak of the sinusoidal curves.

The source incident wave has a maximum value of 127.27, while for the source reflected wave it is 14.14. The same is the case for the scattering variables of the $R$-element. The values obtained by a traditional analysis do not match these values; hence, the scattering variables are used.

The responses in the conventional variables ($e$ = voltage and $f$ = current) for the MSe source are presented in Figure 6.

The effort and flow responses match exactly the expected values for the voltage ($V_{max} = 100\,V$) and current ($I_{max} = 80\,A$) if a traditional bond graph is used.

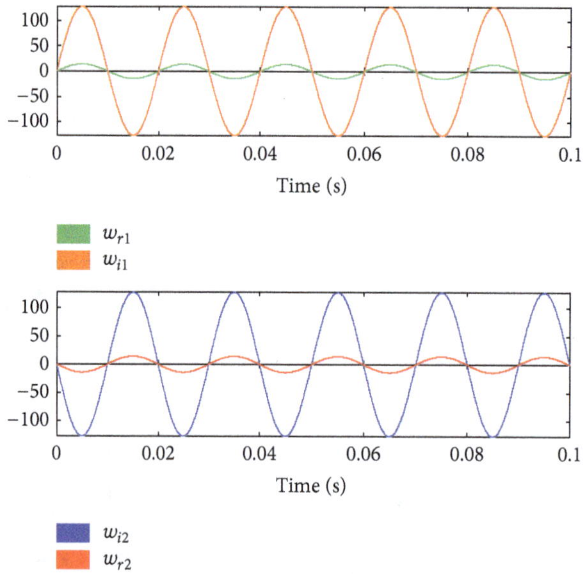


FIGURE 5: Reflected and incident waves for the *VR* circuit.

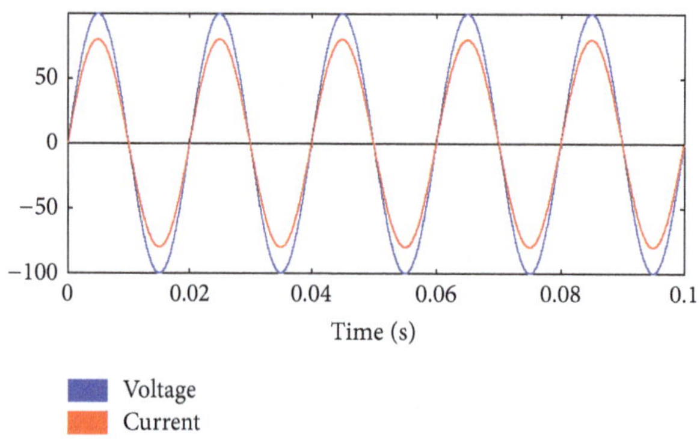


FIGURE 6: Effort and flow variables responses.

As mentioned earlier, the objective to use the scattering variables is to deal with the active and reactive power directly in the StBG. Figure 7 shows the power responses for the *R*-element.

The instantaneous power is presented in Figure 7. As expected, the frequency of the instantaneous power is twice the frequency of the input source. The reflected power $P_{rR}$ is smallest compared with the incident power $P_{iR}$. Actually, the total power is slightly lower than the incident power because the reflected power is subtracted.

Generally, the average value of the instantaneous power is considered (Figure 8) as the power delivered or absorbed by each element.

In conclusion, it is observed that the incident power is slightly higher than the total power and the reflected power has a very small value.

From this result, it can be concluded that the incident power has a correspondence with the active power in an electrical circuit, the reflected power has a correspondence with the reactive power, and the apparent power can be related to the total power.

It is important to note that the *VR* circuit does not have a dynamic element; thus, only active power is present in the circuit. Nevertheless, in the example presented earlier, incident (active) power, reflected (reactive) power, and total (apparent) power are present. This discrepancy can be interpreted on the basis of bond graph methodology as follows: to any action (effort) there is a reaction (flow); this is the principle of causality. Extrapolating this concept to the scattering variables, for any inflow wave, there is an outflow wave.

Despite the presence of reflected power in the StBG, its contribution is minimum and is compensated with the slightly higher value of the incident power. Actually, the difference is the same: reflected power is equal to 50 and the slight value of incident power is equal to 50. Their sum gives the total power (apparent) in the element.

*4.2. Circuit VRL.* Circuit *VRL* is composed of a voltage source, a resistance, and an inductor, all connected in series. Figure 9 shows the StBG of this circuit.

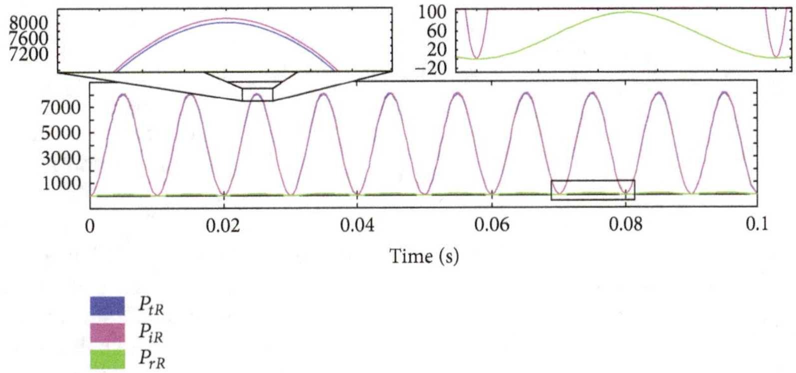


FIGURE 7: Power waves responses for the $R$-element.

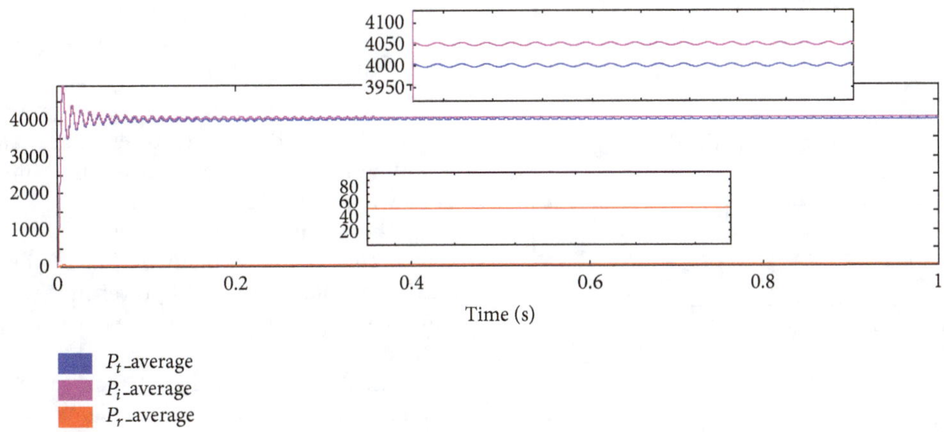


FIGURE 8: Average power for the $VR$ circuit.

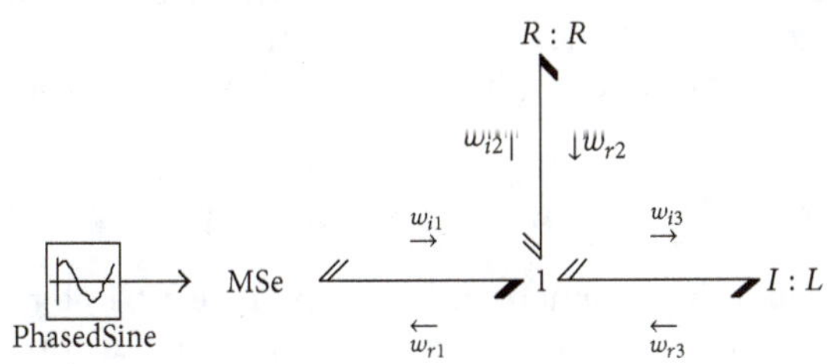


FIGURE 9: StBG of a $VRL$ electrical circuit.

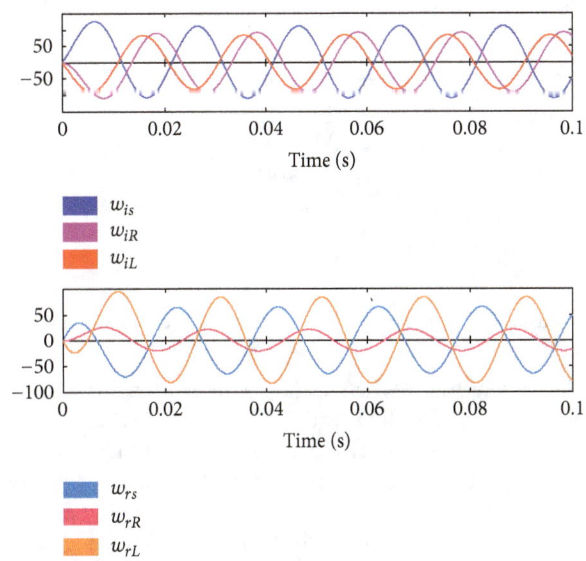


FIGURE 10: Scattering wave's responses of the $VRL$ circuit.

There is a dynamical element in this StBG, so a voltage and current phase shift will be present.

As in the case of circuit $VR$, the StBG of Figure 9 is simulated. Figure 10 shows the scattering waves' responses. The numerical values for the simulation are $R = 0.625\,\Omega$, $L = 0.0034457$ H, and $V = 100 * \sin(2\pi 50)$.

Incident and reflected waves in each element have different amplitudes and their own phases. It is important to note that $R$ and $L$ incident waves ($w_{iR}$, $w_{iL}$) have a phase shift around 60°. This phase shift is actually the value that can be obtained mathematically and corresponds to the angle of the impedance load ($z = 1.25, 60°$). The same phase shift is visualized between the $w_{rR}$ and $w_{rL}$.

The efforts and flows can be obtained from the final results. Figure 11 shows the power presented in each wave.

As in the case of $VR$ circuit, the instantaneous curves of power are obtained. It is observed that the source presents

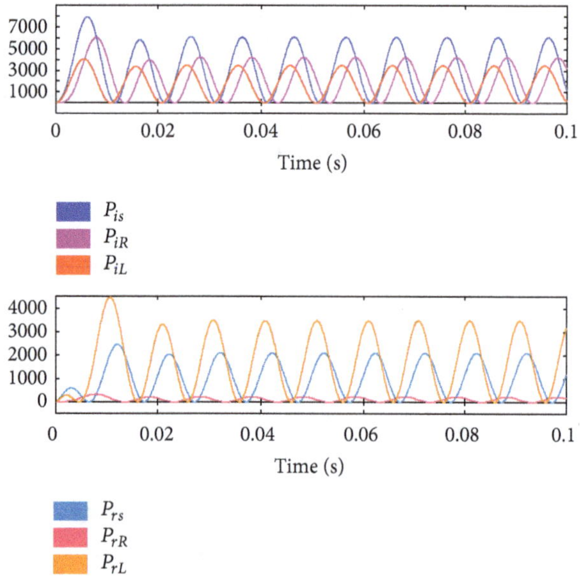


FIGURE 11: Power waves responses in the *VRL* circuit.

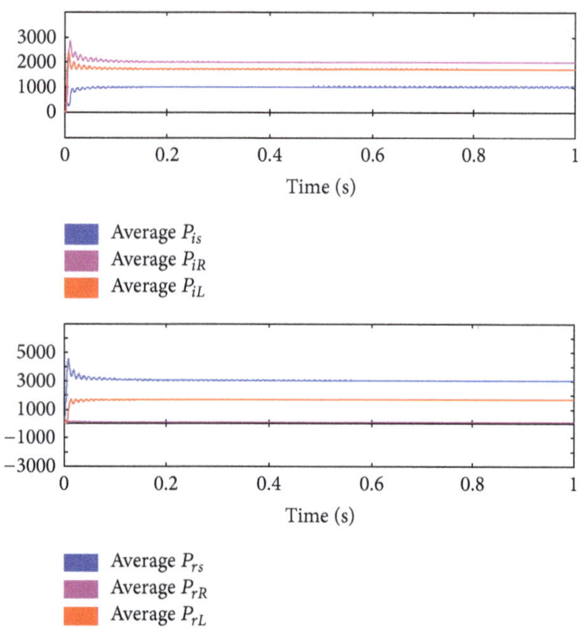


FIGURE 12: Average power responses for the *VRL* circuit.

the larger amplitude when dealing with incident power ($P_{is}$). However, when dealing with reflected power, the larger amplitude is presented by the inductor element ($P_{rL}$). This is normal for this circuit because the reactive power is delivered by the inductor.

If the total power in each element is calculated, from Figure 12 it is found that $P_{tR} = 1890$, $P_{tL} = 6$, and $P_{ts} = -2009$, which correspond to the total power in the $R$ element, the $L$ element, and the source, respectively. Some interesting observations can be made with these results.

(i) The fact that the total power in the $L$ element is close to zero is because this element does not contribute any usable power. Actually, this element contributes only reflected power. In the language of electrical systems, it means that the inductor contributes only with reactive power.

(ii) The total power in the $R$ element and the total power in the source have a contrary sign, and they are almost the same.

(iii) There must be a balance in the power delivered or absorbed for each element. If all the total power is added up, it can be shown that it is not balanced. Nevertheless, there is a balance in power, but, for showing this balance, all the power needs to be taken into consideration (incident power and reflected power). In this case, $P_{tR} + P_{tL} + P_{ts} = -113$, which is equal to the value of $P_{rR}$ (with a positive sign). With this result, is verified the same behavior as presented in the *VR* circuit (Figure 4).

The active and reactive power values can be obtained by regarding the total power waves in the sinusoidal reference, as shown in Figure 13.

In the $R$ element, the total power has a maximum value of 4000. Active power is obtained by taking its mean value. The maximum value of the total power in the $L$ element is 3450, which corresponds directly to the reactive power value in the circuit.

Thus, it can be concluded that, for a given StBG, the mean total power waves in the $R$ elements correspond to the electrical active power, and the total power waves in the $L$ or $C$ elements correspond to the electrical reactive power.

Other circuit configurations, that is, capacitance and resistance in parallel with a source of current (*IRC*), *VRLC*, or a mix of different basic circuits using a single-phase electrical system, can be done simply by using the basic StBG elements.

## 5. Power Flow in a Single-Phase System

Power flow is an important concept in electrical power systems. This concept is used in transmission, distribution and also when the power electronic converters are connected into the electrical network.

Here, the basic concept of power flow in a single-phase electrical system is considered.

Consider two ideal voltage sources connected by a transmission line as shown in Figure 14.

It is known that, in this circuit, small changes in phase shift between the two sources will have a significant effect on the real power flow, while small changes in the voltages' amplitudes will not have an appreciable effect on it. Therefore, the flow of real power in a transmission line is governed by the angle difference between both sources. For maintaining transient stability, the power system is usually operated with small angle difference. Moreover, the reactive power flow is determined by the magnitude difference of terminal voltages.

The traditional analysis (using the phasor concept) of this circuit is given here by considering the numerical values $V_1 = 120\angle-5^\circ$; $V_2 = 100\angle 0^\circ$; impedance line $z = 1 + j7\,\Omega$. Then,

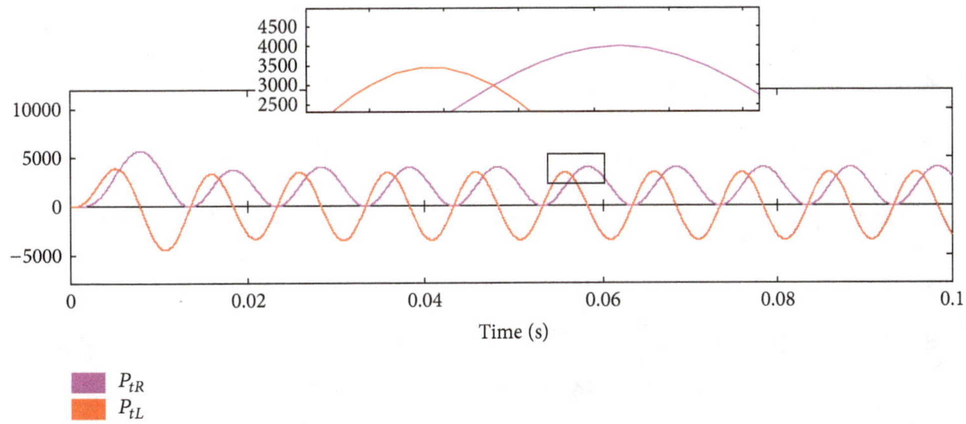


FIGURE 13: Power waves in the *VRL* StBG.

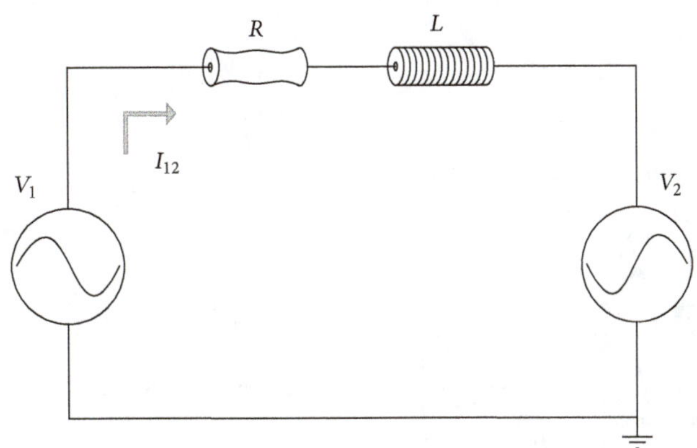


FIGURE 14: Two interconnected voltages sources.

the circulating current from source 1 ($V_1$) to source 2 ($V_2$) is given by $I_{12}$, and $I_{21}$ is the reverse current. Consider

$$I_{12} = \frac{120\angle -5^\circ - 100\angle 0^\circ}{1 + j7} = 3.135\angle -110.02^\circ,$$
$$I_{21} = \frac{100\angle 0^\circ - 120\angle -5^\circ}{1 + j7} = 3.135\angle 69.98^\circ. \tag{9}$$

The apparent power delivered or absorbed by each source is given by

$$A_{12} = V_1 I_{12}^* = 376.2\angle 105.02^\circ = -97.5\,\text{W} + j363.3\,\text{VAR}, \tag{10}$$

$$A_{21} = V_2 I_{21}^* = 313.5\angle -69.98^\circ = 107.3\,\text{W} - j294.5\,\text{VAR}. \tag{11}$$

Equation (10) means that source 1 absorbs 97.5 W of active power and delivers 363.3 VAR. From (11), we can see that the source 2 provides 107.3 W and absorbs 294.5 VAR.

Also, the loss in the transmission line can be calculated by

$$A_L = A_{12} + A_{21} = 9.8\,\text{W} + j68.8\,\text{VAR}. \tag{12}$$

The corresponding StBG for the interconnected voltage sources is shown in Figure 15.

The StBG presented in Figure 15 has the same structure (as the examples presented earlier), but it can be seen that the scattering bonds 4 and 5 are added. The scattering bond 4 is kept intentionally taking the convention current into consideration. This means that the power is transferred from source $V_1$ to source $V_2$.

The parameters taken in the simulation are the same as those taken in the electrical system of Figure 14: MSe: $V_1 = 120\sin(\omega t - 5^\circ)$; MSe: $V_2 = 100\sin(\omega t)$; $R = 1\,\Omega$ and $L = 22.28$ H; where $\omega$ is the angular frequency $\omega = 2\pi 50$. Figure 16 shows the scattering variables in each bond.

The incident and reflected waves can been interpreted as follows.

In case of $w_i s_1$ and $w_i s_2$, a phase shift of $-5^\circ$ is visualized (although they have $180^\circ$ phase shift) and the amplitude difference between these two sources is also considered. For the $w_{rR}$, which is the reflected wave in the $R$ element, it can be seen that its value is zero. This is due to the chosen $R$ element value, which is equal to $1\,\Omega$; then, the scattering relation inside this element is $S = 0$ (the element is matched).

The incident power and reflected power in the sources are given in Figure 17.

From Figure 17, the average total power in the source 1 is −96.5, while it is 106.5 for the source 2. These two values

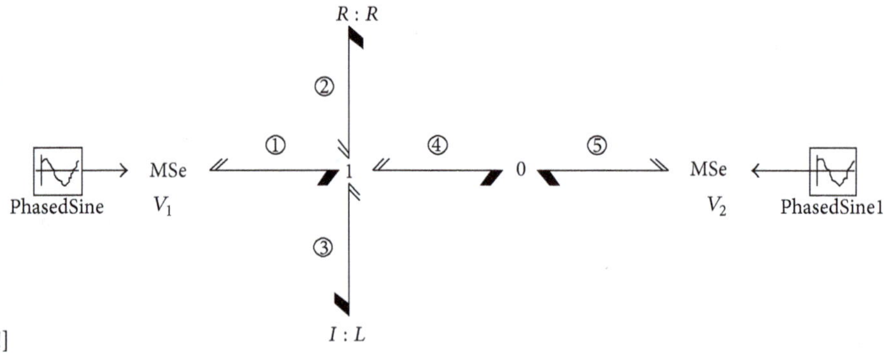


[t!]

FIGURE 15: StBG of two interconnected sources.

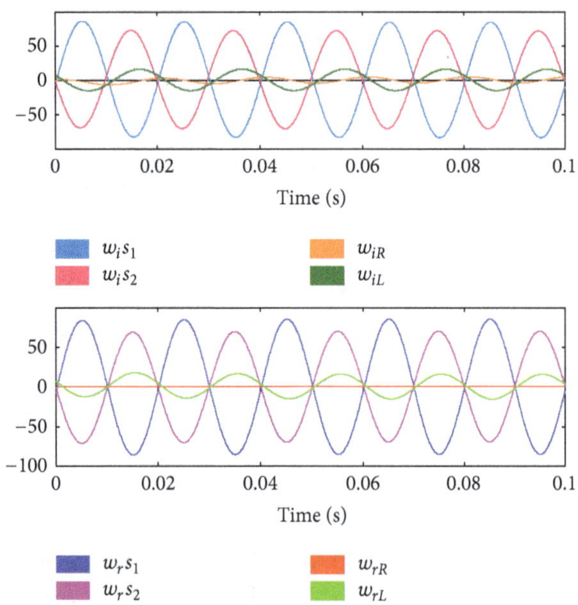


FIGURE 16: Power waves for the StBG of Figure 15.

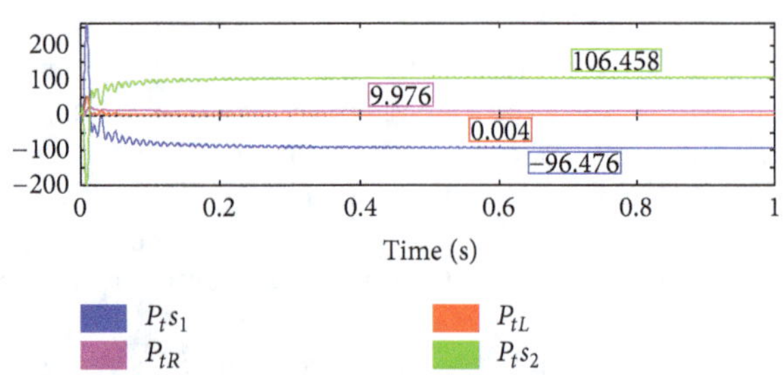


FIGURE 17: Sources average power responses.

match closely the active power values calculated earlier ((10) and (11)).

When the power flow is analyzed, it is important to know the active power and reactive power delivered or absorbed by each source. The two sources are not dynamic elements; however, they can deliver or absorb reactive power. In order to get the values of this reactive power it is necessary to see the response of the reflective power waves in these elements. Figure 18 shows the reflected power waves.

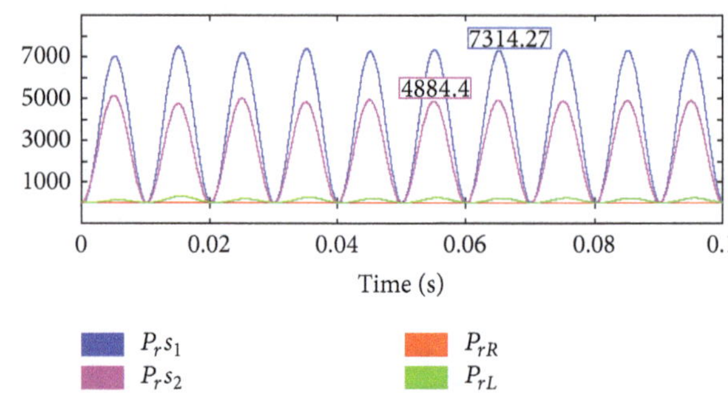


FIGURE 18: Reflected waves in the StBG of Figure 15.

As earlier, the reactive power value is taken directly from the peak value of the reflected power waves. For this case, it can be seen that the peak value of the reflected power waves for sources $V_1$ and $V_2$ is $P_r s_1 = 7314.27$ VAR and $P_r s_2 = 4884.4$ VAR, respectively. These values do not match (10) and (11); nevertheless, this can be interpreted as follows: the two sources have a difference of 20 volts in their amplitudes, and as mentioned before, the difference between two sources has a direct impact on the reactive power interchanged between them. Then, if the peak value of these two reflected waves is divided by this difference (20 volts), the reactive power can be obtained ($P_r s_1 = 365.7$ VAR and $P_r s_2 = 244.2$ VAR).

It is evident that, in this part, the absorbed or generated reactive power in each source cannot be differentiated. It means that the sign cannot be obtained directly. In order to solve this disadvantage, it is necessary to look at the phase shift between these two sources and then assign the corresponding sign.

The power losses in the elements $R$ and $L$ are shown in Figure 19.

Same as before, the active power ($P = 10$ W) is obtained by taking the mean value of the total power wave in the $R$ element. The maximum value of the total power in the $L$ element is 69.9 VAR, which corresponds directly to the reactive power loss in the transmission line.

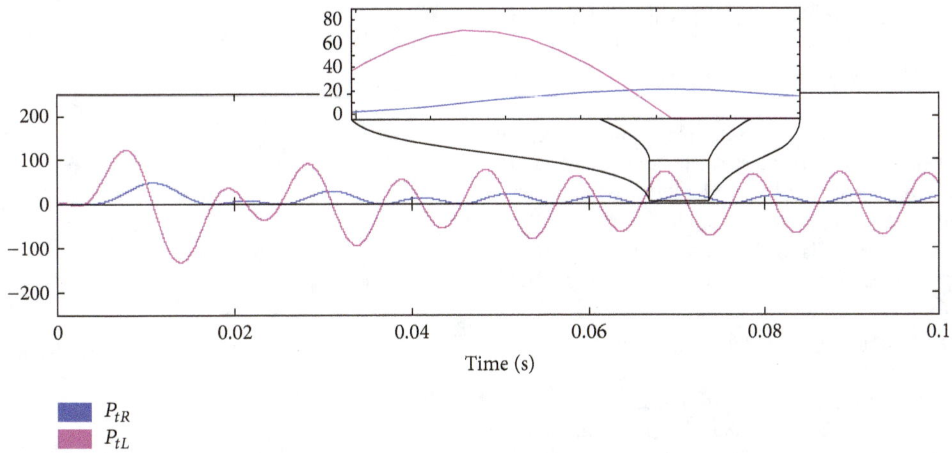


FIGURE 19: Losses power in the StBG of Figure 15.

## 6. Conclusions and Future Work

The active power and reactive power were obtained from the scattering bond graph. The problem of visualizing the reactive power in a bond graph was resolved. A library developed in the software 20Sim allowed simulating the single-phase electrical circuits.

The exposed scattering bond graph analysis represents an alternative to the classical phasor analysis. This has been shown through different examples.

The use of the StBG allowed obtaining an online reactive and active power calculation, when the system is simulated. As the proposed analysis is based on a graphical methodology, the power interaction can be easily shown in a system. Besides, the different techniques already developed by the traditional bond graph methodology can be applied to this StBG formulation.

The scattering variables have allowed obtaining the active and reactive power directly by regarding the incident and reflected waves. This means that the system has been treated in a different reference frame (power frame) in which the systems interaction was described by a duplex pair of oppositely directed signals. Since different variables have been used, the traditional variables (i.e., voltage, current) can be obtained by using the scattering and normalized relationship.

In the next study, the balanced and unbalanced three-phase circuits will be presented using the concept exposed here. Also, the use of incident and reflected variables is visualized in order to derive the control laws necessary in the electrical systems. It is important to note that the StBG does not apply exclusively to electrical systems; it can also be applied to other systems where it could be necessary to interpret the results.

## Conflict of Interests

The author declares that there is no conflict of interests regarding the publication of this paper.

## References

[1] G. Dauphin-Tanguy, *Les Bond Graphs*, Hermès Science, 2000.

[2] D. Karnopp, D. Margolis, and R. Rosenberg, *System Dynamics: A Unified Approach*, John Wiley & Sons, New York, NY, USA, 1990.

[3] H. M. Paynter and I. J. Busch-Vishniac, "Wave-scattering approaches to conservation and causality," *Journal of the Franklin Institute*, vol. 325, no. 3, pp. 295–313, 1988.

[4] M. Amara and S. Scavarda, "A procedure to match bond graph and scattering formalisms," *Journal of the Franklin Institute*, vol. 328, no. 5-6, pp. 887–899, 1991.

[5] A. Kamel and G. Dauphin-Tanguy, "Power transfer in physical systems using the scattering bond graph and a parametric identification," *Systems Analysis Modelling Simulation*, vol. 27, pp. 1–13, 1996.

[6] R. G. Longoria, "Wave-scattering formalisms for multiport energetic systems," *Journal of the Franklin Institute*, vol. 333, no. 4, pp. 539–564, 1996.

[7] W. V. Lyon, "Reactive power and unbalanced circuits," *Electrical World*, vol. 75, no. 25, pp. 1417–1420, 1920.

[8] C. I. Budeanu, *Puissances réactives et fictives*, Institut Romain de l'Energie, Bucarest, Romania, 1927.

[9] C. I. Budeanu, *The Different Options and Conceptions Regarding Active Power in Non-Sinusoidal Systems*, vol. 4, Institute Romain de l'Energie, 1927.

[10] S. Fryze, "Wirk-, Blind- und Scheinteistung in elektrischen Stromkreisen mit nicht-sinusformigen Verlauf von Strom und Spannung," *Przegląd Elektrotechniczny* (Polish), no. 7, pp. 193–203, no. 8, pp. 225–234, no. 22, 673–676, ETZ, no. 25–29, 1931 (German).

[11] H. Sasaki and T. Machida, "A New method to eliminate AC harmonic currents by magnetic flux compensation-considerations on basic design," *IEEE Applications on Power Apparatus and Systems*, vol. 90, no. 5, pp. 2009–2019, 1971.

[12] M. S. Erlicki and A. Emanuel-Eigeles, "New aspects of power factor improvement part I—theoretical basis," *IEEE Transactions on Industry and General Applications*, vol. 4, no. 4, pp. 441–446, 1968.

[13] T. Fukao, H. Lida, and S. Miyairi, "Improvements of the power factor of distorted waveform by thyristor based switching filter," *Transaction of the IEE-Japan Part B*, vol. 92, no. 6, pp. 342–349, 1972.

[14] H. Akagi, Y. Kanazawa, and A. Nabae, "Instantaneous reactive power compensator comprising switching devices without energy storage components," *IEEE Transactions on Industry Applications*, vol. 20, no. 3, pp. 625–630, 1984.

[15] H. J. Carlin and A. B. Giordano, *Network Theory: An Introduction to Reciprocal and Nonreciprocal Circuits*, Prentice-Hall, Englewood Cliffs, NJ, USA, 1964.

[16] R. W. Newcomb, *Linear Multiport Synthesis*, McGraw-Hill, New York, NY, USA, 1966.

[17] K. Kurokawa, "Power waves and scattering matrix," *IEEE Transactions on Microwave Theory and Techniques*, vol. 13, no. 2, pp. 194–202, 1965.

[18] H. J. Carlin and A. B. Giordano, *Networks Theory: An Introduction to Reciprocal and Non-reciprocal Circuits*, Prentice Hall, Englewood Cliffs, NJ, USA, 1964.

[19] K. Kurokawa, *An Introduction to the Theory of Microwave Circuits*, Academic Press, New York, NY, USA, 1969.

# 17

# Prediction of Neutron Yield of IR-IECF Facility in High Voltages Using Artificial Neural Network

**A. Sadighzadeh,[1] A. Salehizadeh,[1,2] M. Mohammadzadeh,[3] F. Shama,[4] S. Setayeshi,[5] S. A. H. Feghhi,[3] S. M. Sadati,[3] M. Rezaei,[3] E. Haji Ebrahimi,[5] and G. H. Roshani[1,3]**

[1]*Plasma and Fusion Research School, Nuclear Science and Technology Research Institute, Tehran, Iran*
[2]*Faculty of New Science and Technology, University of Isfahan, Isfahan, Iran*
[3]*Department of Radiation Application, Shahid Beheshti University, GC, Tehran, Iran*
[4]*Young Researchers and Elite Club, Islamic Azad University, Kermanshah Branch, Kermanshah, Iran*
[5]*Department of Energy Engineering and Physics, Amirkabir University of Technology, Tehran, Iran*

Correspondence should be addressed to G. H. Roshani; hosseinroshani@yahoo.com

Academic Editor: Jyh-Hong Chou

Artificial neural network (ANN) is applied to predict the number of produced neutrons from IR-IECF device in wide discharge current and voltage ranges. Experimentally, discharge current from 20 to 100 mA had been tuned by deuterium gas pressure and cathode voltage had been changed from −20 to −82 kV (maximum voltage of the used supply). The maximum neutron production rate (NPR) of $1.46 \times 10^7$ n/s had occurred when the voltage was −82 kV and the discharge current was 48 mA. The back-propagation algorithm is used for training of the proposed multilayer perceptron (MLP) neural network structure. The obtained results show that the proposed ANN model has achieved good agreement with the experimental data. Results show that NPR of $1.855 \times 10^8$ n/s can be achieved in voltage and current of 125 kV and 45 mA, respectively. This prediction shows 52% increment in maximum voltage of power supply. Also, the optimum discharge current can increase 1270% NPR.

## 1. Introduction

Inertial electrostatic confinement fusion (IECF) device is a compact and simple structure for nuclear fusion researches by electrical discharge, which can operate in the pulsed or continuous mode. It consists of two concentric (or coaxial) electrodes in which usually the central one (cathode) is negatively high-voltage-biased and the outward electrode (anode) is grounded. In this configuration, strong electric fields between electrodes lead to iodinating the filling gas and then accelerate the created ions toward the center, where the electrons are placed in the opposite direction. As a result, rather hot and dense plasma is formed in the center of cathode. In this situation, the continuous nuclear fusion reactions occur, which are the results of the beam-target and beam-beam interactions depending on the working conditions, although the beam-beam interactions will be negligible compared to beam-target reactions in the plasma conditions in this study. Therefore IECF is considered as a source of hot and dense plasma, highly energetic ions, and fast neutrons (when using deuterium or mixture of deuterium-tritium gas). IECF device is an excellent apparatus because of its ability to generate fast neutrons with high-flux from a small source for many applications, such as medicine (e.g., boron neutron capture cancer therapy) [1–3], radiography or tomography of thick materials, space propulsion system [4], inspection system and explosive landmine detection [5], neutron activation analysis, mine and petroleum exploration, and security screening. Therefore neutron production rate optimization in this device will increase its efficiency for the above applications.

IEC researches have been performed since the 1950s decade. Recently, over the last 10 years, many researches and developments on IECF neutron source have been done [2, 4, 6–8]. Most of these developments are created by the usage of the different ion sources and enhancement of voltage and current [2, 4, 6]. IR-IECF is the prototype IEC device that is structured at the Atomic Energy Organization of Iran (AEOI) and the preliminary experiments on neutron yield were performed [7]. D-D neutrons have lower energy (2.5 MeV) and yield compared with D-T neutrons (14 MeV) and therefore they can be moderated within a thinner moderation layer, and thermal neutron flux can be obtained with D-D neutrons. The mentioned benefit and other advantages of D-D neutrons caused that deuterium gas to be used in IECF device.

Reported results in previous works suggest that using the D-T gas mixture and increasing the ion current via ion source would give a neutron rate of $7.3 \times 10^{12}$ n/sec at 1.2 mTorr, 75 kV, and 1.5 A ion current [6, 8, 9]. Researchers in Wisconsin University, in order to increase and optimize the neutron production rate in the UW IECF device, studied effect of cathode's size (diameter), geometry, and material composition on NPR [10]. Dietrich in MIT University investigated multigrid IEC device theoretically and experimentally for improving particle confinement [11]. In this study, for the first time, the ability of artificial neural network for modeling the IECF device in order to optimize the neutron production rate was investigated and the maximum neutron yield was obtained. In the recent work [12], dependence of neutron production rate (NPR) on cathode voltage and discharge current in IR-IECF device for deuterium gas has been studied.

Objective of this work is to train an ANN based on the data presented in [12] to predict NPR at high voltages that cannot be achieved experimentally by the current facilities. The use of ANN is discussed in detail in the following sections to predict NPR for voltages up to 130 kV.

## 2. Experimental Setup

IR-IECF device consists of a spherical grounded anode, which serves as a chamber, and a grid cathode that are set concentrically as shown in Figure 1. This device consists of 13.5 cm diameter stainless steel cathode, 41 cm diameter anode with a 60 cm diameter, and 60 cm height vacuum chamber [7].

The IR-IECF facility accompanied with the cathode and the anode geometry is shown in Figure 2. In the experiments, facility is adjusted in a desired gas pressure and then by changing the cathode negative voltage glow plasma inside the cathode grid is produced. Produced ions are accelerated toward the cathode, and most of them penetrate the hollow cathode spherical wire net. In this situation, fusion reactions through beam-beam and beam-target collisions occurred. By increasing the cathode voltage and decreasing gas pressure, desired dense plasma is obtained.

In order to measure the neutron production rate (NPR), a calibrated neutron probe LB6411, which consists of a cylindrical $^{3}$He proportional counter tube, was placed in 60 cm distance from cathode center of IR-IECF device [12]. The cathode voltage and discharge current have a significant impact on the NPR. The experiments had been carried out with the tunable 82 kV DC high voltage. Maximum NPR was $1.46 \times 10^{7}$ n/s and had occurred when the voltage was 82 kV and the discharge current was 48 mA. In the next sections

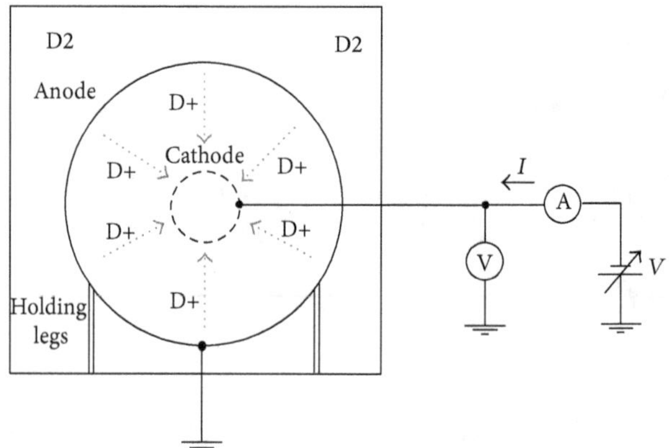


FIGURE 1: Schematic diagram of IR-IECF device and parameters used in this study.

using artificial neural network and these experimental data the NPR has been predicted for these ranges: current discharge = 40 mA–120 mA and voltage = 50 kV–130 kV. The extrapolation beyond the range of experimental data is one of the most important applications of using ANNs to model the nonlinear unknown-response systems such as the studied IECF. This method has been used in many published works such as [13–15].

## 3. Neural Network Computations

Artificial neural networks (ANNs) are mathematical models inspired in the human brain. Ability of learning to use experimental data is the main characteristic of this technique. This model can predict behaviors and patterns from a finite set of experimental data, called the "training set" of the ANN [16]. In general, the structure of ANN typically is comprised of three layers including input, hidden, and output layers. First primary data is collected in input layer and then sent to different processing units (neuron), which constitute the hidden layer of the networks [17]. The structure of a neuron consists of two major terms of weight and activation function as shown in Figure 3.

Activation function receives an argument $n$ and generates an output $a$. The input of network for each neuron is sum of all input values in which each is multiplied by its weight and a bias term. The output value can be calculated by feeding the network input into the activation function of the neuron. Many activation functions are applied in ANN model such as triangular basis, pure line, soft max, log sigmoid, tan sigmoid, and hard limit. To achieve an acceptable prediction, the ANN is trained to minimize the error between ANN output and experimental data.

In this research, multilayer perceptron (MLP) neural networks were used. MLP is a feed-forward artificial neural network model that maps input data sets onto a set of appropriate outputs. What makes a multilayer perceptron a different tool to predict the nonlinear unknown-response systems is that each neuron uses a nonlinear activation function (such as tangent sigmoid), which was developed to model

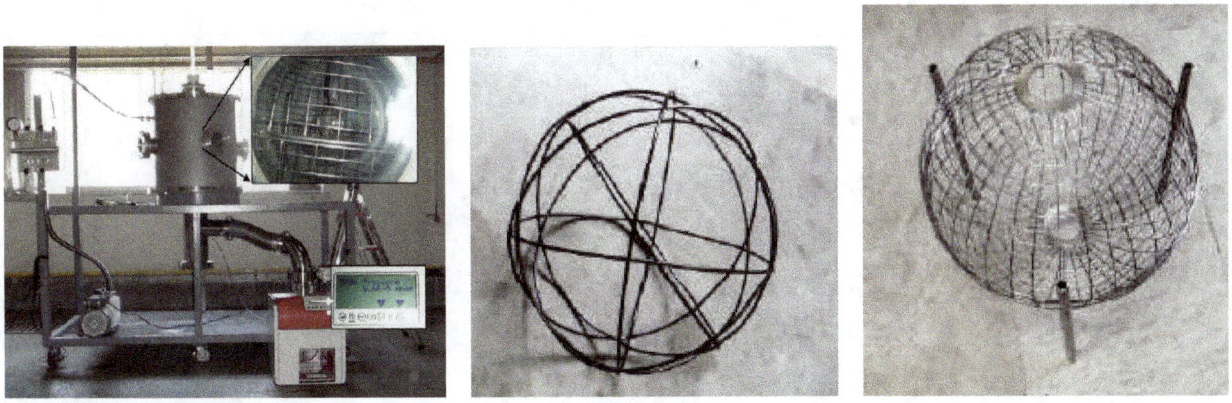

(a) (b) (c)

FIGURE 2: (a) IR-IECF device, (b) cathode (13.5 cm diameter), and (c) anode (41 cm diameter).

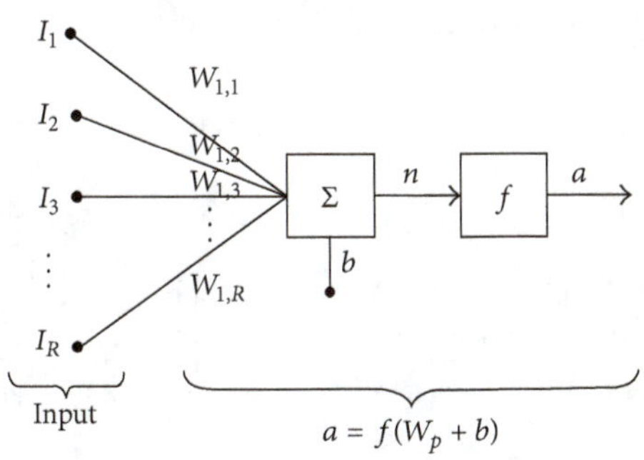


FIGURE 3: The neuron mathematical model.

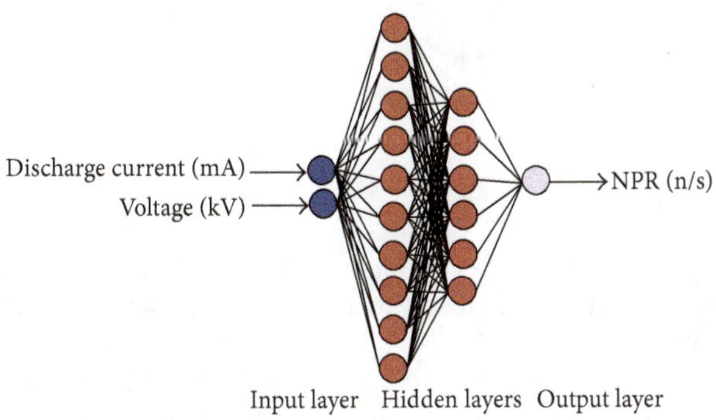


FIGURE 4: Architecture for the proposed MLP model.

the frequency of action potentials, or firing, of biological neurons in the brain [18, 19]. The proposed MLP model is shown in Figure 4, where the inputs are discharge current (mA) and voltage (kV) and the output is NPR (n/s).

The input to the node $t$ in the first hidden layer is given by [20, 21]

$$\eta_t = \sum_{u=1}^{2} (X_u W_{ut}) + \theta_t, \quad t = 1, 2, \ldots, 10. \tag{1}$$

TABLE 1: Specification of the proposed ANN model.

| Neural network | MLP |
|---|---|
| Number of neurons in the input layer | 2 |
| Number of neurons in the first hidden layer | 10 |
| Number of neurons in the second hidden layer | 6 |
| Number of neurons in the output layer | 1 |
| Number of epochs | 1200 |
| Activation function | tansig |

The output from the $t$th neuron of the second hidden layer is given by

$$O_t = f(\eta_t), \tag{2}$$

where $X$ is the input variables, $\theta$ is the bias term, $W$ is the weighting factor, and $f$ is the activation function of the hidden layer. The output of the neuron in the output layer is given by

$$Y = \sum_{u=1}^{6} (O_u W_u) + b. \tag{3}$$

Required data set for training the network is obtained by use of IR-IECF experimental data. The presented MLP network is trained by Levenberg-Marquardt (LM) algorithm [22]. In this method, first derivative and second derivative (Hessian) are used for network weight correction [23]. The numbers of samples for training and testing data are 36 (about 70%) and 15 (about 30%), respectively. In this study, for optimizing the ANN configuration many different structures with one hidden layer and two, three, and four hidden layers were tested by applying different number of neurons in each layer and different epochs. MATLAB 7.0.4 software was used for training the ANN model. Table 1 shows the specification of the proposed ANN model being used in this study. In this table, an epoch represents completion of processing on the entire training data set [23].

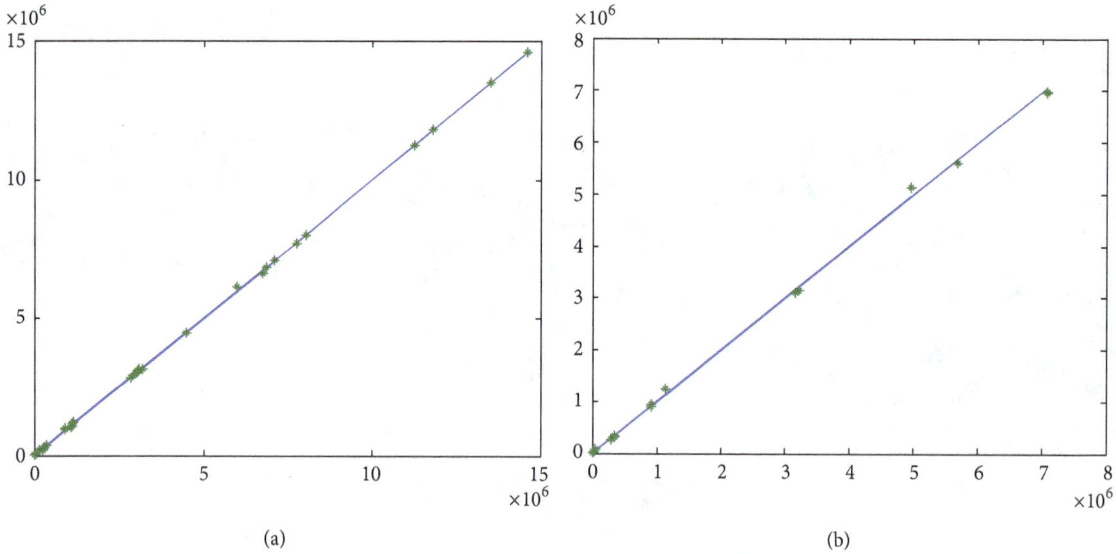


FIGURE 5: Comparison of experimental and predicted results for (a) training data and (b) testing data.

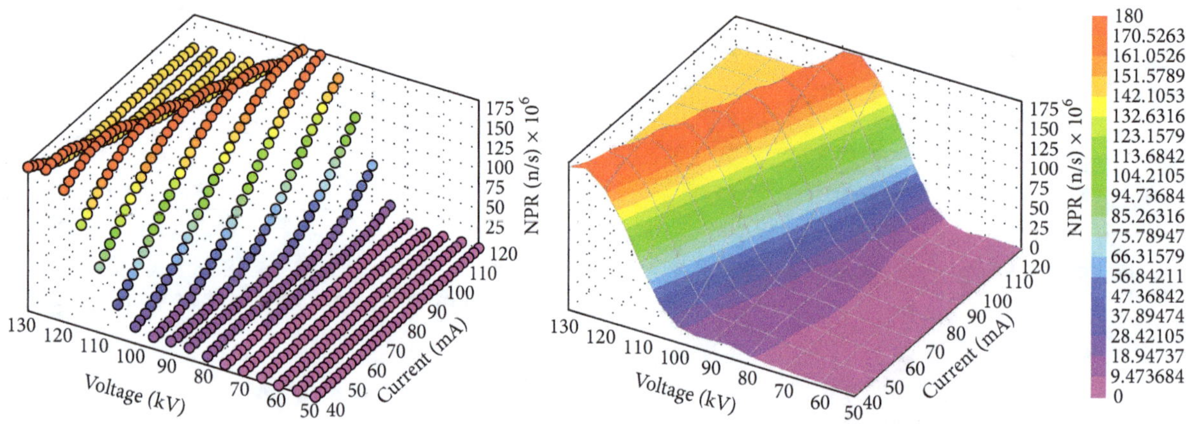


FIGURE 6: Prediction of NPR for current discharge = 40 mA–120 mA and voltage = 50 kV–130 kV.

TABLE 2: Obtained errors for training and testing results of the proposed ANN model.

| Error | Train | Test |
|---|---|---|
| MRE% | 0.0072 | 0.6563 |
| RMSE | 0.0445 | 0.0719 |

## 4. Results and Dissection

Table 2 shows the obtained errors for the proposed ANN model, where the mean relative error percentage (MRE%) and the roots mean square error (RMSE) are calculated by

$$\text{RMSE} = \left[ \frac{\sum_{m=1}^{N} \left( X_m(\text{Exp}) - X_m(\text{Pred}) \right)^2}{N} \right]^{0.5},$$
$$\text{MRE\%} = 100 \times \frac{1}{N} \sum_{m=1}^{N} \left| \frac{X_m(\text{Exp}) - X_m(\text{Pred})}{X_m(\text{Exp})} \right|, \quad (4)$$

where $N$ is the number of data and "$X$(Exp)" and "$X$(Pred)" stand for experimental and predicted (ANN) values, respectively.

Figure 5 shows the comparison of the experimental and predicted results using the proposed ANN model for training and testing data. The comparisons between experimental and predicted (ANN) results for training and testing data are listed in Tables 2 and 3, respectively.

For training results, variance and standard deviation of differences are obtained as $1.98E{-}03$ and 0.0445, respectively.

For testing result, the variance and standard deviation of differences are obtained as 0.005088 and 0.07134, respectively. From Tables 3 and 4 and Figure 5, clearly the predicted produced neutrons by ANN model are close to the experimental results. These results show the applicability of ANN as an accurate and reliable model for the prediction of NPR in IECF device according to the discharge current and voltage. Figure 6 shows the prediction of NPR in the range of current discharge = 40 mA–120 mA and voltage = 50 kV–130 kV.

TABLE 3: The data that were used for training the network and predicted number of produced neutrons by ANN.

| Discharge current (mA) | Voltage (kV) | Produced neutron ($\times 10^6$) | Predicted neutron ($\times 10^6$) | Difference |
|---|---|---|---|---|
| 20 | 30 | 0.055164 | 0.0552 | $-3.648E-05$ |
| 20 | 40 | 0.364079 | 0.3641 | $-2.08E-05$ |
| 20 | 60 | 2.846438 | 2.8464 | $3.8E-05$ |
| 20 | 70 | 4.468245 | 4.4682 | $4.5E-05$ |
| 30 | 40 | 0.358563 | 0.3667 | −0.0081371 |
| 30 | 50 | 1.152918 | 1.2269 | −0.073982 |
| 30 | 60 | 3.033994 | 3.0024 | 0.031594 |
| 40 | 30 | 0.052405 | 0.0584 | −0.0059947 |
| 40 | 50 | 1.130852 | 1.1437 | −0.012848 |
| 40 | 60 | 3.072608 | 3.0582 | 0.014408 |
| 50 | 30 | 0.042476 | 0.0469 | −0.0044241 |
| 50 | 50 | 1.125336 | 1.0738 | 0.051536 |
| 50 | 70 | 5.95766 | 6.1068 | −0.14914 |
| 60 | 40 | 0.292367 | 0.2832 | 0.0091667 |
| 60 | 50 | 1.10327 | 1.0192 | 0.08407 |
| 60 | 70 | 6.729949 | 6.5857 | 0.144249 |
| 70 | 30 | 0.027582 | 0.029 | −0.0014182 |
| 70 | 50 | 0.926747 | 0.9802 | −0.0534529 |
| 70 | 60 | 3.199484 | 3.155 | 0.044484 |
| 80 | 30 | 0.017652 | 0.0214 | −0.0037477 |
| 80 | 40 | 0.248236 | 0.2421 | 0.0061358 |
| 80 | 60 | 3.089157 | 3.1426 | −0.053443 |
| 80 | 70 | 7.060931 | 7.0959 | −0.034969 |
| 90 | 40 | 0.24272 | 0.2172 | 0.0255195 |
| 90 | 50 | 0.910198 | 0.9312 | −0.0210019 |
| 90 | 60 | 3.056059 | 3.0759 | −0.019841 |
| 100 | 30 | 0.012136 | 0.0121 | $3.597E-05$ |
| 100 | 40 | 0.171007 | 0.1823 | −0.0112931 |
| 100 | 60 | 2.923667 | 2.8965 | 0.027167 |
| 100 | 70 | 6.840276 | 6.8374 | 0.002876 |
| 50 | 75 | 7.99871 | 7.988 | 0.01071 |
| 96 | 77 | 11.80499 | 11.8062 | −0.001207 |
| 55 | 77 | 11.25336 | 11.2547 | −0.001342 |
| 26 | 81 | 7.722893 | 7.7191 | 0.003793 |
| 75 | 82 | 13.51506 | 13.513 | 0.002062 |
| 48 | 82 | 14.61833 | 14.6195 | −0.001167 |

In this range, maximum number of produced neutrons happened in $I = 45$ mA and $V = 125$ kV. In this situation, IECF can produce $1.855 \times 10^8$ neutrons per second. In another optimum point where $I = 72.5$ mA and $V = 115$ kV, the IECF can produce $1.854 \times 10^8$ neutrons per second. These results show significant NPR increase in the IECF device according to increasing voltage. 52% increment in maximum voltage of power supply (82 kV to 125 kV) was found and, in optimum operational discharge current (in this case $I = 45$ mA), the neutron produced rate can be increased to 1270% NPR ($1.46 \times 10^7$ n/s to $1.855 \times 10^8$ n/s).

## 5. Conclusion

In this paper, the development of an ANN model for prediction of the neutron yield in IECF devices is presented. The comparison between experimental and predicted results by ANN model shows that there is a good agreement between them with mean relative error (MRE) less than 0.66%. Therefore, the MLP network can be used as an efficient tool to predict the NPR in the IECF device. For developing the model, the input parameters are voltage (kV) and discharge current (mA) and the output is neutron yield (n/s). Results

TABLE 4: The data that were used for testing the network and predicted number of produced neutrons by ANN.

| Discharge current (mA) | Voltage (kV) | Produced neutron (×10$^6$) | Predicted neutron (×10$^6$) | Difference |
|---|---|---|---|---|
| 20 | 50 | 1.130852 | 1.2495 | −0.11865 |
| 30 | 30 | 0.052957 | 0.0723 | −0.01934 |
| 30 | 70 | 4.964717 | 5.131 | −0.16628 |
| 40 | 40 | 0.34753 | 0.3329 | 0.01463 |
| 40 | 70 | 5.681843 | 5.6062 | 0.075643 |
| 50 | 60 | 3.166386 | 3.1047 | 0.061686 |
| 50 | 40 | 0.342014 | 0.3058 | 0.036214 |
| 60 | 30 | 0.035856 | 0.0373 | −0.00144 |
| 60 | 60 | 3.210517 | 3.1387 | 0.071817 |
| 70 | 40 | 0.28685 | 0.2629 | 0.02395 |
| 70 | 70 | 7.060931 | 6.955 | 0.105931 |
| 80 | 50 | 0.915714 | 0.9536 | −0.03789 |
| 90 | 30 | 0.014894 | 0.0133 | 0.001594 |
| 90 | 70 | 7.060931 | 6.9842 | 0.076731 |
| 100 | 50 | 0.904682 | 0.8952 | 0.009482 |

show that in optimum discharge current when maximum voltage of power supply increases 1.5 times, NPR can increase more than 12 times.

## Conflict of Interests

The authors declare that there is no conflict of interests regarding the publication of this paper.

## References

[1] G. L. Kulcinski, "Non-electric applications of fusion energy—an important precursor to commercial electric power," *Fusion Technology*, vol. 34, no. 3, pp. 477–483, 1998.

[2] R. P. Ashley and the UW IEC Team, "Experimental progress in 2003 of the UW IEC facility," in *Proceedings of the 6th U.S.-Japan IEC Workshop*, Tokyo, Japan, October 2003.

[3] R. P. Ashley, G. L. Kulcinski, J. F. Santarius, S. K. Murali, G. Piefer, and R. Radel, "Steady-state D3He proton production in an IEC fusion device," *Fusion Technology*, vol. 39, no. 2, pp. 546–551, 2001.

[4] C. C. Dobson and I. Hrbud, "Electron density and two-channel neutron emission measurements in steady-state spherical inertial-electrostatically confined plasmas, with review of the one-dimensional kinetic model," *Journal of Applied Physics*, vol. 96, no. 1, pp. 94–108, 2004.

[5] K. Yoshikawa, K. Masuda, T. Takamatsu et al., "Research and development on a compact discharge-driven D-D fusion neutron source for explosive detection," in *Proceedings of the 2nd Joint International Conference on Sustainable Energy and Environment (SEE '06)*, Bangkok, Thailand, November 2006.

[6] T. Takamatsu, K. Masuda, T. Kyunai, H. Toku, and K. Yoshikawa, "Inertial electrostatic confinement fusion device with an ion source using a magnetron discharge," *Nuclear Fusion*, vol. 46, no. 1, pp. 142–148, 2006.

[7] V. Damideh, A. Sadighzadeh, A. Koohi et al., "Experimental study of the Iranian inertial electrostatic confinement fusion device as a continuous neutron generator," *Journal of Fusion Energy*, vol. 31, no. 2, pp. 109–111, 2012.

[8] G. H. Miley, Y. Yang, J. Webber, Y. Shaban, and H. Momota, "RF ion source-driven IEC design and operation," *Fusion Science and Technology*, vol. 47, no. 4, pp. 1233–1237, 2005.

[9] G. H. Miley, Y. Shaban, and Y. Yang, "RF ion gun injector in support of fusion ship II research and development," *AIP Conference Proceedings*, vol. 699, no. 1, p. 406, 2004.

[10] A. L. Wehmeyer, R. F. Radel, and G. L. Kulcinski, "Optimizing neutron production rates from DD fusion in an inertial electrostatic confinement device," in *Proceedings of the 6th ANS Topical Meeting on Fusion Energy*, Madison, Wis, USA, September 2004.

[11] C. C. Dietrich, *Improving particle confinement in inertial electrostatic fusion for spacecraft power and propulsion [Ph.D. thesis]*, MIT University, 2007.

[12] E. H. Ebrahimi, R. Amrollahi, A. Sadighzadeh et al., "The influence of cathode voltage and discharge current on neutron production rate of inertial electrostatic confinement fusion (IR-IECF)," *Journal of Fusion Energy*, vol. 32, no. 1, pp. 62–65, 2013.

[13] H. K. Cigizoglu, "Estimation, forecasting and extrapolation of river flows by artificial neural networks," *Hydrological Sciences Journal*, vol. 48, no. 3, pp. 349–362, 2003.

[14] Z. Liu and J. Liu, "Seismic-controlled nonlinear extrapolation of well parameters using neural networks," *Geophysics*, vol. 63, no. 6, pp. 2035–2041, 1998.

[15] Ö. Kişi, "Streamflow forecasting using different artificial neural network algorithms," *Journal of Hydrologic Engineering*, vol. 12, no. 5, pp. 532–539, 2007.

[16] C. M. Salgado, L. E. B. Brandão, R. Schirru, C. M. N. A. Pereira, A. X. da Silva, and R. Ramos, "Prediction of volume fractions in three-phase flows using nuclear technique and artificial neural network," *Applied Radiation and Isotopes*, vol. 67, no. 10, pp. 1812–1818, 2009.

[17] F. Rahimi-Ajdadi and Y. Abbaspour-Gilandeh, "Artificial Neural Network and stepwise multiple range regression methods for prediction of tractor fuel consumption," *Measurement*, vol. 44, no. 10, pp. 2104–2111, 2011.

[18] F. J. de Cos Juez, M. A. Suárez-Suárez, F. Sánchez Lasheras, and A. Murcia-Mazón, "Application of neural networks to the study of the influence of diet and lifestyle on the value of bone mineral density in post-menopausal women," *Mathematical and Computer Modelling*, vol. 54, no. 7-8, pp. 1665–1670, 2011.

[19] A. Ebrahimzadeh and A. Khazaee, "Detection of premature ventricular contractions using MLP neural networks: a comparative study," *Measurement*, vol. 43, no. 1, pp. 103–112, 2010.

[20] A. R. Gallant and H. White, "On learning the derivatives of an unknown mapping with multilayer feedforward networks," *Neural Networks*, vol. 5, pp. 129–138, 1992.

[21] J. G. Taylor, *Neural Networks and Their Applications*, John Wiley & Sons, West Sussex, UK, 1996.

[22] M. T. Hagan and M. B. Menhaj, "Training feedforward networks with the Marquardt algorithm," *IEEE Transactions on Neural Networks*, vol. 5, no. 6, pp. 989–993, 1994.

[23] B.-H. Juang and S. Katagiri, "Discriminative learning for minimum error classification," *IEEE Transactions on Signal Processing*, vol. 40, no. 12, pp. 3043–3054, 1992.

# 18

# Performance Evaluation of Antlion Optimizer Based Regulator in Automatic Generation Control of Interconnected Power System

**Esha Gupta and Akash Saxena**

*Department of Electrical Engineering, Swami Keshvanand Institute of Technology, Management and Gramothan, Office No. AC-201, Ramnagaria, Jagatpura, Jaipur, Rajasthan 302017, India*

Correspondence should be addressed to Esha Gupta; esha.gupta@outlook.com

Academic Editor: Chung-Liang Chang

This paper presents an application of the recently introduced Antlion Optimizer (ALO) to find the parameters of primary governor loop of thermal generators for successful Automatic Generation Control (AGC) of two-area interconnected power system. Two standard objective functions, Integral Square Error (ISE) and Integral Time Absolute Error (ITAE), have been employed to carry out this parameter estimation process. The problem is transformed in optimization problem to obtain integral gains, speed regulation, and frequency sensitivity coefficient for both areas. The comparison of the regulator performance obtained from ALO is carried out with Genetic Algorithm (GA), Particle Swarm Optimization (PSO), and Gravitational Search Algorithm (GSA) based regulators. Different types of perturbations and load changes are incorporated to establish the efficacy of the obtained design. It is observed that ALO outperforms all three optimization methods for this real problem. The optimization performance of ALO is compared with other algorithms on the basis of standard deviations in the values of parameters and objective functions.

## 1. Introduction

With the increase in the interconnection of the utilities, complexity in power system operation and control has emerged as a challenging problem in front of design engineers. Variation of the system parameters (voltage and frequency) from their nominal values can present a potential threat to the system stability. To control these deviations, intelligent design is required at generation, transmission, and distribution end. Hence, Automatic Generation Control (AGC) of interconnected power network is a major thrust area of research. To keep frequency and tie-line power exchanges in a nominal range AGC of the interconnected power system is required [1, 2]. The prime objective of the successful power system operation is to transmit, distribute, and utilize the electrical power within the nominal range of frequency and terminal voltage. Normally, the interconnection of different power plants (nuclear, thermal, and hydro) introduces different complexities in the operation of power system. Hydro power plants are less operative in developing countries like India, as the availability of the water for irrigation purpose is a critical issue. Moreover, the constraints related to regulations hinder the participation of hydro power plants in the AGC. High efficiency of nuclear units prevents the participation of these units in AGC. The role of AGC is prominent in thermal power plants. Control of generator consists in functioning of two major loops: Automatic Voltage Regulator (AVR) loop to maintain the nominal voltage and Load Frequency Control (LFC) loop to maintain the system frequency.

In 1970, the concept of modern optimal control was introduced by Elgerd [1]. In the modern optimal control, the determination of the parameters of primary governor loop is performed to enhance the system's damping performance. In recent years, this field has emerged as a potential area of research. Approaches employed for AGC can be subdivided into two categories. First, the application of supervised learning models and expert systems was employed to obtain an intelligent control of interconnected power system. Second, the applications of metaheuristic techniques to obtain the controller settings were explored. A rich literature survey on

AGC is provided in [3]. A philosophy of AGC is explained in [4, 5]. Some of the approaches for effective AGC were based on Pole Placement Technique [6], Coefficient Diagram Method (CDM) [7, 8], Neural Networks (NN) [9–11], Fuzzy Logic (FL) [12–15], and Super Magnetic Energy Storage (SMES) Devices [16]. Calculation of CDM coefficients for large interconnected units involves heavy computations. These approaches require large data sets and observations for training and rule formation. Moreover, fuzzy approaches are based on approximations. AGC of a power system is a responsible operation; minute changes in the controller settings can put a question mark on the reliable operation of the power system. Hence, Neural Network and Fuzzy Logic based approaches are not suitable in modern power system's context. As "nature is the best teacher," by mimicking the biological behavior of plants, insects, and species, some beautiful analogies were developed and simulated by researchers in the form of mathematical paradigms. Some of these approaches include Gravitational Search Algorithm (GSA) [17] based on Newton's law of attraction, Particle Swarm Optimization (PSO) [18] based on the behavior of flock and swarm fishes, Genetic Algorithm (GA) [19] based on Darwinian's survival of the fittest and natural evolution, Bacterial Foraging (BF) [20] based on foraging behavior of bacteria, Differential Evolution (DE) [21], Artificial Bee Colony (ABC) [22] based on the behavior of bees, Firefly Algorithm (FA) [23], and Cuckoo Search (CS) [24]. Some hybrid approaches are also reported in which the Fuzzy Logic is combined with the PI controllers. Majorly, the proportional and integral gains of a controller were considered as parameters of interest in the optimization process. However, in the literature, parameters of primary governor loop were also optimized in [20]. The effect of speed regulation on the performance of the regulator was also discussed in [25]. Recently, Teaching Learning Based Optimization (TLBO) is applied to find the scaling factors and integral gains for two thermal units' interconnected power systems in [26]. Recently, Grey Wolf Optimizer (GWO) is applied to find out the optimal settings of PID controller for three thermal units by Sharma and Saikia [27]. In the work, frequency droop was observed in the presence of solar power plants. Bat algorithm is applied to find the regulator settings of multiarea thermal power system in [28]. The author employed PD-PID cascade controllers to obtain the AGC. From the literature review, it is clear that application of metaheuristic algorithm in AGC regulator design is a potential area. The search of a proper set of parameters (integral and differential gains, primary loop parameters) by which Area Control Error (ACE) can be reduced to zero is a major objective to solve AGC problem. In the literature, two objective functions (design criteria) were employed to carry out the estimation process. To find the minima of these functions by considering the gains and other parameters as variables is the essence of AGC problem. Evolutionary algorithms search for global optima of the function by the combined actions of agents and deciding operator in a predefined search space. Hence, the quality of exploration and exploitation is a major deciding factor in the performance of the algorithm. Another noteworthy feature of these algorithms is randomness; although randomness gives different results in each run, they can yet be able to avoid the local minima trap.

Recently, Mirjalili proposed an Antlion Optimizer (ALO) algorithm on the behavior inspired from antlions [29]. ALO has been successfully applied over 19 benchmark functions along with four classical engineering problems. This algorithm is based on the foraging behavior of antlions. Moreover, salient features of algorithms are the effective exploration of the search space by random walk and random selection of agents. Similarly, exploitation of the search space is assured by adaptive boundaries of traps. Since it is a population based algorithm, the avoidance of local optima is indispensable. Fewer parameters, gradient-free structure, and adaptive intensity with iterations are some salient features of the algorithm. In view of the above literature survey, salient features and computational efficacy of the ALO motivated us to employ ALO in AGC regulator design for the very first time. The following are the objectives of this research work:

(1) To solve the optimization process by ISE and ITAE objective functions to find out the parameters of primary governor loop, that is, speed regulation constant ($R$), frequency bias ($D$), and integral gains ($K_I$).

(2) To test the efficacy of the objective functions with the help of damping performance obtained by ALO regulators.

(3) To test the robustness and efficacy of the proposed design with other recently employed regulators and test the design for various types of perturbations and topological changes.

This paper is organized as follows. The details of system modeling are presented in brief in Section 2. In Section 3, details of ALO along with the functioning of operators are explained in a lucid manner. Section 4 discusses the simulation results and analysis. Section 5 provides the comparative analysis of the optimization process for all the algorithms. And finally, in Section 6, conclusions and future scope of the work are pointed out.

## 2. System Modeling

*2.1. AGC Model.* The two-area nonreheat thermal interconnected power system is shown in Figure 1. The main components of the power system include speed governor, turbine, rotating mass, and load. The inputs of the power system are controller output $u$, load disturbance $\Delta P_L$, and tie-line power $\Delta P_{\text{tie}}$ and the outputs are frequency deviations $\Delta f$ and Area Control Error (ACE). The ACE signal controls the steady state errors of frequency deviation and tie-power deviation. Mathematically, ACE can be defined as

$$\text{ACE} = B\Delta f + \Delta P_{\text{tie}}, \quad (1)$$

where $B$ indicates the frequency bias parameter.

The operating behavior of the power system is dynamic so it must be assumed that the parameters of the system are linear. For mathematical modeling, the transfer function is used.

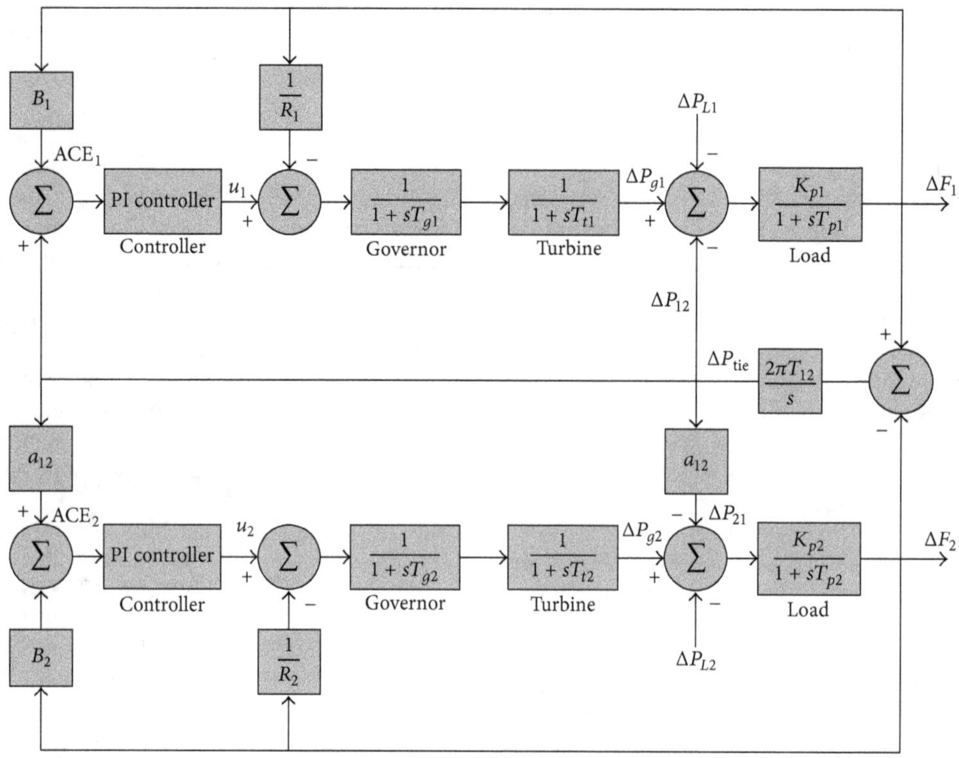


FIGURE 1: Transfer function model of two-area nonreheat thermal interconnected system.

The transfer function of a governor is represented by [1]

$$G_g(s) = \frac{1}{1 + sT_g}. \tag{2}$$

The turbine is represented by the transfer function as [1]

$$G_t(s) = \frac{1}{1 + sT_t}. \tag{3}$$

The transfer function of rotating mass and load [1] is as follows:

$$G_L(s) = \frac{K_p}{1 + sT_p}, \tag{4}$$

where $T_p = 2H/fD$ and $K_p = 1/D$.

$\Delta P_G$ and $\Delta P_L$ are the two inputs of rotating mass and load and $\Delta f(s)$ is the output and is represented by [1]

$$\Delta f(s) = G_L(s)\left[\Delta P_G(s) - \Delta P_L(s)\right]. \tag{5}$$

*2.2. The System Investigated.* The system was investigated on two equal thermal areas connected by a weak tie line having the same generation capacity of 1000 MVA. The parameters of the system are taken from [4]. A sudden step perturbation of 0.1875 p.u. occurs in area 1 and another one of 0.1275 p.u. occurs in area 2. The transfer function model of the two-area thermal system is shown in Figure 1. The system is implemented using MATLAB 2013 and run on a Pentium IV CPU, 2.69 GHz, and 1.84 GB RAM computer [30].

*2.3. The Proposed Approach.* The controller used in AGC system is PI controller as it determines the difference between set point and reference point and removes the steady state error. For the design of PI controller, the parameters proportional gain ($K_P$) and integral gain ($K_I$) are essential. However, in this work, for the ease and simplicity of the optimization process, we consider proportional gain 1. Area Control Errors are the input of the controllers for area 1 and area 2 and are defined as

$$\begin{aligned} \mathrm{ACE}_1 &= B_1 \Delta f_1 + \Delta P_{\mathrm{tie}}, \\ \mathrm{ACE}_2 &= B_2 \Delta f_2 + \Delta P_{\mathrm{tie}}, \end{aligned} \tag{6}$$

where $B_1 = 1/R_1 + D_1$ and $B_2 = 1/R_2 + D_2$.

The outputs of the controllers are $u_1$ and $u_2$ and are obtained as follows:

$$\begin{aligned} u_1 &= K_{P1}\mathrm{ACE}_1 + K_{I1}\int \mathrm{ACE}_1, \\ u_2 &= K_{P2}\mathrm{ACE}_2 + K_{I2}\int \mathrm{ACE}_2. \end{aligned} \tag{7}$$

In this paper, the estimation of integral gains and parameters of primary governor loop is based on two objective functions (ITAE and ISE) which are given in (8). These objective functions aim to reduce the steady state error to zero and maximize the damping ratio of the system. Hence,

$$J_1 = \text{ITAE} = \int_0^T \left(|\Delta f_1| + |\Delta f_2| + |\Delta P_{\text{tie}}|\right) \cdot t\, dt,$$
$$J_2 = \text{ISE} = \int_0^T \left(|\Delta f_1|^2 + |\Delta f_2|^2 + |\Delta P_{\text{tie}}|^2\right) dt. \tag{8}$$

The problematic constraints are the parameters of AGC regulator which contains integral gains, speed regulations, and the frequency sensitivity coefficients as they are bounded with the limits. These parameters are system specific. Hence, the design problem can be formulated as follows:

$$\begin{aligned} &\text{Minimize} \quad J \\ &\text{Subjected to} \quad K_{I_{\min}} \le K_I \le K_{I_{\max}} \\ &\qquad R_{\min} \le R \le R_{\max} \\ &\qquad D_{\min} \le D \le D_{\max}. \end{aligned} \tag{9}$$

$J$ is the objective function ($J_1$ and $J_2$).

## 3. Antlion Optimizer

A novel algorithm inspired by nature named Antlion Optimizer (ALO) is presented in this section. This technique was proposed by Mirjalili [29] in 2015. In ALO, the hunting mechanism of antlions is mimicked. Antlions belong to Myrmeleontidae family of class net winged insects. ALO employs five main steps of hunting, that is, random walk of ants, building trap, entrapment of ants in trap, catching prey, and rebuilding traps. The ALO algorithm is a gradient-free algorithm which also provides greater exploration and exploitation of search space. Exploration is guaranteed by the random selection of antlions and random walks of ants around them whereas exploitation is guaranteed by adaptive shrinking boundaries of antlion's trap. With the help of roulette wheel and random walks, ALO has high probability to resolve local optima stagnation. The life cycle of antlions consists of two main phases: larvae and adults. Total natural lifespan can take up to 3 years which mostly occurs in larvae and only 3–5 weeks in adulthood. Antlions undergo metamorphosis in a cocoon to become adult. They mostly hunt in larvae and the adulthood period is for reproduction. An antlion larva digs a cone shaped pit in sand by moving along a circular path and throwing out sand with its massive jaw. After digging the trap, the larvae hide underneath the bottom of the cone and wait for the insect (preferably ant) to be trapped in the pit. The edge of the cone is sharp enough for insects to fall to the bottom of the trap easily. Figure 2 illustrates the hunting behavior in which antlions wait for the ants to be trapped in the cone shaped pit.

Once the antlion realizes that the prey is in the trap, it tries to catch it. Another interesting behavior in the lifestyle of ant behavior is the relevancy of size of the trap, level of hunger, and shape of the moon. Antlions dig out larger traps as they become more hungry and when the moon is full. And in this way they improve their chance of survival.

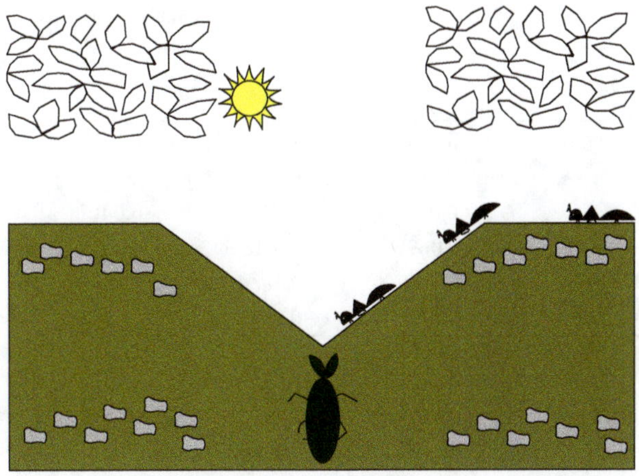

FIGURE 2: The hunting behavior of antlion.

*3.1. Mathematical Modeling of the ALO Algorithm*

*(a) Random Walks of Ants.* Random walks of ants are given in

$$\begin{aligned} X(t) = [&0, \text{cumsum}(2r(t_1) - 1), \\ &\text{cumsum}(2r(t_2) - 1), \dots, \text{cumsum}(2r(t_n) - 1)], \end{aligned} \tag{10}$$

where $n$ is the maximum number of iterations, cumsum calculates the cumulative sum, and $t$ is the step of random walk. Hence,

$$r(t) = \begin{cases} 1 & \text{if rand} > 0.5 \\ 0 & \text{if rand} < 0.5. \end{cases} \tag{11}$$

Here, $r(t)$ is a stochastic function and rand is a random number generated with uniform distribution in the interval of $[0, 1]$.

The positions of ants are saved and utilized during optimization in the matrix:

$$M_{\text{Ant}} = \begin{bmatrix} A_{1,1} & A_{1,2} & \cdots & \cdots & A_{1,d} \\ A_{2,1} & A_{2,2} & \cdots & \cdots & A_{2,d} \\ \vdots & \vdots & \vdots & \vdots & \vdots \\ \vdots & \vdots & \vdots & \vdots & \vdots \\ A_{n,1} & A_{n,2} & \cdots & \cdots & A_{n,d} \end{bmatrix}, \tag{12}$$

where $M_{\text{Ant}}$ is the matrix for saving the position of each ant, $A_{i,j}$ shows the value of the $j$th variable of $i$th ant, $n$ is the number of ants, and $d$ is the number of variables.

At each step of optimization, ants update their position according to random walk. Equation (10) cannot be directly used for updating position of ants. The random walks are normalized using the following equation (min-max normalization):

$$X_i^t = \frac{(X_i^t - a_i) \times (d_i - c_i^t)}{(d_i^t - a_i)} + c_i, \tag{13}$$

TABLE 1: Optimized parameters of AGC regulator.

| Parameters | ALO | | GSA [17] | | PSO [18] | | GA [19] | |
|---|---|---|---|---|---|---|---|---|
| | $J_1$ | $J_2$ | $J_1$ | $J_2$ | $J_1$ | $J_2$ | $J_1$ | $J_2$ |
| $K_{I1}$ | 0.3260 | 0.4002 | 0.3817 | 0.4171 | 0.3131 | 0.4498 | 0.3031 | 0.6525 |
| $K_{I2}$ | 0.2135 | 0.2010 | 0.2153 | 0.2028 | 0.1091 | 0.2158 | 0.3063 | 0.7960 |
| $R_1$ | 0.0491 | 0.0404 | 0.0401 | 0.0435 | 0.0581 | 0.0201 | 0.0794 | 0.0503 |
| $R_2$ | 0.0699 | 0.0509 | 0.0657 | 0.0635 | 0.0531 | 0.03 | 0.0737 | 0.0609 |
| $D_1$ | 0.4457 | 0.4884 | 0.5889 | 0.4778 | 0.4756 | 0.5910 | 0.7591 | 0.7216 |
| $D_2$ | 0.8770 | 0.8975 | 0.8946 | 0.8744 | 0.6097 | 0.8226 | 0.8950 | 0.8984 |

where $a_i$ is the minimum of random walk of $i$th variable, $d_i$ is the maximum of random walk of $i$th variable, $c_i^t$ is the minimum of $i$th variable at $t$th iteration, and $d_i^t$ is the maximum of $i$th variable at $t$th iteration.

*(b) Trapping in Antlion's Pit.* Random walks of ants are affected by antlions' trap. Mathematical modeling of trapping in antlion's pit is proposed by the following equations:

$$c_i^t = \text{Antlion}_j^t + c^t, \quad d_i^t = \text{Antlion}_j^t + d^t, \tag{14}$$

where $c^t$ represents the minimum of all variables at $t$th iteration, $d^t$ indicates the vector including the maximum of all variables at $t$th iteration, $c_i^t$ is the minimum of all variables for $i$th ant, $d_i^t$ is the maximum of all variables for $i$th ant, and $\text{Antlion}_j^t$ shows the position of the selected $j$th antlion at $t$th iteration.

*(c) Building Trap.* For building trap, a roulette wheel is employed to model the hunting capability of antlions. The ALO algorithm is required to utilize a roulette wheel operator for selecting antlions based on their fitness during optimization. This mechanism provides high chances to the fitter antlions for catching ants.

*(d) Sliding Ants towards Antlion.* Antlions are able to build traps which are proportional to their fitness and ants are required to move randomly. Once the antlion realizes that an ant is in the trap, it shoots sand out the centre of the pit. The ants which are trying to escape slide down the trap. The radius of the ant's random walks hypersphere is decreased adaptively in the mathematical modeling. The following equations are proposed for this:

$$c^t = \frac{c^t}{I}, \quad d^t = \frac{d^t}{I}, \tag{15}$$

where $I$ is a ratio, $c^t$ is the minimum of all variables at $t$th iteration, and $d^t$ indicates the vector including the maximum of all variables at $t$th iteration.

*(e) Catching Prey and Rebuilding the Pit.* This is the final stage of hunt. At this stage, an ant reaches the bottom of the pit and is caught in the antlion's jaw. After this stage, the antlion pulls the ant inside the sand and consumes its body. Catching the prey occurs when the ant goes inside the sand and becomes fitter than its corresponding antlion. According to the position of the latest hunted ant, the antlions update their position to enhance the chances of catching new prey. Mathematically, the following equations can be proposed in this regard:

$$\text{Antlion}_j^t = \text{Ant}_i^t \quad \text{if } f\left(\text{Ant}_i^t\right) > f\left(\text{Antlion}_j^t\right), \tag{16}$$

where $t$ represents the current iteration, $\text{Antlion}_j^t$ is the position of the selected $j$th antlion at $t$th iteration, and $\text{Ant}_i^t$ represents the position of $i$th ant at $t$th iteration.

*(f) Elitism.* For any evolutionary algorithm, elitism is an important feature that allows antlions to maintain the best solution obtained at any stage of optimization process. In this algorithm, the best obtained antlion during the entire iteration is saved and is considered as an elite. Since the fittest antlion is elite, it affects the movement of all the ants during iteration. Hence, it is assumed that every ant walks randomly around a selected antlion by roulette wheel and the elite simultaneously as follows:

$$\text{Ant}_i^t = \frac{R_A^t + R_E^t}{2}, \tag{17}$$

where $R_A^t$ is the random walk around the antlion selected by the roulette wheel at $t$th iteration, $R_E^t$ is the random walk, and $\text{Ant}_i^t$ represents the position of $i$th ant at $t$th iteration.

The following section presents analysis of simulation results.

## 4. Results and Analysis

This section presents simulation results and analysis of AGC regulator performance on two-area thermal interconnected power system with different step perturbations and loading conditions. Different AGC regulator settings are obtained with the application of four algorithms (GA, PSO, GSA, and ALO) on two standard objective functions (ISE and ITAE). Table 1 shows the values of optimized parameters of regulator with the application of the abovementioned algorithms on two objective functions.

Table 2 shows the values of system's minimum damping ratio and eigenvalues after the application of these AGC

Table 2: Eigenvalues and minimum damping ratio.

| Parameter | ALO | | GSA [17] | | PSO [18] | | GA [19] | |
|---|---|---|---|---|---|---|---|---|
| | $J_1$ | $J_2$ | $J_1$ | $J_2$ | $J_1$ | $J_2$ | $J_1$ | $J_2$ |
| System modes | −5.8548 | −5.9604 | −5.8468 | −5.976 | −5.846 | −6.5657 | −5.6586 | −5.808 |
| | −4.2219 | −4.4165 | −4.313 | −4.4257 | −4.4443 | −4.8155 | −4.2083 | −4.2168 |
| | −0.3805 ± 1.7191$i$ | −0.2885 ± 1.8854$i$ | −0.3994 ± 1.7029$i$ | 0.2511 ± 1.9124$i$ | **−0.4010 ± 1.7004$i$** | −0.0030 ± 2.6953$i$ | −0.4925 ± 1.3799$i$ | −0.2024 ± 1.6817$i$ |
| | −0.3007 ± 1.4854$i$ | −0.2088 ± 1.7320$i$ | −0.2606 ± 1.6066$i$ | −0.1924 ± 1.7420$i$ | **−0.2406 ± 1.7718$i$** | −0.0220 ± 2.1889$i$ | −0.2491 ± 1.4729$i$ | **0.0361 ± 1.5786$i$** |
| | −0.3716 | −0.4624 | −0.3395 | −0.5169 | **−0.0983 ± 0.0157$i$** | −0.4666 | −0.1353 | −0.1058 |
| | −0.1179 | −0.0910 | −0.1102 | −0.0884 | −0.3521 | −0.0494 | −0.3294 | −0.7991 |
| | −0.2256 | −0.2351 | −0.2061 | −0.2416 | | −0.2144 | −0.3712 | −0.9209 |
| Minimum damping ratio | **0.1984** | **0.1197** | 0.1601 | 0.1098 | 0.1345 | 0.0011 | 0.1668 | 0.0229 |

regulators. Eigenvalue analysis plays an important role in stability studies. Complex conjugate eigenvalues are also known as swing modes and these eigenvalues are responsible for oscillatory instability, when the real part of the eigenvalue is positive. From Table 2, it is observed that when optimization process is carried out with the application of GA on $J_2$ system mode contains a swing mode with positive eigenvalue (0.361). Real positive part of eigen is the indication of oscillations of growing amplitude. The minimum damping ratios obtained from the application of different regulators with $J_1$ and $J_2$ criteria are shown in Table 2. For PSO regulator, minimum damping ratios obtained from these criteria are (0.1345, 0.0011); similarly, the ratios for GA are (0.1668, 0.0229), for GSA are (0.1601, 0.1098), and for ALO are (0.1984, 0.1197). It can be said that a considerable amount of damping is enhanced in each case when the regulator parameters are obtained with criterion $J_1$. Overall damping of the system is the highest with ALO regulator ($J_1$) (0.1984). Prima facie, it can be concluded that the regulator design obtained from criterion $J_1$ is more effective as the damping enhanced by this regulator is higher. In eigenvalue analysis, both real and imaginary parts have their interpretation and physical significance. The real part of the complex conjugate eigenvalue shows the damping behavior which represents the damp oscillations, which means the larger the magnitude, the higher the rate of decay. Imaginary components show the frequency of oscillations. It can be observed from Table 2 that high frequency oscillations are associated with setting $J_2$. Higher frequency oscillations are bad for equipment health and often cause the damage of physical structure of controllers. In this case for $J_2$, GA frequency of oscillations is (1.57, 1.68) for PSO, (2.69, 2.18) for GSA (1.91, 1.74), and (1.88, 1.73) for ALO. It is observed that although frequency of oscillations is in moderate range for GA regulator, the amplitude of the oscillation is growing with time as it has a positive real part of eigenvalue. However, the other regulators have high frequency of oscillations modes as compared with ALO. To show this analysis in a more prominent way, AGC regulators are designed with ALO algorithm and tested for different levels of perturbations. Figures 3(a) and 3(b) show the dynamic responses of frequency deviations in areas 1 and 2 when area 1 observes a step disturbance of 0.01 p.u. Figures 3(c) and 3(d) show the frequency deviation curves of both areas with both regulator settings $J_1$ and $J_2$ when area 2 is perturbed with 0.02 p.u. Similarly, for both regulator settings, the dynamic responses obtained from both areas are self-explanatory. It is observed that $J_1$ setting is promising. The overshoot and settling time of the frequency deviation curves of both areas are less with $J_1$ regulator. It is also empirical to judge that the variations of tie-line power exchanges are nominal with both types of perturbation with $J_1$ regulator. Hence, it is concluded that $J_1$ optimization criterion is suitable for the designing of the AGC regulator.

To exhibit the comparative performance of the ALO regulator with other regulators, four different loading scenarios are simulated in this work. These loading conditions are summarized below.

*Case 1.* Load changes in area 1 by 10%. The dynamic responses of $\Delta F_1$, $\Delta F_2$, and $\Delta P_{\text{tie}}$ are given in Figures 4–6 for all the algorithms.

*Case 2.* Load changes in area 2 by 20%. Figures 7–9 show the dynamic responses of the system.

*Case 3.* Load is increased in area 1 by 25%. In Figures 10–12, the system dynamic responses are shown.

*Case 4.* Load is decreased in area 1 by 25% and the system dynamic responses are given in Figures 13–15.

Dynamic responses along with the system eigenvalues for these conditions are exhibited in Table 3. It is observed that again with setting $J_2$ few eigenvalues possess positive real part when optimized with GA (0.0370, 0.0382, and 0.0368). The real part of swing mode varies from −0.2823 to −0.4567 for ALO regulator, from −0.2541 to −0.4632 for GSA regulator, from −0.0982 to −4.587 for PSO regulator, and from −0.2511 to −0.5411 for GA regulator with criterion $J_1$. It is of note here that the real part of the eigenvalue observes a large variation in case of GA under different loading conditions. This spread put a question mark on the performance of the regulator and robustness of the regulator also. Moderate spread has been observed with ALO regulator. For all cases, higher numeric values of real part of the eigenvalues suggest that the system is more stable. In Case 1, these values are $(-0.4278, -0.2823)$ for ALO, $(-0.4288, -0.2570)$ for GSA, $(-0.4277, -0.2395)$ for PSO, and $(-0.2588, -0.5271)$ for GA. It can be predicted that for Case 1 the robust setting is achieved by ALO. Similarly, in Case 4, the real parts of eigenvalues (swing modes) are $(-0.3276, -0.2879)$ for ALO and $(-0.3106, -0.2589)$ for GSA and an additional swing mode with PSO setting has been observed, ($J_1$) $(-0.3055, -0.2459, 0.0983)$ and $(-0.440, 0.2680)$ for GA. From this, it is also observed that a higher degree of robustness can be achieved by ALO regulator. To understand the dynamic response of the frequency deviation curves, a conventional index Figure of Demerit (FOD) is used in this paper. Figure of Demerit is the summation of the square of the overshoot and settling time of the deviation curves. It is observed that for almost all loading cases the values of settling time, overshoot, and FODs are low for ALO based regulators as compared with other regulator designs. It is observed from Figures 4–6 that ALO based controller exhibits better dynamic performance as compared with others. The percentage of overshoot and settling time is much less in these cases. The low oscillatory response exhibited by ALO is best suited for the equipment's health. FOD values are considered as a close replica of dynamic performance of controller. Higher values of FOD show poor dynamic performance and vice versa. It is also empirical to mention here that for frequency deviation in area 1 the settling time and FOD obtained from ALO are 3.8 and 14.44, respectively, whereas, from GSA, PSO, and GA, the settling time and FOD are 5.6, 5.0, and 4.9 and 31.36, 25, and 24.01, respectively. The frequency deviation in area 2 also shows that the values of settling time and FOD are less when ALO

TABLE [illegible]: System modes for each case of all the algorithms.

| Parameters | ALO | | GSA [17] | | PSO [18] | | GA [19] | |
|---|---|---|---|---|---|---|---|---|
| | $J_1$ | $J_2$ | $J_1$ | $J_2$ | $J_1$ | $J_2$ | $J_1$ | $J_2$ |
| | −5.8014 | −5.8014 | −5.7891 | −5.9112 | −5.7884 | −6.4711 | −5.5532 | −5.752 |
| | −4.2274 | −4.2274 | −4.313 | −4.4257 | −4.4443 | −4.8155 | −4.1792 | −4.2168 |
| | −0.4924 ± 1.6361*i* | −0.4924 ± 1.6361*i* | −0.4288 ± 1.6[illegible]43*i* | −0.2773 ± 1.8079*i* | −0.4277 ± 1.6059*i* | −0.0430 ± 2.5784*i* | −0.2588 ± 1.4307*i* | −0.2211 ± 1.5866*i* |
| Case 1 | −0.2842 ± 1.4933*i* | −0.2842 ± 1.4933*i* | −0.2570 ± 1.6[illegible]85*i* | −0.1941 ± 1.7460*i* | −0.2395 ± 1.7695*i* | −0.0222 ± 2.1888*i* | −0.5271 ± 1.1657*i* | **0.0370** ± 1.5795*i* |
| | −0.1208 | −0.1208 | −0.3454 | −0.5259 | −0.0983 ± 0.0157*i* | −0.4806 | −0.1466 | −0.1058 |
| | −0.2021 | −0.2021 | −0.1101 | −0.0884 | −0.3584 | −0.0494 | −0.3344 | −0.9221 |
| | −0.2229 | −0.2229 | −0.2062 | −0.2416 | | −0.2144 | −0.401 | −0.8182 |
| | −5.8597 | −5.9843 | −5.8468 | −5.976 | −5.846 | −6.564 | −5.5965 | −5.808 |
| | −4.1275 | −4.2672 | −4.2059 | −4.3093 | −4.3269 | −4.6691 | −4.0832 | −4.1155 |
| | −0.4646 ± 1.7341*i* | −0.2906 ± 1.9218*i* | −0.4002 ± 1.7[illegible]63*i* | −0.2534 ± 1.9127*i* | −0.3943 ± 1.7014*i* | **0.0014** ± 2.6941*i* | −0.5108 ± 1.2557*i* | −0.2029 ± 1.6828*i* |
| Case 2 | −0.3315 ± 1.3466*i* | −0.1072 ± 1.5609*i* | −0.3117 ± 1.4[illegible]71*i* | −0.2466 ± 1.5904*i* | −0.3057 ± 1.6218*i* | −0.0944 ± 2.0213*i* | −0.3032 ± 1.2918*i* | −0.0013 ± 1.4380*i* |
| | −0.1204 | −0.0879 | −0.3394 | −0.5169 | −0.0986 ± 0.0155*i* | −0.4774 | −0.1467 | −0.1059 |
| | −0.2047 | −0.4498 | −0.1101 | −0.0884 | −0.3521 | −0.0494 | −0.344 | −0.8003 |
| | −0.2234 | −0.5752 | −0.2096 | −0.245 | | −0.2155 | −0.3879 | −0.9453 |
| | −5.7282 | −5.8373 | −5.7167 | −5.8312 | −5.716 | −6.353 | −5.4991 | −5.6819 |
| | −4.2274 | −4.3812 | −4.313 | −4.4278 | −4.4443 | −4.8155 | −4.1792 | −4.2168 |
| | −0.5297 ± 1.5095*i* | −0.3560 ± 1.6872*i* | −0.4632 ± 1.4[illegible]84*i* | −0.3257 ± 1.6774*i* | −0.4587 ± 1.4794*i* | −0.0990 ± 2.4269*i* | −0.2590 ± 1.4294*i* | −0.2413 ± 1.4636*i* |
| Case 3 | −0.2816 ± 1.4943*i* | −0.0567 ± 1.7151*i* | −0.2541 ± 1.6[illegible]63*i* | −0.1938 ± 1.7505*i* | −0.2396 ± 1.7679*i* | −0.0228 ± 2.1889*i* | −0.5411 ± 1.0386*i* | **0.0382** ± 1.5795*i* |
| | −0.1204 | −0.0878 | −0.355 | −0.5367 | −0.0982 ± 0.0157*i* | −0.4855 | −0.1462 | −0.1058 |
| | −0.2039 | −0.4672 | −0.11 | −0.0886 | −0.3686 | −0.0494 | −0.3357 | −0.9251 |
| | −0.2252 | −0.5609 | −0.2063 | −0.2401 | | −0.2144 | −0.4258 | −0.8475 |
| | −6.0556 | −6.2024 | −6.041 | −6.1949 | −6.0401 | −6.8711 | −5.7445 | −5.9969 |
| | −4.2274 | −4.3812 | −4.313 | −4.4278 | −4.4443 | −4.8155 | −4.1792 | −4.2168 |
| | −0.3695 ± 2.0323*i* | −0.1897 ± 2.2365*i* | −0.3106 ± 2.0[illegible]88*i* | −0.1627 ± 2.2305*i* | −0.3055 ± 2.0076*i* | **0.1510** ± 3.0616*i* | −0.4440 ± 1.5429*i* | −0.1319 ± 1.9807*i* |
| Case 4 | −0.2838 ± 1.4894*i* | −0.0573 ± 1.7122*i* | −0.2589 ± 1.6[illegible]17*i* | −0.1958 ± 1.7442*i* | −0.2459 ± 1.7656*i* | −0.0217 ± 2.1890*i* | −0.2680 ± 1.4339*i* | **0.0368** ± 1.5765*i* |
| | −0.1216 | −0.0879 | −0.3261 | −0.4948 | −0.0983 ± 0.0158*i* | −0.4698 | −0.1478 | −0.1059 |
| | −0.1971 | −0.434 | −0.1105 | −0.0886 | −0.3379 | −0.0494 | −0.3276 | −0.7544 |
| | −0.2194 | −0.5605 | −0.2059 | −0.24 | | −0.2144 | −0.3632 | −0.9192 |

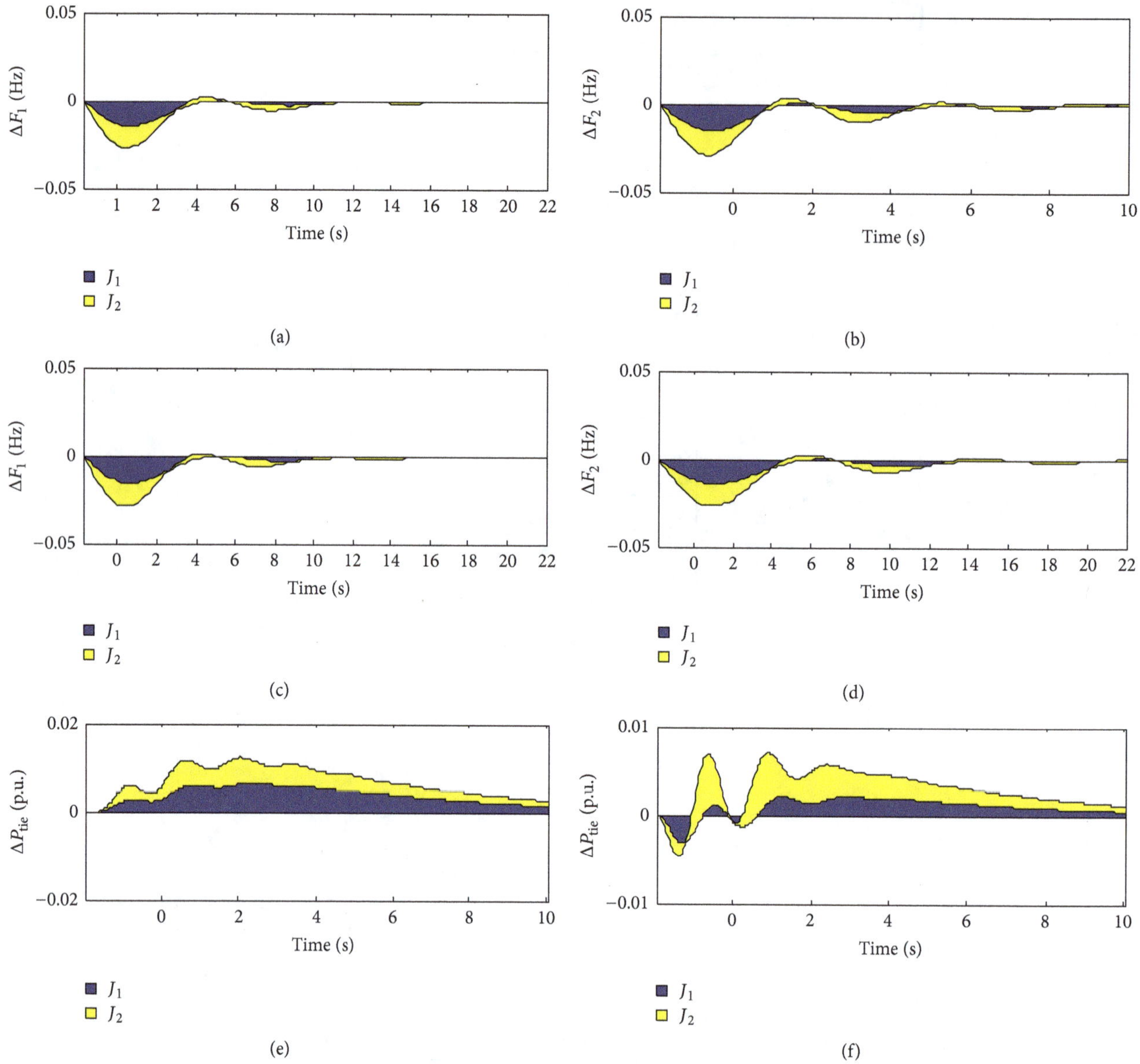


FIGURE 3: Dynamic responses obtained from ALO.

regulator is used. The value of settling time with frequency deviation in area 2 is observed as 6.0 for ALO, 7.2 for GSA, 7.6 for PSO, and 8.0 for GA. It is also interesting to observe that with the 10% increase in the load PSO gives erroneous results and the flow of tie-line power behaves in a different manner. Hence, critical analysis of dynamic responses clearly reveals that better dynamic performance is exhibited by ALO. By examining the responses in Figures 7–9, it is clearly seen that the settling time and peak overshoot are less when load changes in area 2 by 20%. It can be observed from Figure 8 that when area 2 observes 20% increase, GA based controller is not able to mitigate the frequency oscillations. This inculcates oscillatory instability in the system. However, ALO based controller shows a better dynamic response and yields satisfactory performance over a wide range of loading conditions. For Case 2, the settling time of ALO is 3.6 and is 5.3, 4.2, and 5.8 for GSA, PSO, and GA, respectively. Similarly, the FOD is also very low in case of ALO, that is, 12.96, whereas it is 17.64 and 33.64 in case of PSO and GA. Figures 10 and 11 show the frequency deviations of areas 1 and 2. From dynamic responses of overshoot, settling time, and FOD, it is clear that ALO provides competitive results. The dynamic responses for Case 4 are shown in Figures 13–15 and it has been observed that ALO tuned controller yields better dynamic performance. The minimum settling time and FOD obtained from ALO are 5.1 and 26.01 for frequency deviation in area 1 and 6.0 and 36.00 for frequency deviation in area 2. However, in case of GSA, the settling time is 7.9 and FOD is 19.8. An oscillatory response is obtained by the GA, GSA, and PSO tuned controllers.

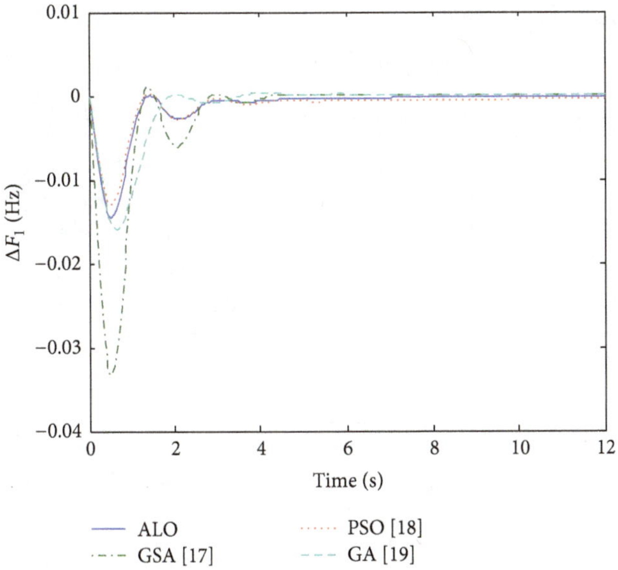


FIGURE 4: Change in frequency of area 1 by 10% load increase in area 1.

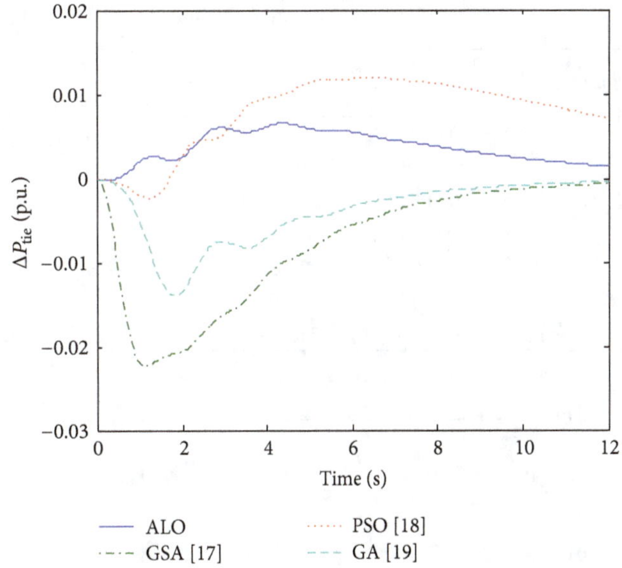


FIGURE 6: Change in tie-line power by 10% load increase in area 1.

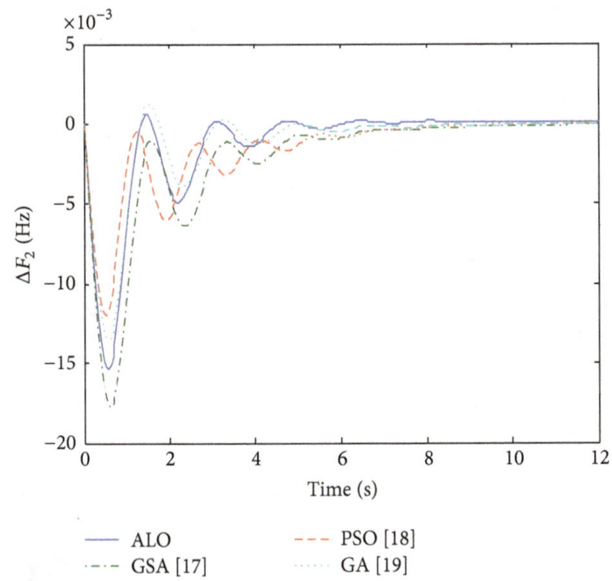


FIGURE 5: Change in frequency of area 2 by 10% load increase in area 1.

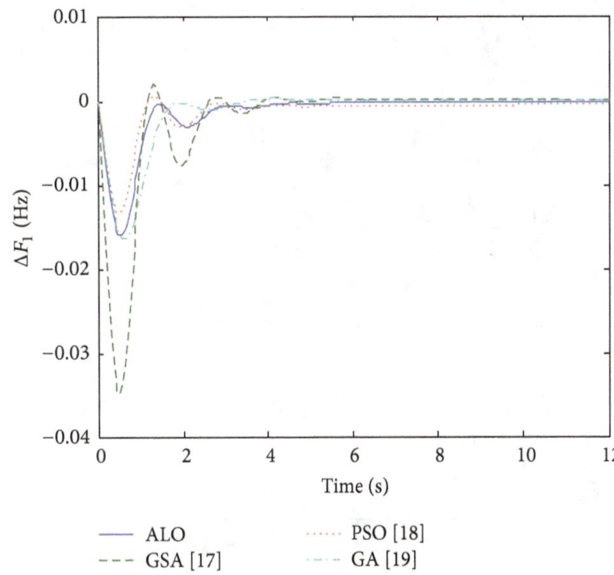


FIGURE 7: Change in frequency of area 1 by 20% load increase in area 2.

## 5. Optimization Performance

To judge the efficiency of the optimization process carried out by all algorithms, 100 trials of optimization are carried out. To provide a fair comparison, the population size (100) and the maximum number of iterations (1000) are kept the same. Stopping criterion for the optimization process is maximum run of the iteration. To observe the optimization process in a critical way, the standard deviations of optimized parameters of the regulator along with the values of objective functions are calculated and shown in Table 4. It is observed that high values of standard deviations are obtained in regulator parameters and values of objective functions when optimization process is handled by GA (0.09729). Comparatively large values of standard deviations are found in GA and PSO when they are compared with ALO. We observed high values of standard deviation in speed regulation parameters after each run of optimization obtained with GA regulators ($J_1$ and $J_2$). Speed regulation parameter is a vulnerable parameter in power system dynamics studies. It affects the system dynamics in a very prominent way. Large values of standard deviations in the calculation of such vulnerable parameters are not acceptable. The values for standard deviations are 0.10256 (GSA, $J_1$) and 0.1550 (GA, $J_2$) and, similarly, for $R_2$, they are 0.0081 (ALO, $J_1$), 0.02 (GSA), 0.38 (PSO), and

TABLE 4: Standard deviation of optimized parameters of the regulator.

| Parameters | ALO | | GSA [17] | | PSO [18] | | GA [19] | |
|---|---|---|---|---|---|---|---|---|
| | $J_1$ | $J_2$ | $J_1$ | $J_2$ | $J_1$ | $J_2$ | $J_1$ | $J_2$ |
| $K_{I1}$ | 0.02923 | 0.03773 | 0.02425 | 0.07720 | 0.02604 | 0.08461 | 0.09729 | 0.20687 |
| $D_1$ | 0.04107 | 0.03315 | 0.04475 | 0.09833 | 0.08416 | 0.04759 | 0.18872 | 0.21313 |
| $R_1$ | 0.00582 | 0.00102 | 0.00045 | 0.00168 | 0.00153 | 0.01403 | 0.00892 | 0.00717 |
| $K_{I2}$ | 0.01763 | 0.04563 | 0.10256 | 0.00769 | 0.09639 | 0.08158 | 0.05800 | 0.15550 |
| $D_2$ | 0.08957 | 0.08807 | 0.04916 | 0.11771 | 0.16726 | 0.08363 | 0.17955 | 0.07945 |
| $R_2$ | 0.00343 | 0.00174 | 0.00173 | 0.00385 | 0.00280 | 0.01080 | 0.00310 | 0.0105937 |
| | 0.00184 | $7.16E-06$ | 0.0212 | 0.000419 | 0.38 | 0.011 | 1.76 | 0.013 |

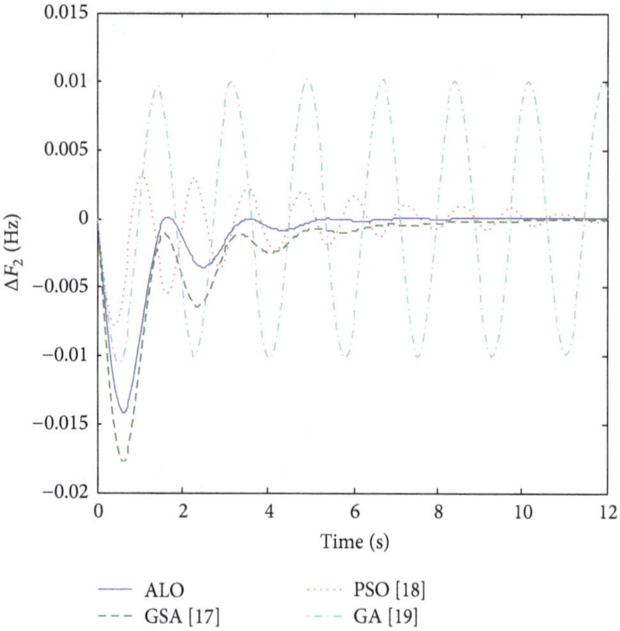


FIGURE 8: Change in frequency of area 2 by 20% load increase in area 2.

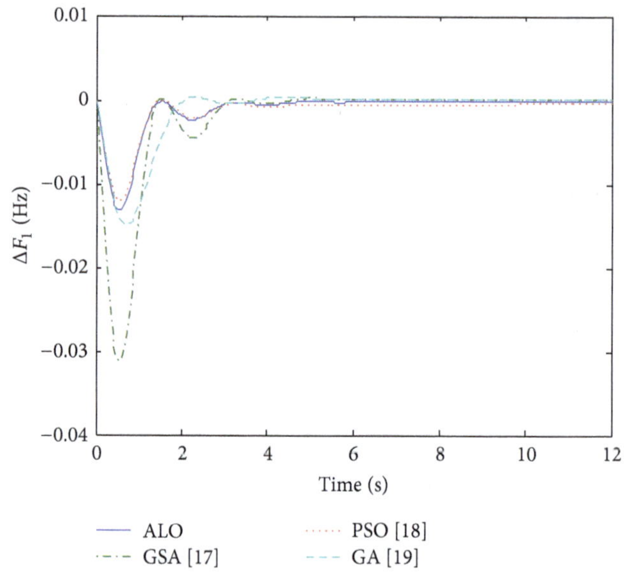


FIGURE 10: Change in frequency of area 1 by 25% load increase in area 1.

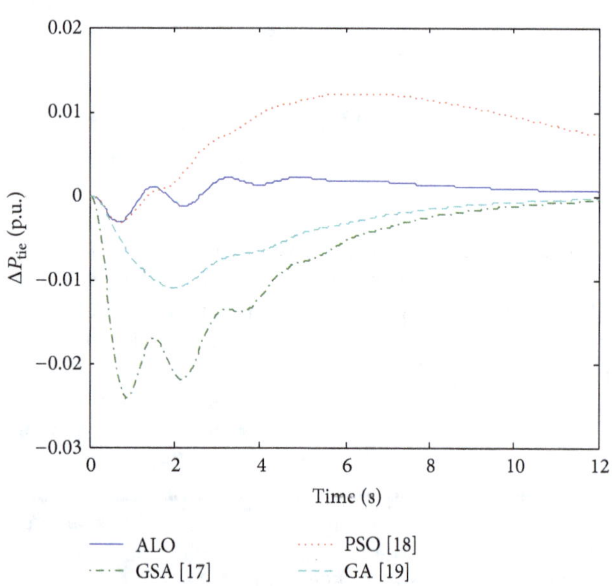


FIGURE 9: Change in tie-line power by 20% load increase in area 2.

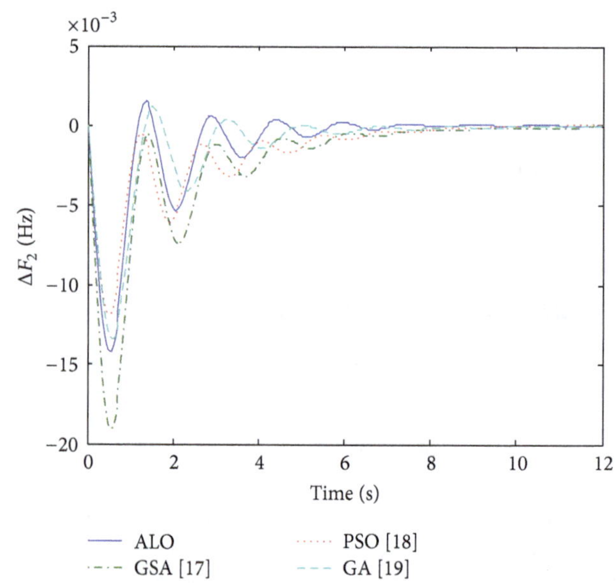


FIGURE 11: Change in frequency of area 2 by 25% load increase in area 1.

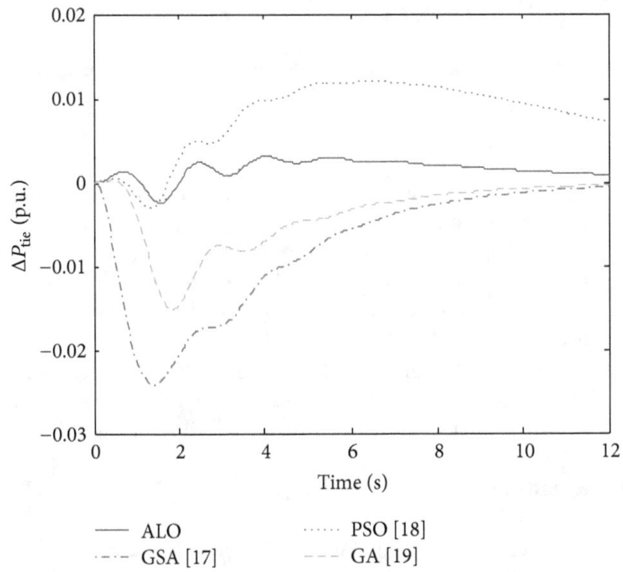


FIGURE 12: Change in tie-line power by 25% load increase in area 1.

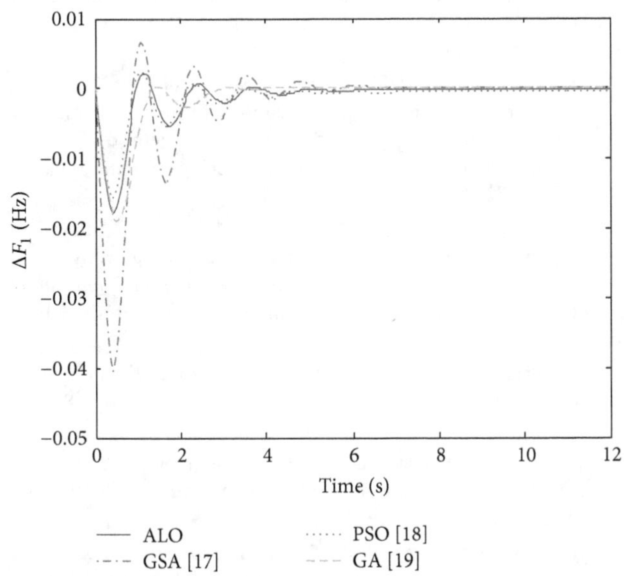


FIGURE 13: Change in frequency of area 1 by 25% load decreases in area 1.

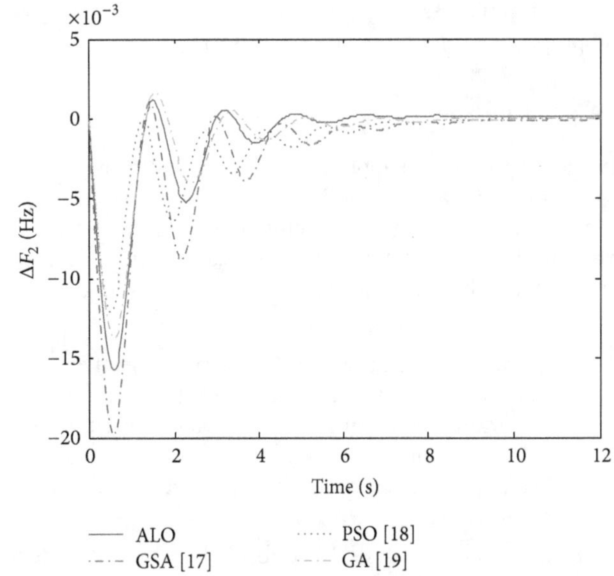


FIGURE 14: Change in frequency of area 1 by 25% load decreases in area 1.

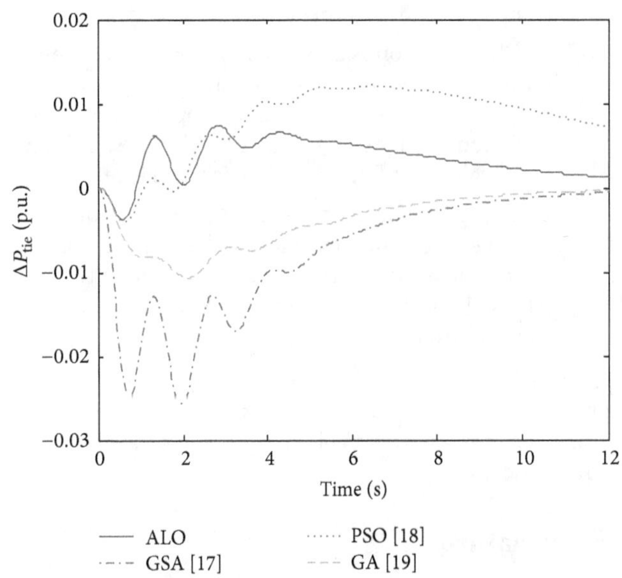


FIGURE 15: Change in tie-line power by 25% load decreases in area 1.

1.76 for GA. The impact of speed regulation parameters on the dynamic response is shown in [25]. The lowest values of standard deviations are observed, when the parameters are optimized by ALO. This basically means that in each run of optimization ALO exhibits precision in computing the parameters. The standard deviations in the values of integral gains for area 1 by $J_1$ and $J_2$ are minimum for ALO (0.041, 0.033); for GA, these values are (0.18 and 0.213) and similarly (0.044, 0.09) for GSA and (0.08 and 0.047) for PSO. It can be concluded that regulator setting integral gain observes the least variation in numerical values when the parameter is optimized through ALO ($J_1$). The values of standard deviations in objective functions $J_1$ and $J_2$ are the lowest for ALO process and the highest for GA. The values of standard deviations in the values of $J_1$ for GA, PSO, GSA, and GWO are 0.0972, 0.0264, 0.0242, and 0.02 and for $J_2$ are 0.2068, 0.0846, 0.0772, and 0.3773. It has been observed that the values obtained by ALO are precise and the optimization processes are reliable enough for obtaining the regulator design. However, high values of standard deviations in parameters of regulator (1.76 with $R_2$) and in the objective functions show that the optimization process loses its relevance when it is handled by GA. Hence, it is concluded from the eigenvalue analysis and dynamic response of the deviation curves that ALO shows a promising response to obtain regulator setting with optimization criterion $J_1$. The following section summarizes the contribution of this research work and proposes a solid milieu for future work.

## 6. Conclusion

This paper presents an application of the recently introduced algorithm ALO to find optimal parameters of the AGC regulator. The ALO regulator is employed on a test system of two thermal units connected with a weak tie line of limited capacity for AGC. The following are the major findings of the work:

(A) Comparison of the application of two objective functions, namely ISE and ITAE, in optimization process for finding the regulator parameters, under different contingencies, is investigated. Results reveal that ITAE is a better choice to optimize the regulator parameters.

(B) Eigenvalue analysis is performed to test the effectiveness of the proposed approach and to compare the results of the proposed approach with recently published approaches. It is observed that the damping obtained from ALO regulator is more positive as compared with the other algorithms.

(C) ALO shows promising results in terms of overshoot, settling time obtained from the frequency responses of both areas under different loading cases, standard deviations in regulator's parameter values, ISE and ITAE values, and optimization performance.

(D) Damping performance is evaluated with different contingencies, load changes, and step disturbances in both areas. PI controller setting obtained through ALO exhibits better dynamic performance and overall low settling time.

Application of other new metaheuristic algorithms in AGC regulator design on different models of the power system considering different renewable energy power sources lies in the future scope.

## Nomenclature

$i$: Subscript referring to area $i$ (1, 2)
$\Delta F_i$: Frequency deviation in area $i$ (Hz)
$\Delta P_{Gi}$: Incremental generation of area $i$ (p.u.)
$\Delta P_{Li}$: Incremental load change in area $i$ (p.u.)
$\mathrm{ACE}_i$: Area Control Error of area $i$
$B_i$: Frequency bias parameter of area $i$
$R_i$: Speed regulation of the governor of area $i$ (Hz/p.u. MW)
$T_{gi}$: Time constant of governor of area $i$ (s)
$T_{ti}$: Time constant of turbine of area $i$ (s)
$K_{pi}$: Gain of generator and load of area $i$
$T_{pi}$: Time constant of generator and load of area $i$ (s)
$\Delta P_{\mathrm{tie}}$: Incremental change in tie line (p.u.)
$T_{12}$: Synchronizing coefficient
$T$: Simulation time (s)
$t$: Current iteration.

## Competing Interests

The authors declare that they have no competing interests.

## References


[1] O. I. Elgerd, *Energy Systems Theory. An Introduction*, McGraw-Hill, New Delhi, India, 1983.

[2] IEEE Standard Committee, "IEEE Standard definitions of terms for automatic generation control on electric power systems," *IEEE Transactions on Power Apparatus and Systems*, vol. 89, no. 6, pp. 1356–1364, 1970.

[3] A. Ibraheem, P. Kumar, and D. P. Kothari, "Recent philosophies of automatic generation control strategies in power systems," *IEEE Transactions on Power Systems*, vol. 20, no. 1, pp. 346–357, 2005.

[4] H. Sadat, *Power System Analysis*, McGraw-Hill, New York, NY, USA, 1999.

[5] J. Nanda and B. Kaul, "Automatic generation control of an interconnected power system," *Proceedings of the Institution of Electrical Engineers*, vol. 125, no. 5, pp. 385–390, 1978.

[6] A. Y. Sivaramakrishnan, M. V. Hariharan, and M. C. Srisailam, "Design of variable-structure load-frequency controller using pole assignment technique," *International Journal of Control*, vol. 40, no. 3, pp. 487–498, 1984.

[7] M. Z. Bernard, T. H. Mohamed, Y. S. Qudaih, and Y. Mitani, "Decentralized load frequency control in an interconnected power system using Coefficient Diagram Method," *International Journal of Electrical Power and Energy Systems*, vol. 63, pp. 165–172, 2014.

[8] R. Ali, T. H. Mohamed, Y. S. Qudaih, and Y. Mitani, "A new load frequency control approach in an isolated small power systems using coefficient diagram method," *International Journal of Electrical Power and Energy Systems*, vol. 56, pp. 110–116, 2014.

[9] I. Vajk, M. Vajta, L. Keviczky, R. Haber, J. Hetthéssy, and K. Kovács, "Adaptive load-frequency control of the hungarian power system," *Automatica*, vol. 21, no. 2, pp. 129–137, 1985.

[10] J. Kanniah, S. C. Tripathy, O. P. Malik, and G. S. Hope, "Microprocessor-based adaptive load-frequency control," *IEE Proceedings C—Generation, Transmission and Distribution*, vol. 131, no. 4, pp. 121–128, 1984.

[11] M. L. Kothari, P. S. Satsangi, and J. Nanda, "Sampled-data automatic generation control of interconnected reheat thermal systems considering generation rate constraints," *IEEE Transactions on Power Apparatus and Systems*, vol. 100, no. 5, pp. 2334–2342, 1981.

[12] S. Wu, M. J. Er, and Y. Gao, "A fast approach for automatic generation of fuzzy rules by generalized dynamic fuzzy neural networks," *IEEE Transactions on Fuzzy Systems*, vol. 9, no. 4, pp. 578–594, 2001.

[13] B. K. Sahu, S. Pati, P. K. Mohanty, and S. Panda, "Teaching-learning based optimization algorithm based fuzzy-PID controller for automatic generation control of multi-area power system," *Applied Soft Computing Journal*, vol. 27, pp. 240–249, 2015.

[14] K. R. Sudha and R. Vijaya Santhi, "Load frequency control of an interconnected reheat thermal system using type-2 fuzzy system including SMES units," *International Journal of Electrical Power & Energy Systems*, vol. 43, no. 1, pp. 1383–1392, 2012.

[15] A. Banerjee, V. Mukherjee, and S. P. Ghoshal, "Intelligent controller for load-tracking performance of an autonomous

power system," *Ain Shams Engineering Journal*, vol. 5, no. 4, pp. 1167–1176, 2014.

[16] S. Padhan, R. K. Sahu, and S. Panda, "Automatic generation control with thyristor controlled series compensator including superconducting magnetic energy storage units," *Ain Shams Engineering Journal*, vol. 5, no. 3, pp. 759–774, 2014.

[17] R. K. Sahu, S. Panda, and S. Padhan, "Optimal gravitational search algorithm for automatic generation control of interconnected power systems," *Ain Shams Engineering Journal*, vol. 5, no. 3, pp. 721–733, 2014.

[18] Y. L. Abdel-Magid and M. A. Abido, "AGC tuning of interconnected reheat thermal systems with particle swarm optimization," in *Proceedings of the 10th IEEE International Conference on Electronics, Circuits and Systems (ICECS '03)*, pp. 376–379, December 2003.

[19] Y. L. Abdel-Magid and M. M. Dawoud, "Optimal AGC tuning with genetic algorithms," *Electric Power Systems Research*, vol. 38, no. 3, pp. 231–238, 1996.

[20] J. Nanda, S. Mishra, and L. C. Saikia, "Maiden application of bacterial foraging-based optimization technique in multiarea automatic generation control," *IEEE Transactions on Power Systems*, vol. 24, no. 2, pp. 602–609, 2009.

[21] U. K. Rout, R. K. Sahu, and S. Panda, "Design and analysis of differential evolution algorithm based automatic generation control for interconnected power system," *Ain Shams Engineering Journal*, vol. 4, no. 3, pp. 409–421, 2013.

[22] H. Gozde, M. C. Taplamacioglu, and I. Kocaarslan, "Comparative performance analysis of Artificial Bee Colony algorithm in automatic generation control for interconnected reheat thermal power system," *International Journal of Electrical Power and Energy Systems*, vol. 42, no. 1, pp. 167–178, 2012.

[23] S. Debbarma, L. Chandra Saikia, and N. Sinha, "Solution to automatic generation control problem using firefly algorithm optimized $I\lambda D\mu$ controller," *ISA Transactions*, vol. 53, no. 2, pp. 358–366, 2014.

[24] P. Dash, L. C. Saikia, and N. Sinha, "Comparison of performances of several Cuckoo search algorithm based 2DOF controllers in AGC of multi-area thermal system," *International Journal of Electrical Power and Energy Systems*, vol. 55, pp. 429–436, 2014.

[25] A. Saxena, M. Gupta, and V. Gupta, "Automatic generation control of two area interconnected power system using Genetic algorithm," in *Proceedings of the 3rd IEEE International Conference on Computational Intelligence and Computing Research (ICCIC '12)*, Coimbatore, India, December 2012.

[26] B. K. Sahu, T. K. Pati, J. R. Nayak, S. Panda, and S. K. Kar, "A novel hybrid LUS-TLBO optimized fuzzy-PID controller for load frequency control of multi-source power system," *International Journal of Electrical Power & Energy Systems*, vol. 74, pp. 58–69, 2016.

[27] Y. Sharma and L. C. Saikia, "Automatic generation control of a multi-area ST—thermal power system using Grey Wolf Optimizer algorithm based classical controllers," *International Journal of Electrical Power and Energy Systems*, vol. 73, pp. 853–862, 2015.

[28] P. Dash, L. C. Saikia, and N. Sinha, "Automatic generation control of multi area thermal system using Bat algorithm optimized PD–PID cascade controller," *International Journal of Electrical Power & Energy Systems*, vol. 68, pp. 364–372, 2015.

[29] S. Mirjalili, "The ant lion optimizer," *Advances in Engineering Software*, vol. 83, pp. 80–98, 2015.

[30] MATLAB, http://www.mathworks.com.

# Routing in Wireless Multimedia Sensor Networks: A Survey of Existing Protocols and Open Research Issues

**Vikas Bhandary,[1] Amita Malik,[1] and Sanjay Kumar[2]**

[1]*Deenbandhu Chhotu Ram University of Science and Technology, Murthal, Sonepat 131039, India*
[2]*SRM University Haryana, Sonepat 131029, India*

Correspondence should be addressed to Amita Malik; amitamalik.cse@dcrustm.org

Academic Editor: Kuei-Ping Shih

With the advancement of wireless sensor networks (WSNs) and technology, applicability of WSNs as a system is touching new heights. The development of multimedia nodes has led to the creation of another intelligent distributed system, which can transfer real-time multimedia traffic, ubiquitously. Wireless multimedia sensor networks (WMSNs) are applicable in a wide range of areas including area monitoring and video surveillance. But due to unreliable error-prone communication medium and application specific quality of service (QoS) requirements, routing of real-time multimedia traffic in WMSNs poses a serious problem. The paper discusses various existing routing strategies in WMSNs, with their properties and limitations which lead to open research issues. Further, detailed classification and analytical comparison of discussed protocols are also presented.

## 1. Introduction

A wireless sensor network (WSN) comprises a substantial number of interconnected sensor nodes (SNs) that have the capability to sense the physical environment attributes such as temperature, pressure, humidity, light, and sound and simultaneously interact with each other over the wireless medium. The advancements in various technologies have led to the creation of tiny and low-cost multimedia devices like video cameras and microphones, which can easily be integrated to form a sensor node. These devices are used in special types of wireless sensor networks, called wireless multimedia sensor networks (WMSNs). WMSN is a special type of WSN which can sense and transfer scalar data as well as multimedia data, that is, image, audio, and video streams, in real-time as well as nonreal-time. Usually these networks are more powerful distributed systems [1]. Akyildiz et al. [2] have stated various applications of WMSNs: multimedia surveillance networks, traffic avoidance, enforcement and control systems, industrial process control, environment monitoring, and so forth. Many applications still require intense efforts in various research areas to implement WMSNs more efficiently.

Figure 1 shows the architecture of WMSNs based on the type of node used. WMSN can be classified into various classes: (a) single-tier flat and homogeneous, (b) single-tier clustered and heterogeneous, and (c) multitier heterogeneous. In single-tier flat and homogeneous architecture, sensors with similar physical capabilities are used [2]. A subset of nodes having data processing capability, called multimedia processing hubs, is used. Data processing is done in a distributed fashion. In single-tier clustered and heterogeneous architecture, there are nodes with different physical capabilities, for example, multimedia node and scalar node [2]. Nodes having more capability, processing power, batteries, and so forth are called cluster heads (CHs). The rest of the nodes communicate with CHs, which then process and transfer the information to sink node. The data is stored in the central storage hub. In multitier clustered and heterogeneous architecture, there are multiple layers of nodes with different type and processing tasks at each layer. Data is processed and stored in a distributed fashion.

Routing is the most challenging problem in wireless communication system. Only a few survey papers have been published on WMSN routing protocols. A survey presented

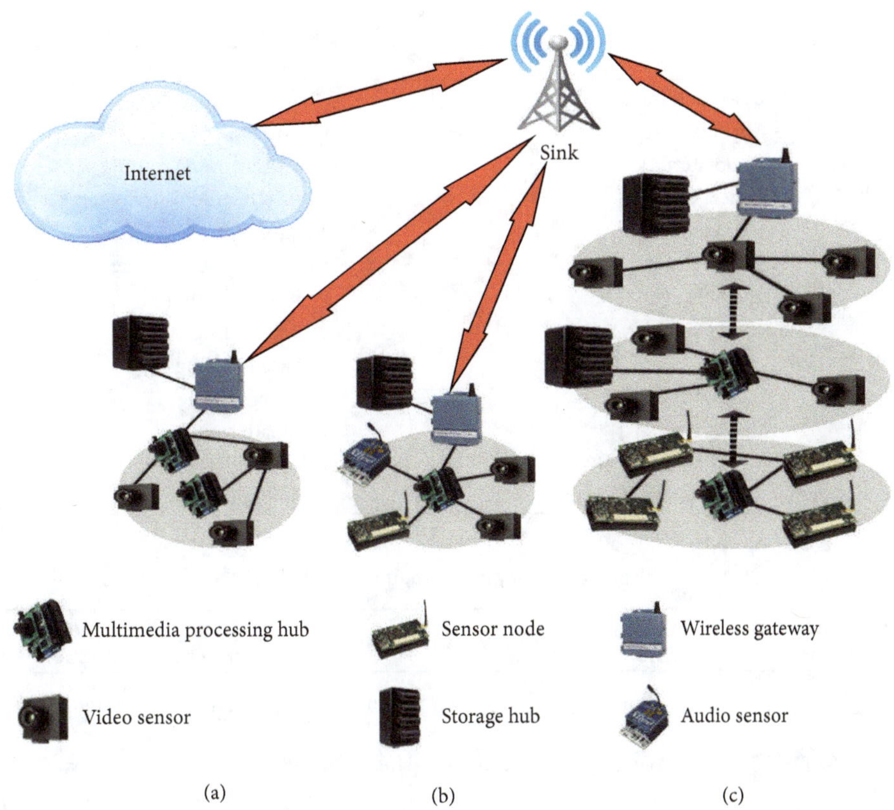


FIGURE 1: Architecture of WMSNs.

by Al-Karaki and Kamal [3] included routing protocols for WSN. The authors presented surveyed protocols in taxonomy and a detailed tabular comparison of all the protocols was also included at the end. Akyildiz et al. [2] gave layerwise description of WMSN protocols and summarized the open issues of WMSNs. In [4], Förster has summarized various machine learning (ML) techniques which can be used in routing in ad hoc networks. He also discussed the importance of ML techniques and summarized their properties in a table. Misra et al. [5] described layerwise requirements, mechanism, and issues of multimedia streaming in WSN. The paper included QoS based and reliability constrained routing protocols. Saleem et al. [6] presented a paper, which included swarm intelligence (SI) based routing protocols for WSN. The authors briefly explained various SI based energy efficient and QoS based routing protocols. Zungeru et al. [7] did an exhaustive survey of classical and SI based routing protocols for WSN. Detailed analytical and experimental comparison of various QoS based energy efficient and reliable routing protocols was done by the authors. Jayashree et al. [8] classified multipath routing protocols and summarized few QoS based routing protocols. Whereas the authors Ehsan and Hamdaoui [1] performed extensive survey of WMSN, which included detailed classification and analytical comparison based on different criteria and protocol's summary and drawbacks. Guo and Zhang [9] described state-of-the-art SI based routing protocols in WSN. The authors have summed up the performance analysis in a tabular format and have given short description of analysis of network lifetime based on different definitions. Abazeed et al. [10] surveyed different routing protocols for WMSNs. The authors performed the analytical comparison of protocols and also considered simulation scenarios and results of surveyed protocols. Masdari et al. [11] surveyed multipath routing protocols. The protocols were analytically compared and classified based on different criteria. Alanazi and Elleithy [12] have studied real-time QoS routing protocols. The authors classified protocols into probabilistic and deterministic protocols. They performed simulation to analyse various metrics and finally summarized the simulation results. Aswale and Ghorpade [13] surveyed state-of-the-art QoS routing protocols and summarized their analysis in table, including each protocol's strengths and weaknesses, and finally summarized various issues in routing in WMSNs.

At the end, previous works related to ours are condensed in Table 1. The paper presents summarized description and key features of routing protocols, which are classified into different categories based on different characteristics. Further, classification-wise analytical comparison of protocols is done at the end of each class (Tables 2, 3, 4, 5, and 6), in which attributes like characteristics features, suitability for WMSNs, performance metrics, type of network, energy efficiency, and performance results are stated. And in the end, surveyed protocol's characteristics are condensed in Table 7.

The remainder of the paper is organized as follows. Section 2 discusses various requirements for WMSN routing protocols. Section 3 discusses various routing techniques,

TABLE 1: Previous surveys.

| Year of survey | Authors | Characteristics | Time period |
|---|---|---|---|
| 2004 | Al-Karaki and Kamal [3] | WSN routing | 1988–2004 |
| 2006 | Akyildiz et al. [2] | WMSN survey | 1993–2006 |
| 2007 | Förster [4] | Machine learning based routing | 1989–2006 |
| 2008 | Misra et al. [5] | Multimedia streaming survey | 1973–2008 |
| 2010 | Saleem et al. [6] | SI based routing | 1989–2010 |
| 2011 | Zungeru et al. [7] | Classical and SI based routing | 1988–2011 |
| 2012 | Jayashree et al. [8] | Multipath routing | 1993–2011 |
| 2012 | Ehsan and Hamdaoui [1] | QoS routing | 1990–2012 |
| 2013 | Guo and Zhang [9] | SI based routing | 1959–2012 |
| 2013 | Abazeed et al. [10] | WMSN routing | 1998–2011 |
| 2013 | Masdari et al. [11] | Multipath routing | 2003–2012 |
| 2015 | Alanazi and Elleithy [12] | Real-time QoS routing | 2000–2015 |
| 2015 | Aswale and Ghorpade [13] | QoS routing | 1988–2015 |

TABLE 2: Latency based routing protocols.

| Author [paper] | Year | Characteristics features | Suitability for WMSN | Performance metrics | Type of network | Energy efficiency | Network lifetime | Performance results |
|---|---|---|---|---|---|---|---|---|
| Zhou and Wang [20] | 2010 | HRT | Medium | Delay, Lifetime | Homogeneous | Medium | High | Suitable for small WMSNs |
| Sohrabi et al. [21] | 2000 | SRT | Low | Delay | Homogeneous | No | Low | Meets SRT |
| Akkaya and Younis [23] | 2003 | SRT | Medium | Delay, RE | Homogeneous | Medium | High | Low throughput |
| Lu et al. [25] | 2002 | SRT | Medium | Delay | Homogeneous | No | Low | Outperforms DSR [26] and GF [27] |
| Karp and Kung [27] | 2000 | SRT | High | Delay, RE | Homogeneous | Medium | Medium | Outperforms DSR [26] and GF [27] |
| Chipara et al. [29] | 2006 | SRT | High | Delay, RE | Homogeneous | Medium | Medium | Outperforms SPEED [28] and EEGR [33] |
| Ratnaraj et al. [34] | 2006 | SRT | High | Delay, RE | Homogeneous | Medium | Medium | Outperforms AODV [36] and DSR [26] |
| Medjiah et al. [38] | 2010 | SRT | High | RE | Homogeneous | High | Medium | Outperforms GPSR [27] |
| Xue et al. [37] | 2011 | SRT | High | Delay, RE | Homogeneous | High | Medium | Outperforms MMSPEED [31] and RAP [25] |
| Ben-Othman and Yahya [39] | 2010 | SRT | High | Delay | Homogeneous | High | Medium | Outperforms MCMP [40] and RAP [25] |
| Li and Kim [15] | 2014 | $(m, k)$ FRT | High | Delay, RE | Homogeneous | High | Medium | Outperforms SPEED [28] |

HRT: hard-real-time, SRT: soft-real-time, FRT: firm-real-time, SD: service differentiation, and RE: residual energy.

which are classified into various categories. Section 4 summarizes open research issues and Section 5 concludes the paper.

## 2. Design Requirements for WMSN Routing

Routing governs the performance of WMSNs in terms of quality of service and energy efficiency. Routing policies accommodate multiple conflicting objectives and constraints, imposed by technologies and user requirements [14]. Real-time multimedia traffic in WMSNs raises many design challenges for routing protocols:

(i) *QoS Requirements.* WMSNs are used in different types of applications. These applications need different requirements. For example, real-time multimedia applications require strictly less end-to-end delay and jitter but high bandwidth and reliability. Various QoS requirements are given below:

*Latency.* End-to-end delay requirements can be classified into soft latency bounded systems, hard delay bounded systems, and firm delay

TABLE 3: Multiconstrained routing protocols.

| Author [paper] | Year | Characteristics features | Suitability for WMSN | Performance metrics | Type of network | Energy efficiency | Network lifetime | Performance results |
|---|---|---|---|---|---|---|---|---|
| Felemban et al. [31] | 2006 | Multiconstrained | High | Delay and reliability | Homogeneous | High | Low | Outperforms SPEED [28] |
| Razzaque et al. [41] | 2008 | Multiconstrained | High | Delay and reliability | Homogeneous | High | High | Outperforms MMSPEED [31] and MCMP [40] |
| Ahmed and Fisal [42] | 2008 | Multiconstrained | High | Delay and LQ | Homogeneous | High | Medium | Outperforms MMSPEED [31] and RTPC [43] |
| Yan et al. [48] | 2009 | Multiconstrained | High | Delay, PLR | Homogeneous | Medium | Medium | Outperforms SPEED [28] and DD [49] |
| Shen et al. [44] | 2014 | Multiconstrained | High | Delay and reliability | Homogeneous | High | High | Outperforms CAQR [45] |
| Rao et al. [46] | 2013 | Multiconstrained | High | Delay and reliability | Homogeneous | Medium | Medium | Outperforms VIBE [47] |

LR: packet loss ratio; LQ: link quality.

bounded systems. In soft latency bounded systems, the system needs to maintain probabilistic delay guarantee; that is, some delay in packet delivery can be tolerated. But in hard latency bounded system, the system needs to maintain deterministic end-to-end delay, which is also denoted as deadline [15, 16]. The entire system is said to fail, if the system is not able to provide service in the deadline guarantee period.

*Reliability*. It is defined as ability to deliver the data to sink node with minimum packet loss [1, 2]. The unreliability of a system is due to congestion and interference. So it is also regarded as reciprocal of packet loss [1, 2].

*Bandwidth*. Multimedia traffic requires huge bandwidth. In order to fulfill bandwidth requirements and utilize the resources to its maximum available capacity, routing protocols must employ multipath, multichannel, or stream division method.

*Jitter*. It is defined as the accepted variability of delay between received packets [1]. If the requirement is not met, then it may result in error in audio or video data stream, glitches, and discontinuity.

*Lifetime*. The lifetime of a system is an important factor for deciding the performance of system.

(ii) *Energy Efficiency [1, 2]*. Multimedia sensor nodes are battery constrained devices just like scalar sensor nodes. So energy is one of the major concerns in WMSNs too. Due to network congestion, packet interference, retransmission, and too much concentration on QoS requirements, battery of sensor nodes drains quickly.

(iii) *Hole Detection [1, 2]*. Real-time applications require high bandwidth and low end-to-end delay. As a result, the energy of the routes that are used repeatedly exhausts quickly. This leads to the creation of special scenarios called dynamic holes. An efficient routing protocol must facilitate balanced energy usage in the whole network.

(iv) *Traffic Classes [1, 2]*. Admission control in WMSNs must be based on QoS requirements of the requested service or application as there is a need to provide differentiated service between different applications. Various traffic classes that are defined in [2] are as follows:

*Real-Time, Loss-Tolerant Multimedia Streams*. Usually, this traffic class requires high bandwidth and strict delay bounds. This class includes audio, video stream, multilevel stream, scalar data, and their metadata.

*Delay-Tolerant, Loss-Tolerant Multimedia Streams*. This class includes traffic that is intended for offline storing or processing. This traffic must be transmitted in almost real-time to avoid packet loss.

*Delay-Tolerant, Loss-Intolerant, Scalar Data*. This class includes data from monitoring process and data which are intended for offline storing or processing. The bandwidth requirement is low.

*Delay-Tolerant, Loss-Tolerant, Scalar Data*. This may include any scalar sensor data, non-time critical multimedia snapshot whose bandwidth requirement is low.

*Real-Time, Loss-Tolerant Scalar Data*. This class includes data from monitoring process or loss-tolerant snapshot multimedia. Bandwidth demand is moderate to low.

*Real-Time, Loss-Intolerant Scalar Data*. This class includes data from time critical monitoring process and bandwidth requirement is moderate to low.

Table 4: SI based routing protocols.

| Author [paper] | Year | Characteristics features | Suitability for WMSN | Performance metrics | Type of network | Energy efficiency | Network lifetime | Performance results |
|---|---|---|---|---|---|---|---|---|
| Di Caro and Dorigo [14] | 1998 | Multipath (ACO) | No | Delay | Homogeneous | No | Medium | Slow convergence, high overhead |
| Rahman et al. [51] | 2008 | Flat (ACO) | Low | Delay and jitter | Homogeneous | Low | Low | Limited-scalability, high overhead |
| Sun et al. [53] | 2008 | Hierarchical (ACO) | Medium | Delay, PLR | Homogeneous | Medium | High | Outperforms Dijkstra and DD [49] |
| Bi et al. [54] | 2010 | Hierarchical (ACO) | Medium | Delay | Homogeneous | Medium | High | Outperforms AGRA [55] and M-IAR [51] |
| Cobo et al. [57] | 2010 | Hierarchical (ACO) | High | RE, delay, and PDR | Homogeneous | High | High | Outperforms T-ANT [58] |
| Yu et al. [56] | 2011 | QoS based (ACO) | High | Delay, PLR, and RBW | Homogeneous | High | High | Limited-scalability |
| Nayyar et al. [60] | 2011 | Energy aware (ACO) | High | RE, LB | Homogeneous | Medium | Medium | Outperforms Tiny AODV [36] |
| Kumar et al. [61] | 2014 | QoS based (ACO) | High | RBW, RE | Heterogeneous | Medium | Medium | Outperforms EEABR [62] |
| Liu et al. [63] | 2012 | QoS based (PSO) | High | Delay, RBW, and jitter | Homogeneous | Medium | Low | Outperforms EEABR [62] and AODV [36] |
| Huang et al. [64] | 2014 | Hierarchical (ACO) | Medium | RE, delay, and HC | Homogeneous | Medium | High | Outperforms Dijkstra and IPACR [64] |
| Förster and Murphys [65] | 2007 | RL based | Medium | Lifetime, percentage gain over DD | Homogeneous | Medium | High | Outperforms DD [49] |
| Kotecha and Popat [66] | 2007 | GA based | Medium | Delay | Homogeneous | Medium | Medium | Outperforms AODV [36] and QoS AODV [67] |
| Saleem et al. [69] | 2012 | HBO based | Medium | Routing overhead, PLR, and latency | Homogeneous | Medium | Medium | Outperforms FP-Ant [70], EEABR [62], and AODV [36] |
| Cai et al. [72] | 2015 | Hierarchical (HBO) | High | PDR, ESD, and latency | Homogeneous | High | High | Outperforms IEEABR [73], FF-Ant [70], and BeeSensor [69] |
| Jafarzadeh and Moghaddam [74] | 2014 | RL | Medium | Delay, PDR, and traffic load | Homogeneous | High | High | Outperforms QoS-AODV [67], RL-QRP [75], and RSSI [76] |
| Kiani et al. [77] | 2015 | Hierarchical (RL) | High | Lifetime, PLR | Homogeneous | High | High | Outperforms LEACH [79], HEED-NPF [80], and EECS [81] |

ACO: ant colony optimization, PSO: particle swarm optimization, PLR: packet loss ratio, RBW: residual bandwidth, RE: residual energy, HC: hop-count, RL: reinforcement learning, GA: genetic algorithm, HBO: honey bee optimization, PDR: packet delivery ratio, and ESD: energy standard deviation.

TABLE 5: Flat routing protocols.

| Author [paper] | Year | Characteristics features | Suitability for WMSN | Performance metrics | Type of network | Energy efficiency | Network lifetime | Performance results |
|---|---|---|---|---|---|---|---|---|
| Lan et al. [82] | 2008 | Flat | Medium | Delay, RE | Homogeneous | Medium | Low | Outperforms SAR [21] |
| Nayyar et al. [83] | 2011 | Flat (Multipath) | Medium | Delay, RE, and RBW | Homogeneous | High | Low | Outperforms AODV [36] |

RE: residual energy; RBW: residual bandwidth.

TABLE 6: Hierarchical routing protocols.

| Author [paper] | Year | Characteristics features | Suitability for WMSN | Performance metrics | Type of network | Energy efficiency | Network lifetime | Performance results |
|---|---|---|---|---|---|---|---|---|
| Kandris et al. [85] | 2011 | Hierarchical | Low | RE | Homogeneous | Medium | High | Outperforms TEEN [87] |
| Lin et al. [88] | 2011 | Hierarchical | Medium | Delay, LQ | Homogeneous | Medium | High | Outperforms ARCH [89] |

RE: residual energy; LQ: link quality.

TABLE 7: Location based routing protocols.

| Author [paper] | Year | Characteristics features | Suitability for WMSN | Performance metrics | Type of network | Energy efficiency | Network lifetime | Performance results |
|---|---|---|---|---|---|---|---|---|
| Karp and Kung [27] | 2000 | Location based | No | Shortest path | Homogeneous | No | Low | Outperforms DSR [26] |
| Chen et al. [93] | 2007 | Location based | High | Delay, RE, and RBW | Homogeneous | Medium | Medium | Outperforms GPSR [27] |
| Shu and Hauswirth [59] | 2008 | Location based | Low | Shortest path | Homogeneous | Low | Medium | Outperforms GPSR [27] and less throughput |
| Medjiah et al. [94] | 2009 | Location based | Low | Shortest path, RE | Homogeneous | Medium | Medium | Outperforms GPSR [27] |
| Bennis et al. [95] | 2013 | Location based | High | Shortest path, RE | Homogeneous | High | Medium | Outperforms TPGF [59] |
| Wang et al. [96] | 2015 | Location based | High | Delay, RE, and AHC | Homogeneous | High | Medium | Outperforms DGR [93] |

RE: residual energy, RBW: residual bandwidth, and AHC: average hop-count.

## 3. WMSN Routing Techniques

Many WMSN routing protocols have been proposed in previous literature. There are many ways to classify routing protocols depending on different parameters. The classes and subclasses are not mutually exclusive as many protocols belong to more than one class or subclass. According to current research trend, categorization of routing protocols is based on number and type of QoS constraints [1]. Hence, QoS based routing protocols are mainly divided into latency constrained and multiconstrained routing protocols [1]. Again, based on deadline delays, latency constrained routing protocols are further classified into protocols providing hard real time (HRT), soft-real-time (SRT), and firm-real-time (FRT) QoS guarantees [15, 16]. Another category called "swarm intelligence (SI) based" routing protocol is included in classification. Finally, based on the network structure, routing protocols are further classified into flat routing, hierarchical routing, and location based routing. Classification of different routing protocols for WMSNs is shown in Figure 2.

*3.1. QoS Based Routing.* In order to guarantee qualitative or quantitative performance parameters, the notion of QoS has been proposed [17]. QoS requirement for WMSNs is a set of constraints like reliability, latency, bandwidth, and jitter. Based on number and type of constraints, QoS based routing protocols are classified into latency constrained and multiconstrained routing protocols.

*3.1.1. Latency Based Constrained Routing.* In HRT based systems, if packets arrive after the predefined delay deadline, then this is considered as a system failure [15, 16]. But due to lossy links and error-prone medium, it is almost

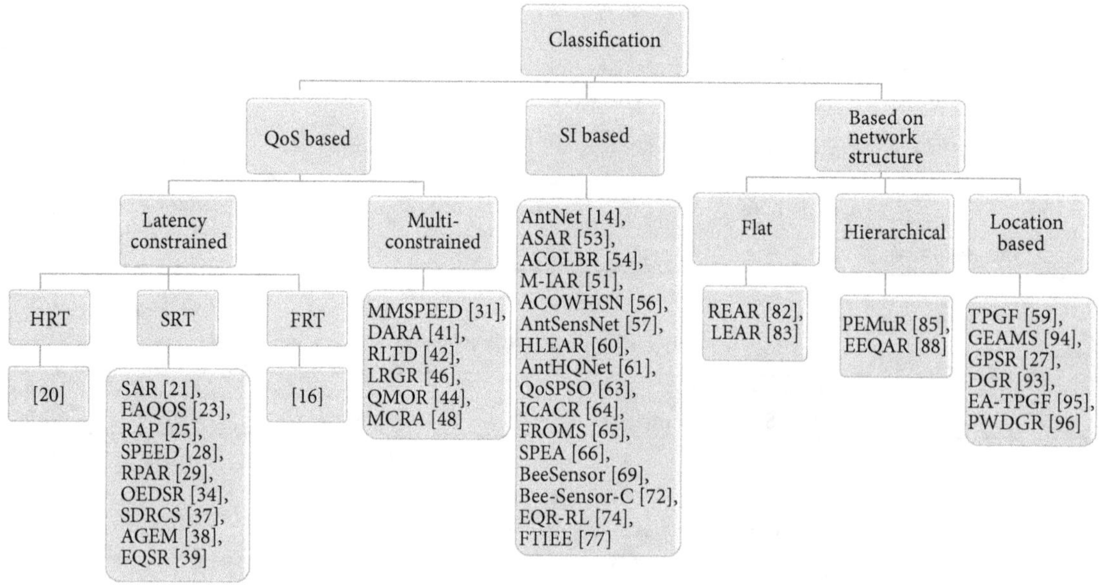


FIGURE 2: Classification of routing protocols in WMSNs.

impossible to guarantee HRT in WMSNs. In SRT based systems, some packet missing the delay deadline is tolerable, but a probabilistic delay guarantee is required in order to prevent the system from crashing. There are various protocols that can provide required delay deadline guarantees and deliver the data packets in SRT. In FRT based systems, $(m, k)$-firm concept is introduced, which demands at least $m$ out of any $k$ consecutive messages must deliver before their delay deadline; otherwise, it is considered as a system failure. Therefore, it is necessary to implement priority based scheduling of stream packets. In [18], distance based Priority (DBP) policy is used to choose between streams. The system maintains DBP state for each stream. DBP is usually calculated at the sink and is used to evaluate the stream not for priority assignment. Recent work in [19] tires to implement cross layer solutions to improve routing.

*(1) HRT Based Protocols.* The paper [20] proposed delay-constrained routing protocol. The main goal of the protocol is to provide hard real-time delay guarantees and minimize the energy consumption within WMSNs. The proposed algorithm uses global link-state routing algorithm, which needs complete knowledge of network topology in order to compute best possible paths. Global link-state routing algorithms can provide hard real-time delay guarantees. The proposed algorithm is divided into three parts. The first part removes the node with less residual energy from the graph $G(V, E)$, in order to avoid hole problem. The second part uses Dijkstra algorithm in order to find energy efficient routes. Then, the third part ensures that all the paths meet delay upper bound. If a path does not meet delay constraints, then a new path is computed using Dijkstra algorithm and delay as a performance metric.

Key features are as follows:

(i) It uses a global state routing algorithm.

(ii) It focuses on energy efficiency and QoS.

(iii) It computes delay-constrained least energy consumption tree using the Dijkstra algorithm.

The proposed protocol guarantees hard-real-time delay bounds. However, the state information exchanged between nodes consumes a huge amount of memory and bandwidth. It is difficult to store and maintain state information in large WMSNs. Therefore, it is only suitable for small WMSNs.

*(2) SRT Based Protocols.* Sequential assignment routing (SAR) [21] considers both QoS and energy efficiency in route selection process. SAR assigns a priority level to each packet. The main objective of SAR algorithm is to minimize the average weighted QoS metric throughout the lifetime of the network. Weighted QoS metric is computed as the product of additive QoS metric and weight coefficient associated with a priority level. SAR algorithm elects special node called the central node (CN) to perform more sophisticated information processing. It uses Single Winner Election (SWE) algorithm and spanning tree (ST) algorithm. ST algorithm computes a minimum hop spanning tree rooted at CN. Failure recovery in SAR uses a handshaking procedure which enforces table consistency among neighbors. If there is a path failure, then a recomputation procedure is triggered automatically. SNR uses two cooperative signal processing techniques: noncoherent and coherent. Extensive simulation done in [22] shows that SAR has better performance than the minimum metric algorithm.

Key features are as follows:

(i) It is a table driven routing protocol.

(ii) It is both a QoS based and an energy efficient protocol.

(iii) It is a multipath and fault-tolerant routing protocol.

The major drawback of SAR is that it is a table driven protocol and hence it is nonscalable. If the number of nodes increases, then the overhead of maintaining the routing table and status information increases drastically [1].

Energy-aware QoS (EAQoS) [23] finds the optimal path to the destination in terms of energy consumption and error rate while meeting the end-to-end delay requirements. EAQoS specifies a queuing model which can handle both nonreal-time and real-time traffic. The implemented queuing model is inspired from class based queuing model [24]. A cost function is associated for each link. This cost function is defined to capture various parameters such as residual energy, transmission energy, and end-to-end delay. EAQoS finds $K$ least cost paths using an extended version of Dijkstra's algorithm. Then, only those paths are chosen that meet the required QoS requirement and maximize network throughput. The cost of each path between nodes is stored. For every discovered path, a tree-like structure is created and total cost of the path between the nodes is recorded. The algorithm terminates when $K$ shortest paths are discovered. If a new path is searched for a particular destination, then only node-disjoint paths are considered. This helps in reducing the congestion over newly found path due to former path.

Key features are as follows:

(i) It is a cluster based energy efficient and delay-constrained routing protocol.

(ii) It facilitates both real-time and nonreal-time best effort traffic simultaneously.

(iii) It ensures guaranteed bandwidth and most energy efficient paths.

(iv) The use of node-disjoint paths helps in reducing congestion over the newly found path.

EAQoS meets various QoS and energy efficiency requirements. However, EAQoS does not consider MAC and other delays while calculating end-to-end delay and it requires knowledge of the whole network. EAQoS does not treat different real-time traffic separately; therefore, it cannot be used for applications that coexist with other multimedia traffic.

Real-Time Communication Architecture for Large-Scale WSN (RAP) [25] is a priority based, localized routing protocol, which provides soft-real-time delay bounds. RAP introduced a novel packet scheduling policy, Velocity Monotonic Scheduling (VMS) [25]. VMS schedules packets according to their requested velocity. VMS improves deadline miss ratio by giving priority to packets based on requested velocity. VMS is capable of reducing deadline miss ratio by as much as 72.1% [25]. RAP assumes that each node knows its own location, making it location dependent protocol. The combination of VMS and geographic forwarding (GF) makes RAP more scalable and more suitable for multimedia traffic. RAP supports both query oriented and event oriented services. Extensive simulation shows that RAP outperforms DSR [26] and GF [27].

Key features are as follows:

(i) It is a location dependent routing protocol, which implements both VMS and GF.

(ii) It provides soft-real-time delay bound guarantees.

(iii) It provides both query and event based service APIs.

(iv) It is a scalable protocol for multimedia applications.

There is no need to maintain neighbor database and hence the routing overhead is very little as compared to other classical routing algorithms. RAP uses GF, which can deplete the bandwidth quickly. Also, it does not consider hop-count as routing metric and there is no direct metric to show energy conservation [1].

Stateless Protocol for Real-Time Communication (SPEED) [28] is a stateless, localized, soft-real-time delay bounded protocol, which requires minimum MAC layer support for QoS based routing. Stateless Nondeterministic Geographic Forwarding (SNGF) method is used for routing in SPEED. SNGF maintains desired delivery speed and traffic balance, to reduce the network congestion. MAC layer adaptation of SNGF drops packets in case of congestion. If congestion remains for a long time, then traffic is redirected to a less congested area. SNGF uses four modules.

*Beacon Exchange.* SNGF uses two extra beacons. One beacon is for delay estimation and the other is for back pressure.

*Delay Estimation.* Round trip single hop delay for each packet is calculated at the sender's site after receiving ACK.

*Neighborhood Feedback Loop (NFL).* NFL maintains system performance by maintaining single hop relay speed above a fixed value.

*Back Pressure Rerouting.* This module is used to reduce congestion in the network.

Key features are as follows:

(i) It is an energy efficient and localized routing protocol.

(ii) It handles congestion and provides soft-real-time delay bounds.

(iii) It avoids the void problem.

(iv) It also supports load balancing.

SPEED uses nondeterministic geographic forwarding and it maintains only localized database. As a result, the routing overhead is minimized. But SPEED is nonscalable as it uses the idea of per-flow reservation [1], and its performance degrades when it encounters a large hole and sudden congestion.

Real-time Power-Aware Routing (RPAR) [29] protocol can achieve application specific delay requirements by dynamically adjusting transmission power and other routing metrics. RPAR looks into various serious issues such as network scalability and memory and bandwidth constraints. RPAR have four components: neighborhood manager, delay estimator, forwarding policy, and velocity assignment policy. The neighborhood manager finds eligible forwarding choices that meet both end-to-end delay and energy efficiency requirements. The delay estimator supports real-time multimedia traffic. It further supports two mechanisms: (1) Power Adaptation and (2) neighbor discovery. The delay estimator

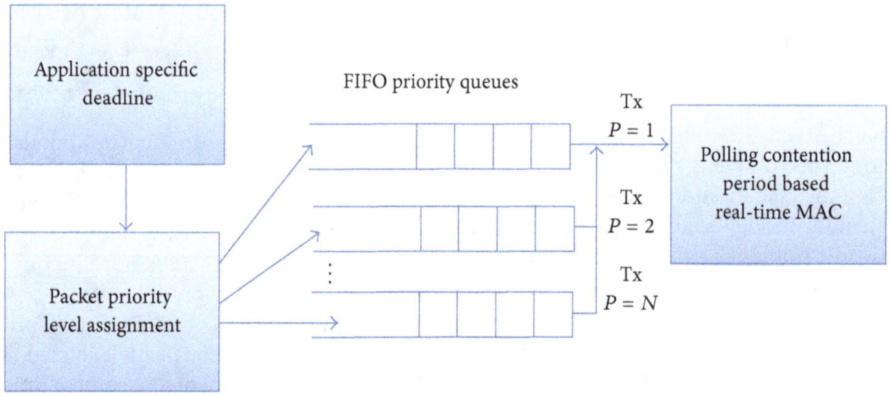


FIGURE 3: Traffic classification using priority queues.

uses Jacobson's algorithm [30] to estimate contention delay and transmission counts. Jacobson's algorithm computes the retransmission timeout by considering both the average and the estimated variable. RPAR is very good at handling congestion and Holes. RPAR is related to SPEED [28], MMSPED [31], and LAPC [32]. But it is different from these because it integrates power efficiency and supports real-time communication, it handles lossy links, and it employs novel neighborhood management mechanism. Simulation results show that RPAR outperforms SPEED [27] and EEGR [33].

Key features are as follows:

(i) It is a power-aware, location dependent, and query based routing protocol.

(ii) It can handle holes and congestion.

(iii) It is a scalable routing protocol, which can deliver data within delay bounds.

(iv) It looks into bandwidth and memory constraints.

Neighborhood management and forwarding policy further reduce power consumption. But RPAR's performance degrades when it encounter large holes and sudden congestion.

Optimized energy-delay subnetwork routing (OEDSR) [34] protocol is cluster based and event driven routing protocol where the information from CHs to the base station (BS) is routed in a multihop fashion. OEDSR protocol borrows the idea of an energy-delay metric from OEDR [35]. Routing decision is based on factors such as: available energy, end-to-end delay and distance from BS. The proposed protocol ensures that routes are loop-free. In OEDSR, total numbers of active nodes are reduced by forming a subnetwork with nodes having significant information regarding the event. A temporary header (TH) is selected based on maximum available energy, which calculates and selects required number of CHs. After the selection of CHs, TH broadcasts all the information about CHs and then it becomes a regular node. It does not consider hop-count as one of routing metrics. Simulation results show that OEDSR outperforms AODV [36] and DSR [26].

Key features are as follows:

(i) It is a proactive cluster based routing algorithm.

(ii) It supports event driven services.

(iii) It is multipath and energy efficient and delivers data within delay deadlines.

OEDSR minimizes the product of available energy and average end-to-end delay, and it also minimizes the total number of collisions by restricting the number of nodes accessing the channel. OEDSR does not consider hop-count for making routing decisions and average delay is increased if the packets are transmitted between sensor node, THs, and CHs.

Service differentiated real-time communication scheme (SDRCS) [37] is an event driven routing scheme, which implements cross layer functionality in order to route real-time traffic. SDRCS functionally integrates both dynamic forwarding technique and CSMA/CA based prioritization technique [37]. SDRCS uses a novel polling contention period based prioritized MAC. The advantage of the proposed design is that it improves bandwidth utilization even when the total numbers of categories are greater than four. SDRCS also proposes light-weight packet schedulability estimation mechanism, which uses polling contention period and received signal strength based grouping technique. The protocol operation can be categorized in five components: (0) RSS based grouping, (1) admission control, (2) prioritized queuing, (3) real-time MAC, and (4) dynamic forwarding. The grouping mechanism uses RSS to strip the sensor field to layers. Accuracy of distance is maintained using factor called grouping granularity (GRA).

The queuing policy in SDRCS uses per-hop deadline based queues. Every node uses FIFO priority queues for packet scheduling. As shown in Figure 3, per-hop deadline requirements are mapped into $N$ priority levels. Polling contention period based MAC can only provide a limited number of traffic classes. Early deadline miss (EDM) policy is employed in order to drop any unscheduled packets. Simulation results show that SDRCS outperforms MMSPEED [31] and RAP [25], in terms of throughput, on-time delivery ratio, and energy efficiency.

Key features are as follows:

(i) It is an event driven routing protocol.

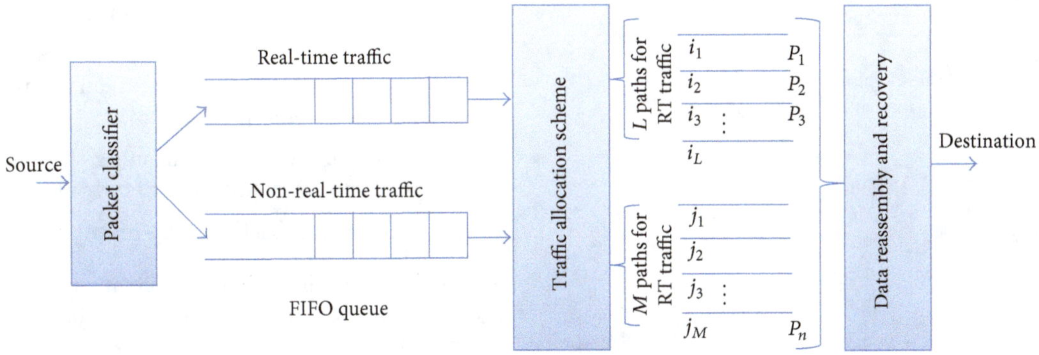


FIGURE 4: Functional diagram of EQSR.

(ii) It requires no extra hardware for localization.

(iii) It can adapt to network dynamics (such as channel, void, and channel quality).

(iv) It uses cross-layer packet forwarding design for routing multimedia traffic.

(v) It can avoid voids very efficiently.

SDRCS can avoid false packet drops due to EDM, because it adapts better to change in a network environment.

Adaptive Greedy-Compass Energy-Aware Multipath (AGEM) [38] routing protocol is location based routing protocol for WMSNs which can be considered as an enhanced version of GPSR [27] protocol. AGEM has load-balancing feature to increase the lifetime of the network and to reduce queue size of nodes. It uses adaptive greedy compass policy. Adaptive compass policy selects only a subset of neighbors. Its forwarding policy uses four parameters: remaining energy, hop-count, distance between the current node and its neighbors, and previous history of forwarded packets of the same stream. AGEM protocol has two modes: (1) smart greedy forwarding and (2) walking back forwarding. It is assumed that nodes gather information regarding their position using GPS or by using distributed localization techniques. Each node maintains database containing information regarding one-hop neighbors. Simulation results show that AGEM performs better than GPSR [27], in terms of packet-loss, transmission delay, and energy efficiency.

Key features are as follows:

(i) It is a location based, multipath routing protocol, which focuses on load balancing.

(ii) It uses adaptive greedy compass policy.

(iii) It can be used for both static and mobile sensor networks.

(iv) It uses the shortest path which has minimum end-to-end delay.

Unlike GPSR [27], data streams in AGEM are routed by different paths; hence, mostly used nodes do not die rapidly because of energy depletion. If network size is low, then mean energy consumption is greater for AGEM as compared to GPSR [27].

Energy efficient and QoS aware multipath routing (EQSR) [39] protocol tries to minimize energy consumption rate to increase the network lifetime. EQSR uses residual energy, available buffer size, and signal-to-noise ratio (SNR) as routing metrics. EQSR increases reliability factor by using XOR-based FEC (forward error correction) technique. As shown in Figure 4, EQSR implements a queuing model, which enables EQSR to handle both real-time and nonreal-time traffic. EQSR discovers node-disjoint paths, which helps in utilizing maximum available network resources. Path discovery phase of EQSR is divided into initialization phase, primary path setup phase, and alternate path setup phase. Each node finds preferred next hop by using the link cost function as defined in [39]. Path discovery phase finds primary and alternate paths. Updated values of metrics are appended on data packets to update previously established paths. This eliminates refreshment overhead.

The segmentation of data packets is done before data transmission, in order to split the data packets into equal sized segments. Then, these segments are scheduled for transmission. Before scheduling process, traffic allocation scheme adds error correction code, which helps in recovering from node failures. Efficient data distribution among the available paths is done using weighted traffic allocation strategy. Simulation results show that EQSR outperforms MCMP [40] and RAP [25].

Key features are as follows:

(i) It is an energy efficient multipath routing protocol.

(ii) It can handle both real-time and nonreal-time traffic.

(iii) It uses FEC codes to increase reliability.

EQSR avoids network-wide flooding of route discovery in case of node failures. It also reduces overhead due to route update messages. EQSR uses FEC codes for increasing reliability, but it increases data redundancy.

*(3) FRT Based Protocols.* The paper [15] defines a real-time fault-tolerant mechanism, which tries to achieve application QoS requirement by utilizing Local Status Indicator (LSI). LSI oversees and assesses the local condition of the nodes and is used to detect network faults. The proposed mechanism enables each node to dynamically adjust its capabilities in order to attenuate performance degradation. As shown in

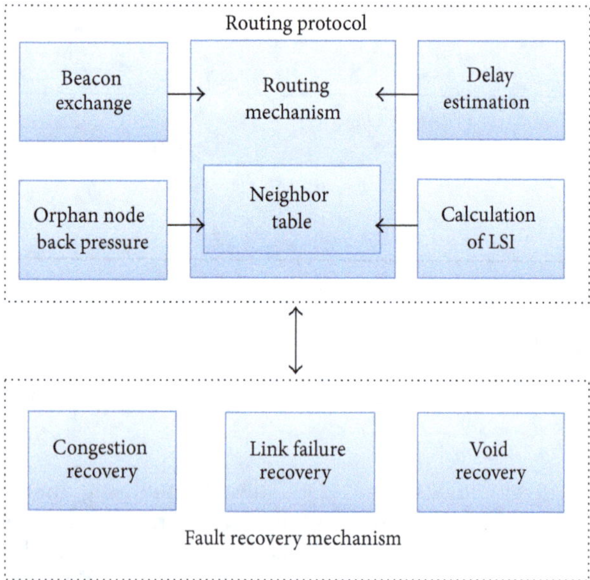


FIGURE 5: Components of scheme proposed in [11].

Figure 5, the working of the proposed mechanism is further divided into four submechanisms: (1) neighbor beacon exchange, (2) single hop delay estimation, (3) LSI calculation, and (4) fault recovery mechanisms.

In order to share location information among neighbors and to prolong the lifetime of overloaded nodes and of the whole network, periodic beaconing is used. Therefore, the beacon messages contain location information and residual energy of nodes. Three on-demand beacons, single hop delay estimation beacon, stream DBP beacon, and orphan node removal beacon, are used by the proposed mechanism in order to implement different functionalities. At a regular interval, stream DBP beacons are sent by the sink node to source node as a feedback. The value of stream DBP is propagated to the intermediate nodes, which help them in the fault recovery mechanism. The proposed mechanism calculates DBP at the sink node, by using past packet deadline misses. The factor "distance to failure" of a particular node is known after calculating LSI by using $m$, $k$, congestion, and link failure level. After calculating LSI, a node can locally decide to implement recovery mechanisms. The paper [15] proposes a routing protocol, which makes optimal forwarding decisions based on LSI, packet deadline, and remaining power. The proposed protocol implements the mechanism defined in [16]. The protocol has five components: routing mechanism, beacon exchange, orphan node back pressure, delay estimation, and calculation of LSI. Each node maintains a neighbor table, which stores neighbor ID, energy level, estimated delay, expiration time, $(m, k)$, deadline, and DBP. The expiration time field is set to standard round trip time (RTT). Simulation results show that the proposed routing protocol outperforms SPEED [28].

Key features are as follows:

(i) It is a robust routing protocol, which can fulfill required latency and reliability requirements.

(ii) It uses local system evaluation that reduces routing overhead.

(iii) It ensures timeliness of the stream transfer by choosing a node with qualified LSI.

(iv) It uses fault recovery mechanisms to handle problems like congestion, link failure, and void.

(v) Orphan node back pressure prevents void occurrence.

The proposed routing protocol efficiently routes stream packets within delay constraints and provides better fault recovery mechanism.

In Table 2, latency constrained routing protocols are compared with each other, where suitability for WMSN for each protocol is based on delay, energy efficiency, network lifetime, and performance of routing protocol.

*3.1.2. Multiple QoS-Constrained Routing.* Multiple QoS-constrained routing protocols focus not only on latency constraint, but also on other constraints like load balancing, reliability, congestion, bandwidth, and so forth. Many protocols were proposed by researchers in previous literature.

Multipath Multi-SPEED (MMSPEED) [31] is a location based routing protocol, which improves the timeliness and reliability of sensor networks communicating over wireless channels. Reliability requirements are met by using probabilistic multipath forwarding mechanism. The main goals of MMSPEED are as follows: (1) localized packet routing decision making and (2) providing QoS in timeliness and reliability domains. MMSPEED assumes that each node is aware of its location, which makes it more scalable and self-adaptive and removes setup and recovery latency. The virtual isolation of speed layers helps in providing service differentiation. Virtual isolation is maintained by classifying packets according to their speed classes and placing them in priority queues. Probabilistic multipath forwarding mechanism differentiates packets with different requirements. MMSPEED makes proper use of paths without much energy consumption. The back pressure mechanism is employed to remove the void area problem and reduce the incoming traffic. MMSPEED greatly relies on MAC protocol for priority based access to medium depending on speed layer, reliable multicast delivery of packets, and calculation of loss rate and average delay to individual neighbors. MMSPEED relies on EDCF mode of IEEE 802.11e standard.

Key features are as follows:

(i) It is a location based and query driven routing protocol.

(ii) It focuses on both timeliness and reliability.

(iii) It is a multipath routing protocol.

(iv) It can solve the void area problem.

MMSPEED uses redundant path selection techniques and prevents unnecessary flooding to improve reliability, network lifetime, and load balancing. But MMSPEED does not consider hop-count as one of the routing metrics and it does not consider energy consumption directly. Moreover, the multicast capability of EDCF results in an increment in total size of packets, which in turn increases power consumption.

Distributed aggregate routing algorithm (DARA) [41] is a multisink, multipath, and location aware protocol, which implements the model for achieving multiconstrained QoS (MCQoS). The main goal of [41] is to improve path reliability and delay guarantee while preserving the battery power. MCQoS is a location aware routing scheme which defines three metrics: (1) residual energy, (2) expected sojourn time of packet at the receiving node, and (3) geographic progress towards the destination. It proposes selective greedy forwarding (SGF) to reduce energy consumption and interference [41]. Delay-differentiated packet scheduling (DDPS) reduces packet drop due to deadline failure [36]. DDPS separates queues for time-constrained and non-time-constrained packets in the MAC layer. Further MAC protocol is modified to give higher priority to time-constrained data packets in two levels: (1) internode and (2) intranode. This modification also allows us to control transmission power and save battery and reduce interference. Nodes in DARA algorithm dynamically decide when to broadcast BEACON messages. Simulation results show that DARA outperforms MMSPEED [31] and MCMP [40].

Key features are as follows:

(i) It is a fully localized algorithm which means there is no need to maintain any global state information.

(ii) It supports both static and dynamic network topologies.

(iii) It increases reliability by sending duplicate packets from the source node.

(iv) It modifies MAC to give more priority to real-time traffic.

DARA reduces energy by using the power control transmission scheme and reducing retransmissions; it also optimizes the trade-off between reliability and delay. But DARA does not provide efficient solution to void problem and it increases routing overhead due to duplicate packet transmission.

Real-time routing protocol with load distribution (RTLD) [42] ensures high packet throughput with minimal packet overhead and prolongs the network lifetime. A novel communication method called "geodirection-cast forwarding" is proposed in [38], which is a combination of both geocast and directional forwarding. It also introduced a mechanism to compute optimal forwarding based on packet reception rate (PRR), remaining power of sensor nodes, and packet velocity over one hop. This optimal forwarding helps in balancing the load of real-time traffic. There are four modules in RTLD: (1) location management, (2) routing management, (3) power management, and (4) neighborhood management. Location management module calculates the location of nodes using the location of predetermined neighbor nodes. The power management module determines the state of the transceiver and the transmission power of sensor nodes. Neighborhood management module discovers a subset of forwarding candidate nodes and maintains a neighbor table of forwarding candidate nodes. Routing management module computes optimal forwarding choices and makes all routing decisions. It further contains submodules: forwarding metrics calculation, forwarding mechanism, and routing problem handler. RTLD outperforms SPEED [28] and RTPC [43].

Key features are as follows:

(i) It is a location based routing protocol, which combines geocast forwarding with link quality.

(ii) It maximizes the velocity and remaining energy to achieve delay bounds.

(iii) It uses neighborhood and power management policies.

(iv) It can efficiently detect and bypass holes.

(v) It supports query based services and focuses on load balancing.

RTLD can deliver data within delay bounds and minimize miss ratio and energy consumption. The only drawback of RTLD is its high cost of deployment.

QoS aware multisink opportunistic routing (QMOR) [44] is an efficient routing protocol, especially designed for delivering real-time multimedia streams on time. The main objective of QMOR is to minimize the energy consumption and meeting the delay and reliability requirements [44]. The use of opportunistic routing scheme enhances the performance of communication by utilizing the broadcast nature of wireless communication. The nodes maintain a forwarder list, which contains a list of nodes that are allowed to forward the broadcasted packets. QMOR focuses on selection and prioritization of the forwarder list to meet application specific QoS requirements.

High level of correlation between video information received from adjacent multimedia sensor nodes results in redundant traffic. QMOR uses a correlation-aware differential coding scheme to reduce redundancy. For reducing redundancy and maximizing energy gains, it is very much important to determine source node and its destination node in the same correlation group. This problem is solved using differential coding-based source and intermediate selection (DCISIS) protocol. Simulation results show that QMOR outperforms CAQR [45].

Key features are as follows:

(i) It is a QoS based multisink routing protocol, which implements an opportunistic routing scheme to reduce energy consumption.

(ii) It implements priority based forwarding.

(iii) It can evaluate expected communication cost.

(iv) It meets the reliability and delay requirements.

QMOR greatly improves video transmission quality and energy efficiency. But QMOR does not provide service differentiation in heterogeneous networks.

Latency and Reliability-Aware Geographic Routing (LRGR) [46] is a hierarchical routing protocol, which focuses on mobility prediction and energy-aware cluster formation. Each node stores information about two-hop neighbor nodes and adjacency table. Working of LRGR protocol is divided into several rounds. Each round consists of two phases:

clustering phase and routing and data transmission phase. Clustering phase is further divided into energy exchange, CH setup, and cluster formation. Cluster maintenance is implemented as required during routing and transmission phase. Energy efficient neighbor discovery protocol (ENDP) is used at MAC layer because it is a low duty beacon-enabled MAC protocol.

In clustering phase, each node broadcasts its residual energy. After energy exchange, sensor nodes calculate average residual energy of their neighbors. The node with a maximum ratio of residual energy/average residual energy becomes CH. For mobility prediction, link expiration time (LET) is calculated, which is computationally less complex. Another factor called linear quality estimation (LQE) is calculated at each sensor node, which is based on previous receptions and transmissions. Both LET and LQE are used by nodes to choose between the CHs. In order to use the most stable path, the main focus is on factors like balanced energy usage, connectivity, position, and congestion. LRGR uses connection reliability metric, forwarding progress metric, and aggregated path metric. Simulation results show that LRGR outperforms VIBE [47].

Key features are as follows:

(i) It is a location based hierarchical routing protocol.

(ii) It uses energy-aware cluster setup phase.

(iii) It uses mobility prediction based clustering, which estimates link quality and improves reliability.

(iv) It defines forwarding progress metric to evaluate end-to-end delay and packet loss.

LRGR guarantees low latency and maintains a balance between energy efficiency and reliability. But in case of highly mobile environments, LRGR suffers because of limited bandwidth, error-prone wireless channel.

Multiconstrained routing algorithm (MCRA) [48] is based on query-flooding and query driven data delivery models. In MCRA, special messages called "interest" are used to query an event. The MCRA borrows the idea of the logical coordinate system. Logical Coordinate System is a localized approach which calculates the coordinates of nodes using hop-count information. This approach drastically reduces message overhead. Both geographic information based routing and virtual coordinate system based routing algorithms sometimes result in route failure due to inaccurate node coordinate information. MCRA implements message suppression techniques to reduce redundant transmission and retransmission of messages caused by collisions. In MCRA message suppression is done using restraining forwarding and deferring forwarding. Both try to reduce the amount of interests by restraining and deferring forwarding node. Extensive simulation performed in [48] shows that MCRA outperforms agent based DD [49] and SPEED [28].

Key features are as follows:

(i) It is a logical location based multiconstrained routing protocol which supports query based services.

(ii) It focuses on reliability and energy efficiency.

(iii) It implements message suppression techniques to reduce redundant packets.

By using the logical location based technique, MCRA reduces routing overhead and it also minimizes redundant messages. But MCRA does not consider reliability and bandwidth.

Multiconstrained routing protocols are analytically compared in Table 3, which contains characteristics features, suitability for WMSNs, performance metrics, type of network, energy efficiency, network lifetime, and performance results of routing protocols.

*3.2. SI Based Routing Protocols.* SI based routing protocols are inspired by the collective behavior of intelligent biological species (e.g., ant colonies). SI based routing protocols simulate the behavior of these species by using mobile software agents. These agents intelligently perform problem solving tasks, in coordination with other agents. Stigmergy plays an important role by supporting communication between the agents, through the environment. SI based routing protocols are scalable, fault tolerant, adaptive, fast, and autonomous. In addition to this, these protocol's operations are modular in nature and they can be executed in parallel. Due to these reasons, SI based routing protocols have become a better alternative to the routing problem. For more than two decades, ant colonies have motivated the design of various novel algorithms [6].

Distributed Stigmergic Control for Communication Networks (AntNet) [14] is an adaptive, distributed, and mobile agent based Monte Carlo system which is used for routing of multimedia traffic. AntNet implements adaptive learning of routing tables using agents called forward ants and backward ants. At fixed intervals forward ants are sent from any network nodes towards a destination to discover feasible, low-cost path. Routing decisions are based on pheromone trails left by ants. QoS Parameters such as residual energy, delay, and bandwidth are used to calculate forwarding probability, which makes the process of exploration adaptable according to traffic distribution. When a forward ant reaches the destination, backward ants are generated. Backward ants update routing tables.

Key features are as follows:

(i) It is an adaptive and hybrid routing protocol.

(ii) It uses ACO algorithm and reinforcement learning to solve routing problem.

(iii) It is a multipath, multiobjective, and multiconstrained routing algorithm.

(iv) It is stochastic in nature.

(v) It delivers the data in real-time latency constraints.

AntNet algorithm provides better throughput as compared to classical routing protocols. But AntNet is very slow as it requires ant agents to reach at the destination without updating routing tables [50] and its routing overhead is very high.

Multimedia-Enabled Improved Adaptive Routing (M-IAR) [51] is an extension of Improved Adaptive Routing (IAR) algorithm [52]. M-IAR is an ant-based flat multihop routing protocol which incorporates some extra multimedia QoS parameters (end-to-end delay and jitter) as compared with IAR. It also supports fault tolerance and self-organization without any need for global route information. Novelty of M-IAR protocol is that it considers the closeness of next-hop neighbors towards the sink but also takes into account the distance of the neighbor from the sender node [46]. This protocol gives a reliable solution as it piggybacks end-to-end acknowledgement packets. M-IAR assumes that each node in the network knows its own position and all other nodes through GPS. In 98% of test cases, M-IAR successfully found the shortest path within the first three route discovery attempts by forward ants [51].

Key features are as follows:

(i) It is a localized and ant-based routing protocol.

(ii) It is an improved version of IAR algorithm.

(iii) It focuses on energy efficiency and end-to-end delay, jitter.

(iv) It is self-organized, adaptive, and fault tolerant in nature.

M-IAR provides reliable solution and it can be configured for both acknowledgement-based and non-acknowledgement based approach. But M-IAR needs high cost of deployment and it does not concentrate on load balancing between nodes and scalability. Moreover, M-IAR's performance degrades due to hole creation and exponential increase in overhead, in case of high network load.

Ant-based service-aware routing (ASAR) [53] is an ant-based service-aware hierarchical routing protocol for wireless multimedia sensor networks. It facilitates three types of basic services (query driven, data driven, and stream driven). This protocol uses the concept of "pheromone quantization" to improve ant-based optimization algorithm. This quickens the convergence of ASAR. The proposed algorithm is made to run in all CHs. It finds three different routes for each type of services by positive feedback mechanism used in ACO and the results are stored on CHs, respectively. QoS parameters such as delay, packet loss rate, network bandwidth, energy consumption, and network lifetime are used as routing metric. Extensive simulation verified that ASAR performs better than Dijkstra and DD [49]. This algorithm is often used for optimized routing and load balancing in telecommunications and networks.

Key features are as follows:

(i) It is a clustered, ant colony optimization based routing protocol.

(ii) It supports both query and data driven services.

(iii) It uses pheromone quantization to accelerate the convergence of the algorithm and optimize the network resources.

(iv) It supports three kinds of services in WMSNs and meets QoS requirements and improves network performance.

ASAR converges more quickly and it supports three types of services and meets diverse QoS requirements. But ASAR's performance degrades due to bottleneck problem and repetitive usage of optimal paths [10].

Ant Colony Optimization-based Load-Balancing Routing Algorithm (ACOLBR) [54] is a biologically inspired hierarchical routing algorithm. ACOLBR is further divided into intercluster routing and intracluster routing. Intracluster routing uses minimal spanning tree (MST) algorithm to construct MST with CHs as the root. Intercluster routing finds a path with minimal cost from source node to the destination node and some suboptimal paths using ACO algorithm. ACOLBR enhances the exploring ability of ants by introducing maximum and minimum value of pheromone density values. It also improves transmission efficiency by splitting the stream, when data amount is greater than the path's maximum flow. Simulation results and analysis show that ACOLBR has better adaptability, can achieve load balancing, reduces the end-to-end delay, and prolongs the network lifetime as compared to AGRA [55] and M-IAR [51].

Key features are as follows:

(i) It is a swarm-based, hierarchical, and multipath routing protocol.

(ii) It focuses on congestion control.

(iii) It uses the minimum spanning tree algorithm to create MST.

(iv) It has enhanced the exploring ability as compared to other ACO techniques.

ACOLBR can recover from path failure by setting pheromone value to zero and then sending an error message to sender. But ACOLBR requires more complex calculations to find optimal route and heavy load results in a bottleneck situation over CHs.

Ant Colony Optimization-Based QoS Routing Algorithm (ACOWMSN) [56] is a reactive routing protocol, which focuses on four QoS metrics. ACOWMSN finds routes which meet application specific QoS requirements. Artificial ants wander around the network in order to find paths which meet the requirements. ACOWMSN uses two types of artificial ants: forward ants and reverse ants. Forwarding ants store information fields like minimum residual energy and cumulative queue delay and packet loss and memory of node visited, in order to evaluate the route in terms of QoS. ACOWMSN protocol finds minimum cost route using heuristics. At each node, next forwarding node is found by calculating the probability. Routing in ACOWMSN is divided into two stages: route discovery and routing confirm. In route discovery stage, forward ants are sent towards the destination node using broadcast or unicast approach. Broadcast approach is used only when current node does not know about the destination node. In routing confirm stage, forward ants are converted into backward ants. Backward ants travel from the destination node to source node. Backward ants update the pheromone table of every intermediate node.

Key features are as follows:

(i) It is an ant colony based routing protocol.

(ii) It is reactive in nature; therefore, it finds route only when it is needed.

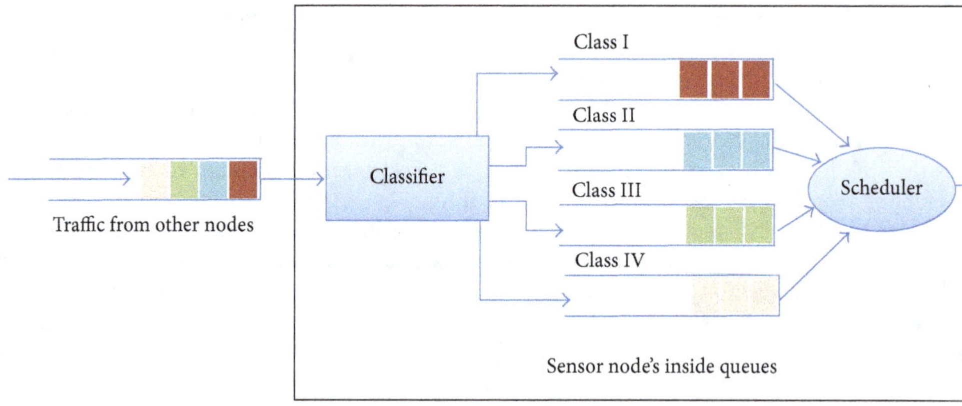


FIGURE 6: Traffic classification process used in AntSensNet.

(iii) It is adaptable to any change in network conditions.

(iv) It is an energy-aware and QoS based routing protocol.

ACOWMSN can satisfy bandwidth, delay, and packet loss rate requirements and increase the network lifetime. But in large-scale WMSNs, it is very much difficult to determine the protocol parameters and therefore algorithm converges very slowly.

AntSensNet [57] combines features of both hierarchical network structure and Ant colony optimization based routing. It is specifically designed for WMSNs and hence it meets application specific requirements. It is based on T-ANT [58]. Just like T-ANT [58] algorithm, its operation is split into rounds. Each round is further divided into $t$ cluster setup phase and steady phase. It uses special types of agents for cluster setup called cluster ants (CANTS). The node with CANT becomes CH. HELLO packets broadcasted by each node constantly contain information regarding its ID, clustering pheromone value, and its state. The clustering pheromone value decides if the current node is suitable to become CH. A tunable factor called "cluster radius" determines minimum distance between two CH nodes.

Routing tables in AntSensNet are updated in three phases: forward ant phase, backward ant phase, and route maintenance phase. Forward ants are generated if there is no satisfactory route to the sink node. Forward ants carry information regarding routes they travel. The routes are evaluated when they reach the sink node. If the evaluated route does not meet application specific requirements, then forward ants are discarded; otherwise, backward ants are generated in order to update the pheromone values. Backward ants travel the reverse path. Maintenance phase handles the congestion and lost link problems. Figure 6 shows diagram of traffic classifier used by AntSensNet to handle and schedule four different types of traffic classes.

Key features are as follows:

(i) It is a hybrid routing protocol, and hence it comprises reactive as well as proactive components.

(ii) It implements both ACO based routing and clustering.

(iii) It supports a power efficient multipath video packet scheduling in order to minimize the distortion in video transmission.

(iv) It uses completely distributed clustering algorithm.

(v) It can handle different types of traffic classes.

AntSensNet can handle congestion problem. It is scalable and multipath and it provides better video quality in terms of distortion as compared to ASAR [53] and TPGF [59].

Hop and load based energy-aware routing protocol (HLEAR) [60] is a swarm-based, reactive, self-organizing routing protocol, which avoids hole and increases network lifetime. HLEAR's main aim is to find completely or partially disjoint path with relative minimum distance. It uses hop-count, rate of forwarding nodes, and residual energy as metric, to find an efficient path for routing multimedia streams. Based on these values, another routing metric $\beta$ is calculated. Path with the lowest value of $\beta$ is considered as best path. The protocol operation is further divided into route discovery and path establishment phases. In route discovery phase, RREQ is broadcasted in a controlled manner. When RREQ reaches the sink node, it does not need to analyse path selection, as the path has been already scanned by intermediate nodes. In path establishment phase, destination node sends a route reply message RREP via reverse path. Simulation results show that HLEAR performs better than Tiny AODV in terms of energy efficiency and throughput [60].

Key features are as follows:

(i) It is a reactive, swarm-based, and energy efficient routing protocol.

(ii) It avoids holes by stopping nodes from taking part in route discovery phase, if their energy is less than 0.5 J.

(iii) It finds the minimum distance to destination by finding disjoint paths.

(iv) Unlike LEAR, it considers hop-count as routing metric.

HLEAR provides better throughput and energy efficiency than Tiny AODV [60], but it cannot handle congestion efficiently due to which its throughput is very low when frame rate is high.

AntHQSeN [61] is a reactive ant-based QoS enabled routing protocol, which works in two phases: (1) route discovery phase and (2) route maintenance phase. The agents called forward ants are created by the source node to find multiple paths to the destination node. They also carry network information in order to evaluate various QoS parameters. AntHQSeN forwards the data packets stochastically. Each forwarding node decides next node based on pheromone values, which makes the protocol adaptable to any change in the network. The route setup phase is completed when destination nodes generate backward ants to update the pheromone tables. Route maintenance phase is activated whenever a node failure occurs. For pheromone calculation, various factors like residual bandwidth, residual energy, and route cost are considered. In AntQHSeN protocol hello ants play a major role in immediate neighbor discovery, exchanging information about the bandwidth, timestamp, energy, and pheromone concentration. Extensive simulation results show that AntQHSeN performs better than AODV [36] and EEABR [62].

Key features are as follows:

(i) It is a reactive ant-based routing protocol for heterogeneous sensor network.

(ii) It can work fine in both static and dynamic topologies.

(iii) It treats multimedia and scalar traffic differently.

(iv) It uses a special flag to give an option to use the admission control scheme or not.

AntQHSeN provides a more consistent solution due to its reactive route discovery mechanism as compared to EEABR. But AntQHSeN lacks sufficient information at the beginning to find routes. Its setup time is directly proportional to the time needed for convergence of the algorithm. But still the discovered routes are better in terms of quality than those found by AODV [36].

An agent-assisted QoS-based routing (QoS-PSO) [63] is an agent based routing protocol, which focuses on delivering application based QoS services by using practical swarm optimization (PSO) technique. The synthetic QoS parameters are used as parameters of the objective function of PSO. QoS-PSO uses two different types of agents called forward agent and reverse agent in order to find out the optimized route. The synthetic QoS metrics are stored in agents. Forward agent's aim is to find a new route to the destination nodes. If the new route can provide required QoS, then reverse agents are created. All the information stored in forward agents is copied to reverse agents. Reverse agents travel in the reverse direction and update routing tables of all the nodes lying on the newly found path. The algorithm firstly selects the source and destination nodes and uses forward agent and reverse agent to establish a new path to destination. The QoS of each path is evaluated to obtain local best values. Iteration formulae are used to find a new routing path and its QoS is evaluated. If the QoS of a particular path is better, then it is made global best solution. This process is iterated until required values are achieved. After this, global best solution is used to determine the path and update routing table.

Key features are as follows:

(i) It is an agent based routing protocol, which uses PSO for optimizing synthetic QoS.

(ii) It uses agents which participate in the routing path establishment and maintenance.

(iii) It focuses on improving QoS metrics like delay, bandwidth, packet loss, and energy.

QoS-PSO shows improvement in QoS metrics by more than 60% and 20%, if compared with AODV [31] and EEABR [57], respectively.

A QoS-aware routing algorithm based on ant-cluster (ICACR) [64] is a hierarchical routing protocol for WMSNs, which integrates clustering and IPACR [64]. IPACR is a flat routing protocol, which outperforms standard ant colony routing algorithm (SACR). The principle disservice of SACR is that it considers the performance of every link between nodes and their neighbors as equivalent and sets the same initial pheromone value. IPACR protocol maintains a neighbor-list, which stores information regarding adjacent nodes. IPACR ensures that there is at least one feasible path. Based on the assumption "if a hello packet arrives earlier at a node, then it can be implied that the path is optimal," the optimality of paths depends on log SEQ. Therefore, IPACR sets more initial pheromone value of nodes with smaller log SEQ. The overall performance of routing protocol depends on the pheromone update procedure followed by forward ants and backward ants.

ICACR uses "divide and conquer" methodology to improve the scalability of the network. The network is divided into a number of clusters. The total number of subnetworks in the original network is one more than the total number of clusters. The additional subnetwork consists of sink node and all the CHs. IPACR protocol finds local optimal paths in each cluster.

Key features are as follows:

(i) It is an ant-based hierarchical routing protocol.

(ii) It uses IPACR routing in order to find local optimal paths.

(iii) It is a multipath routing protocol.

(iv) It is a scalable protocol, which can meet QoS requirements for WMSNs.

IPACR converges faster than SACR. ICACR is more scalable, energy efficient routing algorithm as compared to IPACR, SACR, and Dijkstra. But if the source node is placed near the sink node, then the performance of ICACR goes down.

Feedback-Enhanced Learning Approach (FROMS) [65] is an in-network protocol which learns by using local information about network topology and tries to converge to optimal routing paths. It assumes that each sensor has to route sensed data to multiple number of sink nodes.

Key features are as follows:

(i) It is an intelligent routing algorithm based on reinforcement learning.

(ii) It uses local information and optimizes route using feedback mechanism.

(iii) It considers equal cost paths to minimize energy consumption.

The technique uses three main phases, namely, sink announcement, exploration, and stable data gathering.

In Phase I, each sink broadcasts request message to entire network which contains time to live (TTL), sink ID, and number of hops. When a node receives this message, it verifies if the message has been already received. If entry is already there in routing table and it has less hop-count, received message is not processed. If there is no routing table entry or the entry has more hop-count, routing table is updated and message is broadcasted with decreased TTL and increased hop-count.

In Phase II, all best possible paths are discovered. This phase initiated, when a data packet is received. Fitness function gives the route cost based on single or multiple variables for a set of sinks. The proposed algorithm uses Route Length Estimate (RLE) as metric to decide the fitness of a path, which is equal to total number of hops. Each node organizes the available options for next hop and makes selection task manageable using Path Sharing Tree (PST). Nodes can manage the complex task of data delivery to multiple sinks by using PST. Using PST gives an extra advantage of escaping the path with loops. Updated RLE are shared in form of feedback when the package is rebroadcasted by upstream node. This feedback mechanism results in better decision-making over time. According to selected exploration strategy, Phase II ends and stable data gathering phase starts where sender node starts transmission to set of sinks. Whenever the PST changes Phase II is reinitiated.

FROMS minimizes the energy consumption by considering equal cost paths and switching between them to increase the lifetime of overall network. FORMS is compared with implementation version of directed diffusion [49]. Simulation results show FORMS performs better.

Strength Pareto Evolutionary algorithm (SPEA) [66] uses genetic algorithm (GA) and tries to optimize various QoS parameters such as bandwidth, delay, traffic, and hop-count. GA is best for multiobjective problems where the values to be optimized are more than one. There is no ideal "optimal" solution for such problems. There are sets of solutions, in which improvement in one objective leads to degradation in other objectives. The main objectives of SPEA are to optimize the QoS parameters and provide adaptive route with optimized delay.

Key features are as follows:

(i) It is an intelligent routing algorithm based on GA.

(ii) It tries to optimize four QoS parameters.

(iii) It quickly adapts to changes in network topology.

Route discovery in SPEA is carried out at MAC layer of TCP/IP model; therefore, the MAC is redesigned to support the algorithm. To make the process clear, it is designed into two modules: QoS routing module and GA module. Data exchange between the modules is bidirectional. GA generates new routes by combining the existing routes with common node and destination.

QoS routing module on receiving a data transfer request checks for a route to the destination. If route to the destination is not available, then it initiates route discovery procedure by transmitting smart agents. When source node receives RREP, it follows the reverse path and updates the table. When RERR message is received by source node, it uses GA module. Available entries in buffer are used as initial population and fitness function is applied and optimally finds the destination node with desired QoS requirements. If there is no RERR then the buffer is used to find route to destination. The buffer is evaluated periodically to maintain the efficiency of system.

The proposed protocol is compared with AODV [36] and QoS-AODV [67] routing and simulation results show it outperforms both.

Further work done in [68] has proved that using implementation of SPEA and changing the objective functions has increased the performance of system 15 times [68]. The paper has used similar approach and used expected transmission count and delay as objective functions. The expected transmission count (ETX) metric is the expected total number of transmissions. It is calculated using link quality and neighbor link quality. Link quality can be seen as probability that packet originated at node successfully arrives at a subsequent node and neighbor link quality is a measure of quality in opposite direction.

Beesensor [69] is an on-demand SI based reactive, event driven, energy efficient, and scalable routing protocol, which is inspired from behavior of honey bees.

Key features are as follows:

(i) It is an intelligent routing algorithm based on Honey bee optimization.

(ii) It minimizes the communication and processing cost.

(iii) It is energy-aware, scalable, and efficient.

Honey bee optimization uses recruitment and agent communication energy efficient foraging and multiflower-site foraging mechanism found in bee colony [69]. There are four types of agents: packers, scouts, foragers, and swarms. Packers are like food storer bees in hive. They receive packets from upper layer and find appropriate forager for them. Scouts explore the network in search of a potential sink node. Foragers play an important role in carrying packets to the sink using predetermined path. They also evaluate the quality of the path and give updates to other foragers. Foragers are piggy-backed in the link layer acknowledgement packets to source node. Foragers wait for certain amount of time at sink node and build a swarm of waiting foragers, which is transported back to source node. The algorithm is divided into four phases: scouting, foraging, swarming and routing loops, and path maintenance.

Scouting is further divided into forward scouting and backward scouting. Packers look for the appropriate forager to transfer an event at a sensor node. If a packer fails to find forager it launches forward scout with source ID, scout ID minimum remaining energy, and number of hops fields in header. This scout is then broadcasted to neighbor.

Interested sink converts the forward scout into backward stable. The algorithm maintains three tables: routing table, probability-distribution table, and a forwarding table. Source node maintains routing table and probability-distribution table. Forwarding table is maintained by sink and forwarding nodes. Sink node creates forwarding table entry from the received scout header information and replies with backward scout with same information fields. After discovering the route, foragers transport events to sink node. Packer selects a forager stochastically based on probability-distribution table. Intermediate node does not make routing decisions; next hop is found using path ID.

Swarming encapsulate foragers with identical path ID at sink node. The swarm is routed towards the source node using forward table entries. Swarming also helps in path maintenance. The path becomes invalid when the dance number becomes zero for a particular entry in routing table, which is deleted after some waiting period. Simulation results show that BeeSensor outperforms FP-Ant [70], EEABR [71], and AODV [36].

Bee-Sensor-C [72] is an on-demand event driven multipath routing algorithm based on dynamic clustering scheme and BeeSensor [69]. Bee-Sensor-C reduces the routing overhead and scalability as compared to BeeSensor. It follows multipath construction methodology to increase the overall lifetime of network.

Bee-Sensor-C is divided into three phases: cluster formation, multipath construction, and data transmission. In the second phase, various improvements are made to BeeSensor for finding multipath between CH and relative sink and the path is selected stochastically for data transmission.

BeeSensor is poor in overall performance, when all the nodes near the event need to send data to sink node. Bee-Sensor-C adds a new agent called HiveHeader to bee hive for each sensor. Its major responsibility is to claim the node wants to be cluster head node.

Nodes having information about the cluster join the cluster and the clustering region is called event-subnetwork. Any node can decide to join a cluster based on received signal strength. If a node receives multiple HiveHeader from different node it chooses the node with higher residual energy. The node, which sends the HiveHeader first, becomes the CH node. The node with higher residual energy and closer to event sends HiveHeader earlier and is more likely to become CH node. Each cluster has its TTL; if remaining energy of CH gets lower than 60%, CH selection starts again.

Like BeeSensor [69], Bee-Sensor-C utilizes forward scouts to explore the network. CH broadcasts a forward scout to its neighbors. The scouts are self-destructed when the hop-count is more than H_max, to avoid the communication overhead. Backward scout travels back to CH node using previous hop information in cache. Bee-Sensor-C also modifies the reward mechanism to include the average residual energy factor. When the backward scout arrives at CH with new path, it recruits foragers using "waggle dance" [72]. The simulation results show that Bee-Sensor-C outperforms FF-Ant [70], BeeSensor [69], and IEEABR [73].

EQR-RL [74] is energy-aware QoS routing which uses reinforcement learning. EQR-RL uses control flooding algorithm for updating the latest changes to the network. Each node uses header of its neighbor data packet to maintain up-to-date information about the routes.

Key features are as follows:

(i) It is an intelligent dynamic routing algorithm based on reinforcement learning.

(ii) It focuses on various QoS requirements like latency hop-count and so forth.

(iii) It utilizes data packets to get latest information of neighbors.

If a node does not hear any packet from a neighbor, it removes it from its routing table. The routing packet contains neighbor ID, timestamp, number of packets sent, end-to-end delay, and minimum residual energy in route.

When a node sends a packet to its neighbor, it starts the timer to calculate the round trip time. The receiver node finds the next node and updates the receiver address in data packet. After that, it updates minimum residual energy field. The feedback information is piggybacked in the header of data packets.

Exploration strategy provides the trade-off between exploring the low-cost routes and all possible routes. So it uses $\varepsilon$-greedy method for random selection of next hop for exploration [74].

When a node receives a data packet, it accesses various parameters from the packet and compares those parameters with the routing table entries. If it does not find any route, it sends the packet with best effort; otherwise, it selects next hop using roulette wheel. Simulation results show that ERQ-RL outperforms QoS-AODV [36], RL-QRP [75], and RSSI [76].

FTIEE [77] is an intelligent clustering based routing protocol, which aims to reduce energy consumption and make the system more reliable. It focuses on three parameters, namely, network lifetime, packet delay, and delivery. In this routing algorithm, learning and routing phase are realized at the same time, which reduces overall energy and overhead of the system.

Key features are as follows:

(i) It is an intelligent cluster based routing algorithm based on reinforcement learning.

(ii) It makes system more reliable and reduces packet delay.

(iii) It increases the lifetime of system.

In this algorithm, any node can act as CH and CH is chosen by election done by means of learning machine. Number of clusters is constant and clusters closer to sink are smaller as compared to the other clusters, size of cluster increases with increase in distance between the clusters and sink. CH nodes have multihop communication to the sink, and sensor nodes do not need to know about the CH.

FTIEE uses Q-learning technique [78] to select the optimal CH node and reduce the cost of CH-sink data transmission to minimum. Each sensor node is independent learning agent and therefore chooses to forward data to neighbor or select itself as CH node. The reward system in protocol is calculated using distance between node and sink and residual energy of node. In simulation, results show that it outperforms LEACH [79], HEED-NPF [80], and EECS [81].

SI based routing protocols are analytically compared in Table 4, which contains characteristics features, suitability for WMSNs, performance metrics, type of network, energy efficiency, network lifetime, and performance results of routing protocols.

*3.3. Based on Network Structure.* Based on network structure, routing protocols are classified into three different categories: (1) flat routing; (2) hierarchical routing; and (3) location based routing.

*3.3.1. Flat Routing Protocols.* In flat network architecture, all the nodes have the same responsibilities. Flat routing protocols are also called data-centric routing protocol. A BS is responsible for sending requests and queries. However, the sensor nodes can sense the physical environment and transfer the data to BS. Flat routing protocols use data aggregation to reduce redundant data, which ultimately reduces energy consumption. But flat routing protocols are not able to fulfill application specific latency requirements. These protocols suffer due to high routing overhead, delay, congestion, and low versatility.

Real-Time and Energy-Aware Routing (REAR) [82] protocol is a multipath routing algorithm which also reduces energy consumption of links and trades off the relationship of energy and delay. In REAR, metadata is used to find routing paths. Use of metadata stops repeated retransmission of exactly same packets; thus, huge bandwidth is saved. Metadata is used for routing discovery processes. REAR uses Optimized Dijkstra Algorithm which is used to select a link with lowest weight among marked nodes. For QoS determination factors like time delay, bandwidth, and remaining energy are included. In order to guarantee real-time transmission and reduce queue delay, a "classified queue model" is introduced to each node. For link cost evaluation, estimated distance, remaining energy, and queuing delay are used.

Key features are as follows:

(i) It is a metadata based routing protocol, which supports query based services.

(ii) It focuses on energy efficiency and end-to-end delay.

(iii) It uses modified Dijkstra algorithm.

(iv) It implements "classified queue model" to reduce queuing delay.

REAR reduces repeated retransmissions and energy consumption by using meta-data for route setup. But REAR does not consider reliability and bandwidth requirements. The concept of metadata in streaming applications consumes a lot of energy and bandwidths.

Load Based Energy-Aware Multimedia Routing (LEAR) [83] protocol is a reactive, self-organizing, and energy efficient protocol. LEAR is AODV [49] based protocol. Here active routes refer to the number of routes for different sources in which a node is participating. For route selection, a special factor ("route selection factor $\beta$") is calculated by forwarding nodes. If node's power reaches 25% of its total power, then it is labeled as "Swap Node." Swap Nodes do not take part in routing; they transmit packets only in certain critical conditions. So LEAR also has the ability of hole avoidance. LEAR finds disjoint path, if it is available. If disjoint path is not available, then it finds a partial disjoint path. Creating a partial disjoint path increases the hop distance, but it ensures high throughput, reduces congestion, and lowers the latency. This makes LEAR perfectly suitable for routing of multimedia traffic. Simulation results show that "the number of paths selected increases with the increase in the number of multimedia nodes within the event region" [61].

Key features are as follows:

(i) It is a reactive, multipath, self-organizing, and event driven routing protocol.

(ii) It focuses on energy efficiency, reliability, and load balancing.

(iii) It can avoid holes efficiently.

(iv) It finds disjoint paths in order to reduce congestion and meet delay bounds.

LEAR can find disjoint path while maintaining load balance between the nodes and increase network lifetime. The main drawback of LEAR is that it does not consider delay, reliability, network lifetime, and bandwidth requirements. Table 5 shows comparison of flat routing protocols.

*3.3.2. Routing Protocols.* Initially, hierarchical protocols were proposed for routing in wired networks [84]. However, they are also suitable for WMSNs. In hierarchical routing protocols, the entire network is divided into different levels. Nodes at different levels have different responsibilities. Generally, the protocol consists of two routing layers only, and data flows from a low level layer to a higher level layer. The nodes with the higher capabilities and responsibilities are known as CHs. Each CH can directly communicate with the BS. The rest of the nodes are called child nodes or cluster members, which associate with CHs to form clusters. Child nodes can only communicate with their CH. The advantages of these protocols over flat routing protocols are as follows: they are more scalable, energy efficient, and reliable. Overall performance of protocol depends on the process of formation of clusters.

Energy efficient and perceived QoS aware video routing (PEMuR) [85] is a novel energy-aware hierarchical routing protocol for WMSNs, which uses intelligent video packet scheduling scheme and predicts the video distortion. It is based on scalable power efficient routing (SHPER) protocol [86]. CHs are further categorized into upper level CHs and lower level CHs. Upper level CHs are near BS and can transmit data directly to the BS. Whereas lower level CHs are far from the BS, distant nodes route their messages to the BS by multihop routing. PEMuR is divided into two phases: (1) initialization phase and (2) steady state phase. The initialization phase starts with the creation of TDMA schedules for all nodes to advertise themselves and calculate relative distances between them. Random numbers of upper and lower CHs are elected by the BS. Non-CH nodes associate with CH having maximum signal strength. In steady state phase, BS elects a node with maximum residual energy in each cluster as CH.

In case of nonavailability of sufficient bandwidth, each sensor node drops video packets to reduce the retransmission rate. The packets to be dropped are chosen wisely in order to minimize overall video distortion using video distortion model [85]. The video distortion is measured in terms of Mean Square Error (MSE). The model also takes into account the random behavior of wireless medium; hence, it can predict video distortion accurately. Simulation result shows PEMuR outperforms TEEN [87].

Key features are as follows:

(i) It is an energy-aware hierarchical routing protocol.

(ii) It uses an intelligent video scheduling scheme.

(iii) It is based on SHPER protocol.

(iv) It intelligently handles low bandwidth situation.

PEMuR ensures that minimum power is consumed and required QoS is maintained for the transmission of video streams. Also, it can maintain a high level of perceived video quality (PSNR) for uniform and nonuniform energy distributions. But PEMuR is fully dependent on BSs for carrying out the routing process; hence, it is not fully distributed.

Energy efficiency QoS assurance routing (EEQAR) [88] protocol is a hierarchical routing protocol, which implements cellular topology to form clusters. EEQAR uses a QoS trust estimation model, which is based on social network analysis. On a social network, the relationship between two individuals is determined by the trust relationship between them. The concept of social network analysis is used by EEQAR in order to ensure energy efficiency and determine trust relationship. For building trust relationships, distributed metrics are introduced in EEQAR. Direct behavior monitoring and indirect information gathering are used in order to calculate trust value.

Trust value is calculated by calculating various other factors like transmission delay, frame rate, image quality, and audio quality. On the basis of the geographical position of agents and multimedia sensor nodes, the cellular topology is formed. Agent nodes are always deployed in the center of the hexagon. Agents send advertisement messages with their ID, after deployment. Multimedia nodes join agent node with maximum signal strength. If multimedia nodes have no task, then they can go into sleeping mode. But agent nodes must remain awake.

Each node maintains optimization factor table, which stores trust value, energy level, and correlation for each neighbor. The correlation factor estimates correlation between two nodes, which directly affects the fusion process. Each node finds a next forwarding node using optimization factor table. EEQAR operation is divided into various rounds. Where each round is divided into three phases: cluster building, routing probe, and steady state. Cellular clusters are formed in cluster building phase. Intracluster and intercluster routes are built in routing probe phase. Data is collected in steady phase. Simulation results in [88] show that EEQAR outperforms ARCH [89], in terms of delay, energy efficiency, and reliability.

Key features are as follows:

(i) It is an energy efficient hierarchical routing protocol.

(ii) It uses "data fusion" process in order to reduce unnecessary data transmission.

(iii) It uses trust analysis to fulfill application specific QoS requirement.

(iv) It uses TDMA at MAC layer to make efficient use of energy resources.

EEQAR efficiently improves performance by decreasing delay and energy utilization. But trust relationship is slowly varying and it is not error-free. Direct monitoring increases the energy consumption rate. Table 6 shows comparison of hierarchical routing protocols.

*3.3.3. Location Based Routing Protocol.* The central idea of location based routing protocols is to exploit the knowledge of location of nodes in order to route data packets. The location of the node can be determined using global positioning system (GPS) or other localization techniques like DVHop [90], Amorphous [91], and so forth.

Few advantages of location based routing protocols over other routing techniques are as follows:

(i) Minimized routing overhead.

(ii) Efficient bandwidth utilization.

(iii) Fully distributed and stateless nature.

(iv) Scalable and tolerant to route faults.

(v) Less memory and maintenance requirement.

However, these routing protocols suffer from "energy hole" problem. Greedy nature of these protocols results in repetitive usage of the same path, if the destination lies in the same region. The batteries of nodes lying on that path drain quickly, resulting in the creation of energy hole. An efficient location based routing protocol must handle and avoid hole problem efficiently.

Greedy Perimeter Stateless Routing (GPSR) [27] is a novel routing protocol, which uses the positions of the router and the destination node to make packet forwarding decisions. GPSR uses two modes: (1) greedy forwarding and (2) perimeter forwarding. GPSR packet headers include a flag to indicate the current forwarding mode of the packet. If a packet reaches a region where greedy forwarding is not possible, GPSR recovers by routing the packet around the perimeter of the region. In perimeter mode GPSR forwards packets using planar graph traversal. Further, a packet can again reenter the greedy forwarding mode if any neighbor is closer to the destination node.

Key features are as follows:

(i) It is a location based routing algorithm.

(ii) It can detect and avoid the hole problem.

(iii) It supports MAC-layer failure feedback.

(iv) It uses promiscuous mode.

(v) It supports queue traversal.

(vi) It implements planarization of graphs.

GPSR has a good packet delivery ratio even when the network density is too high. It requires less memory because there is no need to store full knowledge of the network at each node. There is no need for periodic updating of the table. It is an energy efficient routing protocol. End-to-end delay increases when mobility is high, which in turn makes it unsuitable for handling multimedia traffic [92].

Directional Geographic Routing (DGR) [93] protocol is a location based routing protocol, which gives better video transmission quality and prolongs the network lifetime. DGR assumes that any node can send video packets to sink at any instance. It forms multiple application specific disjoint paths to transmit real-time traffic over low bandwidth and unreliable network. Forward error correction (FEC) coding is used in order to make data delivery more reliable. A single real-time video stream is divided into multiple substreams. These substreams are transferred in parallel through disjoint paths, in order to make full utilization of resources. DGR follows the right hand rule to recover from the void problem [27].

Key features are as follows:

(i) It is a location based routing protocol.

(ii) It uses disjoint paths to optimize bandwidth and energy and reduce delays.

(iii) It is a fault-tolerant routing protocol.

(iv) It supports various features such as load balancing, fast packet delivery, and good peak signal-to-noise ratio (PSNR).

DGR utilizes resources very efficiently, increases the delivery ratio, and reduces end-to-end delay requirements. But DGR is not suitable for WMSNs because it requires additional overhead and hence results in transmission delay in case of fault, and it is impractical to implement for large networks.

Two-phase geographic greedy forwarding (TPGF) [59] routing algorithm overcomes various shortcomings of flat routing protocols. TGPF can be efficiently used in multimedia applications. TGPF avoids "Local Minimum Problem" by adopting face routing (right hand rule) as adopted in GPSR. The Label Based Optimization method is adopted to eliminate path circles and optimize the routing path found by TPGF. It is different from existing geographic routing algorithms as it does not adopt face routing to bypass holes. For hole-bypassing, it follows two approaches: (1) hole-bypassing by computing the planar graph in advance without knowing holes or boundary nodes; (2) hole-bypassing by identifying the holes or boundary node information in advance.

Key features are as follows:

(i) It is a location based routing protocol which supports event driven services.

(ii) It is a multipath routing protocol.

(iii) It implements GF algorithm.

(iv) It can bypass holes using two different methods.

TPGF provides paths with minimum number of hop-counts, but it requires full knowledge of the network; therefore, it is not a scalable protocol.

Geographic Energy-Aware Multipath Stream-based (GEAMS) [94] routing protocol focuses on load-balancing and minimizing energy consumption. In GEAMS routing protocol, routing decisions are made without global topology knowledge and there is no need for maintenance of topology changes. Here each node stores information about its one-hop neighbors, which is updated by beacon messages. The information includes estimated distance to its neighbors, the distance of neighbor to the sink, data rate of link, and remaining energy. GEAMS uses two policies: (1) Smart Greedy Forwarding and (2) Walking Back Forwarding. Walking Back Forwarding policy is used when there are no neighbors closer to sink. Otherwise, GEAMS uses the Smart Greedy Forwarding policy.

Key features are as follows:

(i) It is a geographical multipath routing protocol.

(ii) It enforces uniform energy consumption and meets delay requirement.

(iii) It works on two policies, Smart Greedy Forwarding and Walking Back Forwarding.

(iv) It can detect and avoid hole problems.

GEAMS does not require global topology information and its maintenance and it maximizes network lifetime and guarantees QoS requirements. But in sparse network, GEAMS can cause congestion and packet loss.

Energy-Aware TPGF (EA-TPGF) [95] is a routing protocol which is an enhanced version of TPGF [45]. TPGF is a routing protocol, which implements greedy forwarding and step back and mark methods. The first phase in TPGF consists of two methods, greedy forwarding and step back and mark. The greedy forwarding algorithm selects a neighbor closest to the BS as the next node for routing. Step back and mark method is used to handle the situations, when there is not even a single node in the neighborhood that is closer to BS. The second phase of TPGF deals with label based optimizations of path. EA-TPGF defines distance-energy formula, which is used to calculate the score of neighbors; node with minimum value is selected. EA-TPGF also defines the energy cost transmission formula to calculate the cost of energy consumed by the transmitter.

Key features are as follows:

(i) It is an enhanced version of TPGF.

(ii) It focuses on energy efficiency.

(iii) It uses distance-energy formula and energy cost transmission formula to evaluate the path.

(iv) It also uses GF algorithm to route the data packets.

EA-TPGF routing protocol greatly reduces the energy consumption rate in WMSNs as compared to TPGF. But EA-TPGF increases the average delay of nodes and it always uses the same discovered paths again and again.

Pairwise directional geographical routing (PWDGR) [96] is a multipath routing protocol, which solves the energy bottleneck problem. PWDGR is proposed based on DGR [93]. It makes full use of nodes around the sink. PWDGR finds pairwise nodes around the sink for data transmission. After the data is transferred to pairwise nodes, pairwise node sends data to sink.

Routing in PWDGR is divided into three stages:

(I) In first stage video data is collected and packed, and then the data is broadcasted. Broadcasting saves energy and reduces delay.

(II) After receiving neighbor nodes check if they are in the cooperative node list. If yes, then they calculate coordinates of the next neighbor node.

(III) In the third stage, source cooperative node is selected. Two rules are used to select a source cooperative node: (i) based on cumulative numbers to be selected for cooperative node and (ii) based on distance between ideal point and neighbor node.

Key features are as follows:

(i) It is a multipath, location based routing algorithm.

(ii) It uses the concept of pairwise nodes to fully utilize the nodes around the sink.

(iii) It increases the network lifetime by nearly 70%.

(iv) It is reliable and fault tolerant and improves network bandwidth.

Use of pairwise nodes in PWDGR around the sink makes full use of nodes within 360° so the energy of nodes around the sink is utilized, to increase the network lifetime. DGR scheme suffers due to energy hole around the sink. If the sink is at the edge of the network, then PWDGR degenerates into DGR algorithm. Also, it increases the delay by 8% as compared to similar protocol DGR.

In Table 7, location based routing protocols are compared with each other, where suitability for WMSN for each protocol is based on delay, energy efficiency, network lifetime, and performance of routing protocol.

In Table 8, aforementioned routing protocols are analytically compared, based on architecture, location awareness, multipath, energy efficiency, real-time nature, congestion control, load balancing, reliability, hole detection/bypassing, and data delivery model as attributes.

## 4. Open Research Issues

WMSN has got huge attention in the past few years, but still it is in evolving stage [13]. Research papers [1, 2, 13] have summed up current issues in QoS and energy efficient routing for WSNs and WMSNs. This paper has already mentioned various design requirements. This section is outlining common problems for future exploration.

*Selection of Optimal Metrics.* QoS routing being the multiobjective problem, it is very difficult to decide between the metrics, because the optimization of one metric leads to degradation of another. So there is need of designing and developing routing protocols which can focus more and more on multiple metrics. The protocols following SI based methodologies like ACO, PSO, RL, and GA are becoming prevalent as they are adaptive and multiobjective in nature.

*Cross Layer Functionality.* In typical networks, each layer is assigned its functionality and service. One layer can only interact with adjacent layers only. Various factors in wireless communication make layered approach worse, resulting in degradation of overall system performance. For optimal performance, routing protocols in WMSNs share QoS parameters among the layers. Papers [19, 97] propose cross layer solutions for efficient routing.

*Multiple Sources and Multiple Sink.* In large scale WMSNs, if a single sink is placed there are the chances that the performance of system depletes because of the presence of multiple sources and single sink. Moreover, sink's neighbor nodes get drained much faster and the sink gets isolated. Therefore in real practice, multiple sinks are used, there are some protocols which support multiple source and multiple sinks, but still there is need to make system more efficient.

*Duty Cycle Based Routing.* Sensors can be in active or in sleep mode; these modes are called duty cycle. Most of the energy is utilized in communication with neighboring nodes. Switching between the states and making the node least idle saves a considerable amount of energy. But to achieve duty cycling in WMSNs is a very complex problem due to volatility of video traffic.

*Integration with Other Networks.* Sensors sense the environment and transmit the observed data to sink. As shown in Figure 1, sink communicates the data to internet or other network in order to make it useful to the end users. There is need to develop new architecture in order to provide guarantee of QoS, energy efficiency, and other requirements.

*Security.* Due to various features of WMSNs and unreliable wireless communication channel. WMSNs are always vulnerable to various attacks such as wormhole, sinkhole, and Sybil. Low computation power makes it difficult to implement strong security protocols.

## 5. Conclusions

This paper presents survey of various existing routing protocols for WMSNs, with summarized description and properties of each algorithm. Also, categorywise analytical comparison of protocols is done. QoS routing algorithms can be classified using different metrics but that could make classification more complex so to maintain simplicity, protocols are classified into only main three categories. Tables 2–7 give summarized comparison of protocols based on intrinsic characteristics. Finally, Table 8 gives the overall comparison of all the protocols.

Owing to the increase in applicability of WMSNs, researchers are more inclined to persuade research on WMSN routing protocols, in order to solve various problems

Table 8: Comparison of protocols.

| Protocol | Architecture | | Location aware | Multipath | Energy efficiency | Real-time | Congestion control | Load balancing | Reliability | Hole detection/bypassing |
|---|---|---|---|---|---|---|---|---|---|---|
| | Flat | Clustered | | | | | | | | |
| Zhou and Wang [20] | √ | | | | √ | √ | | | | |
| SAR [21] | √ | | | √ | √ | | | | | |
| EAQoS [23] | | √ | | | √ | √ | | | | |
| RAP [25] | | | √ | | | √ | | | | |
| SPEED [28] | | | √ | | √ | √ | √ | √ | | √ |
| RPAR [29] | | | √ | | √ | √ | √ | | | √ |
| OEDSR [34] | | √ | | √ | √ | √ | | | | |
| SDRC [37] | | | | | √ | √ | | | | |
| AGEM [38] | | | √ | | √ | √ | | | | √ |
| EQSR [39] | | | | √ | √ | | | | √ | |
| Li and Kim [16] | | | | | √ | √ | | | √ | |
| MMSPEED [31] | | | √ | √ | | √ | | | √ | |
| DARA [41] | | | √ | √ | √ | √ | | | √ | |
| RLTD [42] | | | √ | √ | √ | √ | | √ | | √ |
| LRGR [46] | | | √ | | √ | √ | | | √ | |
| QMOR [44] | | | | | √ | √ | | | √ | |
| ASAR [53] | | √ | | √ | √ | √ | | | √ | |
| M-IAR [51] | | | √ | | √ | | | | | |
| ACOLBR [54] | | √ | | √ | | √ | √ | | | √ |
| AntNet [14] | √ | | | √ | | | | | | |
| ACOWMSN [56] | | | | | √ | √ | | | | |
| AntSensNet [57] | | √ | | √ | √ | √ | √ | √ | | |
| HLEAR [60] | | | | | √ | | | √ | | |
| AntHQSeN [61] | | √ | | | √ | | | √ | | |
| QOSPSO [63] | | | | | √ | √ | | | | |
| ICACR [64] | | √ | | | √ | √ | | | | |
| FROMS [65] | √ | | | | | | | | | |
| SPEA [66] | √ | | | | √ | | | | | |
| BeeSensor [69] | √ | | | | | | | | | |
| Bee-Sensor-C [72] | | √ | | √ | √ | √ | | √ | | |
| EQR-RL [74] | √ | | | | | | | | | |
| FTIEE [77] | | √ | | | √ | √ | | √ | √ | |
| REAR [82] | √ | | | √ | √ | √ | | | | |
| LEAR [83] | √ | | | √ | √ | | | √ | √ | √ |
| PEMuR [85] | | √ | | | √ | | | √ | | |
| EEQAR [88] | | √ | | | √ | | | √ | | |
| MCRA [48] | | | √ | | √ | | | | √ | |
| TPGF [59] | | | √ | √ | | | | | | √ |
| GEAMS [94] | | | √ | √ | √ | | | | | √ |
| GPSR [27] | | | √ | | √ | | | | | |
| DGR [93] | | | √ | | √ | √ | | | | √ |
| EA-TPGF [95] | | | √ | √ | √ | | | | | √ |
| PWDGR [96] | | | √ | √ | √ | √ | | | | |

in routing in WMSNs. The paper aims to give the introduction to current WMSN routing protocols and improve the understanding of basic concepts. We hope it will encourage them to design an efficient routing protocol.

## Competing Interests

The authors declare that they have no competing interests.

## References


[1] S. Ehsan and B. Hamdaoui, "A survey on energy-efficient routing techniques with QoS assurances for wireless multimedia sensor networks," *IEEE Communications Surveys and Tutorials*, vol. 14, no. 2, pp. 265–278, 2012.

[2] I. F. Akyildiz, T. Melodia, and K. R. Chowdhury, "A survey on wireless multimedia sensor networks," *Computer Networks*, vol. 51, no. 4, pp. 921–960, 2007.

[3] J. N. Al-Karaki and A. E. Kamal, "Routing techniques in wireless sensor networks: a survey," *IEEE Wireless Communications*, vol. 11, no. 6, pp. 6–28, 2004.

[4] A. Förster, "Machine learning techniques applied to wireless ad-hoc networks: guide and survey," in *Proceedings of the 3rd International Conference on Intelligent Sensors, Sensor Networks and Information (ISSNIP '07)*, pp. 365–370, Melbourne, Australia, December 2007.

[5] S. Misra, M. Reisslein, and G. Xue, "A survey of multimedia streaming in wireless sensor networks," *IEEE Communications Surveys and Tutorials*, vol. 10, no. 4, pp. 18–39, 2008.

[6] M. Saleem, G. A. Di Caro, and M. Farooq, "Swarm intelligence based routing protocol for wireless sensor networks: survey and future directions," *Information Sciences*, vol. 181, no. 20, pp. 4597–4624, 2011.

[7] A. M. Zungeru, L.-M. Ang, and K. P. Seng, "Classical and swarm intelligence based routing protocols for wireless sensor networks: a survey and comparison," *Journal of Network and Computer Applications*, vol. 35, no. 5, pp. 1508–1536, 2012.

[8] A. Jayashree, G. Biradar, and V. Mytri, "Review of multipath routing protocols in wireless multimedia sensor network—a survey," *International Journal of Scientific & Engineering Research*, vol. 3, no. 7, pp. 1–9, 2012.

[9] W. Guo and W. Zhang, "A survey on intelligent routing protocols in wireless sensor networks," *Journal of Network and Computer Applications*, vol. 38, no. 1, pp. 185–201, 2014.

[10] M. Abazeed, N. Faisal, S. Zubair, and A. Ali, "Routing protocols for wireless multimedia sensor network: a survey," *Journal of Sensors*, vol. 2013, Article ID 469824, 11 pages, 2013.

[11] M. Masdari, M. Tanabi, and W. Azarbayjan, "Multipath routing protocols in wireless sensor networks: a survey and analysis," *International Journal of Future Generation Communication and Networking*, vol. 6, no. 6, pp. 181–192, 2013.

[12] A. Alanazi and K. Elleithy, "Real-time QoS routing protocols in wireless multimedia sensor networks: study and analysis," *Sensors*, vol. 15, no. 9, pp. 22209–22233, 2015.

[13] S. Aswale and V. R. Ghorpade, "Survey of QoS routing protocols in wireless multimedia sensor networks," *Journal of Computer Networks and Communications*, vol. 2015, Article ID 824619, 29 pages, 2015.

[14] G. Di Caro and M. Dorigo, "AntNet: distributed stigmergetic control for communications networks," *Journal of Artificial Intelligence Research*, vol. 9, pp. 317–365, 1998.

[15] B. Li and K.-I. Kim, "A novel real-time scheme for (m, k)-firm streams in wireless sensor networks," *Wireless Networks*, vol. 20, no. 4, pp. 719–731, 2014.

[16] B. Li and K.-I. Kim, "An ($m$, $k$)-firm real-time aware fault-tolerant mechanism in wireless sensor networks," *International Journal of Distributed Sensor Networks*, vol. 2012, Article ID 905740, 12 pages, 2012.

[17] S. Chen and K. Nahrstedt, "An overview of quality of service routing for next-generation high-speed networks: problems and solutions," *IEEE Network*, vol. 12, no. 6, pp. 64–79, 1998.

[18] M. Hamdaoui and P. Ramanathan, "A dynamic priority assignment technique for streams with (m, k)-firm deadlines," *IEEE Transactions on Computers*, vol. 44, no. 12, pp. 1443–1451, 1995.

[19] C. Lee, B. Shah, and K.-I. Kim, "An architecture for (m, k)-firm real-time streams in wireless sensor networks," *Wireless Networks*, vol. 22, no. 1, pp. 69–81, 2016.

[20] L. Zhou and J.-X. Wang, "Delay-constrained maximized lifetime routing algorithm in wireless multimedia sensor networks," in *Proceedings of the 2nd International Conference on Future Computer and Communication (ICFCC '10)*, pp. V1215–V1219, IEEE, Wuhan, China, May 2010.

[21] K. Sohrabi, J. Gao, V. Ailawadhi, and G. J. Pottie, "Protocols for self-organization of a wireless sensor network," *IEEE Personal Communications*, vol. 7, no. 5, pp. 16–27, 2000.

[22] K. Sohrabi, J. Gao, and V. Ailawadhi, "A self organizing wireless sensor network," in *Proceedings of the Annual Allerton Conference on Communication Control and Computing*, vol. 37, pp. 1201–1210, 1999.

[23] K. Akkaya and M. Younis, "An energy-aware QoS routing protocol for wireless sensor networks," in *Proceedings of the 23rd International Conference on Distributed Computing Systems Workshops (ICDCSW '03)*, pp. 710–715, Providence, RI, USA, May 2003.

[24] S. Floyd and V. Jacobson, "Link-sharing and resource management models for packet networks," *IEEE/ACM Transactions on Networking*, vol. 3, no. 4, pp. 365–386, 1995.

[25] C. Lu, B. M. Blum, T. F. Abdelzaher, J. A. Stankovic, and T. He, "RAP: a real-time communication architecture for large-scale wireless sensor networks," in *Proceedings of the 8th IEEE Real-Time and Embedded Technology and Applications Symposium (RTAS '02)*, pp. 55–66, September 2002.

[26] D. B. Johnson and D. A. Maltz, "Dynamic source routing in ad hoc wireless networks," in *Mobile Computing*, T. Imielinski and H. F. Korth, Eds., vol. 353 of *The Kluwer International Series in Engineering and Computer Science*, pp. 153–181, 1996.

[27] B. Karp and H. T. Kung, "GPSR: greedy perimeter stateless routing for wireless networks," in *Proceedings of the 6th Annual International Conference on Mobile Computing and Networking (MOBICOM '00)*, pp. 243–254, August 2000.

[28] T. H. T. He, J. A. Stankovic, C. L. C. Lu, and T. Abdelzaher, "SPEED: a stateless protocol for real-time communication in sensor networks," in *Proceedings of the 23th IEEE International Conference on Distributed Computing Systems*, pp. 46–55, IEEE, May 2003.

[29] O. Chipara, Z. He, G. Xing et al., "Real-time power-aware routing in sensor networks," in *Proceedings of the 14th IEEE International Workshop on Quality of Service (IWQoS '06)*, pp. 83–92, New Haven, Conn, USA, June 2006.

[30] V. Jacobson, "Congestion avoidance and control," *ACM SIGCOMM Computer Communication Review*, vol. 18, no. 4, pp. 314–329, 1988.

[31] E. Felemban, C.-G. Lee, and E. Ekici, "MMSPEED: multipath multi-SPEED protocol for QoS guarantee of reliability and timeliness in wireless sensor networks," *IEEE Transactions on Mobile Computing*, vol. 5, no. 6, pp. 738–753, 2006.

[32] M. R. Fouad, S. Fahmy, and G. Pandurangan, "Latency-sensitive power control for wireless ad-hoc networks," in *Proceedings of the 1st ACM International Workshop on Quality of Service & Security in Wireless and Mobile Networks (Q2SWinet '05)*, p. 31, 2005.

[33] K. Seada, M. Zuniga, A. Helmy, and B. Krishnamachari, "Energy-efficient forwarding strategies for geographic routing in lossy wireless sensor networks," in *Proceedings of the 2nd International Conference on Embedded Networked Sensor Systems (SenSys '04)*, pp. 108–121, November 2004.

[34] S. Ratnaraj, S. Jagannathan, and V. Rao, "OEDSR: optimized energy-delay sub-network routing in wireless sensor network," in *Proceedings of the IEEE International Conference on Networking, Sensing and Control (ICNSC '06)*, pp. 330–335, IEEE, April 2006.

[35] N. Regatte and S. Jagannathan, "Optimized energy-delay routing in ad hoc wireless networks," in *Proceedings of the World Wireless Congress*, San Francisco, Calif, USA, 2005.

[36] C. E. Perkins and E. M. Royer, "Ad-hoc on-demand distance vector routing," in *Proceedings of the 2nd IEEE Workshop on Mobile Computing Systems and Applications (WMCSA '99)*, pp. 90–100, New Orleans, La, USA, February 1999.

[37] Y. Xue, B. Ramamurthy, and M. C. Vuran, "SDRCS: a service-differentiated real-time communication scheme for event sensing in wireless sensor networks," *Computer Networks*, vol. 55, no. 15, pp. 3287–3302, 2011.

[38] S. Medjiah, T. Ahmed, and F. Krief, "AGEM: adaptive greedy-compass energy-aware multipath routing protocol for WMSNs," in *Proceedings of the 7th IEEE Consumer Communications and Networking Conference (CCNC '10)*, pp. 1–6, IEEE, January 2010.

[39] J. Ben-Othman and B. Yahya, "Energy efficient and QoS based routing protocol for wireless sensor networks," *Journal of Parallel and Distributed Computing*, vol. 70, no. 8, pp. 849–857, 2010.

[40] X. Huang and Y. Fang, "Multiconstrained QoS multipath routing in wireless sensor networks," *Wireless Networks*, vol. 14, no. 4, pp. 465–478, 2008.

[41] M. A. Razzaque, M. M. Alam, M. Mamun-Or-rashid, and C. S. Hong, "Multi-constrained QoS geographic routing for heterogeneous traffic in sensor networks," *IEICE Transactions on Communications*, vol. 91, no. 8, pp. 2589–2601, 2008.

[42] A. A. Ahmed and N. Fisal, "A real-time routing protocol with load distribution in wireless sensor networks," *Computer Communications*, vol. 31, no. 14, pp. 3190–3203, 2008.

[43] O. Chipara, Z. He, G. Xing, Q. Chen, and X. Wang, "Real-time power control in wireless sensor networks,," Tech. Rep. WUCSE-2005-31, Washington University in St. Louis, St. Louis, Mo, USA, 2005.

[44] H. Shen, G. Bai, Z. Tang, and L. Zhao, "QMOR: QoS-aware multi-sink opportunistic routing for wireless multimedia sensor networks," *Wireless Personal Communications*, vol. 75, no. 2, pp. 1307–1330, 2014.

[45] R. Dai, P. Wang, and I. F. Akyildiz, "Correlation-aware QoS routing with differential coding for wireless video sensor networks," *IEEE Transactions on Multimedia*, vol. 14, no. 5, pp. 1469–1479, 2012.

[46] Y. Rao, C. Yuan, Z. Jiang, L. Fu, and J. Zhu, "Latency and reliability-aware geographic routing for mobile wireless sensor networks," *Advances in information Sciences and Service Sciences*, vol. 5, no. 8, pp. 738–748, 2013.

[47] A. Papadopoulos, A. Navarra, J. A. McCann, and C. M. Pinotti, "VIBE: an energy efficient routing protocol for dense and mobile sensor networks," *Journal of Network and Computer Applications*, vol. 35, no. 4, pp. 1177–1190, 2012.

[48] X. Yan, L. Li, and F. J. An, "Multi-constrained routing in wireless multimedia sensor networks," in *Proceedings of the International Conference on Wireless Communications & Signal Processing (WCSP '09)*, pp. 1–5, IEEE, November 2009.

[49] C. Intanagonwiwat, R. Govindan, D. Estrin, J. Heidemann, and F. Silva, "Directed diffusion for wireless sensor networking," *IEEE/ACM Transactions on Networking*, vol. 11, no. 1, pp. 2–16, 2003.

[50] A. A. A. Radwan, T. M. Mahmoud, and E. H. Hussein, "AntNet-RSLR: a proposed Ant routing protocol for MANETs," in *Proceedings of the Saudi International Electronics, Communications and Photonics Conference (SIECPC '11)*, pp. 1–6, Riyadh, Saudi Arabia, April 2011.

[51] A. Rahman, R. Ghasemaghaeil, A. El Saddikl, and W. Gueaieb, "M-IAR: biologically inspired routing protocol for wireless multimedia sensor networks," in *Proceedings of the Instrumentation and Measurement Technology Conference (IMTC '08)*, pp. 1823–1827, Victoria, Canada, May 2008.

[52] R. G. Aghaei, M. A. Rahman, W. Gueaieb, and A. El Saddik, "Ant colony-based reinforcement learning algorithm for routing in wireless sensor networks," in *Proceedings of the IEEE Instrumentation & Measurement Technology Conference (IMTC '07)*, 6, 1 pages, IEEE, Warsaw, Poland, May 2007.

[53] Y. Sun, H. Ma, L. Liu, and Y. Zheng, "ASAR: an ant-based service-aware routing algorithm for multimedia sensor networks," *Frontiers of Electrical and Electronic Engineering in China*, vol. 3, no. 1, pp. 25–33, 2008.

[54] J. Bi, Z. Li, and R. Wang, "An ant colony optimization-based load balancing routing algorithm for wireless multimedia sensor networks," in *Proceedings of the IEEE 12th International Conference on Communication Technology (ICCT '10)*, pp. 584–587, Nanjing, China, November 2010.

[55] Z. Ke, L. Li, Q. Sun, and N. Chen, "Ant-like game routing algorithm for wireless multimedia sensor networks," in *Proceedings of the 4th International Conference on Wireless Communications, Networking and Mobile Computing (WiCOM '08)*, pp. 1–4, Dalian, China, October 2008.

[56] X. Yu, J. Luo, and J. Huang, "An ant colony optimization-based QoS routing algorithm for wireless multimedia sensor networks," in *Proceedings of the 3rd International Conference on Communication Software and Networks (ICCSN '11)*, pp. 37–41, IEEE, Xi'an, China, May 2011.

[57] L. Cobo, A. Quintero, and S. Pierre, "Ant-based routing for wireless multimedia sensor networks using multiple QoS metrics," *Computer Networks*, vol. 54, no. 17, pp. 2991–3010, 2010.

[58] S. Selvakennedy, S. Sinnappan, and Y. Shang, "A biologically-inspired clustering protocol for wireless sensor networks," *Computer Communications*, vol. 30, no. 14-15, pp. 2786–2801, 2007.

[59] L. Shu, Y. Zhang, L. T. Yang, Y. Wang, M. Hauswirth, and N. Xiong, "TPGF: geographic routing in wireless multimedia sensor networks," *Telecommunication Systems*, vol. 44, no. 1-2, pp. 79–95, 2010.

[60] A. Nayyar, F. Bashir, Ubaid-Ur-Rehman, and Z. Hamid, "Intelligent routing protocol for multimedia sensor networks," in *Proceedings of the 5th International Conference on Information Technology and Multimedia (ICIM '11)*, pp. 1–6, Kuala Lumpur, Malaysia, November 2011.

[61] S. Kumar, M. Dave, and S. Dahiya, "ACO based QoS aware routing for wireless sensor networks with heterogeneous nodes," in *Emerging Trends in Computing and Communication*, vol. 298 of *Lecture Notes in Electrical Engineering*, pp. 157–168, Springer, Berlin, Germany, 2014.

[62] T. Camilo, C. Carreto, J. Silva, and F. Boavida, "An energy-efficient ant-based routing algorithm for wireless sensor networks," in *Ant Colony Optimization and Swarm Intelligence*, vol. 4150 of *Lecture Notes in Computer Science*, pp. 49–59, Springer, Berlin, Germany, 2006.

[63] M. Liu, S. Xu, and S. Sun, "An agent-assisted QoS-based routing algorithm for wireless sensor networks," *Journal of Network and Computer Applications*, vol. 35, no. 1, pp. 29–36, 2012.

[64] H. Huang, X. Cao, R. Wang, and Y. Wen, "A QoS-aware routing algorithm based on ant-cluster in wireless multimedia sensor networks," *Science China Information Sciences*, vol. 57, no. 10, pp. 1–16, 2014.

[65] A. Förster and A. L. Murphy, "FROMS: feedback routing for optimizing multiple sinks in WSN with reinforcement learning," in *Proceedings of the 3rd International Conference on Intelligent Sensors, Sensor Networks and Information (ISSNIP '07)*, pp. 371–376, IEEE, Melbourne, Australia, December 2007.

[66] K. Kotecha and S. Popat, "Multi objective genetic algorithm based adaptive QoS routing in MANET," in *Proceedings of the IEEE Congress on Evolutionary Computation (CEC '07)*, pp. 1423–1428, Singapore, September 2007.

[67] I. Gerasimov and R. Simon, "A bandwidth-reservation mechanism for on-demand ad hoc path finding," in *Proceedings of the 35th Annual Simulation Symposium*, pp. 27–34, 2002.

[68] N. Magaia, N. Horta, R. Neves, P. R. Pereira, and M. Correia, "A multi-objective routing algorithm for Wireless Multimedia Sensor Networks," *Applied Soft Computing*, vol. 30, pp. 104–112, 2015.

[69] M. Saleem, I. Ullah, and M. Farooq, "BeeSensor: an energy-efficient and scalable routing protocol for wireless sensor networks," *Information Sciences*, vol. 200, pp. 38–56, 2012.

[70] F. Çelik, A. Zengin, and S. Tuncel, "A survey on swarm intelligence based routing protocols in wireless sensor networks," *International Journal of Physical Sciences*, vol. 5, no. 14, pp. 2118–2126, 2010.

[71] T. C. Camilo, C. Carreto, J. S. Silva, and F. Boavida, "An energy-efficient ant-based routing algorithm for wireless sensor networks," in *Ant Colony Optimization and Swarm Intelligence*, M. Dorigo, L. M. Gambardella, M. Birattari, A. Martinoli, R. Poli, and T. Stützle, Eds., vol. 4150 of *Lecture Notes in Computer Science*, pp. 49–59, 2006.

[72] X. Cai, Y. Duan, Y. He, J. Yang, and C. Li, "Bee-sensor-C: an energy-efficient and scalable multipath routing protocol for wireless sensor networks," *International Journal of Distributed Sensor Networks*, vol. 2015, Article ID 976127, 14 pages, 2015.

[73] Y.-L. Jin, Y. Zhang, Y. Xue, C. Guo, and L.-N. Xu, "Ant colony optimization routing based on congestion control in WSNs," *Journal of Shanghai University (Natural Science Edition)*, vol. 18, no. 6, pp. 551–554, 2012.

[74] S. Z. Jafarzadeh and M. H. Y. Moghaddam, "Design of energy-aware QoS routing protocol in wireless sensor networks using reinforcement learning," in *Proceedings of the IEEE 27th Canadian Conference on Electrical and Computer Engineering (CCECE '14)*, pp. 1–5, Toronto, Canada, May 2014.

[75] X. Liang, I. Balasingham, and S.-S. Byun, "A reinforcement learning based routing protocol with QoS support for biomedical sensor networks," in *Proceedings of the International Symposium on Applied Sciences on Biomedical and Communication Technologies (ISABEL '08)*, pp. 1–5, IEEE, Aalborg, Denmark, October 2008.

[76] M. Maalej, S. Cherif, and H. Besbes, "QoS and energy aware cooperative routing protocol for wildfire monitoring wireless sensor networks," *The Scientific World Journal*, vol. 2013, Article ID 437926, 11 pages, 2013.

[77] F. Kiani, E. Amiri, M. Zamani, T. Khodadadi, and A. A. Manaf, "Efficient intelligent energy routing protocol in wireless sensor networks," *International Journal of Distributed Sensor Networks*, vol. 2015, Article ID 618072, 13 pages, 2015.

[78] A. Förster, "Machine learning techniques applied to wireless ad-hoc networks: guide and survey," in *Proceedings of the International Conference on Intelligent Sensors, Sensor Networks and Information Processing (ISSNIP '07)*, pp. 365–370, Melbourne, Australia, December 2007.

[79] W. R. Heinzelman, A. Chandrakasan, and H. Balakrishnan, "Energy-efficient communication protocol for wireless microsensor networks," in *Proceedings of the 33rd Annual Hawaii International Conference on System Sciences*, January 2000.

[80] H. Taheri, P. Neamatollahi, M. Naghibzadeh, and M.-H. Yaghmaee, "Improving on HEED protocol of wireless sensor networks using non probabilistic approach and fuzzy logic (HEED-NPF)," in *Proceedings of the 5th International Symposium on Telecommunications (IST '10)*, pp. 193–198, Tehran, Iran, December 2010.

[81] M. Ye, C. Li, G. Chen, and J. Wu, "EECS: an energy efficient clustering scheme in wireless sensor networks," in *Proceedings of the 24th IEEE International Performance, Computing, and Communications Conference (IPCCC '05)*, pp. 535–540, IEEE, Phoenix, Ariz, USA, April 2005.

[82] Y. Lan, W. Wenjing, and G. Fuxiang, "A real-time and energy aware QoS routing protocol for multimedia wireless sensor networks," in *Proceedings of the 7th World Congress on Intelligent Control and Automation (WCICA '08)*, pp. 3304–3309, IEEE, Chongqing, China, June 2008.

[83] A. Nayyar, F. Bashir, and Ubaid-Ur-Rehman, "Load based energy aware multimedia routing protocol-(LEAR)," in *Proceedings of the 3rd International Conference on Computer Research and Development (ICCRD '11)*, vol. 2, pp. 427–430, Shanghai, China, March 2011.

[84] S.-H. Yang, "Routing technologies in WSNs," in *Wireless Sensor Networks: Principles, Design and Applications*, pp. 101–128, Springer, London, UK, 2014.

[85] D. Kandris, M. Tsagkaropoulos, I. Politis, A. Tzes, and S. Kotsopoulos, "Energy efficient and perceived QoS aware video routing over wireless multimedia sensor networks," *Ad Hoc Networks*, vol. 9, no. 4, pp. 591–607, 2011.

[86] D. Kandris, P. Tsioumas, A. Tzes, G. Nikolakopoulos, and D. D. Vergados, "Power conservation through energy efficient routing in wireless sensor networks," *Sensors*, vol. 9, no. 9, pp. 7320–7342, 2009.

[87] A. Manjeshwar and D. P. Agrawal, "TEEN: a routing protocol for enhanced efficiency in wireless sensor networks," in *Proceedings of the 15th International Parallel and Distributed Processing*

*Symposium (IPDPS '00)*, pp. 2009–2015, San Francisco, Calif, USA, April 2000.

[88] K. Lin, J. J. P. C. Rodrigues, H. Ge, N. Xiong, and X. Liang, "Energy efficiency QoS assurance routing in wireless multimedia sensor networks," *IEEE Systems Journal*, vol. 5, no. 4, pp. 495–505, 2011.

[89] K. Lin, M. Chen, and X. Ge, "Adaptive reliable routing based on cluster hierarchy for wireless multimedia sensor networks," *EURASIP Journal on Wireless Communications and Networking*, vol. 2010, Article ID 567952, pp. 1–11, 2010.

[90] D. Niculescu and B. Nath, "DV based positioning in Ad Hoc networks," *Telecommunication Systems*, vol. 22, no. 1–4, pp. 267–280, 2003.

[91] R. Nagpal, H. Shrobe, and J. Bachrach, "Organizing a global coordinate system from local information on an ad hoc sensor network," in *Proceedings of the 2nd International Conference on Information Processing in Sensor Networks (IPSN '03)*, pp. 333–348, Palo Alto, Calif, USA, April 2003.

[92] M. A. Zayene, N. Tabbane, and R. Elidoudi, "Performance evaluation of greedy perimeter stateless routing protocol in ad hoc networks," in *Proceedings of the 4th International Conference on Computer Sciences and Convergence Information Technology (ICCIT '09)*, pp. 907–912, Seoul, South Korea, November 2009.

[93] M. Chen, V. C. M. Leung, S. Mao, and Y. Yuan, "Directional geographical routing for real-time video communications in wireless sensor networks," *Computer Communications*, vol. 30, no. 17, pp. 3368–3383, 2007.

[94] S. Medjiah, T. Ahmed, and F. Krief, "GEAMS: a geographic energy-aware multipath stream-based routing protocol for WMSNs," in *Proceedings of the Global Information Infrastructure Symposium (GIIS '09)*, pp. 1–8, Hammamet, Tunisia, June 2009.

[95] I. Bennis, O. Zytoune, D. Aboutajdine, and H. Fouchal, "Low energy geographical routing protocol for wireless multimedia sensor networks," in *Proceedings of the 9th International Wireless Communications and Mobile Computing Conference (IWCMC '13)*, pp. 585–589, July 2013.

[96] J. Wang, Y. Zhang, J. Wang, Y. Ma, and M. Chen, "PWDGR: pair-wise directional geographical routing based on wireless sensor network," *IEEE Internet of Things Journal*, vol. 2, no. 1, pp. 14–22, 2015.

[97] L. Pradittasnee, "Predicting path quality with cross-layer information in multi-hop wireless networks," in *Proceedings of the 7th International Conference on Information Technology and Electrical Engineering (ICITEE '15)*, pp. 464–469, IEEE, October 2015.

# Permissions

All chapters in this book were first published in JE, by Hindawi Publishing Corporation; hereby published with permission under the Creative Commons Attribution License or equivalent. Every chapter published in this book has been scrutinized by our experts. Their significance has been extensively debated. The topics covered herein carry significant findings which will fuel the growth of the discipline. They may even be implemented as practical applications or may be referred to as a beginning point for another development.

The contributors of this book come from diverse backgrounds, making this book a truly international effort. This book will bring forth new frontiers with its revolutionizing research information and detailed analysis of the nascent developments around the world.

We would like to thank all the contributing authors for lending their expertise to make the book truly unique. They have played a crucial role in the development of this book. Without their invaluable contributions this book wouldn't have been possible. They have made vital efforts to compile up to date information on the varied aspects of this subject to make this book a valuable addition to the collection of many professionals and students.

This book was conceptualized with the vision of imparting up-to-date information and advanced data in this field. To ensure the same, a matchless editorial board was set up. Every individual on the board went through rigorous rounds of assessment to prove their worth. After which they invested a large part of their time researching and compiling the most relevant data for our readers.

The editorial board has been involved in producing this book since its inception. They have spent rigorous hours researching and exploring the diverse topics which have resulted in the successful publishing of this book. They have passed on their knowledge of decades through this book. To expedite this challenging task, the publisher supported the team at every step. A small team of assistant editors was also appointed to further simplify the editing procedure and attain best results for the readers.

Apart from the editorial board, the designing team has also invested a significant amount of their time in understanding the subject and creating the most relevant covers. They scrutinized every image to scout for the most suitable representation of the subject and create an appropriate cover for the book.

The publishing team has been an ardent support to the editorial, designing and production team. Their endless efforts to recruit the best for this project, has resulted in the accomplishment of this book. They are a veteran in the field of academics and their pool of knowledge is as vast as their experience in printing. Their expertise and guidance has proved useful at every step. Their uncompromising quality standards have made this book an exceptional effort. Their encouragement from time to time has been an inspiration for everyone.

The publisher and the editorial board hope that this book will prove to be a valuable piece of knowledge for researchers, students, practitioners and scholars across the globe.

# List of Contributors

**Chuanxue Song, Yulong Shao, Silun Peng and Feng Xiao**
State Key Laboratory of Automotive Simulation and Control, Jilin University, Changchun 130022, China

**Shixin Song**
School of Mechanical Science and Engineering, Jilin University, Changchun, China

**Abeer Alhashash**
Department of Mathematics, Faculty of Science, Jouf University, 24241 Sakaka Aljawf, Saudi Arabia

**Kai Kordelin, Jaana Kordelin, Johanna Virkki, Leena Ukkonen and Lauri Sydänheimo**
BioMediTech Institute and Faculty of Biomedical Sciences and Engineering, Tampere University of Technology, Tampere, Finland

**Markku Johansson**
Aitosolutions Oyj, Kangasala, Finland

**Vikram C. Patil and Paul I. Ro**
North Carolina State University, Raleigh, NC, USA

**Doniazed Sioud, Raoudha Garma and Ahmed Bellagi**
Energy Engineering Department, Ecole Nationale d'Ing´enieurs de Monastir (ENIM), University of Monastir, Tunisia

**Julius Mwaniki and Hui Lin**
Department of Electrical Engineering, School of Automation, Northwestern Polytechnical University, Xi'an 710129, China

**Zhiyong Dai**
Automatic Control Department, School of Mechano-Electronic Engineering, Xidian University, Xi'an 710071, China

**Xiaochen Chen, Leena Ukkonen and Johanna Virkki**
Department of Electronics and Communications Engineering, Tampere University of Technology, Tampere, Finland

**Aruhan Liu and Zhigang Wei**
Department of Electronic Engineering, City University of Hong Kong, 83 Tat Chee Avenue, Kowloon, Hong Kong

**Zhongfu Tan, Qingkun Tan and Yuwei Wang**
School of Economics and Management, North China Electric Power University, Beijing 102206, China

**Amnart Boonloi**
Department of Mechanical Engineering Technology, College of Industrial Technology, King Mongkut's University of Technology North Bangkok, Bangkok 10800,Thailand

**Withada Jedsadaratanachai**
Department of Mechanical Engineering, Faculty of Engineering, King Mongkut's Institute of Technology Ladkrabang, Bangkok 10520, Thailand

**Julius Mwaniki and Hui Lin**
Department of Electrical Engineering, School of Automation, Northwestern Polytechnical University, Xi'an 710129, China

**Zhiyong Dai**
Automatic Control Department, School of Mechano-Electronic Engineering, Xidian University, Xi'an 710071, China

**X. H. Zhang**
Department of Electrical and Electronic Engineering, Southern University of Science and Technology, Shenzhen 518055, China

**H. Li**
Department of Electrical and Electronic Engineering, Southern University of Science and Technology, Shenzhen 518055, China
Institute of Applied Physics and Materials Engineering, University of Macau, Macau

**Z. K. Tang**
Institute of Applied Physics and Materials Engineering, University of Macau, Macau

**Baseem Khan**
School of Electrical and Computer Engineering, Hawassa University Institute of Technology, Hawassa, Ethiopia

**Pawan Singh**
School of Informatics, Hawassa University Institute of Technology, Hawassa, Ethiopia

**Yusuke Hayashi, Yoshikatsu Matsugaki and Tamotsu Ninomiya**
Green Electronics Research Institute, Kitakyushu (GRIK), 1-8-409 Hibikino, Wakamatsu-ku, Kitakyushu, Fukuoka 808-0135, Japan

**Abdel-Rahman Al-Qawasmi**
Electrical Engineering Department, College of Engineering, Majmaah University, Majmaah 11952, Saudi Arabia

**I. Tlili**
Energy and Thermal Systems Laboratory, National Engineering School of Monastir, Street Ibn El Jazzar, 5019 Monastir, Tunisia

**Nizar Al Bassam**
Middle East College, Muscat, Oman

**Oday D. Jerew**
Asia Pacific International College, Sydney, NSW 2008, Australia

# Index

**U**



**V**



**W**

www.ingramcontent.com/pod-product-compliance
Lightning Source LLC
LaVergne TN
LVHW081020210926
848485LV00004B/912